Failur

Industri

Materials

Failure Analysis of Industrial Composite Materials

E. E. Gdoutos   Editor
School of Engineering, Democritus University of Thrace
Xanthi, Greece

K. Pilakoutas   Editor
Department of Civil Engineering and Structural Engineering
University of Sheffield, Sheffield, United Kingdom

C. A. Rodopoulos   Editor
Structural Integrity Research Institute of Sheffield and
Department of Mechanical and Process Engineering
University of Sheffield, Sheffield, United Kingdom

McGraw-Hill

New York   San Francisco   Washington, D.C.   Auckland   Bogotá
Caracas   Lisbon   London   Madrid   Mexico City   Milan
Montreal   New Delhi   San Juan   Singapore
Sydney   Tokyo   Toronto

Library of Congress Cataloging-in-Publication Data

Failure analysis of industrial composite materials/E. E. Gdoutos editor, K. Pilakoutas, editor, C. A. Rodopoulos, editor.
      p.   cm.
   ISBN 0-07-134517-5
      1. Composite materials–Fracture. 2. Fracture mechanics. 3. Structural failures–Case Studies. I. Gdoutos, E. E., 1948-II. Pilakoutas, K. III. Rodopoulos, C. A.

   TA418.9.C6 F324 2000
   620.1′186–dc21   99-056278

McGraw-Hill

A Division of The McGraw-Hill Companies

1 2 3 4 5 6 7 8 9 0 AGM/AGM 0 6 5 4 3 2 1 0

ISBN 978-0-07-173790-6

The sponsoring editor for this book was Robert Esposito and the production supervisor was Sherri Souffrance. It was set in Century Schoolbook by Techset Composition Limited.

This book is printed on recycled, acid-free paper containing a minimum of 50% recycled, de-inked fiber

To those who believe in and search for a better world ahead

Contents

Part 2 Case Studies: Aerospace, Construction, and Medical Engineering

Contents

Introduction

Composite materials are the new products of the so-called advanced engineering age. Their unique integration of mechanical properties along with their customized fabrication provide the building blocks of our modern infrastructure. From the aerospace to the automotive industry, from household products to optical devices, composite materials have just started to dominate the future engineering reality of the twenty-first century. Even in the era of decreasing research budgets for frontier materials research, the discovery of new composite materials with unique characteristics moves aggressively. Unfortunately, their high discovery rate does not come without cost. For the first time in our engineering history, structural engineers and designers are incapable of assimilating such a huge volume of new materials. The result in some cases is devastating. Aerospace companies are becoming day by day more reluctant in adopting these new materials due to lack of airworthiness standards, and construction companies are facing difficulties in persuading their customers to accept that today's high cost will be compensated for by their long term durability.

This book does not attempt to solve those questions, but to provide an introductory background in the understanding of failure of various composite material applications. Based on real life problems, derived from the use of composite materials in modern engineering applications, a collection of European specialists offer new and authoritative solutions to the problems. The book itself is divided into two sections. The *Fracture and Failure Mechanisms* section, provides an extensive background on various failure topics. To maintain a progressive degree of difficulty, the section starts with basic failure analyses and moves forward to more advanced problems like notch sensitivity and fatigue. Each of these topics is designed in a way of providing updated information about failure methodologies and solving techniques. Some of these topics also include comparisons between different theories allowing readers to develop their own conceptions. The second section, *Case Studies*, is divided into three parts: a) aerospace engineering; b) medical engineering and c) construction engineering. From impact and crashworthiness to bridge repair, the section provides all-around solutions of realistic problems faced by engineers everyday.

This book is intended for practicing engineers, academics, and research students who have some background and familiarity with composite materials and failure analysis and who wish to upgrade and review their techniques.

The editors and authors are indebted to an endless list of people, especially our teachers who have polished our minds, and our students who, by their continuous quest for knowledge, forced us to write and publish this book. In particular, we would like to thank our editor, B. Esposito, and McGraw-Hill for their support beyond all expectation.

Unfortunately, prior to publication of this book, Prof. P. S. Theocaris, one of the greater contributors to applied mechanics, died. Fortunately, being the mentor to many of the authors of this book, he managed to convey his academic quests to them, thus allowing his legacy to continue, and the science to which he dedicated himself to progress into the new millennium.

E. E. Gdoutos
K. Pilakoutas
C. A. Rodopoulos

Fracture and Failure Mechanisms

Theoretical and Experimental Composite Strength Evaluation

Chris A. Rodopoulos

SIRIUS-Department of Mechanical Engineering,
University of Sheffield, Sheffield, U.K.

1.1   Introduction

Accurate and reliable prediction of the mechanical properties of a composite system is undoubtedly not an easy task even for the most experienced engineer. Under the application of load, the system has the tendency to produce complex distributions of stress and strain. This is basically because the system is heterogeneous at the constituent materials level(reinforcement, matrix, and interface)with mechanical properties of the phases (phase denotes one of the constituent materials) that may remain constant or in some cases degrade (owing to fatigue, micro-cracking, etc.) during service.

The subject of pursuing the correct mechanical properties of a composite system can be approached from several different directions. We can rely on extensive experimental data to produce master curves of nondimensional variables, or we can seek analytical functions. In one or the other way, the price to be paid is enormous and in most cases unwanted. Therefore, the goal of obtaining a more fundamental or probably an effective approach is more than necessary even if assumptions, generalities, and compromises are employed.

In the first section of this chapter we demonstrated the best-known theoretical methods for evaluating the basic mechanical properties of composite materials. The second section is denoted to experimental techniques currently used for testing composites.

1.2   The Basic Mechanics of a Composite System—Simple Analytical Models

1.2.1   The netting analysis

In 1952 Cox[1] published the first analytical model (**netting analysis**) to predict the stress–strain behavior of a typical composite system. Definitely influenced by the theory of elasticity, he based his mathematical analysis on the following assumptions: (a) the fibers are thin, long, and straight (the effect of the fiber diameter is minimal); (b) the load is applied at the fiber ends (no interfacial stresses); and (c) the bending stiffness of the fibers is negligible. It is perhaps obvious that Cox tried to avoid any contribution of the matrix phase.

Let us now consider a planar mat of fibers subjected to tensile strains ε_{11}, ε_{22} and to shear strain γ_{12} (Fig. 1.1). According to the **netting analysis**, the strain of a fiber at an arbitrary angle θ is given by[2]

$$\varepsilon_{11} \cos^2 \theta + \varepsilon_{22} \sin^2 \theta + \gamma_{12} \sin \theta \cos \theta \qquad (1.1)$$

and the stress at the fiber is assumed to be proportional to this strain. If now a load Q is applied at the fiber, then the contribution of the fiber to the loads in direction X_1 and X_2 will be $Q \cos \theta$ and $Q \sin \theta$, respectively.

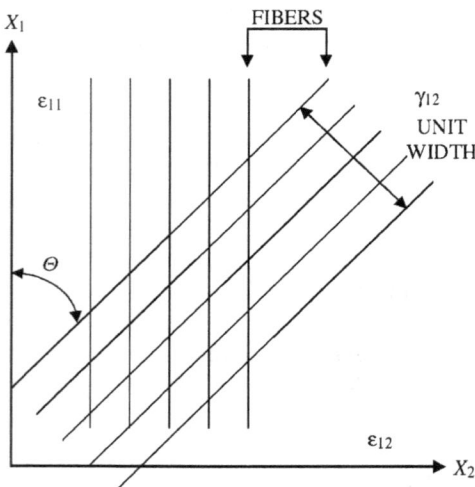

Figure 1.1 The planar mat of fibers.

For all the fibers inclined at angle to the direction X_1, there is a distribution function $f(\theta)$ to represent unit width transverse to their direction

$$\int_0^\pi f(\theta)\, d\theta = 1 \tag{1.2}$$

Therefore, the fractions of fibers intersecting lines perpendicular to the directions X_1 and X_2 are $f(\theta)\cos\theta$, $f(\theta)\sin\theta$ respectively. Consequently, the stresses at the mat are

$$\sigma_{11} = E_f V_f \int_0^\pi (\varepsilon_{11}\cos^2\theta + \varepsilon_{22}\sin^2\theta + \gamma_{12}\cos\theta\sin\theta)\cos^2\theta f(\theta)\, d\theta$$

$$\sigma_{22} = E_f V_f \int_0^\pi (\varepsilon_{11}\cos^2\theta + \varepsilon_{22}\sin^2\theta + \gamma_{12}\cos\theta\sin\theta)\sin^2\theta f(\theta)\, d\theta \tag{1.3}$$

$$\tau_{12} = E_f V_f \int_0^\pi (\varepsilon_{11}\cos^2\theta + \varepsilon_{22}\sin^2\theta + \gamma_{12}\cos\theta\sin\theta)\sin\theta\cos\theta f(\theta)\, d\theta$$

where E_f and V_f are the elastic modulus (Young's modulus) and the volume fraction of the fibers, respectively. Using the elastic constants, C_{ij}, Eq. (1.3) can alternatively be written as

$$\sigma_{11} = C_{11}\varepsilon_{11} + C_{12}\varepsilon_{22} + C_{16}\gamma_{12}$$
$$\sigma_{22} = C_{12}\varepsilon_{11} + C_{22}\varepsilon_{22} + C_{26}\gamma_{12} \tag{1.4}$$
$$\tau_{12} = C_{16}\varepsilon_{11} + C_{26}\varepsilon_{22} + C_{66}\gamma_{12}$$

where

$$C_{11} = E_f V_f \int_0^\pi \cos^4\theta f(\theta)\, d\theta$$

$$C_{16} = E_f V_f \int_0^\pi \cos^3\theta\sin\theta f(\theta)\, d\theta$$

$$C_{22} = E_f V_f \int_0^\pi \sin^4\theta f(\theta)\, d\theta \tag{1.5}$$

$$C_{26} = E_f V_f \int_0^\pi \sin^3\theta\cos\theta f(\theta)\, d\theta$$

$$C_{12} = C_{66} = E_f V_f \int_0^\pi \cos^2\theta\sin^2\theta f(\theta)\, d\theta$$

Assuming that Eq. (1.2) is a rectangular probability density function (the fibers are uniformly distributed within the interval $0 < \theta < \pi$)

$$f(\theta) = \frac{1}{\pi}, \qquad 0 \le \theta \le \pi \tag{1.6}$$

the elastic constants are

$$C_{11} = C_{22} = \tfrac{3}{8} E_f V_f, \qquad C_{12} = \tfrac{1}{8} E_f V_f, \qquad C_{16} = C_{26} = 0 \qquad (1.7)$$

Therefore the elastic properties of the composite system are:

Young's Modulus	Shear Modulus	Poisson's Ratio
$E_c = C_{11} - \left(\dfrac{C_{12}^2}{C_{22}}\right) = E_f\left(\dfrac{V_f}{3}\right)$	$G_c = C_{12} = E_f\left(\dfrac{V_f}{8}\right)$	$\nu_c = \dfrac{E_c}{2G_c} - 1 = \dfrac{1}{3}$

The netting analysis for a three-dimensional system is similar to the above (Fig. 1.2). The strain ε on a fiber at polar angles θ, φ is

$$\varepsilon = \varepsilon_{11} \cos^2 \theta + \varepsilon_{22} \cos^2 \varphi \sin^2 \theta + \varepsilon_{22} \sin^2 \varphi \sin^2 \theta$$
$$+ \gamma_{23} \sin \varphi \cos \varphi \sin^2 \theta + \gamma_{31} \sin \varphi \sin \theta \cos \theta$$
$$+ \gamma_{12} \cos \phi \sin \theta \cos \theta \qquad (1.8)$$

Assuming that a distribution function $f(\theta, \varphi)$ exists in a way similar to Eq. (1.2), then the stresses can be written as[3]

$$
\begin{Bmatrix} \sigma_{11} \\ \sigma_{22} \\ \sigma_{33} \\ \tau_{23} \\ \tau_{31} \\ \tau_{12} \end{Bmatrix}
=
\begin{bmatrix}
C_{11} & C_{12} & C_{13} & C_{14} & C_{15} & C_{16} \\
 & C_{22} & C_{23} & C_{24} & C_{25} & C_{26} \\
 & & C_{33} & C_{34} & C_{35} & C_{36} \\
 & & & C_{44} & C_{45} & C_{46} \\
 & & & & C_{55} & C_{56} \\
\text{SYM} & & & & & C_{66}
\end{bmatrix}
\begin{Bmatrix} \varepsilon_{11} \\ \varepsilon_{22} \\ \varepsilon_{33} \\ \gamma_{23} \\ \gamma_{31} \\ \gamma_{12} \end{Bmatrix}
\qquad (1.9)
$$

where C_{ij} are the elastic constants of the three-dimensional system, i.e.,

$$C_{22} = E_f V_f \int_0^{2\pi} \int_0^{\pi/2} \sin^4 \theta f(\theta, \varphi)\, d\theta\, d\varphi \qquad (1.10)$$

Similarly to Eq. (1.6), the distribution function for the three-dimensional system is taken as

$$f(\theta, \varphi) = \frac{\sin \theta}{2\pi}, \qquad 0 \le \theta \le \frac{\pi}{2}, \qquad 0 \le \varphi \le 2\pi \qquad (1.11)$$

and the elastic properties are:

Young's Modulus	Shear Modulus	Poisson's Ratio
$E_c = E_f\left(\dfrac{V_f}{6}\right)$	$G_c = E_f\left(\dfrac{V_f}{15}\right)$	$\nu_c = \dfrac{1}{4}$

It is quite obvious that the netting analysis has a number of drawbacks, especially by not considering the contribution of the matrix phase. By doing so, Cox managed to neglect stress transfers from the matrix to the fibers and vice versa, while by assuming that the load is exclusively carried by the fiber ends, he disregarded the effect of the fiber length (also known as gauge length) even if he knew that fiber ends can only carry a small portion of load.[1] Additionally, it predicts zero strength transverse to the direction of the fiber. However, the netting analysis is still used for the design of filament-wound pressure vessels.[4]

1.2.2 The law of mixtures (LOM)

The flaws of the netting analysis were recognized by Gordon[5] and Arridge.[6] They proposed several corrections to include the stress interactions between the fiber and the matrix. Consider a block of unidirectional aligned continuous fibers as shown in Fig. 1.3.

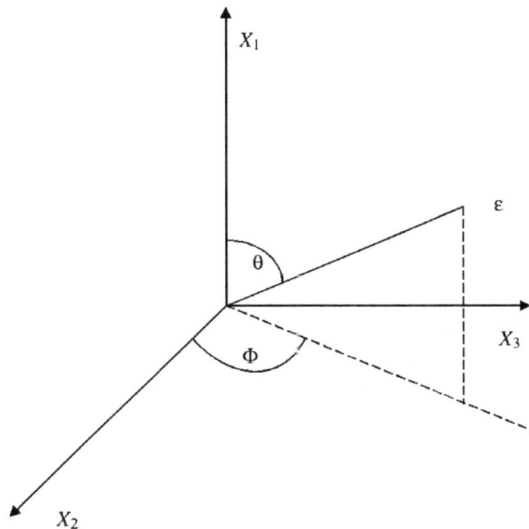

Figure 1.2 The three-dimensional coordinate system.

If a stress σ_1 is applied parallel to axis 1, while a mechanical coupling exists between the fiber and the matrix to allow them to elongate equally, we can then assume strain compatibility between the two phases (often mentioned as the **Voigt model** or **isostrain analysis**),

$$\varepsilon_c = \varepsilon_f = \varepsilon_m \tag{1.12}$$

where c, f, m subscripts denote composite, fiber, and matrix, respectively.

The fundamental principle of Eq. (1.12) is based on the assumption that when an external shear strain ε_a is applied to a multiphase system, and all the phases respond evenly by producing shear strain equal to the applied, then ε_a can be accepted as average, $\bar{\varepsilon}$. Considering that the stress in each phase is given by $G_i\bar{\varepsilon}$, then the average stress of the composite system can be written as

$$\bar{\sigma} = \sum_{i=0}^{N} V_i\sigma_i = \sum_{i=0}^{N} V_iG_i\bar{\varepsilon} \qquad \text{where} \qquad \sum_{i=0}^{N} V_i = 1 \tag{1.13}$$

Additionally, if all phases follow a generalized Hooke's law, then the average stress can be approximated as

$$\bar{\sigma} = G_c\bar{\varepsilon} \tag{1.14}$$

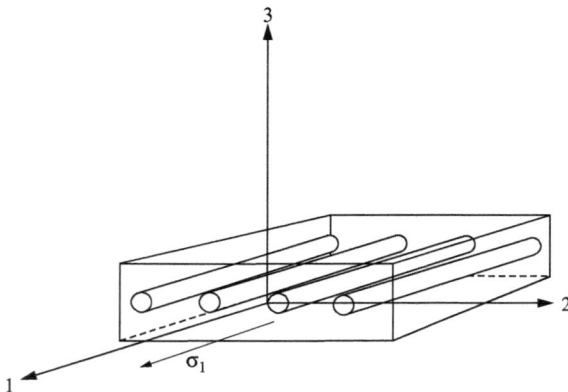

Figure 1.3 Block of unidirectional aligned continuous fibers.

From Eq. (1.13) and Eq. (1.14), G_c is equal to the weighted sum of shear moduli of each phase,

$$G_c = \sum_{i=0}^{N} V_i G_i \qquad (1.15)$$

Equation (1.15) can also predict an average value of the stress of the composite by replacing the shear moduli,

$$\sigma_c = \sum_{i=0}^{N} V_i \sigma_i \qquad (1.16)$$

For the case of a two-phase system, Eq. (1.16) can be rewritten as

$$\sigma_c = \sigma_f V_f + \sigma_m V_m \qquad \text{where} \qquad V_m = (1 - V_f) \qquad (1.17)$$

Accordingly, the tension stress σ_1 in Fig. 1.3 is distributed to the phases as

$$c_1 = \sigma_{f1} V_f + \sigma_{m1}(1 - V_f) \qquad (1.18)$$

where σ_{m1}, σ_{f1} are the corresponding fractions of stress on the matrix and fibers, respectively, in direction 1.

If the matrix is considered isotropic and the fiber orthotropic, then we can relate the stresses and strains of each phase to yield an elastic modulus formula:

$$E_{1c} = E_{1f} V_f + E_{1m}(1 - V_f) \qquad (1.19)$$

Equation (1.19) is the well known **Law of Mixtures equation**. It is obvious that is directly based on the assumption that strain compatibility stands between the phases. One of the drawbacks of Eq. (1.19), is that Eq. (1.15) has an upper bound approximation of the exact value of shear modulus.[7] This is basically because the phases are mainly characterized by different Poisson's contractions ($v_f \neq v_m$). Such difference results in additional radial stresses which are not considered by Eq. (1.19). In addition, Eq. (1.19) could lead to erroneous predictions, especially when phenomena of matrix viscoelasticity and yielding are present (when a polymer or ductile matrix composite is examined). The validity of Eq. (1.19) has been examined extensively throughout the years and, despite its simple structure, it was found adequate to predict experimental results.[8] In Fig. 1.4 the effect of volume fraction on the longitudinal elastic modulus according to Eq. (1.19) is presented.

However, when a stress acts in planes 2–3, the load-bearing capacity of the matrix is significant. As shown in Fig. 1.5 in plane 2, fibers and

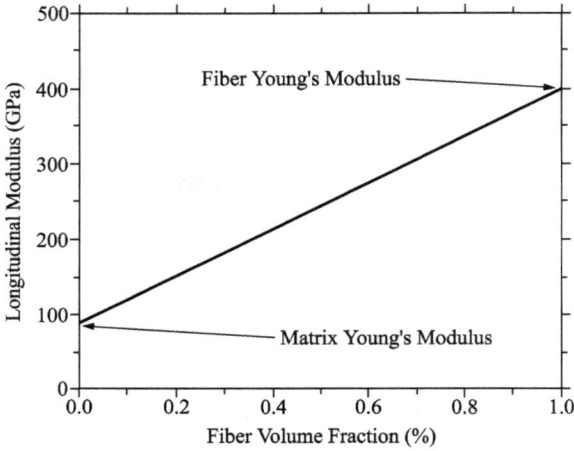

Figure 1.4 The effect of fiber volume fraction on E_c for a SiC/Ti-15-3 composite.

matrix are coupled in "series" and therefore the applied tensile stress is equal in both phases,

$$\sigma_{f2} = \sigma_{m2} = \sigma_2 \qquad (1.20)$$

Because the phases possess different elastic moduli, displacements are expected to be different. The total strain can therefore be assumed to be the weighted sum of the strains in each phase. In a similar way to Eq. (1.18), that is,

$$\varepsilon_2 = \varepsilon_{f2} V_f + \varepsilon_{m2}(1 - V_f) \qquad (1.21)$$

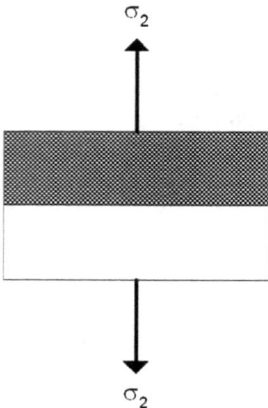

Figure 1.5 Fiber/matrix coupling in plane 2 subjected to tensile stress.

If we invoke once again Hooke's law, the elastic modulus of the composite transverse to the fibers is

$$E_2 = \frac{E_{f2}E_{m2}}{(1 - V_f)E_{f2} + V_f E_{m2}} \tag{1.22}$$

The meaning of Eq. (1.22) is that the fibers play the role of constraining matrix deformation.[9] The assumption of **isostress** condition, Eq. (1.20), is physically unrealistic since the fiber cannot be interpreted as a sheet.

Similarly the Poisson's ratio according to LOM can be written as

$$v_{12} = V_f v_f + (1 - V_f)v_m \tag{1.23}$$

The shear modulus is derived in a way similar to transverse modulus, leading to

$$\frac{1}{G_{12}} = \frac{V_f}{G_{f12}} + \frac{(1 - V_f)}{G_m} \tag{1.24}$$

where G_{f12} is the shear modulus of the fiber in the 1–2 direction and G_m is the shear modulus of the matrix.

1.2.3 The effect of matrix phase yielding on elastic modulus

In the previous section, we mentioned that the basic limitation of the Law of Mixtures is when the matrix material yields. The effect of matrix yielding on elastic modulus was first introduced by Shaffer in 1964.[10] Shaffer considered that in the case of a ductile matrix, there is a point where the elastic-perfectly plastic matrix starts to yield. Consequently, as the stress continue to increase, the stress at the matrix remains constant while the stress at the fiber continues to increase. Let us now consider that a load Q is applied at the end plates of a unit composite area, as shown in Fig. 1.6.

The distribution of load on both phases we assume to be

$$Q = Q_f + Q_m \tag{1.25}$$

To approach a state of matrix yielding, it is necessary to introduce an additional assumption

$$\frac{Y_{f1}}{Y_{m1}} > \frac{E_{f1}}{E_{m1}} \qquad (1.26)$$

where Y_{f1}, Y_{m1} are the yield stress of the fibre and matrix in direction 1, respectively.

Equation (1.26) merely states that the matrix yields before the fiber. In reality, since most types of fiber cannot undergo plastic deformation, Y_{f1} is substituted by the fiber tensile strength, as shown later. Combining Eqs. (1.12), (1.19), and (1.25), we obtain the stress distribution at the instance of matrix yielding (see Fig. 1.7)

$$\sigma_{1c}^{A} = \frac{Y_{m1}}{E_{m1}} E_{f1} V_f + Y_{m1}(1 - V_f) \qquad (1.27)$$

Transformation of Eq. (1.27) in terms of the elastic modulus yields similar results to Eq. (1.19)

$$E_{1c}^{A} = E_{f1} V_f + E_{m1}(1 - V_f) \qquad (1.28)$$

where E_{1c}^{A} represents the first slope in Fig. 1.7.

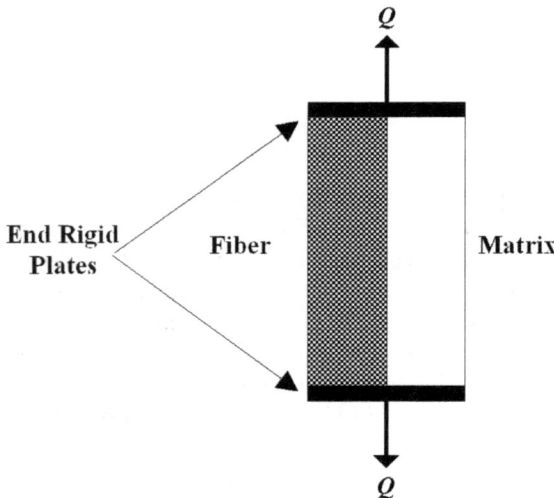

Figure 1.6 Fiber/matrix coupling in plane 1 subjected to load Q parallel to fiber direction. The end rigid plates ensure equal elongation of the phases.

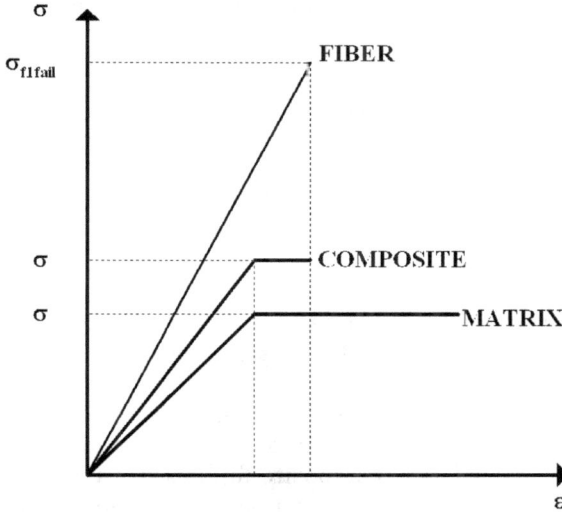

Figure 1.7 Stress–strain diagram of a composite system with ductile matrix phase.

As the stress increases, the stress at the fiber will follow a linear relationship,

$$\sigma_{f1} = \frac{\sigma_{c1}^B - Y_{m1}(1 - V_f)}{V_f} \tag{1.29}$$

where σ_{c1}^B is the stress at the composite after matrix yielding.
   If $Y_{f1} \to \infty$ and therefore $Y_{f1} \gg Y_{m1}$, Eq. (1.29) reduces to

$$\sigma_{c1}^B = \sigma_{f1} V_f \tag{1.30}$$

or $$E_{c1}^B = E_{f1} V_f \tag{1.31}$$

Equation (1.29) represents the second slope of the stress–strain diagram. From Eqs. (1.28) and (1.31) it is obvious that the material stiffness degrades after matrix yielding.
   For the evaluation of the transverse elastic properties, we employ the same configuration as in Fig. 1.6 except that the composite is loaded in plane 2. A typical unit thickness is shown in Fig. 1.8. The fiber diameter is d_f, the matrix phase spacing is S, and the fibers are

equally spaced a distance h apart. If we assume that the triangular area is repeated throughout the composite, then we derive

$$A_{f2} = \left(\frac{\pi}{2\sqrt{3}}\right)\frac{d_f^2}{h^2} \tag{1.32}$$

and
$$S = \left(\frac{\sqrt{3}}{2}\right)h - d_f \tag{1.33}$$

where A_{f2} is the area occupied by the fibers and S is the vertical distance between two fiber rows. Typically, S takes positive values unless A_{f2} is greater than 68%.[11]

In this work we only examine positive values of S, cases of $S = 0$ or $S < 0$ are presented in detail elsewhere.[11] If $S > 0$, then under the application of the external load L, the composite follows two paths of elastic response. The first path represents the response of the matrix strip, S, and the second results from the response of the strip d_f which contains both phases. The problem can be simplified by considering a two-bar model, Fig. 1.9.

The length of the fiber phase, l_f, in the hybrid bar 2 can be simply determined to be

$$\frac{l_f}{h} = \frac{(\pi d_f^2/4)}{h d_f} = \frac{\pi d_f}{4h} \tag{1.34}$$

The hybrid bar could be simplified as two bars in series bonded together by a perfect bond (the concept of perfect bond is used to obviate any debonding phenomena that might cause displacement

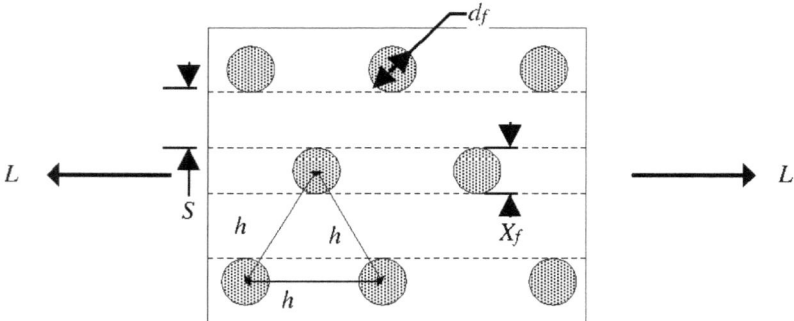

Figure 1.8 Unit thickness of a unidirectional composite system in the transverse direction.

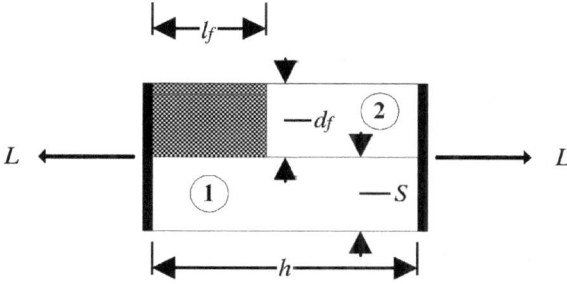

Figure 1.9 The two-bar model simplification. At the end of the bars rigid plates are assumed to ensure equal displacement of both bars. The bar 2 is consisted by a combination of both phases while the bar 1 is consisted only by the matrix phase.

irregularities). In the case of an external stress σ, equally applied at the ends of bar 2, the stress distribution is given as

$$\sigma_m = \sigma_f = \sigma \tag{1.35}$$

while the corresponding displacements are

$$u = u_f + u_m \tag{1.36}$$

where u is the total displacement of the bar and u_f, u_m are the displacements of the fiber and the matrix phase, respectively.

Combining Eq. (1.35) and Eq. (1.36) and introducing the elastic moduli of the phases, we obtain

$$\frac{\sigma}{E_c^h} h = \frac{\sigma}{E_{f2}} l_f + \frac{\sigma}{E_{m2}} (h - l_f) \tag{1.37}$$

where E_c^h is the elastic modulus of the hybrid bar. By substituting l_f, from Eq. (1.34) into Eq. (1.37) we derive

$$\frac{E_{m2}}{E_c^h} = 1 - \left(\frac{\pi d_f}{4h}\right)\left[1 - \left(\frac{E_{m2}}{E_{f2}}\right)\right] \tag{1.38}$$

If $E_{f2} > E_{m2}$ in the hybrid bar, then E_c^h in Eq. (1.38) is greater than the elastic modulus of the pure matrix bar. Consequently, the stress in the hybrid bar will be greater than that of bar 1. It is therefore rational to assume that the pure matrix phase (bar 2) will reach yielding before the matrix phase in bar 1. Therefore, the stress distribution at the instance of yielding will be

$$\sigma_c(d_f + S) = \sigma d_f + \sigma_m^1 S \tag{1.39}$$

where σ_c is the stress at the composite ends, σ is the stress at the matrix phase of the bar 2 $(= Y_{m2})$, and σ_m^1 is the stress of bar 1 $[= (E_{m2}/E_c^h)\sigma]$.

Combining Eq. (1.38) and Eq. (1.39), the stress σ_c is

$$\sigma_c = \frac{Y_{m2}}{df + S}\left\{d_f + S\left[1 - \frac{\pi d_f}{4h}\left(1 - \frac{E_{m2}}{E_{f2}}\right)\right]\right\} \tag{1.40}$$

while the corresponding strain of the composite is

$$\varepsilon_c = \frac{Y_{m2}}{E_c^h} = \frac{Y_{m2}}{E_{m2}}\left[1 - \frac{\pi d_f}{4h}\left(1 - \frac{E_{m2}}{E_{f2}}\right)\right] \tag{1.41}$$

Combining Eqs. (1.40) and (1.41), the elastic modulus of the composite transverse to the fiber direction is

$$E_c = \frac{\sigma_c}{\varepsilon_c} = \frac{E_{m2}}{d_f + S}\left\{\frac{d_f + S\left[\left(1 - \frac{\pi d_f}{4h}\right)\left(1 - \frac{E_{m2}}{E_{f2}}\right)\right]}{1 - \left(\frac{\pi d_f}{4h}\right)\left(1 - \frac{E_{m2}}{E_{f2}}\right)}\right\} \tag{1.42}$$

1.2.4   The ultimate tensile strength

In Sections 1.1.1 and 1.2.3 we assumed that the matrix material after yielding does not contribute to the stress response of the system. Therefore, according to LOM, the stress of the composite at failure can be written as

$$\sigma_c^f = \sigma_f^f V_f + Y_m(1 - V_f) \tag{1.43}$$

where σ_f^f, is the fracture strength of the fiber phase.

Equation (1.43) states that only the fiber strength dictates the failure stress of the composite. However, many matrices may experience viscoelastic or plastic deformation during loading. In this case the accuracy of Eq. (1.43) is questionable. Since isostrain conditions prevail, the question arising from the above is, Which phase fails first? To answer that, we should consider both possible situations: (a) the fiber fails at strain greater than the matrix failure strain or (b) the fiber fails at lower strain than that of the matrix.

Let us now assume that a unidirectional reinforced lamina is subjected to uniaxial tension and that the failure strain of the fiber is greater than that of the matrix, $\varepsilon_f^f > \varepsilon_m^f$. At the instance of matrix failure, the stress at the composite is given by

$$\sigma_c = \sigma_f^* V_f + \sigma_m^f (1 - V_f) \tag{1.44}$$

where σ_m^f is the matrix strength and σ_f^* is the corresponding stress at the fiber.

When the matrix fails, then the load carried by the matrix is transferred to the fibers, which in turn, depending on their strength and volume fraction, may fail or continue to support full load. This is illustrated in Fig. 1.10.

Generally when V_f is large, the fibers are likely to sustain the extra load due to matrix failure. In this case the ultimate tensile strength of the composite is **fiber dominated** and is given as

$$\sigma_{\text{UTS}} = \sigma_f^f V_f \tag{1.45}$$

However, for small values of V_f, the fibers are unable to support the extra load and consequently they fail simultaneously with the matrix. In that case the ultimate tensile strength is **matrix dominated**

$$\sigma_{\text{UTS}} = \frac{\sigma_m^f}{E_m} E_f V_f + \sigma_m^f (1 - V_f) \tag{1.46}$$

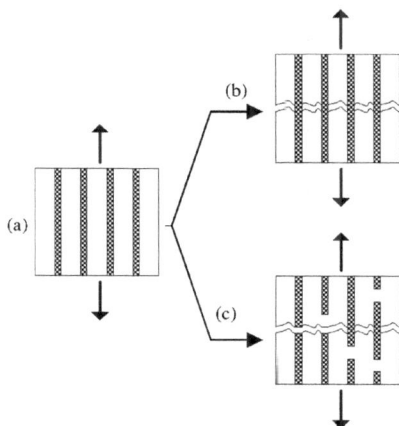

Figure 1.10 Longitudinal tensile failure of a unidirectional lamina subjected to uniaxial tensile load: (a) before matrix failure; (b) the fibers support full load after matrix failure (large V_f); and (c) fibers fail after matrix failure (small V_f).

Combining Eqs. (1.45) and (1.46) we can estimate the value of V_f, where the transition between fiber and matrix dominated strength exists

$$V_f^* = \frac{\sigma_m^f E_m}{\sigma_f^f E_m - \sigma_m^f E_f + E_m \sigma_m^f} \tag{1.47}$$

In Fig. 1.11 the effect of volume fraction on the ultimate tensile strength response of a unidirectional reinforced composite is presented; a unidirectional SiC/Ti-6-4 composite was considered.

A similar analysis could be used in the case of $\varepsilon_f^f < \varepsilon_m^f$. At the instant of fiber failure two different conditions are detected: (a) V_f is low and the extra load transferred by the broken fibers to the matrix is not enough to cause matrix failure; in this case the σ_{UTS} is given by

$$\sigma_{UTS} = \sigma_m^f (1 - V_f) \tag{1.48}$$

The lack of any fiber contribution indicates the reduction of the cross-effective area of the matrix by holes left at fiber breaks. (b) V_f is large and therefore the load carried by the fibers before failure is capable of causing matrix failure just after fiber fracture; therefore, σ_{UTS} is given by

$$\sigma_{UTS} = \sigma_f^f V_f + \frac{\sigma_f^f}{E_f} E_m (1 - V_f) \tag{1.49}$$

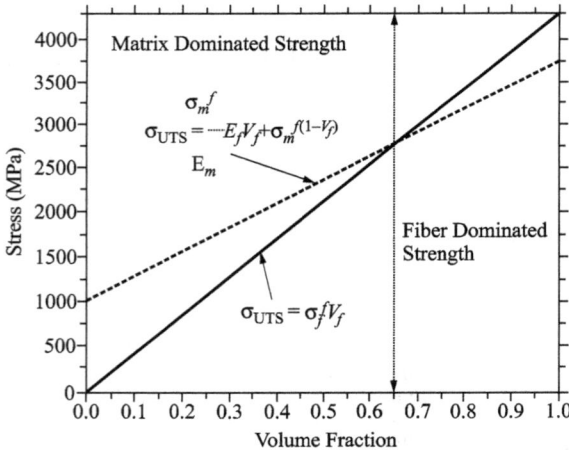

Figure 1.11 The areas of matrix dominated and fiber dominated ultimate tensile strength. For the plotting, the following mechanical properties were used: $\sigma_f^f = 4300\,\mathrm{MPa}$,[12] $\sigma_m^f = 1035\,\mathrm{MPa}$,[13] $E_m = 110\,\mathrm{GPa}$,[14] $E_f = 400\,\mathrm{GPa}$.[13]

It is clear that Eq. (1.48) states the matrix strength dependency, while Eq. (1.49) gives the fiber strength dependency of the σ_{UTS}.

Combination of the last two equations yields a similar conclusion as from Eq. (1.47):

$$V_f^* = \frac{\sigma_f^f E_m}{\sigma_m^f E_f - \sigma_f^f E_f - \sigma_f^f E_m} \qquad (1.50)$$

The two different responses of the material depending on V_f are shown schematically in Fig. 1.12.

It is well known that all materials, but especially brittle ones, are governed by a distribution of strengths. This is basically due to the presence of inherent or manufacturing critical size defects. These defects, depending on their distribution and size, are likely to cause premature fracture (the term premature here defines probability of failure). In monolithic materials, especially ductile ones, the above might not be the most crucial parameters in the design process since strength distribution tolerance already exists in many design methodologies, i.e., the threshold value of the stress intensity factor. In composite materials, however, premature failure of the reinforcing phase may cause unpredictable conditions which mainly result in local overloading of the matrix and of the neighboring fibers. For example, when a fiber fails, then the neighboring fibers, which are now overloaded, might fail as well.

Additionally, the holes left by broken fibers could induce local stress concentrations. These stress concentrations, depending on the matrix toughness, may initiate cracks that could lead to catastrophic failure.

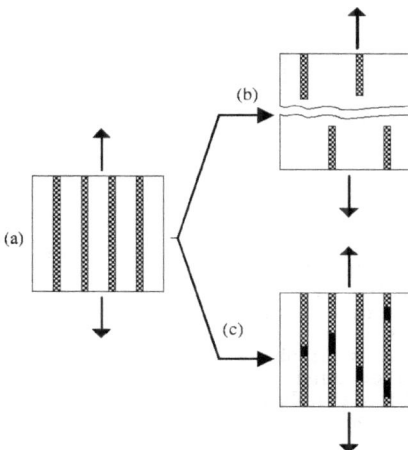

Figure 1.12 Longitudinal tensile failure of a unidirectional lamina subjected to uniaxial tensile load: (a) before fiber failure; (b) the matrix fails after fiber failure (large V_f); and (c) the matrix supports full load after fiber failure (small V_f).

In Fig. 1.13, failures resulting from a single fiber failure are shown schematically.

From the above it is obvious that a rigorous and realistic evaluation of the σ_{UTS} should primarily take into account the strength distribution of the reinforcing phase. For this purpose, we introduce a statistical approach to the fiber strength based on the **Weibull distribution analysis**.[15] Because an extensive analysis of the Weibull distribution is beyond the scope of this book, only the basics will be examined. According to Weibull, the probability of failure of a brittle material under uniform loading at stress σ is given by

$$P_f = 1 - \exp\left[-V\left(\frac{\sigma - \sigma_u}{\sigma_0}\right)^m\right] \tag{1.51}$$

where m is a parameter known as the Weibull modulus, V is the volume of the material, σ_0 is a normalizing factor and σ_u is the stress below which the probability of failure (P_f) is zero.

The Weibull modulus is related to the flaw size distribution of a brittle material. Thus, m is defined as a material parameter that characterizes the **brittleness** of the material. Large values of m indicate materials with a uniform flaw distribution, whereas small values of m relate to materials with flaw sizes described by a wide

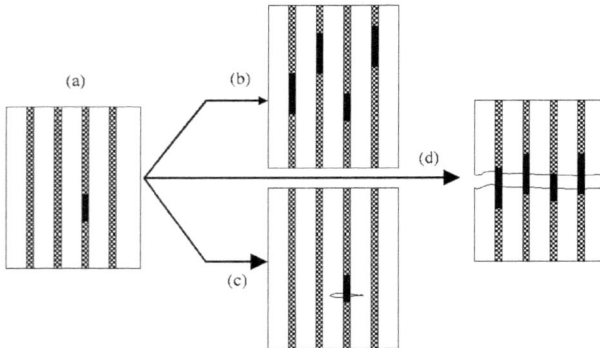

Figure 1.13 Possible results of a single fiber break in a unidirectional composite. (a) Local cracking of the matrix material with possible failure due to general yielding (high matrix toughness). (b) Some of the neighboring fibers fail but the redistribution of stress is not sufficient to cause catastrophic failure. (c) Owing to high density of fiber breaks, extensive micro-cracking may be developed, which could lead to failure of the composite in a brittle manner (low matrix toughness).

scatter of distribution. The mean of the distribution described by Eq. (1.51) is

$$\bar{\sigma} = \sigma_u + \frac{\sigma_0 \Gamma\left[1 + \left(\dfrac{1}{m}\right)\right]}{V^{1/m}} \tag{1.52}$$

where Γ is the tabulated gamma function. Combining Eqs. (1.51) and (1.52) we obtain

$$P_f = 1 - \exp\left\{-\left[\Gamma\left(1 + \frac{1}{m}\right)\left(\frac{\sigma - \sigma_u}{\bar{\sigma} - \sigma_u}\right)\right]^m\right\} \tag{1.53}$$

If the gauge length or the probability of defects in an definite fiber length, L, is involved, Eq. (1.53) can be written in the form[16]

$$P_f(\sigma, L) = 1 - \exp\left[\frac{L}{L_0}\left(\frac{\sigma - \sigma_u}{\sigma_0}\right)^m\right] \tag{1.54}$$

where L_0 is a gauge length for which the distribution is known.

Equation (1.54), can also be used to describe the average fiber strength at a given gauge length considering the fiber diameter[17]

$$\bar{\sigma} = \sigma_0 \left(\frac{L}{d}\right)^{-1/m} \Gamma\left(\frac{m+1}{m}\right) \tag{1.55}$$

where d is the fiber diameter.

Taking the previous considerations into account, Masson and Bourgain[18] proposed the following estimation of the Weibull failure probability: if σ_1, σ_2 are the strengths at a given failure probability for gauge lengths L_1, L_2, then the definition of m may be written as

$$\sigma_2 = \sigma_1 \left(\frac{L_1}{L_2}\right)^{1/m} \tag{1.56}$$

Equation (1.56) implies that the strength variation of brittle materials is an indication of size effects. Therefore, when designing large components using data from small test-pieces, Eq. (1.56) might not be sufficiently accurate.[19] From Eq. (1.55) it is evident that, in the design of composite materials, there is a crucial relationship between the geometrical aspects of the reinforcing phase and the mechanical properties of the component that the design engineer must consider.

Even though Eq. (1.55) provides a good approximation of the fiber strength, an even more accurate approach is to consider the associated strength of a very large number of parallel fibers (**fiber bundle**). If we assume that all fibers have equal diameter d and length L and that they are clamped at their ends in such a way that all the unbroken fibers have the same strain ε, then the bundle will exhibit strength equal to that of the individual fiber.[20] Therefore, as the load carried by the unbroken fibers increases by failure of the weaker fibers, the load-bearing capacity of the bundle is reduced. Hence, the strength of the bundle will be found to be lower than that of an individual fiber. The strength of a bundle, σ_b, has been related to the mean strength of individual fibers, $\bar{\sigma}$[20] by

$$\frac{\sigma_b}{\bar{\sigma}} = \left(\frac{1}{me}\right)^{1/m} \frac{1}{\Gamma\left(1+\frac{1}{m}\right)} \tag{1.57}$$

where e is the base of natural logarithms.

The advantages derived from the use of Eq. (1.57) are that (a) the bundle represents in a more realistic way the composite system and (b) Eq. (1.57) could provide a more reliable approximation of the reinforcement strength, especially in the case of fibers with low value of m (see Fig. 1.14).

However, in some cases the *in situ* average strength of the fibers differs from that calculated from *dry* fiber bundle measurements. It has been understood that this difference arises from the sliding resistance (also known as frictional coupling) of the fiber/matrix interface.[21] Thus, realistic measurements can be achieved by extracting fibers from the composite and measuring their tensile strengths.[22]

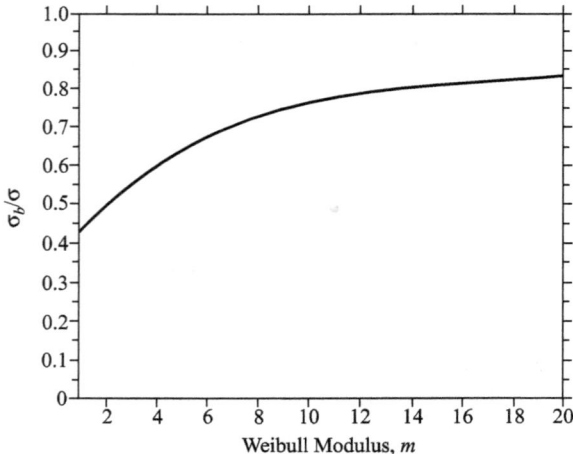

Figure 1.14 The effect of Weibull modulus on the ratio $\sigma_b/\bar{\sigma}$. For large values of m ($m > 14$), the strength ratio tends asymptotically to 0.8.

Traditionally, the approach to evaluating the stress on the fiber has been established considering that no frictional effect exists between the fiber and the matrix.[23] On such basis, the fiber response is given as

$$\sigma_f = (1 - P_f)\varepsilon E_f \tag{1.58}$$

Combining Eqs. (1.54) and (1.58) and considering that $\sigma_u = 0$ as the worst case scenario, we obtain

$$\frac{\sigma_i}{\sigma_0} = \frac{\varepsilon E_f}{\sigma_0} \exp \frac{-L}{L_0} \left(\frac{\varepsilon E_f}{\sigma_0}\right)^m \tag{1.59}$$

Therefore the composite fracture strength and the corresponding fracture strain are evaluated by setting

$$\frac{d\sigma_{\text{CUTS}}}{d\varepsilon} = V_f \frac{d\sigma_f}{d\varepsilon} = 0 \tag{1.60}$$

whereupon $$\sigma_{\text{CUTS}} = V_f \sigma_0 \left(\frac{meL}{L_0}\right)^{-1/m} + (1 - V_f) Y_m \tag{1.61}$$

and $$\varepsilon_{fr} = \frac{\sigma_0}{E_f} \left(\frac{mL}{L_0}\right)^{-1/m} \tag{1.62}$$

The advantages deriving from Eq. (1.61) are that (a) the effect of a single fiber failure on the neighboring fibers can be more realistically approached; (b) the statistical strength of the fiber is taken into account; (c) the matrix part can include phenomena such as residual stresses, straining response (i.e. strain hardening, etc.); and (d) both parts could include environmental and fatigue degradation aspects.

1.3   The Basic Mechanics of a Composite System—Elasticity Models

The need for more sophisticated models to describe the basic mechanical properties of composite materials emanated from the drawbacks deriving from the application of netting analysis and LOM. For example, both models considered properties in one direction by excluding the effect of the other two directions. Hence, models based on elasticity theory should be employed.[24-26] If we consider the three

planes of elastic symmetry of an orthotropic material, then the
elasticity constants required for description are

$$\varepsilon_{11} = \frac{1}{E_1}[\sigma_{11} - (\nu_{12}\sigma_{22} + \nu_{13}\sigma_{33})]$$

$$\varepsilon_{22} = \frac{1}{E_2}[\sigma_{22} - (\nu_{21}\sigma_{11} + \nu_{23}\sigma_{33})]$$

(1.63)

$$\varepsilon_{33} = \frac{1}{E_3}[\sigma_{33} - (\nu_{31}\sigma_{11} + \nu_{32}\sigma_{22})]$$

$$\gamma_{23} = \frac{\tau_{23}}{G_{23}}, \qquad \gamma_{12} = \frac{\tau_{12}}{G_{12}}, \qquad \gamma_{31} = \frac{\tau_{31}}{G_{31}}$$

where
$$\frac{\nu_{13}}{E_1} = \frac{\nu_{31}}{E_3}, \qquad \frac{\nu_{12}}{E_1} = \frac{\nu_{21}}{E_2}, \qquad \frac{\nu_{23}}{E_2} = \frac{\nu_{32}}{E_3}$$

In the case of a unidirectional fiber-reinforced composite, the elastic
constants in the X_2 and X_3 directions will be the same (Fig. 1.15).
Hence Eq. (1.63) reduces to six constants (E_1, E_2, ν_{12}, G_{12}, ν_{23}, G_{23}),[27]

$$\varepsilon_{11} = \frac{1}{E_1}[\sigma_{11} - \nu_{12}(\sigma_{22} + \sigma_{33})]$$

$$\varepsilon_{11} = \frac{1}{E_2}[\sigma_{22} - (\nu_{21}\sigma_{11} + \nu_{23}\sigma_{33})]$$

(1.64)

$$\varepsilon_{11} = \frac{1}{E_2}[\sigma_{33} - (\nu_{21}\sigma_{11} + \nu_{23}\sigma_{22})]$$

$$\gamma_{12} = \frac{\tau_{12}}{G_{12}}, \qquad \gamma_{23} = \frac{\tau_{23}}{G_{23}}, \qquad \gamma_{31} = \frac{\tau_{31}}{G_{12}}$$

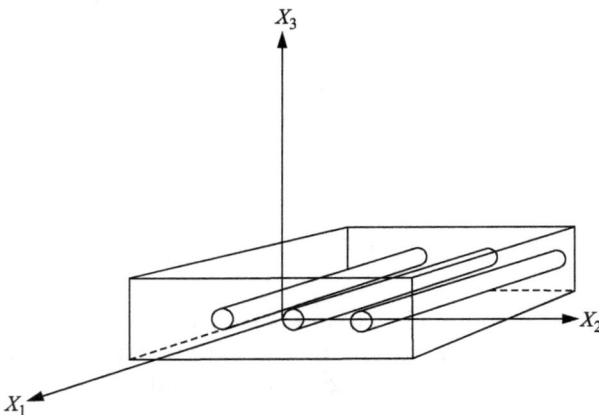

Figure 1.15 The unidirectional configuration with three
planes of reference.

where

$$E_2 = E_3, \qquad \nu_{21} = \nu_{31}, \qquad \nu_{12} = \nu_{13}, \qquad \nu_{23} = \nu_{32}, \qquad G_{12} = G_{31}$$

If planes X_2–X_3 are isotropic then the six independent constants in Eq. (1.64) become five with the introduction of

$$G_{23} = \frac{E_2}{2(1 + \nu_{23})} \tag{1.65}$$

Further reduction into four constants (E_1, E_2, ν_{12}, G_{12}) can be achieved if we considered plane stress conditions (X_3 is very thin). Hence

$$\varepsilon_{11} = \frac{1}{E_1}(\sigma_{11} - \nu_{12}\sigma_{22})$$

$$\varepsilon_{22} = \frac{1}{E_2}(\sigma_{22} - \nu_{21}\sigma_{11}) \tag{1.66}$$

$$\gamma_{12} = \frac{\tau_{12}}{G_{12}}, \qquad E_2\nu_{12} = E_1\nu_{21}$$

Equation (1.66) can be further reduced for specific composite lay-up. For example, in the case of a 0°, 45°, and 90° composite sheet (Fig. 1.16) only two constants are required

$$E_2 = E_1, \qquad \nu_{21} = \nu_{12}, \qquad G_{12} = \frac{E_1}{2(1 + \nu_{12})} \tag{1.67}$$

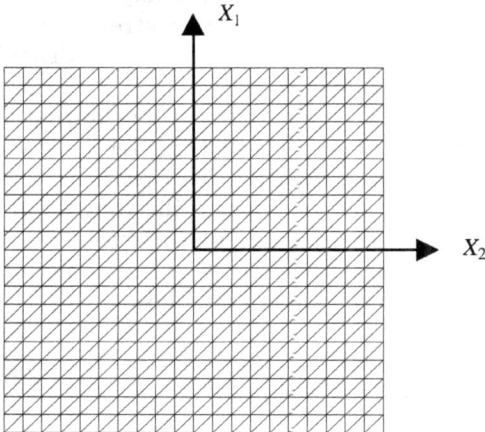

Figure 1.16 The 0°, 45°, and 90° composite lay-up. Such a configuration may be considered as pseudoisotropic.

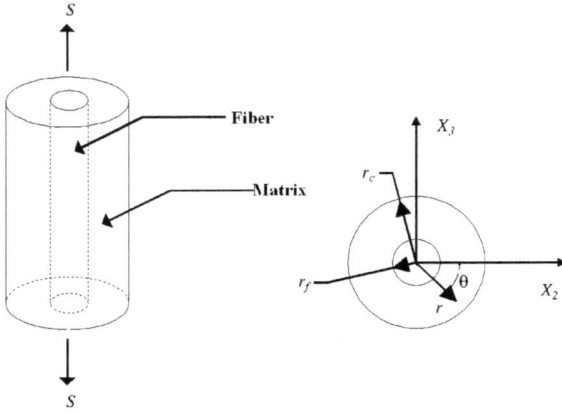

Figure 1.17   The concentric cylinder element introduced by Hashin and Rosen.[26]

1.3.1   Longitudinal modulus

In the previous section it was shown that, depending on the complexity of the composite system (packing configuration of the fibers and volume fraction), the number of the elastic constants required to rigorously describe the elastic properties varies between two and six (more complex systems require greater number of constants). In the case of little regularity, the analysis could be based on the use of **repeating elements**. As repeating element we define the representation of the mechanical system through a set of duplicated elements with identical mechanical responses. The type of repeating element may vary according to the symmetry and regularity of the system from hexagonal elements (regular systems) to concentric cylinder elements (irregular systems).[11,28] Although the repeating elements method is widely used, a typical example being finite element analysis,[29] the application of the method may induce complex boundary stress states due to interactions between the elements. Typical problems and drawbacks of the method are discussed elsewhere.[11]

Let us now consider an axial force S applied on an irregular composite system described by the concentric cylinder element (Fig. 1.17). If we assume that the axial strain ε_1 is constant, then Fig. 1.17 represents a plane strain problem that can be solved by considering the tangential, radial (function of r, θ) and axial (function of X_1) displacements.

If the problem is considered axisymmetric, then we can make use of the Airy* stress function (also called bi-harmonic) as follows[30]

$$\nabla^4 F = \frac{d^4 F}{dr^4} + \frac{2}{r}\frac{d^3 F}{dr^3} - \frac{1}{r^2}\frac{d^2 F}{dr^2} + \frac{1}{r^3}\frac{dF}{dr} = 0 \qquad (1.68)$$

---

* Named after Sir George Bidell Airy (1801–1892), English mathematician and astronomer.

The solution of Eq. (1.68) is

$$F = C_1 \log r + C_2 r^2 \log r + C_3 r^2 + D \tag{1.69}$$

where C_1, C_2, C_3, D are solution constants.

Therefore, the radial and tangential stresses at the fiber and the matrix can be written as

$$
\begin{aligned}
\sigma_{fr} &= \frac{1}{r}\frac{dF_f}{dr} = \frac{A}{r^2} + B(1 + 2\log r) + 2C \\
\sigma_{f\theta} &= \frac{1}{r}\frac{d^2 F_f}{dr^2} = -\frac{A}{r^2} + B(3 + 2\log r) + 2C \\
\sigma_{mr} &= \frac{1}{r}\frac{dF_m}{dr} = \frac{D}{r^2} + J(1 + 2\log r) + 2K \\
\sigma_{m\theta} &= \frac{1}{r}\frac{d^2 F_m}{dr^2} = -\frac{D}{r^2} + J(3 + 2\log r) + 2K
\end{aligned}
\tag{1.70}
$$

The subscripts m and f stand for matrix and fiber respecitvely. A, B, C, D, K are arbitrary constants. If the stress at the center of the fiber ($r = 0$) is finite, then $A = B = 0$. Using the Beltrami–Michell compatibility equation $[\nabla^2(\sigma_{xx} + \sigma_{yy}) = 0]$, it can be easily found that $J = 0$. Therefore, Eq. (1.70) is simplified to

$$
\begin{aligned}
\sigma_{fr} &= \sigma_{f\theta} = 2C \\
\sigma_{mr} &= \frac{D}{r^2} + 2K \\
\sigma_{m\theta} &= -\frac{D}{r^2} + 2K
\end{aligned}
\tag{1.71}
$$

To obtain the axial stresses, we make use of Eq. (1.61) and a stress–strain relationship.[11] Thus

$$
\begin{aligned}
\varepsilon_{f1} = \varepsilon_1 &= \frac{1}{E_f}[\sigma_{f1} - \nu_f(\sigma_{fr} + \sigma_{f\theta})] \\
\varepsilon_{m1} = \varepsilon_1 &= \frac{1}{E_m}[\sigma_{m1} - \nu_m(\sigma_{mr} + \sigma_{m\theta})]
\end{aligned}
\tag{1.72}
$$

From Eqs. (1.71) and (1.72) we derive

$$
\begin{aligned}
\sigma_{f1} &= E_f \varepsilon_1 + 4\nu_f C \\
\sigma_{m1} &= E_m \varepsilon_1 + 4\nu_m K
\end{aligned}
\tag{1.73}
$$

where ν_f and ν_m are the Poisson's ratio of the fiber and matrix, respectively.

To solve Eq. (1.73), we must first determine the arbitrary constants C and K. A typical approach is based on the introduction of additional equations by considering the boundary conditions of the problem. For example, at $r = r_f$ we have

$$\sigma_{fr} = \sigma_{mr}, \qquad u_{fr} = u_{mr} \tag{1.74}$$

where u_{fr}, u_{mr} are radial displacements. Analytically,

$$
\begin{aligned}
u_{fr} &= r\varepsilon_{f\theta} = \frac{r}{E_f}[2C - v_f(2C + 4v_fC + E_f\varepsilon_1)] \\
u_{mr} &= r\varepsilon_{m\theta} = \frac{r}{E_m}\left[2K - \left(\frac{D}{r^2}\right) - v_m\left(2K + 4v_mK + E_m\varepsilon_1 + \frac{D}{r^2}\right)\right]
\end{aligned}
\tag{1.75}
$$

Substituting Eqs. (1.71) and (1.75) into the boundary condition of Eq. (1.74) yields

$$2C = E_f\left[2K(1 - v_m - 2v_m^2) - \frac{D}{r_f^2}(1 + v_m) - v_mE_m\varepsilon_1\right] \tag{1.76}$$

at $r = r_c$ we have

$$\sigma_{mr} = 0 \quad \text{and therefore} \quad 2K = -\frac{D}{2r_c^2} \tag{1.77}$$

According to Reference 11 the elastic modulus, E_1, is obtained by equating the strain energy of the concentric cylinder element to the strain energies of the fiber and the matrix

$$
\frac{1}{2}\int_V \sigma_1\varepsilon_1\,dV = \frac{1}{2}\int_{V_f}(\sigma_{f1}\varepsilon_1 + \sigma_{f\theta}\varepsilon_{f\theta} + \sigma_{fr}\varepsilon_{fr})\,dV_f
$$

$$
+ \frac{1}{2}\int_{V_m}(\sigma_{m1}\varepsilon_1 + \sigma_{m\theta}\varepsilon_{m\theta} + \sigma_{mr}\varepsilon_{mr})\,dV_m \tag{1.78}
$$

and $V = V_f + V_m$. Substituting Eqs. (1.71), (1.73), (1.76), and (1.77) into Eq. (1.78), the elastic modulus is given as

$$
E_1 = \frac{[2(v_f - v_m)^2 E_f E_m(1 - V_f)V_f]}{\{E_m(1 - V_f)(1 - v_f - 2v_f^2) + E_f[V_f(1 - v_m - 2v_m^2) + (1 - v_m)]\}}
$$
$$
+ E_m + (E_f - E_m)V_f \tag{1.79}
$$

It should be noted that for a small difference between the Poisson's ratio of the phases, Eq. (1.79) is simplified to the Law of Mixtures.

1.3.2 Poisson's ratio

For the concentric cylinder element, two values of Poisson's ratio are significant for examination, i.e. v_{12}, v_{23}. To obtain a solution for v_{12}, we consider the radial displacement of the element

$$U_r = r_c v_{12} \varepsilon_1 \qquad (1.80)$$

Equation (1.75) yields a similar equation if we employ the solutions of the arbitrary constants, i.e.

$$U_r = \frac{r_c}{E_m}(\sigma_{m\theta} - v_m \sigma_{m1}) \qquad (1.81)$$

at $r = r_c$.

Combining the previous solution for the tangential and axial matrix stress and Eqs. (1.80) and (1.81), the v_{12} is given by

$$v_{12} = v_m - \frac{[2(v_m - v_f)(1 - v_m^2)E_f V_f]}{\{E_m(1 - V_f)(1 - v_f - 2v_f^2) + E_f[V_f(1 - v_m - 2v_m^2) + (1 + v_m)]\}} \qquad (1.82)$$

The Poisson's ratio v_{23}, can be approximated to follow the LOM[11]

$$v_{23} = v_f V_f + v_m V_m \qquad (1.83)$$

1.3.3 Transverse modulus

If an axial stress and a radial pressure are applied to the concentric cylinder element, the radial pressure is given as

$$p = \frac{\sigma_2 + \sigma_3}{2} \qquad (1.84)$$

If now we assume that the axial strain is zero under the combined loading, the change in volume of the element will be

$$\Delta N = \varepsilon_2 + \varepsilon_3 \qquad (1.85)$$

From Eq. (1.3) the change in volume can be written as

$$\Delta N = \frac{(1 - v_{23})}{E_2}(\sigma_2 + \sigma_3) - \frac{2v_{12}}{E_1}\sigma_1 \qquad (1.86)$$

At this stage we introduce the **Bulk modulus**, K_{23}, which is defined as

$$K_{23} = \frac{p}{\Delta N} \qquad (1.87)$$

Generally the bulk modulus of any material is related to the other elastic constants by

$$K_i = \frac{E_i}{3(1 - 2v_i)} \tag{1.88}$$

By substituting Eqs. (1.84) and (1.86) into Eq. (1.87) we obtain

$$K_{23} = \frac{E_2}{2\left[1 - v_{23} - \left(2v_{12}^2 \dfrac{E_2}{E_1}\right)\right]} \tag{1.89}$$

Using the boundary conditions for equal stress as before, and after some algebraic manipulation, the transverse elastic modulus can be written as

$$E_2 = \frac{2K_{23}E_1(1 - v_{23})}{E_1 + 4v_{12}^2 K_{23}} \tag{1.90}$$

A complete solution of K_{23} can be found elsewhere.[11]

1.3.4 Shear modulus

Let us now assume that a pure shear is applied to the concentric cylinder element. If the element is assumed not to be rotating, then the hypothesis, $\tau_{ij} = \tau_{ji}$ can be assumed. Consequently, $G_{ij} = G_{ji}$ and $\gamma_{ij} = \gamma_{ji}$ can also be assumed. From Fig. 1.17 it is obvious that the element exhibits isotropic characteristics in directions 1–2. Therefore, there are two distinct shear moduli, G_{12} and G_{13}. If we apply some basic mechanics we have

$$G_{13} = \frac{\tau_{13}}{\gamma_{13}} \tag{1.91}$$

Assuming that the shear strain follows a LOM pattern, Eq. (1.91) can be written as[31]

$$G_{13} = \left(\frac{V_f}{G_f} + \frac{1 - V_f}{G_m}\right)^{-1} \tag{1.92}$$

Using displacement solutions, as for the longitudinal modulus, G_{12} is obtained as

$$G_{12} = G_m \left(\frac{(G_f + G_m) + (G_f - G_m)V_f}{(G_f + G_m) - (G_f - G_m)V_f}\right) \tag{1.93}$$

The complete solution of Eq. (1.93) can be found in Reference 26.

1.4 The Basic Mechanics of a Composite System—Semiempirical Models

1.4.1 Halpin–Tsai (H-T) equations

Another approach to estimating the basic mechanical properties of composite materials involves the use of semiempirical and generalized equations to predict experimental results using curve-fitting parameters. The most widely used model has been developed by Halpin and Tsai (H-T).[32] Generally, the H-T equations are based on a single form equation

$$\frac{p}{p_m} = \frac{1 + \xi n V_f}{1 - n V_f} \tag{1.94}$$

where

$$n = \frac{(p_f/p_m) - 1}{(p_f/p_m) + \xi} \tag{1.95}$$

The parameter p represents the composite moduli, i.e., E_{11}, E_{22}, G_{12}, or G_{23}; p_m and p_f are the corresponding moduli of the matrix and fiber, respectively; V_f is the fiber volume fraction; and ξ is an empirical parameter obtained from curve-fitting. The parameter ξ is also a measure of the reinforcement which depends on the fiber geometry, distribution, and loading directions.[33] Equation (1.95) is composed in such a way that when $V_f = 0$, $p = p_m$ and when $V_f = 1$, $p = p_f$.[34] The parameter ξ has two boundary conditions as follows

$$\frac{1}{p} = \frac{V_m}{p_m} + \frac{V_f}{p_f} \qquad \text{for } \xi \to 0 \tag{1.96}$$

and

$$p = p_f V_f + p_m V_m \qquad \text{for } \xi \to \infty \tag{1.97}$$

It is obvious that Eq. (1.96) is a representation of the inverse Law of Mixtures, while Eq. (1.97) represents the normal Law of Mixtures. It should be noted that since Eqs. (1.96) and (1.97) define the minimum and maximum bounds of the elastic moduli, such an approach could be used for selection of composites (Fig. 1.18). A typical example is the H-T equation for the transverse modulus

$$\frac{E_{22}}{E_m} = \frac{1 + \xi n V_f}{1 - n V_f} \qquad \text{and} \qquad n = \frac{(E_f/E_m) - 1}{(E_f/E_m) + \xi} \tag{1.98}$$

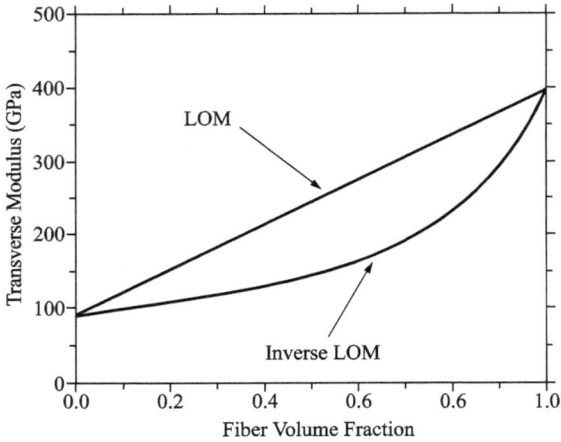

Figure 1.18 The area where the transverse modulus of a SCS/6-Ti-15-3 metal matrix composite is expected according to H-T equations.

The parameter can be obtained by comparing Eq. (1.98) with exact experimental values. Typical values of ξ for some uniaxial composites are shown in Table 1.1.

TABLE 1.1   Typical Values of ξ[34]

Modulus	ξ
E_{11}	$2(l/d)$
E_{22}	0.5
G_{12}	1.0
G_{21}	0.5
K	0

1.4.2   The Nielsen's equations

Nielsen[35] proposed that the H-T equations should take into account the **maximum packing parameter**, Φ_{max}, for the reinforcement, to distinguish between different fiber packing configurations. Therefore, Eqs. (1.94) and (1.95) are modified as

$$\frac{p}{p_m} = \frac{1 + \xi n V_f}{1 - n\Psi V_f} \tag{1.99}$$

$$n = \frac{(p_f/p_m) - 1}{(p_f/p_m) + \xi} \tag{1.100}$$

where

$$\Psi \approx 1 + \left(\frac{1 - \Phi_{max}}{\Phi_{max}^2}\right) V_f \tag{1.101}$$

Square Packing

Hexagonal
Packing

Random
Packing

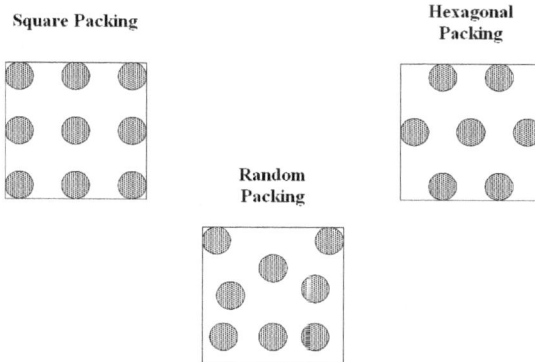

Figure 1.19 The three most common reinforcement configurations.

Nielsen determined different values of the **maximum packing parameter** as a function of the packing configuration, i.e., for square packing of fibers, $\Phi_{max} = 0.785$; for hexagonal arrangement, $\Phi_{max} = 0.907$; and for random packing, $\Phi_{max} = 0.82$.[34] In Fig. 1.19 the three different types of fiber arrangement are shown schematically.

1.5 Methods of Composite Testing

The purpose of this section is to provide the reader with a brief review of the most common but basic testing methods for composite materials. Owing to anisotropy, complicated failure modes and interfacial effects, testing procedures for monolithic materials are not applicable in composite materials. As a result, new test methods and more sophisticated testing equipment have been developed. Although many of these procedures have been accepted by the testing committees, composite testing is still considered in its infancy. A detailed presentation of composite testing techniques can be found in References 36 and 37.

Generally, the mechanical properties of a composite are classified into two categories: (a) thermoelastic constants (linear behavior) and (b) constants for nonlinear behavior.[38] The **thermoelastic constants** comprise information about the Young's modulus, Poisson's ratio, and coefficient of thermal expansion (CTE). The CTE is included in this category because of its relationship with the other two elastic constants. The above three properties provide adequate background for an engineer to perform basic stress analysis and materials selection, especially for unidirectional reinforced composites. However, in the case of whiskers or particulate reinforced composites, we must assume that the material is transversely isotropic, in other words that the elastic properties of the material are isotropic in planes normal to the principal direction (the direction of the reinforcement). For these materials, the thermoeleastic constants required for a basic stress analysis are seven: E_1, E_2, G_2, n_1, n_2, α_1, α_2.

Conversely to thermoelastic constants, the nonlinear behavior constants are of great importance when component design and structural integrity analysis is performed. The properties normally used are: yield stress in tension and compression; proof stress; ultimate tensile strength; compression strength; failure strain; work hardening characteristics; anisotropic yield criteria; fatigue; fracture toughness, etc.

The presentation of all the different techniques currently used to measure the properties of composite materials is beyond the scope of this chapter. However, by presentation measuring techniques for basic properties, the reader will be enabled to understand the major differences between testing a monolithic and a composite material.

1.5.1 Single fiber tests

The major problem confronting scientists when designing testing techniques for fibers was the design of a technique able to operate for all different types of fibers. The two basic parameters of the problem were (a) differences in diameter to modulus ratio and (b) problems regarding the gripping of the fiber to the testing apparatus. Typically, glass, Kevlar, carbon and ceramic fibers have similar small diameters ($6\,\mu m$ to $12\,\mu m$) and therefore can be treated as members of the same team. However, boron and SiC fibers have larger diameters ($100\,\mu m$ to $140\,\mu m$) and higher modulus ($400\,GPa$). Hence, they cannot sustain bending loads without failure. In contrast, polymer fibers, with diameters between $30\,\mu m$ to $40\,\mu m$ and modulus between $50\,GPa$ and $100\,GPa$, can be deformed without damage.

Additionally, it should be noted that testing techniques based on axial tension may only provide adequate amounts of information for isotropic fibers, i.e., glass and boron. For anisotropic fibers like carbon, Kevlar and polyethelene, simple axial tests are far from sufficient in providing information about their transverse properties and Poisson's ratio. For this purpose, estimating techniques using indirect measurement from the axial testing are usually used.

1.5.2 Tensile strength and modulus

Generally, two methods of measuring the tensile properties of fibers are currently in use. These are based on (a) vibration and (b) direct mechanical loading.

1.5.1.1 The vibration technique. The vibration technique uses short fiber lengths (≈ 100 mm) where the one end of the fiber is gripped in a vibration input device, while the other is allowed to vibrate as a cantilever beam. The natural frequency of the fibers is measured by the attainment of a resonant condition. Using the natural frequency, the cross-sectional shape and area, and the fiber density, the elastic modulus is obtained. A schematic setup of the method is shown in Fig. 1.20. There is no standardized version of the vibration technique.

1.5.2.2 ASTM standard. The second method involves the gripping of both fiber ends and pulling in tension until failure. This technique is also known as ASTM D 3379-75 standard test.[39] As shown in Fig. 1.21, the fiber is bonded to a thin paper, metal, or plastic sheet that has a central longitudinal hole or a rhombic hollow. After mounting the sheet in the testing machine, both sides of the sheet are cut or burned, allowing the direct loading of the fiber. A detailed description of the procedure is also given in the standard. As the specimen is pulled to failure, the load and the elongation are recorded, allowing the determination of the tensile strength and modulus. The compliance of the supporting sheet, adhesive, testing machine and end side effects is measured by using specimens of different gage lengths. Hence, the system compliance is determined by plotting the fiber elongation as a function of gage length and then extrapolating the data to zero gage length. The compliance at zero gage length will represent the system compliance. If fiber strain throughout loading is not recorded, then only strength properties can be obtained.

One of the difficulties of ASTM D 3379-75 is obtaining accurate results when fibers with irregular cross-sectional area and shape are in question (e.g., polythylene fibers). This problem can be minimize by testing an entire yarn or tow. However, when a bundle of fibers is tested, only a lower strength limit is determined. This is because, the weaker fibers will fail at lower stresses, forcing the remaining fibers to carry their load. To solve this problem, an impregnating resin is used to force the broken fibers to continue carrying the loading. Of course,

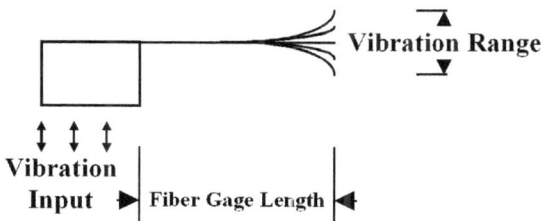

Figure 1.20 The vibration technique.

the properties measured by this testing technique are closer to those of unidirectional composite than to those of a fiber. The impregnating resin testing is classified as ASTM standard D 2343-67[40] for glass fibers and D 4018-81[41] for carbon fibers. The tensile specimen employing the impregnating resin technique is shown in Fig. 1.22.

1.5.3 The coefficient of thermal expansion

The measurement of the coefficient of thermal expansion (CTE), especially for isotropic fibers, is relatively easy. A long fiber is placed in a temperature chamber and length changes as a function of the temperature are measured. The accuracy of the method depends on the gage length selected for the experiment. Usually, long fibers ($\sim 1000\,$mm) are used, which require simpler measurement instrumentation. However, anisotropic fibers require more sensitive measurement instruments, due to their low or even negative CTE, i.e. $-0.1 \times 10^{-6}/°C$ is typical value for Kevlar fibers. A similar method is used to obtain values of the transverse CTE.

1.5.4 Matrix properties testing

Procedures for testing metallic materials are very well established and many of their properties can easily be found in any materials handbooks. However, properties of many ceramics and polymers are still

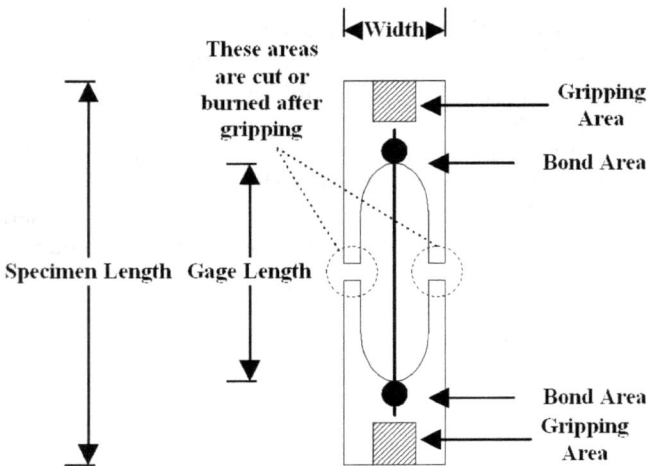

Figure 1.21 The single fiber test specimen according to ASTM D 3379-75.[39] *(Reprinted, with permission from the Annual Book of ASTM Standards, © American Society for Testing and Materials, 100 Barr Harbor Drive, West Conshohocken, PA 19428.)*

5/8 in

13 in   6 in

Cast Resin End
Tabs

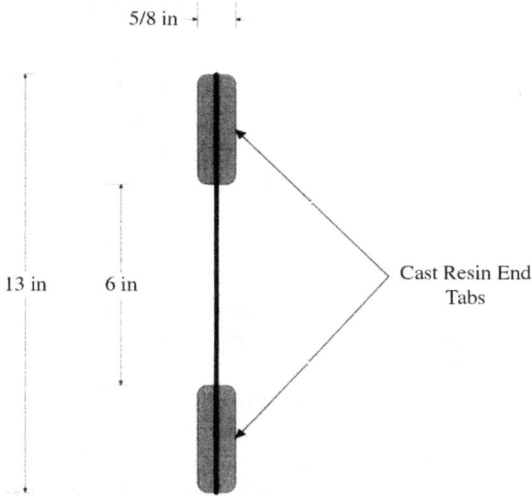

Figure 1.22 Resin-impregnating tensile test specimen according to D 4018-81.[41] (*Reprinted, with permission from the Annual Book of ASTM Standards, © American Society for Testing and Materials, 100 Barr Harbor Drive, West Conshohoken, PA 19428.*)

unknown. This is basically because many of these materials were specifically developed for matrices and therefore mechanical properties exists only for the reinforced form. In addition to the lack of information is that some of these materials exhibit properties incompatible with existing test methods and instruments. For example, many of the epoxies and polyimides are extremely brittle and therefore fabrication of test specimens in the unreinforced form is impossible. Similar problems have been reported for ceramics.[42] Also, the high strain to failure characteristics of thermoplastic materials requires special testing equipment.[42]

Testing of matrix materials is an issue even in our days. The development of new matrices, sometimes monthly, with "exotic" properties has increased their operational cost. New techniques and equipment are not always the desired solution for the industry. To overcome such problems, in some cases we have to assume that these material are characterized by isotropy. Consequently, testing in tension and compression would provide all the necessary information, according to the isotropic relation, $G = E/2(1 + v)$. On this basis, a number of tests known as **neat resin matrix tests** have been developed.

1.5.4.1 Neat resin matrix tests. The tensile properties of plastics (Young's modulus, fracture strength, yield strength, and elongation) can be obtained from the ASTM D 638-90 test method.[43] According to the standard, "dogbone" shaped specimens with several dimensions depending on the thickness of the raw material, can be subjected to tensile loading. The specimens may be machined from sheet or plates or they can be produced by contoured molding. The manufacturing of a mold certainly increases the cost; however, it may be considered a cheap solution when very brittle matrix materials are involved (since the cutting of brittle materials is difficult). It should be noted that during molding many polymer resins may begin to polymerize if held too long at elevated temperatures, making them more viscous. Additionally, air trapped during polymerization may induced areas of stress concentration, leading to premature failure and degraded properties. The testing conditions themselves are also subject to specific standards, since plastics are sensitive to environmental conditions. ASTM standard D 618-81,[44] imposes conditions of **Standard Laboratory Atmosphere** (23°C and 50% relative humidity). In D 638-90, the speed of loading is also specified since many plastics are strain-rate sensitive. Strain gages or extensometers can be used to measure strains. Generally, extensometers are simpler to use, reusable, and less expensive. A typical neat resin tensile specimen is shown in Fig. 1.23.

The compression properties of plastics are obtained according to the ASTM D 695-50 standard.[45] Buckling features of the specimen are

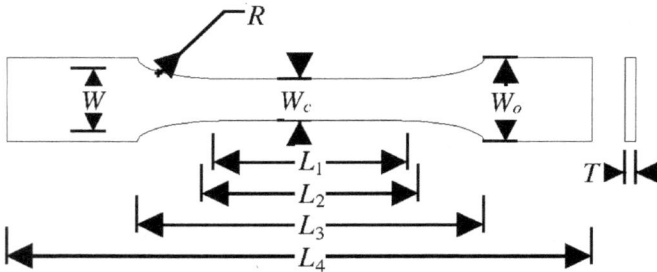

Figure 1.23 Neat resin tensile specimen (Type I) W = width of narrow section, 13 mm; W_c = width of middle section, W_0 = width overall, min, 19 mm; L_1 = gage length, 50 mm; L_2 = length of narrow section, 57 mm; L_3 = distance between grips, 115 mm; L_4 = overall length, min, 165 mm; R = fillet radius, 76 mm; T = thickness, min, 7 mm.[43] (Reprinted, with permission from the Annual Book of ASTM Standards, © American Society for Testing and Materials, 100 Barr Harbor Drive, West Conshohoken, PA 19428.)

avoided by using short testpieces. ASTM D 695-90 also describes a ball-and-socket arrangement to allow compression without bending. Generally, both tensile and compression tests are based on the determination of the elastic properties on the isotropic relationship. No specific procedures exist for ceramic or metal matrices.

1.5.5   Composite properties

Composite properties may differ from those predicted from the individual phases. This is basically due to irregularities in the manufacturing process, reaction kinetics, and thermal expansion. For some scientists, these differences represent the compliance between composite and phase testing. As well as the basic mechanical properties of the composite, properties like density, fiber volume fraction, void volume, etc., are of great importance to designers.

1.5.5.1   Density.   There are three different ways of measuring the density of material: (a) the weighing technique; (b) the displacement technique; and (c) the density gradient technique. The **weighing technique** involves the determination of the (weight) density by measuring the volume and the weight of a sample of simple shape, e.g., cubic. This technique is inaccurate for very brittle composites. In the **displacement technique**, a specimen sample (any shape) is weighed in air and then in a liquid (usually water). The density of the material is then determined by the difference of the weights. The displacement technique is described in detail in ASTM D 792-66.[46] Finally, the **density gradient technique** measures the rate at which a test specimen sinks in a liquid of known density gradient. The procedure has been classified as ASTM D 1505-68.[47]

1.5.5.2   Fiber volume.   A number of different techniques are used to determine the volume fraction of fibers. These techniques are not usually applicable in all kinds of composites. For polymer reinforced composites, a technique known as the **matrix digestion process** is the most widely known. A sample of the composite is weighed and its density is measured. The matrix material is then digested away and the remaining fibers are weighed. If the density of the fibers is known, then the fiber volume can be determined. The matrix digestion technique is described in ASTM D 3171-76.[48] The D 3171-76 standard defines three different procedures depending on the matrix material: (a) epoxy resins require digestion by concentrated nitric acid; (b) polyimides require an aqueous mixture of sulfuric acid and hydrogen peroxide; (c) If procedures (a) and (b) do not apply, then the matrix is digested by mixture of ethylene glygol and potassium hydroxide.

It should be noted that the fibers should not be attacked by the dissolving medium. Although not commonly used, an alternative method to the matrix digestion process is the **electric conductivity method**. The method applies to undirectional composites with electrically conductive fibers (carbon, boron, silicon carbide, or metals) and nonconductive matrices (most polymers). The resistivity of the composite in the direction of the fibers and the resistivity of the fibers are measured. The volume fraction is then determined by the ratio of the fiber resistivity to that of the composite. The basic drawbacks of the method are that (a) the method applies only to undirectional composites with conductive fibers and nonconductive matrices; and (b) broken fibers and fibers with cross-sectional irregularities produce inaccuracies. However, the method is considered as nondestructive. The electric conductivity method is described in ASTM D 3355-74.[49]

The digestion process is also used for metal matrix composites; the process is described in ASTM D 3553-76.[50] As in polymer reinforced composites, the standard suggests three different procedures depending on the matrix material: (a) aluminum, copper, and steel matrices require digestion by nitric acid; (b) for aluminum alloys, where procedure (a) is not applicable, sodium hydroxide can be used as an alternative; (c) magnesium, titanium, steel, and copper may be digested by hydrochloric acid.

Other methods of determining the volume fraction are principally based on optical examination of a given area. Optical microscopy or scanning electron microscopy (SEM) are used to count the numbers of fibers in a specific area of a polished cross section of the composite. These techniques are time consuming and often involve human error. Especially for metal matrix composites, where the fibers do not always follow a uniform distribution, samples of different areas of the composite should be examined to give an average volume fraction.

Automated scanning that records optical density differences is available commercially. According to Cilley, et al.,[51] the digestion process is as accurate as any other technique available. None of the above techniques is covered by ASTM or any other organizational standard.

1.5.5.3 Voids. The percentage of voids in a composite material is vital for high performance engineering, where the component may operate close to failure (e.g., overloads during flight). Generally, voids generate stress concentrations and they represent cracklike defects. ASTM D 2734-70[52] describes a general procedure for determining the void volume. The densities of fiber, matrix, and composite are measured separately. From the densities of the phases, a theoretical composite density is calculated and compared to the true composite density. The difference represents the amount of voids. Void volume of 1% is generally acceptable for most composite systems.

1.5.5.4 Thermal expansion.   Since composites are mainly anisotropic, the thermal expansion coefficient (CTE) should be determined in all three principal directions. For this purpose two techniques originally developed for homogeneous materials are used: dilatometry and optical interferometry. The fact that composite systems can exhibit CTE values from very high to zero or even negative generates problems concerning the accuracy of the technique and the measuring range of the testing equipment. For example, Kevlar and carbon fibers exhibit negative axial expansions. An indication of the required measuring range of the instruments can be obtained by measuring the CTE of the neat matrix.

Measurement by dilatometers is quite common even at industrial scale. Test specimens of length in the range of 50–100 mm are typically used. The specimen is placed in an expansion measuring device and then into temperature chamber. The dilatometer usually consists of a solid push rod inside a tube. One end of the specimen is placed against tube and the other against the push rod. As the temperature increases, the rod moves relative to the tube. Usually, tube and push rod are made of the same material; however, the use of materials with different CTE may provide a scaling factor appropriate for low temperature measurements. Quartz glass is a typical material for the tube and push rod. Dial gages or electronic measuring units (linearly variable differential transformers, LVDT) are used to measure the relative displacement. LVDTs can measure down to $1 \times 10^{-6}/°C$. The dilatometer technique is governed by ASTM D 696-76 standard.[53] A typical dilatometer arrangement is shown in Fig. 1.24.

Optical interferometers are generally specialized devices and not particularly common for industrial use. Interferometers are basically used to measure very low or negative CTEs (composites with very low or "zero" CTEs are used for thermally stable applications). Usually, a laser beam is split and interference occurs after recombination. The accuracy of interferometers is based on the wavelength, λ, of the light beam used to form the interference pattern (typical values are between $\lambda/2$ and $\lambda/3$). With precise arrangement of the mirrors, interferometers can measure CTEs in the range of $\pm 10^{-8}/°C$. Four methods of interferometry are currently used: moiré,[54] Fabry–Perot,[55] Michelson,[56] and Fizeau.[57] Detailed analysis can be found in the

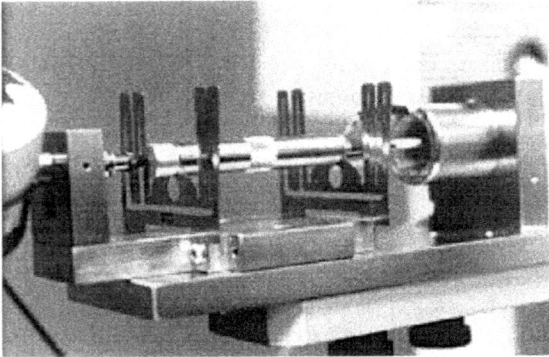

Figure 1.24 A Theta type dilatometer.

corresponding references of each method. Optical interferometers are governed by ASTM E 289-70.[58]

1.5.5.5 Tensile properties. The validity of tensile testing of composite materials, even in our days, is still a great challenge for engineers. This is due to the way we apply load to the specimen. Generally, axial load is applied to the specimen through shear from the testing apparatus. Consequently, shear failure in the gripping area is common, especially when the shear strength of the specimen is much lower than the axial strength. To avoid surface damage, tabs are usually tapered at the specimen ends (especially straight-sided specimens). The design of tabs should ensure uniform distribution of the gripping forces, while the bond area must be large enough to protect the bond from adhesive shear failure. Generally, the possibility of shear failure of the gripping area is reduced when thin specimens are used (specimen thickness close to 0.4 mm is often used).

In general, specimens are cut from bulk material using a diamond saw (ductile matrix), or they can be fabricated into the desired shape using molding (brittle matrix). Tabs are mainly made by printed circuit board techniques to polymer matrix composites or of materials with similar properties to the matrix material for ceramic and metal matrix composites. There is a larger number of different adhesive products with specific characteristics that classify them to meet specific requirements (e.g., Ciba Geigy Redux 403 epoxy paste is recommended for fatigue testing[59]). The bonding process itself may introduce additional problems to the preparation of the specimen. For example, adhesive curing usually requires exposure to high temperatures for a long time; hence polymer matrices may experience polymerization. Furthermore, grinding, cleaning, and degreasing of the

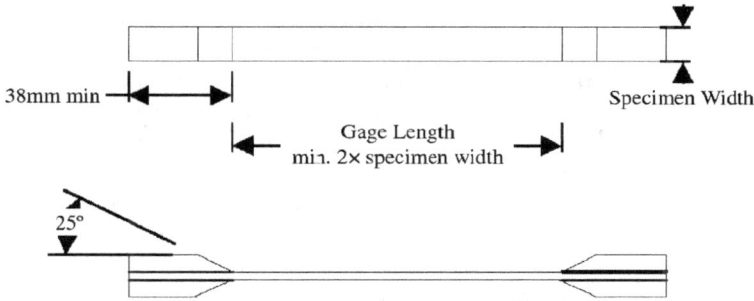

Figure 1.25 Tensile test specimen with tapered tabs, ASTM Standard D 3039-76.[63] (*Reprinted, with permission from the Annual Book of ASTM Standards, © American Society for Testing and Materials, 100 Barr Harbor Drive, West Conshohoken, PA 19428.*)

specimen surface are required prior to bonding; these processes could introduce surface damage and premature failure. Therefore, the selection of the adhesive product itself is a critical task.

The design of the tab ends is another problem and no recommendations are provided by any standardization committee, although ASTM D 3552-77[60] drawings suggests square-ended tabs (Fig. 1.25). The "cut-off" at the end tab and the taper angle (See Fig. 1.26) were found to be the controlling factors of stress concentrations.[61] Parametric finite element studies[61] and evidence from tensile tests on undirectional reinforced polymer composites[62] have shown that by reducing the above factors the rate of load transfer to the tab is reduced and thus the possibility of failure.

Foil resistance strain gages or extensometers may be used to measure strains. The use of extensometers, however, is generally preferable. This is because extensometers are reusable, easier to mount, and accurate at elevated temperatures and humidity percen-

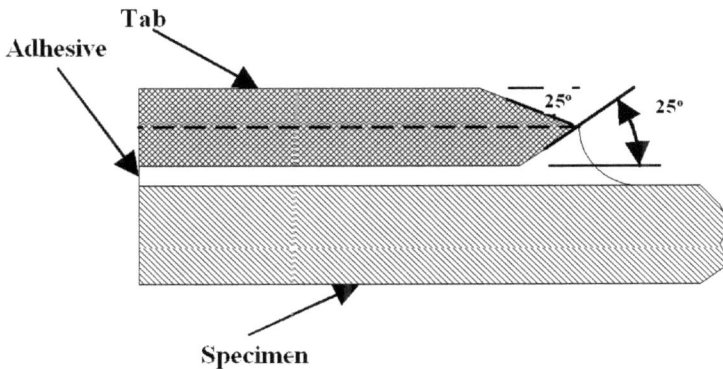

Figure 1.26 Undercut tapered geometry according to Reference 61.

tages, and because they can provide simultaneously axial and transverse measurement necessary for Poisson's ratio calculation.

Using ASTM D 3039-76 (polymer matrix composites) and ASTM D 3552-77 (metal matrix composites), accurate measurements of E_1, ν_{12}, σ_{fr}, etc. can be obtained, especially for orthotropic composites. However, when nonuniformities and multiaxial reinforcements exist, shear forces and bending moments may deform the specimen, leading to inaccurate measurements.[64,64]

1.5.5.6   Transverse properties.   For transverse tensile testing, tabs are not always required owing to the low strength of most composites in the transverse direction, and therefore the test is easier to perform. Care should be taken when mounting the specimen in the testing apparatus. According to ASTM D 3039-76,[63] thicker flat specimens (2 to 3 mm) are used. For extremely weak materials, D 3039-76 prescribes the same kind of cushioning (emery cloth) between the grip face and the specimen.

1.6   Tables of Mechanical Properties for various Fiber and Matrix Materials

See Table 1.2 for mechanical properties of various fiber and matrix materials.

TABLE 1.2 Mechanical Properties for Various Fiber and Matrix Materials

Material	Coefficient of Thermal Expansion ($10^{-6}/°C$)	Tensile Strength (MPa)	Elastic Modulus (GPa)	Poisson's Ratio	Bulk Modulus (GPa)	Shear Modulus (GPa)	Density (g/cm³)
			Fibers				
Boron on tungsten	5.0	3450–3620	400–420	N/A			2.49–2.6
Borsic	5.0	3100	400	N/A			2.71
Boron on carbon	5.0	3450	360	≈ 0.21			?.49
Glass–S	5.0	4140	80–85.5	≈ 0.22			2.49
Glass–E	5.0	2700	70	≈ 0.22			2.49
Steel	13.3	2070	210				7.75
Tantalum	6.5	1089–1520	185–190	0.342	196.5	68.9	16.6–16.88
Tungsten	4.5	3170	390	0.28	311	160	19.38
Ceramic Al_2O_3 whiskers	7.7	42760	450	0.2			3.96
Molybdenum	4.9	660	320	0.293	261		1.02
Beryllium	11.5	970	275–300	0.02–0.075			1.86
SiC on carbon	4.3	3450	400	≈ 0.19			3.04
SiC on tungsten	4.3	3100	430	≈ 0.19			3.32
Nicalon	3.1	2400	190				2.55
Nextel 312 (3M Company)	N/A	1700	150				2.5
SiO_2	0.5	600	70				2.2
Si_3N_4	3	1400–3500	280–350				3.2
Kevlar 49		3024–3414	122–130	0.35			
Graphite PAN HM	–1.1	2210	380				1.86
Graphite PAN HTS	–1.1	2340	210				1.74
Graphite Rayon	–1.1	2170	390				1.66
Graphite Thornel 75	–1.1	2660	520				1.83
Graphite Pitch–P	–1.1	1380	340				1.99
Graphite Pitch UHM	–1.1	2410	690				2.05

(continued)

TABLE 1.2 Mechanical Properties for Various Fiber and Matrix Materials (Continued)

Material	Coefficient of Thermal Expansion (10^{-6}/°C)	Tensile Strength (MPa)	Elastic Modulus (GPa)	Poisson's Ratio	Bulk Modulus (GPa)	Shear Modulus (GPa)	Density (g/cm^3)
			Metal Matrices				
Aluminum	23.4	310	70	0.34	75.1	26	2.8
Beryllium	11.5	620	290				1.9
Copper	17.6	340	120	0.34	77.8	27	8.9
Lead	28.8	20	10	0.44	26	3.1	11.3
Magnesium	25.2	280	40	0.291	35.5	17	1.7
Nickel	13.3	760	123–210	0.3–0.32	100–106	42–47	8.9
Niobium	8	28	0	0.39	170	21.2	8.6
Steel	13.3	2.070	210	0.28–0.3	95	46	7.8
Tantalum	6.5	410	190	0.34	110	39	16.6
Tin	23.4	10	40	0.35	32.8	10	7.2
Titanium	9.5	1170	110	0.36	61	25	4.4
Tungsten	4.5	1520	410	0.28	310	160	19.4
Zinc	27.4	280	70	0.25	39	42	6.6
			Polymer Matrices				
Epoxy	28–44	50–130	2.4–4.1	0.34			
Nylon 66		70	2.7	0.33			

Material					
Polyester		55	0.38	3.0	
PA 6		41		1.2	
PBT		120		7.5	
PC		65	0.37	2.3	
PEEK		92		3.6	
PES		84		2.6	
PEI		105		3.0	
PET		137			
PPS		79–134		11.7	
PS		70		2.5	
PSF		70		2.7	
uPVC		55		3.0	
Ceramic Matrices					
Al_2O_3	9	262	0.22	390–420	3.89
MgO	12	140	0.18	280	3.6
SiO_2	0.5			70	
ZrO_2	10			250	
Mullite $3Al_2O_3 \cdot 2SiO_2$				220	
Zircon	5	147		210	
$Zr_2O_2TiO_2$					
ZrO_2TiO_2					
Si_3N_4	3	580		280–310	
SiC	5	299	0.19–0.22	410–450	3.2

47

References

1. H. L. Cox (1952) The elasticity and strength of paper and other fibrous materials, *Br. J. of Appl. Phys.*, **3**, 72–79.
2. A. E. H. Love (1927) *A Treatise on the Mathematical Theory of Elasticity*, 4th ed., Dover Publications, New York.
3. I. S. Sokolnikoff (1956) *Mathematical Theory of Elasticity*, McGraw-Hill, New York.
4. R. C. Stedfeld and C. T. Hoover (1962) *The Filament-Wound Pressure Vessel*, Allison Research and Engineering Co, Report TR-41.
5. J. E. Gordon (1952) On the present and potential efficiency of structural plastics, *J. Royal Aero. Soc.*, **56**, 704–728.
6. B. W. Shaffer (1964) Stress–strain relations of reinforced plastics parallel and normal to their internal filaments, *J. of AIAA*, **2**, 348–352.
7. M. Taya and R. J. Arsenault (1989) *Metal Matrix Composites—Thermomechanical Behavior*, Pergamon Press, New York.
8. R. D. Adams (1987) Damping properties analysis of composites, in *Engineered Materials Handbook, Vol. 1, Composites* (T. J. Reinhart, ed.), ASM International, Materials Part, OH, 206–217.
9. N. G. McCrum, C. P. Buckley, and C. B. Bucknall (1988) *Principles of Polymer Engineering*, Oxford Science Publications, Oxford University Press, New York.
10. B. W. Shaffer (1968) Elastic-plastic stress distribution within reinforced plastics loaded normal to its filaments, *J. of AIAA*, **6**, 2316–2324.
11. S. K. Garg, V. Svalbonas, and G. A. Gurtman (1973) *Analysis of Structural Composite Materials*, Marcel Dekker, New York.
12. S Jansson, H. E. Dve, and A. G. Evans (1991) The anisotropic mechanical properties of a Ti matrix composite reinforced with SiC fibres, *Metall. Trans.*, **22A**, 2975–2984.
13. H. E. Boyer and T. L. Gall (eds.) (1985) *Metals Handbook*, ASM, Metals Park, OH.
14. R. W. Hertzberg (1989) *Deformation and Fracture Mechanics of Engineering Materials*, 3rd ed., Wiley, New York.
15. W. A. Weibull (1951) Statistical distribution function of wide applicability, *J. Appl. Mech.*, **18**, No. 3, 293–297.
16. W. A. Curtin (1993) The "tough" to brittle transition in brittle matrix composites, *J. Mech. Phys. Solids*, **41**, 217–245.
17. D. M. Kotchick, R. C. Hink, and R. E. Tressler (1975) Gauge length and surface damage effects on the strength distributions of silicon carbide and sapphire filaments, *J. Compos. Mater.*, **9**, 327–336.
18. J. J. Masson and E. Bourgain (1992) Some guidelines for a consistent use of the Weibull statistics with ceramic fibres, *Int. J. Fracture*, **55**, 30–319.
19. A. G. Atkins and Y. W. Mai (1985) *Elastic and Plastic Fracture*, Ellis Horwood, London.
20. B. D. Coleman (1958) On the strength of classical fibres and fibre bundles, *J. Mech. Phys. Solids*, **7**, 60–70.
21. C. H. Weber, X. Chen, S. J. Connell, and F. W. Zok (1994) On the tensile properties of a fiber reinforced titanium matrix composite—I. Unnotched behavior, *Acta Metall. Mater.*, **42**, No. 10, 3443–3450.
22. B. W. Boxen (1983) *Mechanics of Composite Materials: Recent Advances*, Pergamon Press, Oxford.
23. K. K. Chawla (1987) *Composite Materials, Science and Engineering*, Springer, New York.
24. C. H. Chen and S. Cheng (1967) Mechanical properties of fiber reinforced composites, *J. Compos. Mater.*, **1**, 30–41.
25. T. S. Chow and J. J. Hermans (1969) The elastic constants of fiber-reinforced materials, *J. Compos. Mater.*, **3**, 382–396.
26. Z. Hashin and B. W. Rosen (1964) The elastic moduli of fiber-reinforced materials, *J. Appl. Mech.*, **31**, 223–232.
27. J. M. Whitney and M. B. Riley (1965) *Elastic Stress–Strain Properties of Fiber-Reinforced Composite Materials*, USAF Materials Laboratory AFML-TR-65-238.
28. C. M. Warwick and T. W. Clyne (1991) Development of a composite coaxial cylinder

stress analysis model and its application to SiC monofilament system, *J. Mater. Sci.*, **26**, 3817–3827.
29. O. C. Zienkiewicz and R. L. Taylor (1989) *The Finite Element Method, Vol. 1, Basic Formulation and Linear Problems*, 4th ed., McGraw-Hill, New York.
30. S. P. Timoshenko and J. N. Goodier (1951) *Theory of Elasticity*, 2nd ed., McGraw-Hill, New York.
31. T. W. Clyne and P. J. Withers (1993) *An Introduction to Metal Matrix Composites*, Cambridge University Press, Cambridge.
32. J. C. Halpin and S. W. Tsai (1967) *Environmental Factors in Composite Design*, USAF Materials Laboratory, AFML-TR-67-423.
33. J. M. Whitney (1967) Elastic moduli of unidirectional composites with anisotropic fibers, *J. Compos. Mater.*, 1, 188–193.
34. K. K. Chalwa (1987) *Composite Materials—Science and Engineering*, Springer-Verlag, London.
35. L. E. Nielsen (1974) *Mechanical Properties of Polymers and Composites*, vol. 2, Marcel Dekker, New York.
36. *ASTM Standards and Literature for Composite Materials* (1990), 2nd ed., American Society for Testing and Materials, Philadelphia.
37. R. L. Pendleton and M. E. Tuttle (1989) *Manual on Experimental Methods for Mechanical Testing of Composites*, Society of Experimental Mechanics, Bethel, CT.
38. L. N. McCartney (1990) Testing of metal matrix composites—An overview, in *Test Techniques for Metal Matrix Composites* (N. D. R. Goddard, ed.), The Institute of Physics, London.
39. D 3379-75, Standard Test Method for Tensile Strength and Young's Modulus of High-Modulus Single-Filament Materials (1990), *ASTM Standards and Literature for Composite Materials*, 2nd ed., American Society for Testing and Materials, Philadelphia, 34–37.
40. D 2343-67, Tensile Properties of Glass Fiber Strands, Yarns and Roving Used in Reinforced Plastics (1986), American Society for Testing and Materials, Philadelphia.
41. D 4018-81, Tensile Properties of Continuous Filament Carbon and Graphite Yarn, Strands, Rovings, and Tows (1986), American Society for Testing and Materials, Philadelphia.
42. D. F. Adams (1989) Properties characterization—Mechanical/physical/hygrothermal properties test methods, in *Reference Book for Composites Technology* (S. M. Lee, ed.), Technomic Publishing, Lancaster, 49–79.
43. D 638-90, Standard Test Method for Tensile Properties of Plastic (1990), 2nd ed., American Society for Testing and Materials, Philadelphia.
44. D 618-81, Standard Practice for Conditioning Plastics and Electrical Insulating Materials for Testing (1990), 2nd. ed., American Society for Testing and Materials, Philadelphia.
45. D 695-90, Standard Test Method for Compressive Properties of Rigid Plastics (1990), 2nd ed., American Society for Testing and Materials, Philadelphia.
46. D 792-66, Specific Gravity and Density of Plastics by Displacement (1986), American Society for Testing and Materials, Philadelphia.
47. D 1505-68, Density of Plastics by the Density-Gradient Technique (1986), American Society for Testing and Materials, Philadelphia.
48. D 31271-76, Fiber Content of Resin-Matrix Composites by Matrix Digestion (1986), American Society for Testing and Materials, Philadelphia.
49. D 3355-74, Fiber Content of Unidirectional Fiber–Resin Composites by Electrical Resistivity (1986), American Society for Testing and Materials, Philadelphia.
50. D 3553-76, Fiber Content by Digestion of Reinforced Metal Matrix Composites (1986), American Society for Testing and Materials, Philadelphia.
51. E. Cilley, D. Roylance, and N. Schneider (1974) Methods of fiber and void measurement in graphite/epoxy composites, in *Composite Materials: Testing and Design*, ASTM STP 546, American Society for Testing and Materials, Philadelphia, 237–249.
52. D 2734-70, Void Content of Reinforced Plastics (1986), American Society for Testing and Materials, Philadelphia.

53. D 696-79, Test Method for Coefficient of Linear Thermal Expansion (1986), American Society for Testing and Materials, Philadelphia.
54. D. E. Bowes, D. Post, C. T. Herakovich, and D. R. Tenney (1981) Moiré interferometry for thermal expansion of composites, *Exp. Mech.*, **21**, 441–448.
55. V. E. Botton (1964) Fabry–Perot dilatometer, *Rev. Sci. Instrum.*, **35**, 364–366.
56. E. G. Wolff and S. A. Eselun (1979) Double Michelson interferometer for contactless thermal expansion measurement, *Proc. SPIE*, **192**, Interferometry, 204–208.
57. S. S. Tomkins (1989) Techniques for measurement of the thermal expansion of advanced composite materials, in *Metal Matrix Composites: Testing, Analysis and Failure Modes*, ASTM STP 1032, American Society for Testing and Materials, Philadelphia, 54–67.
58. E 289-70, Test Method for Linear Thermal Expansion of Rigid Solids with Interferometry (1986), American Society for Testing and Materials, Philadelphia.
59. Ciba-Geigy (1993) *Ciba Polymers—Structural Adhesives*, Publication No. A201c-GB, Cambridge.
60. D 3552-77, Tensile Properties of fiber reinforced metal matrix composites (1982), American Society for Testing and Materials, Philadelphia.
61. I. Greaves (1994) The growth of naturally initiating fatigue cracks in titanium-silicon carbide MMCs, Ph.D Thesis, University of Sheffield.
62. M. E. Cunningham, S. V. Schoultz, and J. M. Toth, Jr (1985) Effect of end-tab design on tension specimen concentrations, in *Recent Advances of Composites in the United States and Japan*, ASTM STP 864, American Society for Testing and Materials, Philadelphia, 253–262.
63. D 3039-76, Standard Test Method for Tensile Properties of Fiber Resin Composites (1990), American Society for Testing and Materials, Philadelphia.
64. N. J. Pagano and J. C. Halpin (1968) Influence of end constraint in the testing of anisotropic bodies, *J. Compos. Mater.*, **2**, 18–31.
65. R. M. Jones (1975) *Mechanics of Composite Materials*, Hemisphere Publishing, New York.

Failure Mechanisms and Failure Criteria of Fiber Reinforced Composites

E. E. Gdoutos

*School of Engineering, Democritus University of
Thrace, Xanthi, Greece*

2.1 Introduction

Failure of fiber reinforced composites is generally preceded by an accumulation of different types of internal damage. Failure mechanisms on the micromechanical scale include fiber breaking, matrix cracking, and interface debonding. They vary with type of loading and are intimately related to the properties of the constituents, i.e., fiber, matrix, and interface/interphase. While failure mechanisms are common in most composites, their sequence and interaction depend on the type of the loading and the properties of the constituents. The damage is generally well distributed throughout the composite and progresses with an increasingly applied load. It coalesces to form a macroscopic fracture shorty before catastrophic failure. Study of the progressive degradation of the material as a consequence of growth and coalescence of internal damage is of utmost importance for the understanding of failure.

Stress transfer from the matrix to fibers in a composite takes place by shear at the fiber–matrix interface. Strong interfaces result in high strength and stiffness, but low fracture toughness composites. On the other hand, weak interfaces promote deflection of matrix cracks along the interface and lead to high fracture toughness, but low strength and stiffness of composites. The process of transfer of load between fibers and matrix in the neighborhood of a fiber break or a matrix crack depends on the strength of the interface. Although fiber and matrix can be characterized by conducting simple tests, interface properties are most difficult to determine. Interfacial shear strength is an important parameter that controls the fiber–matrix debonding process and, therefore, the sequence and relative magnitude of the various failure mechanisms in the composite.

The strength of a composite is dictated by the strength of the fibers. The strength of a unidirectional composite is an anisotropic property, i.e., it varies with orientation. For a continuous fiber reinforced lamina, the longitudinal strength is much higher than the transverse strength. Furthermore, the compressive strengths associated with the longitudinal and transverse directions are, generally, different from the corresponding tensile strengths. The in-plane shear strength referred to the principal material axes is another independent material property. It is desirable to correlate the strength along an arbitrary direction or under multiaxial loading to the above five basic lamina strength parameters. For this reason a number of failure criteria have been proposed.

For the analysis of composite strength it is essential to understand first the failure mechanisms and processes within the composite and their effect on the macroscopic lamina strength. In this respect, a micromechanical analysis of failure is first conducted in this chapter. The cases of loading of a lamina under uniaxial and transverse tension or compression and in-plane shear along the principal axes are analyzed. Emphasis is given to the study of the underlying failure mechanisms under longitudinal tension. The cases of fiber- and matrix-dominated failures are studied in detail. Following the analysis of micromechanics of failure, a number of macroscopic failure criteria are discussed. Emphasis is given to the maximum stress theory, the maximum strain theory, the deviatoric strain energy theory for anisotropic materials (Tsai–Hill) and the interactive tensor polynomial theory (Tsai–Wu), which are considered representative and the more widely used of all failure theories available.

2.2 Micromechanics of Failure Under Longitudinal Tension

An elementary mechanics of materials approach can be used to predict the lamina longitudinal tensile strength. Consider a representative volume element subjected to a longitudinal normal stress σ_1. Equilibrium along the longitudinal direction requires that the total force carried by the element must be equal to the sum of the forces acting on the fiber and matrix, that is,

$$\sigma_1 A = \sigma_f A_f + \sigma_m A_m \tag{2.1}$$

where σ_1 = average longitudinal stress on the composite,
$\quad \sigma_f,\ \sigma_m$ = average longitudinal stresses in the fiber and matrix, respectively,
$\quad A_f,\ A_m$ = areas perpendicular to the longitudinal direction related to the fiber and matrix, respectively,
$\quad A = A_f + A_m.$

Equation (2.1) gives a rule of mixtures relation for longitudinal stress

$$\sigma_1 = \sigma_f V_f + \sigma_m V_m \tag{2.2}$$

where V_f, V_m are fiber and matrix volume ratios, respectively.

For a simplified analysis of the composite strength, the deterministic assumption of uniform strength of the fiber and the matrix is invoked. In reality, fiber and matrix strengths are not constant and are dictated by a statistical distribution. For example, fiber strength varies from point to point and from fiber to fiber. Materials failure often initiates at fiber breaks and stress or strain concentrations. Under such circumstances, the simple deterministic mechanics of materials models provide approximate estimates of the lamina strength.

Under longitudinal tension, two types of composite failure can be distinguished. In the first type the ultimate tensile strain of the fiber is lower than that of the matrix. In the second type the ultimate tensile strain of the matrix is lower than that of the fiber. The first type occurs in ductile-matrix composites, such as epoxy–matrix composites. The second occurs in brittle-matrix composites, such as ceramic–matrix composites. In the following, theses two types of composite failure will be dealt with separately in detail.

2.2.1 Composites with ultimate tensile strain of the fiber lower than that of the matrix

2.2.1.1 Mechanics of materials approach. In the case of a composite with ultimate tensile strain of the fiber lower than that of the matrix, i.e., when

$$\varepsilon_{ft}^u < \varepsilon_{mt}^u \tag{2.3}$$

failure of the composite occurs when the applied longitudinal strain becomes equal to the ultimate tensile strain of the fiber. Figure 2.1 shows the longitudinal stress–strain curves for the fiber, matrix, and composite. Note that according to Eq. (2.2) the stress–strain curve of the composite falls between the stress–strain curves of the matrix and fiber. The longitudinal tensile strength of the composite is obtained from Eq. (2.2) as

$$F_{1t} = F_{ft}V_f + \sigma'_m V_m \tag{2.4}$$

where F_{1t} = longitudinal composite tensile strength
$\quad\quad F_{ft}$ = longitudinal fiber tensile strength
$\quad\quad \sigma'_m$ = longitudinal matrix stress at fiber ultimate strain

For linear elastic behavior of the fiber and matrix, Eq. (2.4) is written as

$$F_{1t} = F_{ft}V_f + E_m \varepsilon^u_{ft} V_m = F_{ft}\left(V_f + V_m \frac{E_m}{E_f}\right) \tag{2.5}$$

When $E_f \gg E_m$, that is, for composites with very stiff fibers Eq. (2.5) can be approximated as

$$F_{1t} \simeq F_{ft}V_f \tag{2.6}$$

In Eq. (2.6) it is assumed that the load is carried by the fibers.

2.2.1.2 Stress field around a fiber break. The above mechanics of materials analysis was based on the simplified deterministic assumption of uniform fiber strength. However, fiber strength varies from fiber to fiber and along the fiber length. Fiber breaks occur at weak points. The stress field around fiber breaks is no longer uniform but high stresses develop. Stress is transferred to the fiber through the matrix, and it is important to know how stress is built up in the fiber, particularly near the fiber breaks.

The problem of stress distribution in a composite consisting of a single fiber embedded in a matrix was first studied by Cox[1] for the case when both the fiber and the matrix are linear elastic. The following assumptions were made:

1. The fiber is perfectly bonded to the matrix.

2. The Poisson's ratios of the fiber and the matrix are equal.

Consider a fiber of length l embedded in a matrix subjected to a uniform axial strain ε_0 (Fig. 2.2). For the transfer of load from the

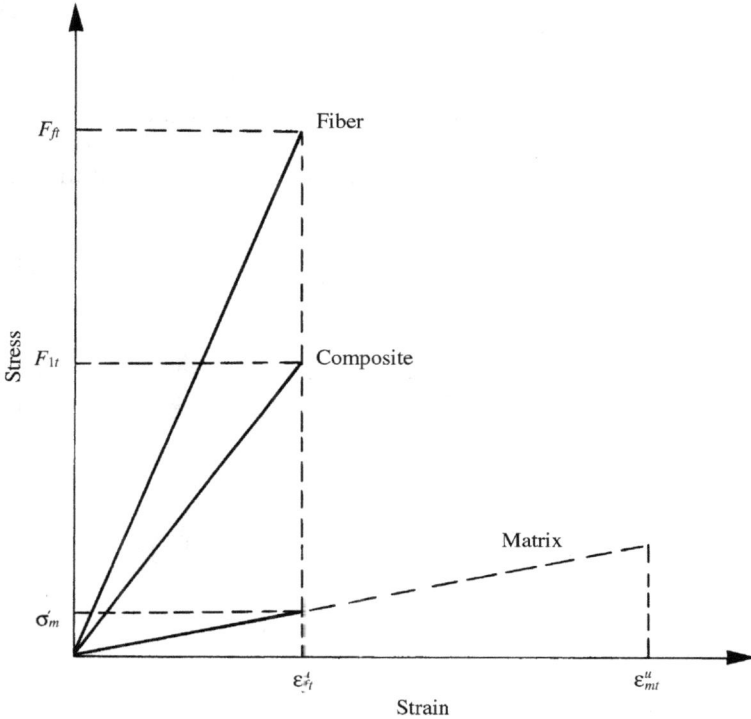

Figure 2.1 Longitudinal tensile stress–strain curves for fiber, matrix and composite when the ultimate tensile strain of the fiber is lower than that of the matrix ($\varepsilon_{ft}^u < \varepsilon_{mt}^u$).

matrix to the fiber, the assumption is made that the rate of change of the load carried by the fiber at a given point is proportional to the difference in displacements of the same point if fiber is present and if fiber is absent, that is,

$$\frac{dP}{dx} = H(u - v) \qquad (2.7)$$

where P = load carried by the fiber

x = distance of a point from the fiber end

u = axial displacement if fiber is present

v = axial displacement if fiber is absent

H = constant (shear lag parameter)

Differentiating Eq. (2.7),

$$\frac{d^2P}{dx^2} = H\left(\frac{du}{dx} - \frac{dv}{dx}\right) \qquad (2.8)$$

We have
$$\frac{du}{dx} = \varepsilon_f, \qquad \frac{dv}{dx} = \varepsilon_0$$

where ε_f, ε_0 are strains in the fiber, matrix, respectively.

The strain in the fiber is created by the transfer of load as the fiber and matrix have different elastic properties. This strain is expressed in terms of the difference of the elastic moduli of the fiber and matrix as

$$\varepsilon_f = \frac{P}{A_f(E_f - E_m)} = \frac{P}{A_f E} \tag{2.9}$$

Substituting in Eq. (2.8),

$$\frac{d^2 P}{dx^2} = H\left(\frac{P}{A_f E} - \varepsilon_0\right) \tag{2.10}$$

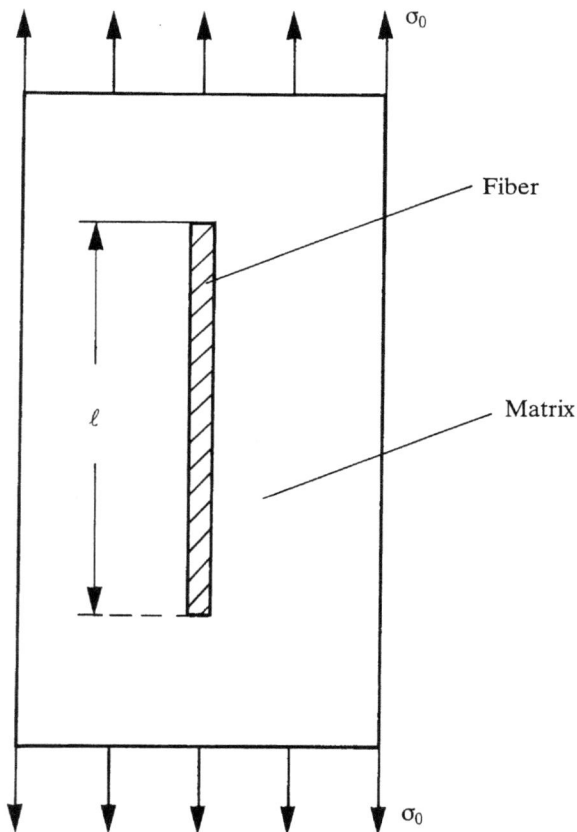

Figure 2.2 A single fiber of length l embedded in a matrix subjected to a uniform axial stress or strain.

The general solution of Eq. (2.10) is

$$P = EA_f\varepsilon_0 + R\sinh\beta x + S\cosh\beta x \qquad (2.11)$$

where $$\beta = \sqrt{\frac{H}{A_f E}} \qquad (2.12)$$

For the determination of R and S the boundary conditions at the fiber ends are used. We have

$$P = 0 \qquad \text{at } x = 0 \text{ and } x = l$$

Thus,

$$P = (E_f - E_m)A_f\varepsilon_0\left[1 - \frac{\cosh\beta\left(\frac{l}{2} - x\right)}{\cosh\beta\frac{l}{2}}\right] \qquad 0 < x < \frac{l}{2} \qquad (2.13)$$

The axial stress in the fiber σ_f is obtained as

$$\sigma_f = \frac{(E_f - E_m)\sigma_0}{E_m}\left[1 - \frac{\cosh\beta\left(\frac{l}{2} - x\right)}{\cosh\beta\frac{l}{2}}\right] \qquad (2.14)$$

where $\sigma_0 \ (=\varepsilon_0 E_m)$ is the applied stress in the matrix.

The critical length of the fiber $l_c/2$ at which the fiber stress takes a maximum value of $\varepsilon_0 E_f$ is obtained by substituting in Eq. (2.14)

$$\sigma_f = \varepsilon_0 E_f \qquad \text{for } x = \frac{l_c}{2} \qquad (2.15)$$

We obtain $$\cosh\beta\frac{l_c}{2} = \frac{E_f}{E_f - E_m} \qquad (2.16)$$

The shear stress τ_i along the fiber–matrix interface is obtained from the equilibrium of the forces acting on a fiber element (Fig. 2.3). For a cylindrical fiber we obtain

$$\frac{dP}{dx}\,dx = 2\pi r_f\,dx\,\tau_i$$

or $$\tau_i = \frac{1}{2\pi r_f}\frac{dP}{dx} \qquad (2.17)$$

Substituting the value of P from Eq. (2.13) we obtain

$$\tau_i = \frac{(E_f - E_m)A_f \beta \varepsilon_0}{2\pi r_f} \frac{\sinh \beta \left(\dfrac{l}{2} - x\right)}{\cosh \beta \dfrac{l}{2}} \qquad (2.18)$$

When the applied stress σ_0 on the matrix is given ($\sigma_0 = \varepsilon_0 E_m$), we have

$$\tau_i = \frac{(E_f - E_m)}{E_m} \frac{A_f \beta \sigma_0}{2\pi r_f} \frac{\sinh \beta \left(\dfrac{l}{2} - x\right)}{\cosh \beta \dfrac{l}{2}} \qquad (2.19)$$

The variation of the normal fiber stress, σ_f, and the interfacial shear stress, τ_i, along the fiber length is shown in Fig. 2.4. Note that the stress τ_i becomes zero at the critical fiber length.

The shear lag constant H is given by

$$H = \frac{2\pi G_m}{\ln \dfrac{r_0}{r_f}} \qquad (2.20)$$

where G_m = shear modulus of the matrix
$\quad r_f$ = radius of the fiber
$\quad r_0$ = radius of the effective matrix surrounding the fiber

Another shear lag model similar to that proposed by Cox was presented by Dow.[2] The form of variation of stresses in both models

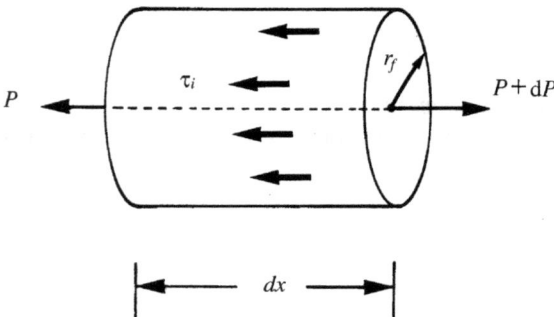

Figure 2.3 Forces acting on a cylindrical fiber element.

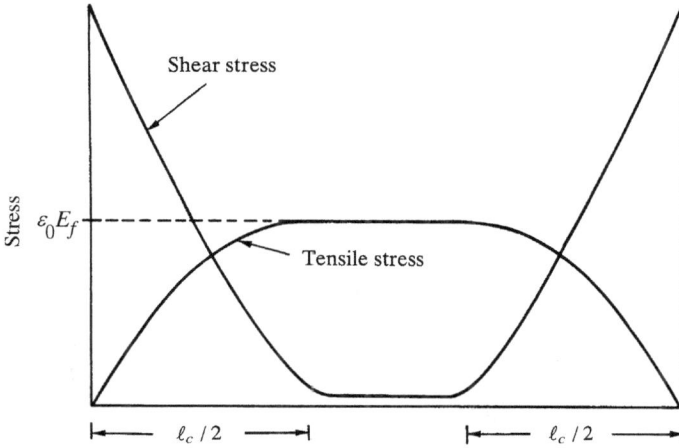

Figure 2.4 Variation of the normal fiber stress and the interfacial shear stress along a fiber of length greater than the critical length.

is the same. The interfacial shear stress expressed from Eq. (2.19) can be put in the form

$$\tau_i = K_1 \sigma_0 \frac{\sinh K_2 \left(\dfrac{l}{2} - x\right)}{\cosh K_2 \left(\dfrac{l}{2}\right)} \tag{2.21}$$

where the expressions for K_1 and K_2 for Cox's and Dow's models are given in Table 2.1. In this table expressions for K_1 and K_2 are also included for a planar model.

A modification of Dow's model was proposed by Rosen.[3] For the case of a matrix in the plastic state, Kelly and Tyson[4] presented an analysis of stress transfer at constant shear stress along the interface.

The fiber tensile stress along the critical length is given by

$$\sigma_f = \frac{2\tau_c \left(\dfrac{l}{2} - x\right)}{r_f} \tag{2.22}$$

The critical length l_c is obtained as

$$l_c = \frac{r_f \sigma_c}{\tau_c} \tag{2.23}$$

where σ_c is the fracture stress of the fiber and τ_c is the interfacial shear strength or the matrix shear yield strength. The variation of σ_f along the fiber is shown in Fig. 2.5.

TABLE 2.1

	Treatment of Cox	Treatment of Dow
	Cylindrical Symmetry	
K_1	$\dfrac{\beta E_f A_f (1 - E_m/E_f)}{2\pi r_f E_m}$	$\dfrac{\lambda}{4}\dfrac{A_m(1 - E_m/E_f)}{A_m(E_f/E_m) + A_f}$
K_2	β	$\dfrac{\lambda}{2r_f}$
	Planar Symmetry	
K_1	$\dfrac{\beta E_f A_f (1 - E_m/E_f)}{2t}$	$\dfrac{\lambda}{2}\dfrac{A_m(1 - E_m/E_f)}{A_m(E_f/E_m) + A_f}$
K_2	β	$\dfrac{\lambda}{2r_f}$

$$\lambda = 2\left(\frac{2\sqrt{2}(G_f/E_f)[1 + (A_f/A_m)(E_f/E_m)]}{(\sqrt{2} - 1) + (G_f/G_m)[\sqrt{A_m/A_t + 2} - \sqrt{2}]}\right)^{1/2} \quad \text{for cylindrical symmetry}$$

$$\lambda = 2\left(\frac{2(A_m/E_f + A_f/E_m)}{(A_m/A_f)(A_f/G_f + A_m/G_m)}\right)^{1/2} \quad \text{for planar symmetry}$$

$$H = \frac{2G_m t}{r_0 - r_f} \quad \text{for planar symmetry}$$

t = thickness (planar model)
G_f = shear modulus of the fiber

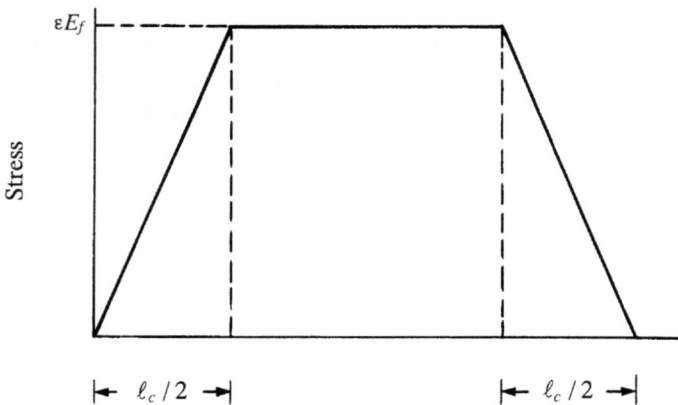

Figure 2.5 Variation of the normal fiber stress along the fiber length for the case of a matrix in the plastic state according to the Kelly and Tyson[4] model.

In the above simplified models only the fiber axial stress and the fiber–matrix interfacial shear stress are determined. An approximate closed form solution that gives the axisymmetric stress distribution in a system consisting of a single broken fiber surrounded by an unbounded matrix was presented by Whitney and Drzal.[5] The fiber and the matrix are assumed to be linear-elastic and perfectly bonded. The fiber is considered to be transversely isotropic. The stresses near the end of the broken fiber are represented by a decaying exponential function multiplied by a polynomial. In the solution, equilibrium and boundary conditions are exactly satisfied in the fiber and matrix, while the compatibility condition is only approximately satisfied. The stress and displacement fields in the fiber and the matrix are given by the following equations:

IN THE FIBER:

$$\sigma_{xf} = [1 - (1 + 4.75\bar{x})e^{-4.75\bar{x}}]A_1\varepsilon_0$$

$$\sigma_{rf} = \sigma_{\theta f} = \left[A_2 - A_1\mu^2\left(2 - \frac{r^2}{R^2}\right)(1 - 4.75\bar{x})e^{-4.75\bar{x}}\right]\varepsilon_0 \qquad (2.24)$$

$$\tau_{xrf} = -4.75\mu A_1\varepsilon_0\left(\frac{r}{R}\right)\bar{x}e^{-4.75\bar{x}}$$

$$u_f = \left[2.375E_{1f}\bar{x} + A_1\left[1 - \left(\frac{r}{R}\right)^2 + 2.375(1 + 4v_{12f}\mu^2)\bar{x}\right]e^{-4.75\bar{x}}\right]\frac{R\varepsilon_0}{\mu E_{1f}}$$

IN THE MATRIX:

$$\sigma_{xm} = \left[E_m(1 - \bar{\varepsilon}_m/\varepsilon_0) - A_1\left(\frac{R}{r}\right)(1 - 4.75\bar{x})e^{-4.75\bar{x}}\right]\varepsilon_0$$

$$\sigma_{rm} = [A_2 - A_1\mu^2(1 - 4.75\bar{x})e^{-4.75\bar{x}}]\left(\frac{R}{r}\right)^2\varepsilon_0$$

$$\sigma_{\theta m} = -[A_2 + A_1\mu^2(1 - 4.75\bar{x})e^{-4.75\bar{x}}]\left(\frac{R}{r}\right)^2\varepsilon_0 \qquad (2.25)$$

$$\tau_{xrm} = -4.75\mu\left(\frac{R}{r}\right)^2\bar{x}A_1\varepsilon_0e^{-4.75\bar{x}}$$

$$u_m = 2.375\frac{R\varepsilon_0}{\mu E_{1f}}\left[E_{1f}\bar{x} + A_1(1 + 4v_{12f}\mu^2)\left(\frac{R}{r}\right)^2\bar{x}e^{-4.75x}\right]$$

where $\varepsilon_0 =$ far-field axial strain
            $\bar{x} = x/L_c$

$L_c =$ critical length defined such that the axial stress becomes equal to 0.95 of its far-field value

$$\mu = \sqrt{\frac{G_m}{E_{1f} - 4v_{12f}G_m}}$$

$E_{1f} =$ axial fiber modulus
$v_{12f} =$ fiber longitudinal Poisson' ratio
$G_m =$ matrix shear modulus

$$A_1 = E_{1f}\left(\frac{\varepsilon_0 - \bar{\varepsilon}_{1f}}{\varepsilon_0}\right) + \frac{4K_f G_m v_{12f}}{K_f + G_m}$$

$$\times \left(v_{12f} - v_m + \frac{1}{\varepsilon_0}[(1 + v_m)\bar{\varepsilon}_m - \bar{\varepsilon}_{2f} - v_{12f}\bar{\varepsilon}_{1f}]\right)$$

$$A_2 = 2\frac{K_f G_m}{K_f + G_m}\left(v_{12f} - v_m + \frac{1}{\varepsilon_0}[(1 + v_m)\bar{\varepsilon}_m - \bar{\varepsilon}_{2f} - v_{12f}\bar{\varepsilon}_{1f}]\right)$$

$$A_3 = E_{1f} + \frac{4K_f v_{12f}G_m(v_{12f} - v_m)}{K_f + G_m}$$

$$A_4 = \frac{-\left(K_f v_{12f} + v_m G_m - \frac{1}{\varepsilon_0}[K_f(\bar{\varepsilon}_{2f} + v_{12f}\bar{\varepsilon}_{1f}) + G_m(1 + v_m)\bar{\varepsilon}_m]\right)}{K_f + G_m}$$

$$A_5 = -v_m + (1 + v_m)\frac{\bar{\varepsilon}_m}{\varepsilon_0}$$

$$A_6 = \frac{R^2 K_f}{K_f + G_m}\left(v_m - v_{12f} + \frac{1}{\varepsilon_0}[\bar{\varepsilon}_{2f} + v_{12f}\bar{\varepsilon}_{1f} - (1 + v_m)]\right)$$

$$K_f = \frac{E_m}{2\left(2 - \frac{E_{2f}}{2G_{2f}} - 2v_{12f}\frac{E_{2f}}{E_{1f}}\right)}$$

$v_m =$ Poisson' ratio of the matrix
$E_{2f} =$ radial modulus of the fiber
$R =$ fiber radius
$\bar{\varepsilon}_{1f} =$ fiber axial expansional strain
$\bar{\varepsilon}_{2f} =$ fiber radial expansional strain
$\bar{\varepsilon}_m = \alpha_m\,\Delta T, \qquad \bar{\varepsilon}_{1f} = \alpha_{1f}\,\Delta T, \qquad \bar{\varepsilon}_{2f} = \alpha_{2f}\,\Delta T$
$\alpha_m, \alpha_{1f}, \alpha_{2f} =$ linear thermal expansion coefficients for the matrix, fiber axial direction, and fiber radial direction
$\Delta T =$ temperature change

Numerical results for the stress components taken from Reference 5 for an AS-4 fiber and a Kevlar 49 fiber embedded in an Epon 828 epoxy matrix are presented in Figs. 2.6–2.9. The material properties of the

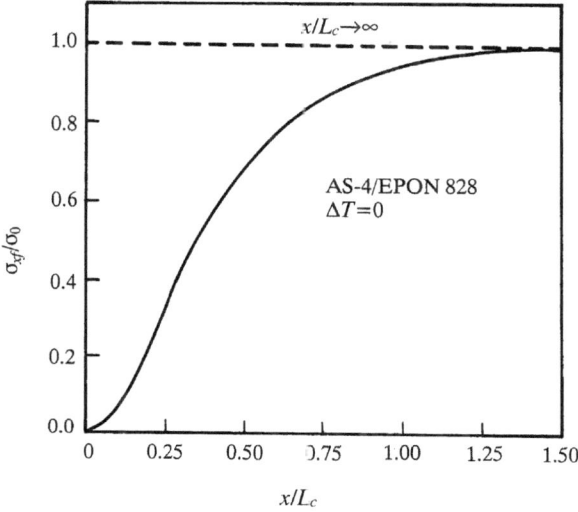

Figure 2.6 Variation of fiber axial stress along fiber length.

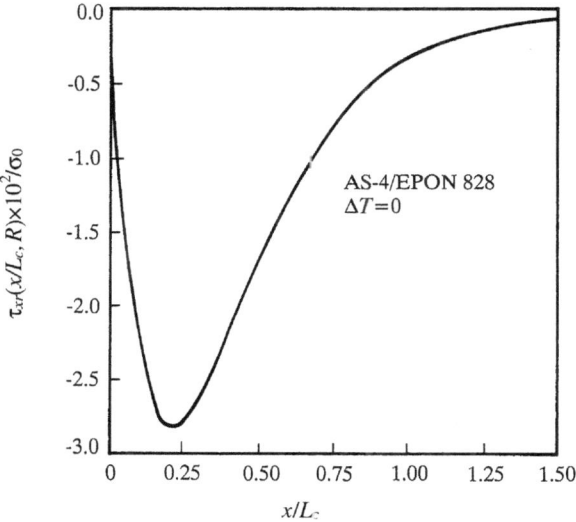

Figure 2.7 Variation of interfacial shear stress along fiber length.

two fibers and matrix appear in Table 2.2. Figures 2.6 and 2.7 present the variation of the normalized fiber axial stress and interfacial shear stress along the fiber length for an AS-4 fiber. The variation of the normalized radial stress along the fiber length for $\varepsilon_0 = 1\%$ and $\Delta T = 0$ and $\Delta T = -75°C$ is presented in Fig. 2.8. Finally, Fig. 2.9 presents results for the radial stress for the AS-4/Epon 828 system and the

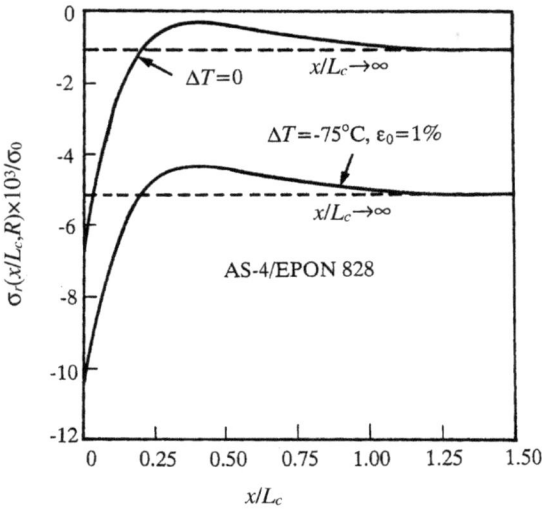

Figure 2.8 Variation of radial stress along fiber length.

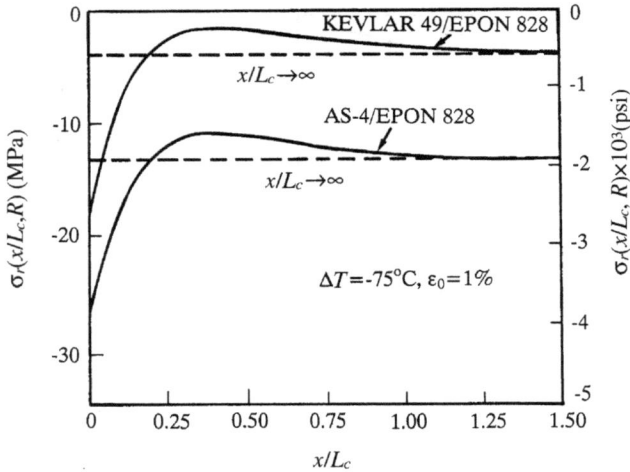

Figure 2.9 Variation of radial stress along fiber length for AS-4 and Kevlar 49 fibers.

TABLE 2.2   Material Properties

Property	Epon 828	AS-4	Kevlar 49
E_1 (GPa [Msi])	3.8 [0.55]	241 [35]	124 [18]
E_2 (GPa [Msi])	3.8 [0.55]	21 [3]	6.9 [1]
v_{12}	0.35	0.25	0.33
G_{23} (GPa [Msi])	1.4 [0.20]	8.3 [1.2]	2.6 [0.38]
α_1 ($10^{-6}/^{\circ}$C [$10^{-6}/^{\circ}$F])	68 [32]	-0.11 [-0.5]	-0.11 [-0.5]
α_2 ($10^{-6}/^{\circ}$C [$10^{-6}/^{\circ}$F])	68 [32]	8.5 [4]	64 [30]

Kevlar 49/Epon 828 system for $\varepsilon_0 = 1\%$ and $\Delta T = -75°C$. Note from Fig. 2.7 that the maximum interfacial shear stress does not occur at the end of the fiber as is predicted by the shear-lag models, but at some distance away from the fiber end. The present solution satisfies the boundary condition of zeroing the shear stress at the free broken end of the fiber.

Axisymmetric elasticity solutions of a cracked fiber in a perfectly bonded composite were presented by McCartney[6] and Nairn.[7]

2.2.1.3 Statistical analysis of failure. The results of the analysis of the stress field around fiber breaks indicate that a nonuniform state of stress is developed around such points. The normal stress along the fiber is not uniform, but increases from a zero value at the fiber break to the far-field value at a critical distance, from the break. An interfacial shear stress with a maximum value near the fiber break develops. This stress becomes zero at the critical distance from the break. In the neighborhood of a fiber break the axial load carried by the fiber is transmitted by shear through the matrix to adjacent fibers. This results in increased axial stresses and development of interfacial shear stresses to adjacent fibers. A portion of the fiber at each break is not fully effective in resisting the applied load.

Fiber failures occur at points of imperfections. As the applied load increases, an increasing accumulation of fiber breaks occurs until the ineffective fiber lengths combine to produce failure of the composite. A model of statistical analysis of composite failure was developed by Rosen.[8] The model expresses the longitudinal tensile strength of unidirectional composites in terms of parameters of the statistical distribution of the fiber strengths. The model is based on the assumption that the composite fails as a result of accumulation of statistically distributed fiber flaws. A Weibull distribution for the fiber tensile stress was assumed in the form

$$P(\sigma) = 1 - \exp\left[-L\left(\frac{\sigma}{\bar{\bar{\sigma}}}\right)^{\beta}\right] \tag{2.26}$$

where $P(\sigma) =$ probability of having a fiber with strength less than or equal to σ
$L =$ total length of a fiber
$\sigma =$ stress along the fiber
$\bar{\bar{\sigma}} =$ scale factor
$\beta =$ shape parameter of fiber strength distribution

The mean stress of an individual fiber is

$$\sigma_{f,0} = \int_0^{\infty} \sigma \, \frac{dP(\sigma)}{d\sigma} \, d\sigma = \left(\frac{1}{L}\right)^{1/\beta} \bar{\bar{\sigma}} \, \Gamma\left(1 + \frac{1}{\beta}\right) \tag{2.27}$$

where δ = ineffective length of the cummulative damage model
  Γ = gamma function

The mean strength of a bundle of fibers of equal length L gathered together and subjected to tension is

$$\sigma^*[1 - P(\sigma^*)]$$

where

$$\left[\frac{d}{d\sigma}\{\sigma[1 - P(\sigma)]\}\right]_{\sigma=\sigma^*} = 0$$

or

$$\sigma^* = \bar{\bar{\sigma}}\left(\frac{1}{\beta L_g}\right)^{1/\beta} \tag{2.28}$$

Thus the mean strength of the bundle of fibers is

$$\sigma^*[1 - P(\sigma^*)] = \bar{\bar{\sigma}}\left(\frac{1}{\beta eL}\right)^{1/\beta} \tag{2.29}$$

where e is the base of natural logarithms.

According to Rosen, the contribution of the fiber strength to the strength of the composite can be obtained by replacing L with δ. Thus the statistical mode of the composite tensile strength σ^* is given by

$$\sigma^* = V_f(\alpha\beta\delta e)^{-1/\beta} \tag{2.30}$$

where α is a constant. The constants α and β are determined by test results of fiber strength versus length. The ineffective length δ is defined by a fiber shear stress analysis.

Figure 2.10 presents the composite failure strength versus the normalized ineffective length for a glass fiber composite. The ineffective length varies in the range of 1 to 100 fiber diameters. The effects of an increase in the dispersion as measured by a 10% change in β and of a decrease in the reference strength as measured by a 10% change in $\alpha^{-1/\beta}$ are shown in Fig. 2.10. As the dispersion in fiber strength is small, the fibers are characterized by a single strength value rather than by a distribution function and the composite will fail when the fibers reach this stress value.

2.2.1.4 Failure mechanisms around a fiber break. The modes of local failure around a fiber break depend on the properties of the fiber and matrix and the interfacial shear strength. To understand the initiation and growth of failure mechanisms, it is essential to establish the state of stress in the matrix. The variation of the fiber normal stress and the interfacial shear stress along the fiber length are shown in Fig 2.4. A photoelastic analysis of the stress field in a two-dimensional model of an aluminum alloy fiber embedded in an Araldite resin matrix was

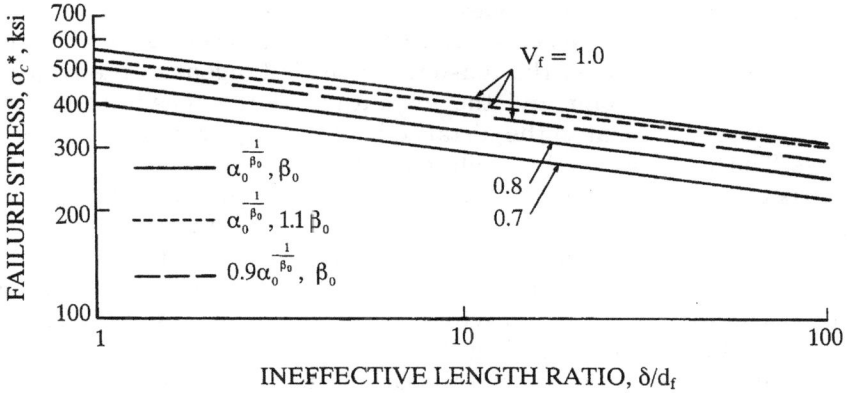

Figure 2.10   Failure stress versus normalized ineffective length for a glass fiber composite according to Rosen.[8]

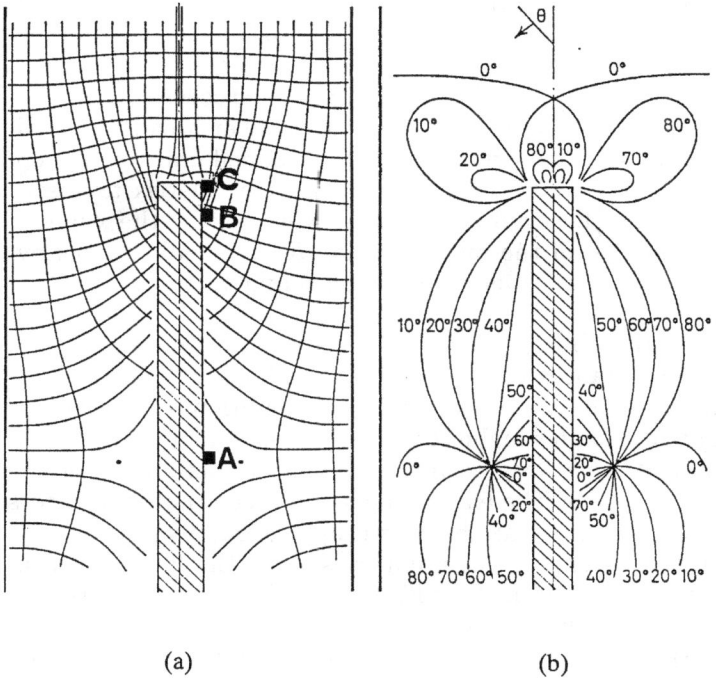

(a)   (b)

Figure 2.11   Stress trajectories (a) and isoclinics (b) of a two-dimensional model of an aluminum alloy fiber embedded in an Araldite resin matrix according to Tyson and Davies.[9]

performed by Tyson and Davies.[9] Figure 2.11 shows the patterns of tensile stress trajectories (Fig. 2.11a) and isoclinics (Fig. 2.11b) in the matrix. The direction of the tensile stresses is tangent to the trajectories at a given point and the magnitude of the tensile stress is inversely proportional to the distance between adjacent stress trajectories. The figure shows that high concentration of stress transfer from matrix to fiber occurs near the fiber end, while at the center of the fiber (element A) the matrix is virtually unstressed, since the load is mostly carried by the stiffer fiber. At element B located along the fiber at a small distance from the fiber end the state of stress is nearly pure shear. The tensile stress trajectory at that point makes an angle of approximately 45° with the fiber. At element C at the fiber end there is a high tensile stress in the matrix parallel to the fiber in addition to shear stress at the interface.

Based on the state of stress in the fiber and the matrix, the possible failure mechanisms can now be defined. The normal stress along most of the fiber length away from the fiber ends (at a distance of almost five fiber diameters) takes a constant maximum value. First fracture of the fiber is expected to occur in the fiber central region. The specific location depends on the variation of the fiber strength along its length. The matrix near the fiber break is suddenly stressed in axial tension to carry the load that was previously carried by the fiber.

When the tensile stress carried by the matrix exceeds its uniaxial strength, a penny-shaped crack is formed at the fiber break perpendicular to the fiber axis. The extent the penny-shaped crack will propagate into the matrix depends on the ability of the matrix to absorb energy along the crack periphery. This is expressed by the value of fracture toughness K_{Ic} of the matrix. When K_{Ic} is small, the penny-shaped crack may grow catastrophically, leading to fracture of the entire specimen.

In some cases the fiber fractures at a weak point under low load. Under such circumstances the energy released during fiber breaking is insufficient to produce a matrix crack. This condition is identified as low energy fracture. Under higher energy released during fiber breaking a disk-shaped crack originating from the circumference of the broken fiber may appear. The elastic energy released that is in excess of the energy required to break the fiber is approximately proportional to the fiber tensile strength. Thus, it varies along the fiber length in accordance with the variation of the fiber strength. The response of the matrix to the propagation of the crack having an area equal to the fiber cross section depends on the behavior of the matrix. When the matrix has a sufficient plastic or viscoelastic response, the excess energy may be absorbed by the matrix. It can also be absorbed by the damping behavior of the specimen. When the value of the released energy exceeds the value that can be absorbed by the matrix, the excess energy is absorbed by the creation of matrix cracks or debonding.

To summarize, the following failure modes can be created at a fiber break in the fragmentation test depending on the amount of the released strain energy, the bond strength of the fiber–matrix interface, and the fracture toughness of the matrix material:

- *Fiber fracture only*: This occurs when low strain energy is released during fiber fracture, the fiber–matrix bond is adequate, and the fracture toughness of the matrix is high.

- *Fiber fracture and fiber–matrix debonding*: This occurs at intermediate strain energy release values, insufficient fiber–matrix bond strength, and high matrix fracture toughness.

- *Fiber fracture with a matrix penny-shaped crack and/or one or two inclined conical cracks*: This occurs at high strain energy release values, high fiber–matrix bond strength, and insufficient matrix fracture toughness.

For more information on the types of failure modes around fiber breaks, the reader is referred to Reference 10. For an increasing applied load the fiber breaks increase in density, the localized failure mechanisms interact and eventually coalesce to produce catastrophic failure.

The interfacial shear strength (ISS) influences the strength and failure mechanisms of unidirectional composites. Low values of the ISS cause failure of the fiber–matrix interface resulting in a cumulative failure process. For high values of the ISS, brittle fractures result because of matrix failure. Intermediate values of the ISS cause a "brushlike" failure, consisting of interfacial debonding and matrix fracture. The effect of ISS on the mechanical and failure properties of graphite/epoxy composites was studied by Madhukar and Drzal.[11]

2.2.2 Composites with ultimate tensile strain of the matrix lower than that of the fiber

2.2.2.1 Mechanics of materials approach. In the case of a composite with ultimate tensile strain of the matrix lower than that of the fiber, i.e., when

$$\varepsilon_{mt}^u < \varepsilon_{ft}^u \tag{2.31}$$

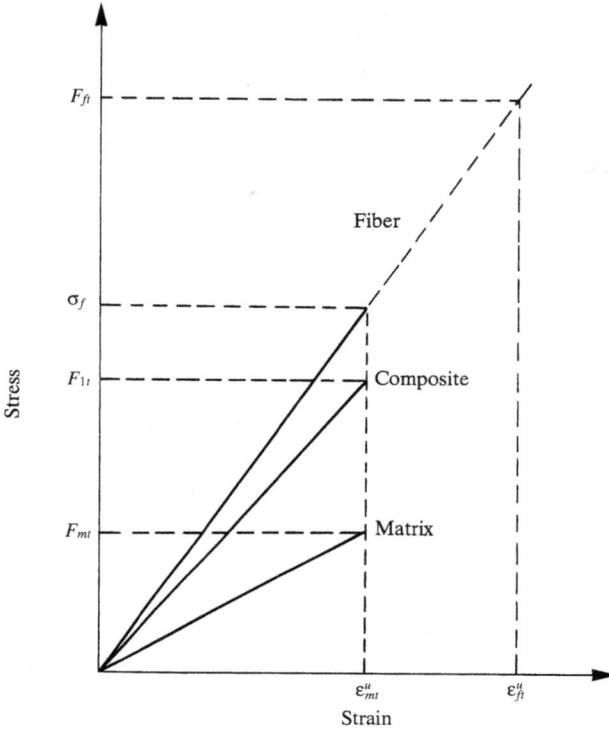

Figure 2.12 Longitudinal tensile stress–strain curves for fiber, matrix and composite when the ultimate tensile strain of the matrix is lower than that of the fiber ($\varepsilon_{mt}^u < \varepsilon_{ft}^u$).

failure of the composite occurs when the applied longitudinal strain becomes equal to the ultimate tensile strain of the matrix. Figure 2.12 shows the longitudinal stress–strain curves for the fiber, matrix, and composite. Note that according to Eq. (2.2) the stress–strain curve of the composite falls between the stress–strain curves of the matrix and fiber. The longitudinal tensile strength of the composite is obtained from Eq. (2.2) as

$$F_{1t} = \sigma_f' V_f + E_{mt} V_m \qquad (2.32)$$

which can be approximated as

$$F_{1t} \cong F_{mt}\left(V_f \frac{E_f}{E_m} + V_m\right) \qquad (2.33)$$

where F_{mt} = matrix tensile strength
       σ_f' = longitudinal fiber stress when ultimate matrix strain is reached

2.2.2.2 Matrix cracking. For a unidirectional composite with ultimate tensile strain of the matrix lower than that of the fiber under long-itudinal tension, matrix cracks will first appear. Near the crack border the axial strain in the matrix is relieved, while the axial fiber stress is increased and a highly localized shear stress develops. As the applied load increases, depending on the relative values of the interfacial shear strength and the matrix tensile strength, debonding may start immediately after the formation of the first matrix cracks or it may initiate at some point in the matrix cracking process. Initiation of debonding occurs when the interfacial shear stress exceeds the fiber bond strength and the maximum axial stress in the matrix remains below its tensile strength. In many cases, multiple matrix cracking occurs first up to a saturation point where a minimum crack spacing is reached. At matrix crack saturation the maximum tensile stress in the matrix is smaller than its tensile strength with any further increase in load. Following matrix crack saturation fiber–matrix debonding occurs. As the applied load is increased, more debonding and fiber failures occur, leading to catastrophic failure.

A number of micromechanical models have been developed for the study of load transfer mechanisms near matrix cracks. The model proposed by Hsueh[12] is based on an axisymmetric element consisting of a cylindrical fiber with a uniform coating of finite thickness in its central portion subjected to an axial tensile stress through the bare fiber ends. A shear lag analysis was performed based on the approx-imation that the shear stress in the fiber is proportional to the distance from the fiber axis and that the axial displacement in the coating is independent of the radial position and identical to the interfacial axial displacement. Under such circumstances, for an applied axial stress σ_a, the axial stress in the matrix σ_{zm} and the interfacial shear stress τ_i are given by

$$\sigma_{zm} = \frac{E_m V_f}{E_1}\left[1 - \frac{\cosh\left[m\left(\frac{l}{2} - z\right)\right]}{\cosh\left(\frac{ml}{2}\right)}\right]\sigma_a \qquad (2.34)$$

$$\tau_i = \left[\frac{E_m V_m}{(1 - v_f)E_1}\right]^{1/2} \frac{\sinh\left[m\left(\frac{l}{2} - z\right)\right]}{\cosh\left(\frac{ml}{2}\right)}\sigma_a \qquad (2.35)$$

where

$$E_1 = E_f V_f + E_m V_m$$

$$m = \frac{2}{r_f}\left[\frac{E_1}{(1 + v_f)E_m V_m}\right]^{1/2}$$

A modified shear-lag analysis of a cylindrical element of matrix with a single fiber and two matrix cracks perpendicular to the fiber axis under longitudinal tensile load was conducted by Daniel and coworkers.[13,14] This analysis is treated here in detail.

Consider a cylindrical element of matrix with a single fiber subjected to an axial load P (Fig. 2.13). Let r_f and r_m represent the radius of the fiber and the outer cylinder, respectively, and l the length of the element. For an increasing applied load P the first set of matrix cracks perpendicular to the loading direction will appear when the axial stress in the matrix σ_{zm} is equal to the tensile strength of the matrix, F_{mT}, or the axial strain of the matrix, ε_{zm}, is equal to the ultimate axial strain ε_{mT}^u, that is,

$$\sigma_{zm} = F_{mT} \quad \text{or} \quad \varepsilon_{zm} = \varepsilon_{mT}^u \tag{2.36}$$

Following the initial matrix cracking, stress redistribution in the matrix and fiber takes place. The equilibrium of forces of the element in the axial direction is obtained as

$$P = P_m + P_f + (\sigma_{rm} V_m + \sigma_{rf} V_f) \pi r_m^2 = P_m + P_f \tag{2.37}$$

where P_m, P_f = forces carried by the matrix and fiber, respectively
σ_{rm}, σ_{rf} = residual stresses in the axial direction in the matrix and fiber, respectively
V_m, V_f = matrix and fiber volume ratios, respectively

The analysis is based on the assumption that the fiber and matrix are perfectly bonded and the rate of change in the load P_m carried by the matrix is proportional to the difference in average displacements of the matrix and fiber:

$$\frac{d(P_m + \pi r_m^2 V_m \sigma_{rm})}{dz} = H(\bar{u}_m - \bar{u}_f) \tag{2.38}$$

where H = the shear-lag parameter, which is a constant
$\bar{u}_m, \bar{u}_f$ = average axial displacements in the radial direction of the matrix and fiber, respectively.

By differentiating Eq. (2.38) we obtain

$$\frac{d^2 P_m'}{dz^2} = H\left(\frac{d\bar{u}_m}{dz} - \frac{d\bar{u}_f}{dz}\right) \tag{2.39}$$

where $$P_m' = P_m + \pi r_m^2 V_m \sigma_{rm}$$

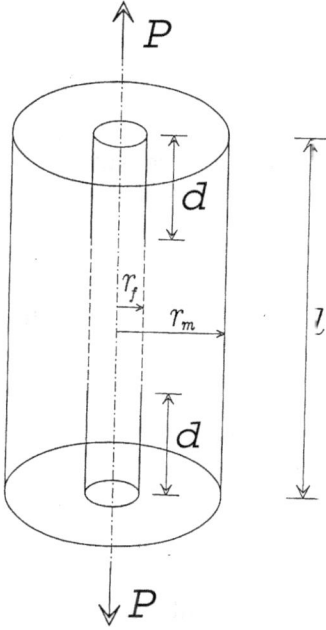

Figure 2.13 Cylindrical element of composite.

From the strain–displacement and linear stress–strain relations we have

$$\frac{d\bar{u}_m}{dz} = \bar{\varepsilon}_m = \frac{P_m}{\pi E_m V_m r_m^2}, \qquad \frac{d\bar{u}_f}{dz} = \bar{\varepsilon}_f = \frac{P_f}{\pi E_f V_f r_m^2} \qquad (2.40)$$

where $\bar{\varepsilon}_m$, $\bar{\varepsilon}_f$ = average axial strains in the radial direction in the matrix and fiber, respectively.

Substituting Eq. (2.40) into Eq. (2.39), we obtain

$$\frac{d^2 P_m'}{dz^2} - \alpha^2 P_m' = -\beta \qquad (2.41)$$

where

$$\alpha^2 = \frac{H}{\pi r_m^2} \left(\frac{1}{E_f V_f} + \frac{1}{E_m V_m} \right)$$

$$\beta = \frac{H}{\pi r_m^2} \left(\frac{P}{E_f V_f} + \left(\frac{1}{E_m V_m} + \frac{1}{E_f V_f} \right) \pi r_m^2 V_m \sigma_{rm} \right)$$

The general solution of Eq. (2.41) is

$$P'_m = \frac{\beta}{\alpha^2} + R\sinh(\alpha z) + S\cosh(\alpha z) \qquad (2.42)$$

where R and S are constants. The first term of the general solution is the force shared by the matrix where there are no matrix cracks in the element and the other two terms are the disturbances due to the existence of matrix cracks.

For the determination of the shear-lag parameter, H, the assumption is made that the variation of the shear stress in the radial direction in the fiber is linear and in the matrix varies according to an inverse polynomial of the second order, that is,

$$\tau_{rzf} = G_f(2C_1 r + C_2), \qquad \tau_{rzm} = G_m\left(-\frac{C_4}{r^2} + C_5\right) \qquad (2.43)$$

where C_i = undetermined parameters that are function of z
        G_m and G_f = shear moduli
        τ_{rzm} and τ_{rzf} = shear stresses in the matrix and fiber, respectively.

From Eq. (2.43) we obtain that the axial displacements in the fiber and matrix are

$$u_f = C_1 r^2 + C_2 r + C_3, \qquad u_m = C_4\frac{1}{r} + C_5 r + C_6 \qquad (2.44)$$

where C_3 and C_6 = undetermined parameters that are functions of z
        u_m and u_f = axial displacements in the matrix and fiber, respectively

The boundary conditions are

$$\begin{aligned}
\tau_{rz} &= 0 &&\text{at } r = 0 \text{ and } r = r_m \\
\tau_{rzf} &= \tau_{rzm} &&\text{at } r = r_f \\
u_f &= u_m &&\text{at } r = r_f
\end{aligned} \qquad (2.45)$$

Using the boundary conditions and introducing the interfacial shear stress τ_i, which will be determined later, we obtain

$$\bar{u}_m - \bar{u}_f = A\tau_i \qquad (2.46)$$

where

$$A = \frac{r_f}{4G_f} + \frac{1}{G_m} \left[\frac{2r_f^2(r_f - r_m)}{3(r_m^2 - r_f^2)^2} (4r_m^2 + r_m r_f + r_f^2) + \frac{r_f(r_m^2 + r_f^2)}{r_m^2 - r_f^2} \right] \qquad (2.47)$$

From equilibrium of axial forces in the matrix of the element, we obtain the relation

$$\tau_i = \frac{1}{2\pi r_f} \frac{dP'_m}{dz} \qquad (2.48)$$

From Eqs. (2.38), (2.46), and (2.48) the shear-lag parameter is obtained as

$$H = \frac{2\pi r_f}{A} \qquad (2.49)$$

Using Eqs. (2.37), (2.42), and (2.48) and the boundary conditions

$$P'_m = 0 \qquad \text{at } z = 0 \text{ and } z = l$$

we obtain for P'_m

$$P'_m = \left(\frac{E_m V_m}{E_1} P + \tau r_m^2 V_m \sigma_{rm} \right) \left[1 - \frac{\cosh\left(\frac{\alpha l}{2} - \alpha z\right)}{\cosh\left(\frac{\alpha l}{2}\right)} \right] \qquad (2.50)$$

The axial stress in the matrix is obtained as follows:

$$\sigma_{zm} = \left(\frac{E_m C_\alpha}{E_1} + \sigma_{rm} \right) \left[1 - \frac{\cosh\left(\frac{\alpha l}{2} - \alpha z\right)}{\cosh\left(\frac{\alpha l}{2}\right)} \right] \qquad (2.51)$$

Using Eqs. (2.37) and (2.50) we obtain the axial stress in the fiber as follows:

$$\sigma_{zf} = \frac{E_f}{E_1} \left[1 + \frac{E_m V_m}{E_f V_f} \frac{\cosh\left(\frac{\alpha l}{2} - \alpha z\right)}{\cosh\left(\frac{\alpha l}{2}\right)} \right] \sigma_a + \left[1 - \frac{\cosh\left(\frac{\alpha l}{2} - \alpha z\right)}{\cosh\left(\frac{\alpha l}{2}\right)} \right] \sigma_{rf} \qquad (2.52)$$

The interfacial shear stress is obtained from Eqs. (2.48) and (2.50) as follows:

$$\tau_i(z) = \frac{1}{2\pi r_f}\frac{dP'_m}{dz} = \frac{\alpha r_m^2 V_m}{2 r_f}\left(\frac{E_m \sigma_a}{E_1} + \sigma_{rm}\right)\frac{\sinh\left(\dfrac{\alpha l}{2} - \alpha z\right)}{\cosh\left(\dfrac{\alpha l}{2}\right)} \qquad (2.53)$$

From the above equations it can be observed that the matrix tensile stress σ_{zm} becomes maximum in the middle ($z = l/2$) and zero at the ends ($z = 0, l$) of the element, whereas the interfacial shear stress τ_i exhibits the opposite trend (it becomes maximum at the ends ($z = 0, l$) and zero in the middle ($z = l/2$) of the element). Figures 2.14 and 2.15 present the variation of the nondimensional stress components σ_{zm}/σ_a, σ_{zf}/σ_a, and τ_i/σ_a along the axis of the cylindrical element according to the shear-lag analysis (SLA) proposed by Hsueh and the modified shear-lag analysis (MSLA) presented by Daniel for a calcium alumi-nosilicate glass ceramic reinforced with silicon carbide fibers. The aspect ratio of the element $k = l/r_f = 10$ and 5, for Figs. 2.14 and 2.15 respectively, $r_m/r_f = 1.6$, and the matrix and fiber elastic modulus and Poisson's ratio take the values $E_m = 98\,\mathrm{GPa}$, $E_f = 179\,\mathrm{GPa}$, $v_m = v_f = 0.16$. The variation of the nondimensional shear stress τ/σ_a along the radial direction for $k = l/r_f = 10$ at various distances from the matrix crack planes $z/r_f = 0.750$, 1.625, and 3.875 is presented in Fig. 2.16. Note that the shear stress τ_i varies linearly in the fiber up to a

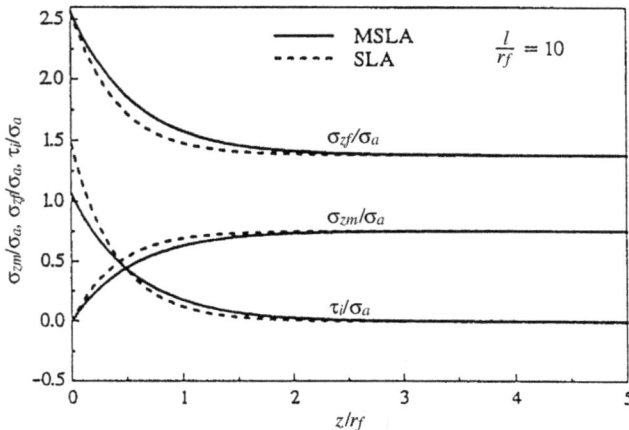

Figure 2.14 Variation of stress components along the axis of the cylindrical element according to the shear lag analysis (dotted lines) and the modified shear lag analysis (continuous lines) for $l/r_f = 10$.

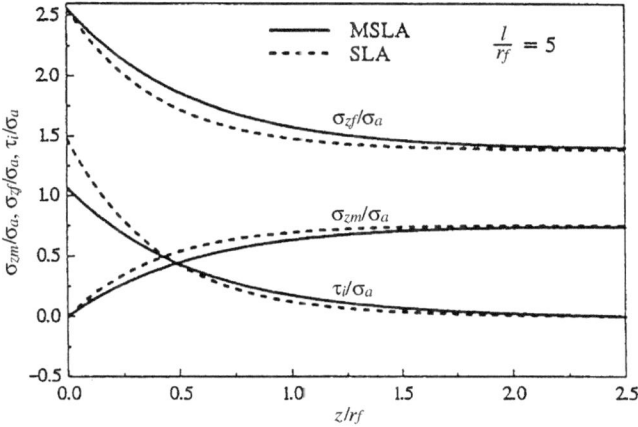

Figure 2.15 Variation of stress components along the axis of the cylindrical element according to the shear lag analysis (dotted lines) and the modified shear lag analysis (continuous lines) for $l/r_f = 5$.

maximum value at the fiber–matrix interface, and in the matrix it decays to zero at the free boundary. The results of the MSLA were verified by a finite element analysis.[15,16]

2.2.2.3 Fiber–matrix debonding. Following matrix crack saturation, fiber–matrix debonding occurs when the interfacial shear stress at the end of the disbond becomes equal to the interfacial shear strength.

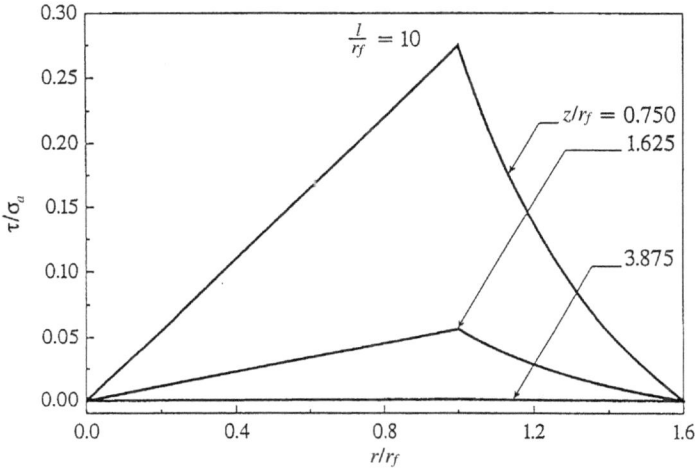

Figure 2.16 Variation of shear stress along the radius of the cylindrical element at various distances from the matrix crack plane for $l/r_f = 10$.

For a debonded length d (Fig. 2.13) the stresses σ_{zf}, σ_{zm}, and τ_i in the debonded area ($0 \leq z \leq d$ and $l - d \leq z \leq l$) and in the bonded area ($d \leq z \leq l - d$) can be approximated as follows according to the modified shear-lag analysis of Daniel that was presented in the previous section:

DEBONDED AREA ($0 \leq z \leq d$ and $l - d \leq z \leq l$)

$$\sigma_{zm} = \frac{2V_f \tau_f z}{V_m r_f} \tag{2.54a}$$

$$\sigma_{zf} = \frac{\sigma_a}{V_f} - \frac{2\tau_f z}{r_f} \tag{2.54b}$$

$$\tau_i = \tau_f \tag{2.54c}$$

BONDED AREA ($d \leq z \leq l - d$)

$$\sigma_{zm} = \left(\frac{E_m}{E_1}\sigma_a + \sigma_{rm}\right) + \left(\frac{2V_f\tau_f d}{V_m r_f} - \frac{E_m}{E_1}\sigma_a - \sigma_{rm}\right)\frac{\cosh \alpha(l/2 - z)}{\cosh \alpha(l/2 - d)} \tag{2.55a}$$

$$\sigma_{zf} = \left(\frac{E_f}{E_1}\sigma_a + \sigma_{rf}\right) + \left(\frac{E_m V_m}{E_f V_f}\sigma_a - \sigma_{rf} - \frac{2\tau_f d}{r_f}\right)\frac{\cosh \alpha(l/2 - z)}{\cosh \alpha(l/2 - d)} \tag{2.55b}$$

$$\tau_i = \frac{\alpha r_f V_m}{2V_f}\left(\frac{E_m}{E_1}\sigma_a + \sigma_{rm} - \frac{2V_f}{V_m}\frac{\tau_f d}{r_f}\right)\frac{\sinh \alpha(l/2 - z)}{\cosh \alpha(l/2 - d)} \tag{2.55c}$$

where τ_f is the interfacial frictional stress in the debonded area. The axial matrix stress and interfacial shear stress distributions in a cracked and partially debonded composite element are shown schematically in Fig. 2.17.

2.2.2.4 Stress–strain curve. The theoretical analysis discussed earlier can predict the stress–strain curve of a unidirectional brittle-matrix composite under longitudinal tensile loading in terms of geometrical elastic and strength parameters of the fiber, matrix and interface. The stress–strain curve is linear up to an applied stress level where the matrix stress (or strain) reaches the ultimate value of the matrix tensile strength (or strain). This stress level is given by

$$\sigma = \frac{E_1}{E_m}(F_{mT} - \sigma_{rm}) \tag{2.56}$$

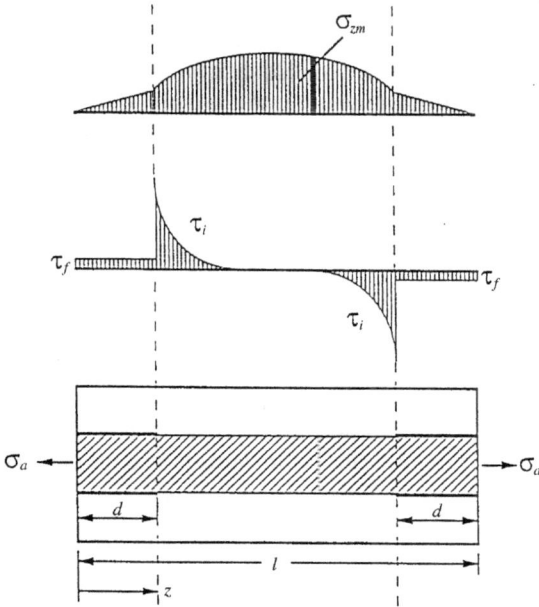

Figure 2.17 Variation of axial matrix stress and interfacial shear stress in a cracked and partially debonded composite element.

Above this applied stress level, matrix cracking takes place. The matrix crack spacing l is obtained from Eq. (2.51) by setting $x = l/2$ and $\sigma_{zm} = F_{mT}$. We obtain

$$l = \frac{2}{\alpha}\cosh^{-1}\left(\frac{E_m\sigma_a + E_1\sigma_{rm}}{E_m\sigma_a + E_1(\sigma_{rm} - F_{mT})}\right) \tag{2.57}$$

The above relation gives the crack spacing as a function of applied stress up to the point of initiation of fiber debonding.

The maximum shear stress at the crack location is calculated from Eq. (2.53) by putting $z = 0$. We obtain

$$\tau_i(0) = \frac{\alpha r_f}{2}\frac{V_m}{V_f}\left(\frac{E_m\sigma_a}{E_1} + \sigma_{rm}\right)\tanh\frac{\alpha l}{2} \tag{2.58}$$

When $\tau_i(0)$ reaches the value of the interfacial shear strength F_{is}, debonding occurs. At debonding initiation we have

$$F_{is} = \frac{\alpha r_f}{2}\left(\frac{V_m}{V_f}\right)\left(\frac{E_m}{E_1}\sigma_a + \sigma_{rm}\right)\tanh\frac{\alpha l}{2} \tag{2.59}$$

The debonded length d is calculated by assuming that the interfacial shear stress at the end of the disbond is equal to the interfacial shear strength. The length d is obtained from Eq. (2.55c) by putting $\tau_i(d) = F_{is}$ and $z = d$. The resulting relation can be solved for d by iteration techniques. If friction is neglected, an explicit expression is obtained for d

$$d = \frac{1}{2}\left[l - \frac{1}{\alpha}\log\frac{1+\xi}{1-\xi}\right] \tag{2.60}$$

where

$$\xi = \frac{2F_{is}}{\alpha r_f}\frac{V_f}{V_m}\frac{E_1}{E_m\sigma_a + E_1\sigma_{rm}} \tag{2.61}$$

Based on the above theory, the stress–strain curve of a unidirectional composite was predicted. Geometrical and material parameters took the values

$$r_f = 8\ \mu\text{m} \qquad V_f = 0.39$$
$$E_m = 98\ \text{GPa}, \qquad E_f = 159\ \text{GPa}$$
$$F_{mT} = 159\ \text{MPa}, \qquad F_{is} = 477\ \text{MPa}$$

Figure 2.18 displays the longitudinal tensile stress–strain curve of the composite up to fracture. The curve presents three distinct characteristic parts that are related to damage development in the

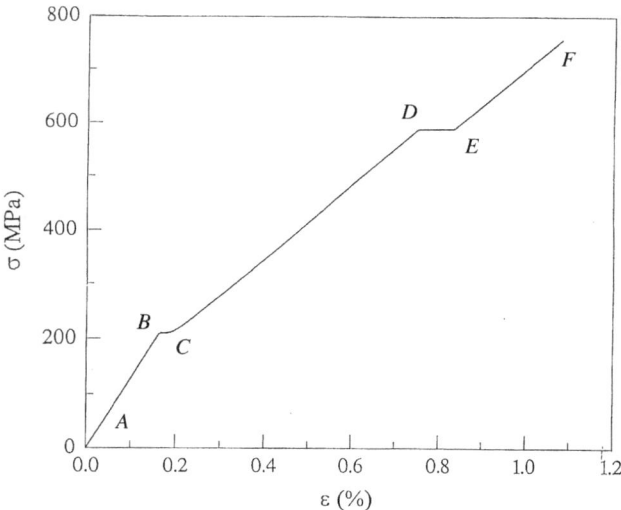

Figure 2.18 Predicted longitudinal stress–strain curve of a SiC/CAS unidirectional composite.

composite. The initial linear part AB corresponds to the elastic linear behavior of the material prior to initiation of damage. The modulus of the material obtained by the rule of mixtures is $E_1 = 130\,\text{GPa}$. At point B the first matrix cracks appear and the stress–strain curve departs from linearity. At this point the stress in the matrix reaches the matrix tensile strength, leading to matrix cracking in a direction perpendicular to the fibers. The stress at point B calculated from Eq. (2.56) is $\sigma_B = 210\,\text{MPa}$; zero residual stresses were assumed in the composite. The second part of the stress–strain curve, BCD, corresponds to transverse crack multiplication up to a saturation crack density at point D. The curve BCD consists of two parts. The first part BC corresponds to the development of the first matrix crack and leads to an abrupt increase in the overall strain of the composite with an almost constant applied stress. After the formation of the first matrix crack, a nonlinear region with decreasing stiffness CD appears in the stress–strain curve, corresponding to transverse cracking in the matrix. The part CD of the curve is almost linear, and the stress–strain curve presents a constant tangent modulus in that region. Following crack saturation, fiber debonding begins. The stress at point D corresponds to termination of crack multiplication and initiation of fiber debonding. As for the case of crack multiplication, the part DEF of the stress–strain curve corresponding to fiber debonding consists of two parts. The first part, DE, corresponds to first fiber debonding and leads to a sharp increase in the overall strain of the composite with an almost constant applied stress. In the second part, EF, fiber debonding progresses gradually up to a limiting debonded length at which fiber fracture occurs, leading to final failure. The part EF of gradual debonding is almost linear, and the stress–strain curve of the composite presents a constant tangent modulus. Parts CD and EF of the stress–strain curve corresponding to progressive matrix cracking and fiber debonding are almost parallel.

To obtain a detailed picture of failure mechanisms in the composite, the stress–strain, the stress versus matrix crack density, and the stress versus debonded length curves are shown in Fig. 2.19. Note the progressive increase of matrix cracking with increased stress after first matrix crack initiation up to debonding initiation (curve $C'D'F'$) and the progressive increase of debond length with increased stress after first debonding up to final fracture (curve $D''E''F'$). The stresses at matrix crack initiation and fiber debonding initiation are clearly identified from the abrupt increase of strain in the stress–strain diagram of the composite at both points.

Theoretical and experimental results for a silicon carbide glass ceramic composite are shown in Fig. 2.20, taken from Reference 13. Note that the predicted transverse matrix crack density increases rapidly after initiation up to a saturation density of $8\,\text{cracks/mm}$ or a

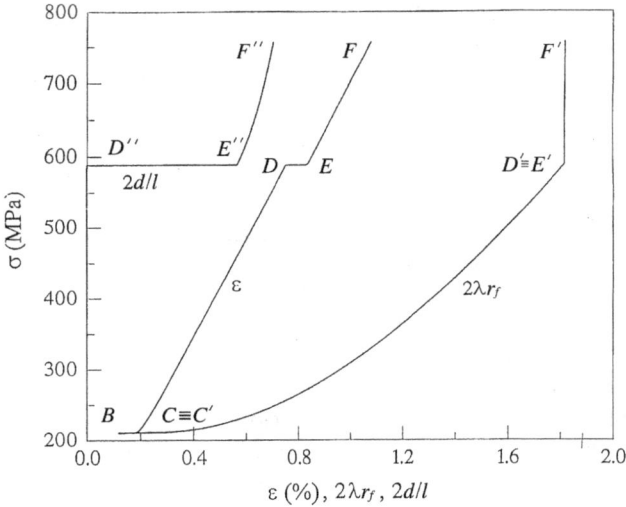

Figure 2.19 Predicted longitudinal stress–strain, stress versus matrix crack density and stress versus debonded length curves for a SiC/CAS unidirectional composite.

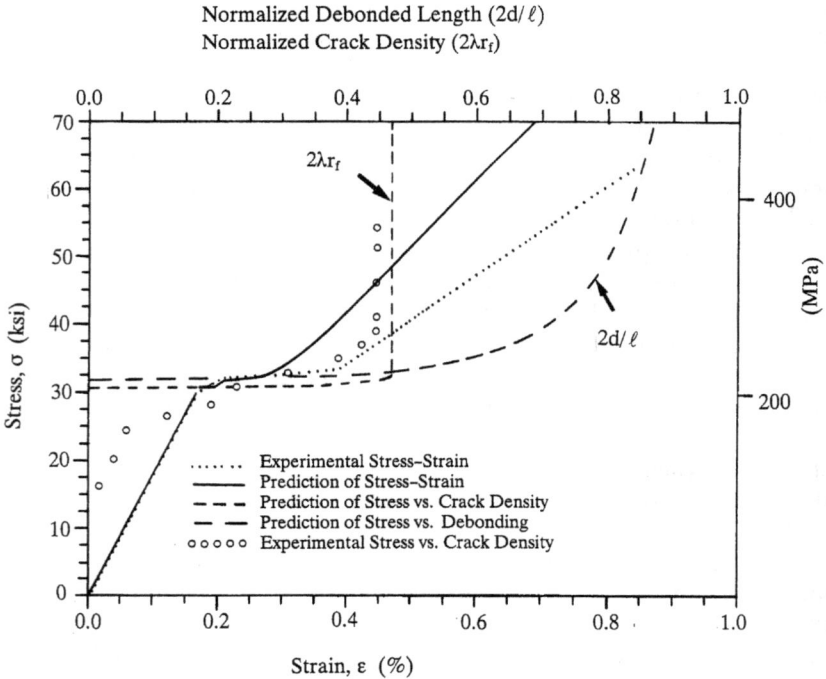

Figure 2.20 Predicted and experimental longitudinal stress–strain, stress versus matrix crack density and stress versus debonded length curves for a SiC/CAS unidirectional composite.

minimum crack spacing of $120\,\mu m$. Although some debonding starts before matrix crack saturation, most of it occurs after crack saturation. Following debonding, fiber breaks occur, leading to catastrophic failure of the composite.

2.3 Micromechanics of Failure under Longitudinal Compression

Experimental results indicate that the compressive strength of unidirectional composites is different from the tensile strength. This is explained by the different failure mechanisms in tension and compression. Thus, the failure modes discussed previously for uniaxial tension do not apply for uniaxial compression. Compression testing of composites is very difficult and test results depend on specimen geometry and/or method of testing. Different compression tests may produce different failure modes and test results are comparable when the failure modes are identical. For a review of the compression test methods, refer to Reference 17.

There are three basic failure modes in longitudinal compression: (1) microbuckling of the fibers within the matrix in extensional (out-of-phase) or shear (in-phase) mode; (2) shear failure of fibers; and (3) transverse tensile failure due to Poisson's effect. These three failure modes of the composite are shown schematically in Fig. 2.21.

Models for microbuckling of fibers in the matrix based on mechanics of materials have been developed by Rosen[3] and Schuerch.[18] In the extensional or out-of-phase mode of microbuckling, the fibers buckle in an out-of-phase pattern and the matrix is extended or compressed. This failure mode applies at low values of the fiber volume ratio. When a sinusoidal buckled shape of the fibers is assumed, the compressive strength (buckling stress of the fibers) predicted by Rosen[3] is

$$F_{1c} = 2V_f \left[\frac{E_m E_f V_f}{3(1 - V_f)} \right]^{1/2} \qquad (2.62)$$

Schuerch[18] gives

$$F_{1c} = 2V_f \left[\frac{V_f}{3(1 - V_f)} \right]^{1/2} E_f \left[1 + \frac{(1 - V_f)}{V_f} \frac{E_m}{E_f} \right] \qquad (2.63)$$

The shear or in-phase mode of microbuckling of fibers applies at higher values of fiber volume ratio. The compressive strength is given by

$$F_{1c} = \frac{G_m}{1 - V_f} \qquad (2.64)$$

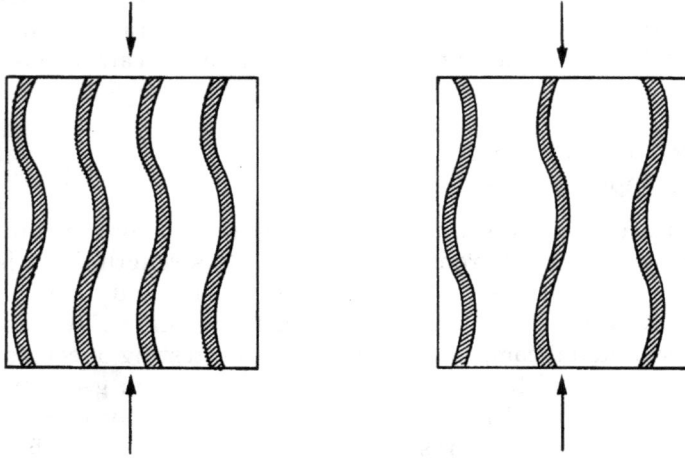

Fiber microbuckling

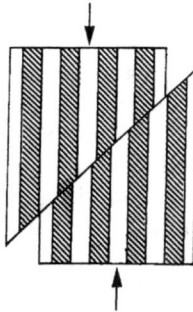

Shear failure

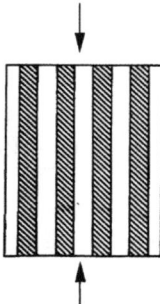

Transverse tensile failure

Figure 2.21   Failure modes of a unidirectional composite in longitudinal compression.

where G_m is the shear modulus of the fiber.

Shear fiber buckling produces tensile and compressive stresses in the fiber that create kink zones. Extensive deformations in ductile fibers or fracture planes in brittle fibers occur. Experimental results indicate that Eq. (2.64) overpredicts the compressive strength. To reduce the shear strength, a lower value of G_m due to inelastic deformation of the matrix can be introduced in Eq. (2.64). It was shown by Creszczuk[19] that for high matrix shear modulus a compressive fiber failure mode occurs. In that case the compressive strength is given by Eq. (2.4), which takes the form

$$\bar{F}_{1c} = F_{fc} V_f + \sigma'_m V_m \tag{2.65}$$

where F_{fc} is the compressive strength of the fiber.

The shear mode of failure of the fibers applies at high fiber volume ratios. The maximum shear stress $\tau_{max} = F_{1c}/2$ acts at an angle 45° with respect to the loading axis. The compressive strength for this mode of failure is

$$F_{1c} = 2F_{6f} \left[V_f + (1 - V_f) \frac{E_m}{E_f} \right] \tag{2.66}$$

where F_{6f} is the shear strength of the fiber.

The third model of transverse tensile failure due to Poisson's effect has been introduced by Agarwal and Broutman.[20] According to the model, failure occurs when the transverse tensile strain due to Poisson's effect becomes equal to its critical value. The transverse strain is

$$\varepsilon_2 = v_{12} \varepsilon_1 = -v_{12} \frac{\sigma_1}{E_1} \tag{2.67}$$

and the resulting compressive strength takes the value

$$F_{1c} = \frac{E_1 \varepsilon_T^u}{v_{12}} \tag{2.68}$$

where E_1 = longitudinal modulus of elasticity of the composite
v_{12} = Poisson's ratio equal to the ratio of strain in the transverse direction to the strain in the longitudinal direction when the applied load is in the longitudinal direction.

2.4 Micromechanics of Failure under Transverse Tension or Compression

The strength of a unidirectional composite under transverse tension is much smaller than under longitudinal tension. Transverse tensile loading is the most critical loading of a lamina. In laminates, first ply failure generally occurs under transverse tension. The low values of transverse tensile strength are due to the high stress and strain concentrations that occur in the matrix and the fiber–matrix interface. Adams and Doner[21] studied the problem of a doubly periodic rectangular array of elastic fibers in an elastic matrix subjected to transverse loading. The maximum principal stress in the matrix is the normal axial stress at the interface along the loading direction. The stress concentration factor, k_σ, is defined as the ratio of the maximum stress to the applied average stress. The variation of k_σ versus the ratio of fiber to matrix elastic modulus for a number of fiber volume ratios or relative fiber spacings δ/r (where δ is fiber spacing and r is fiber radius) is presented in Fig. 2.22. The values of Poisson's ratio for the matrix and fiber were taken $v_m = 0.35$ and $v_f = 0.20$. Note that k_σ increases with increasing fiber stiffness E_f and increasing fiber volume ratio V_f.

The strain concentration factor, k_ε, appears to be a more characteristic quantity than the stress concentration factor because in many cases the stress–strain response of the composite under transverse

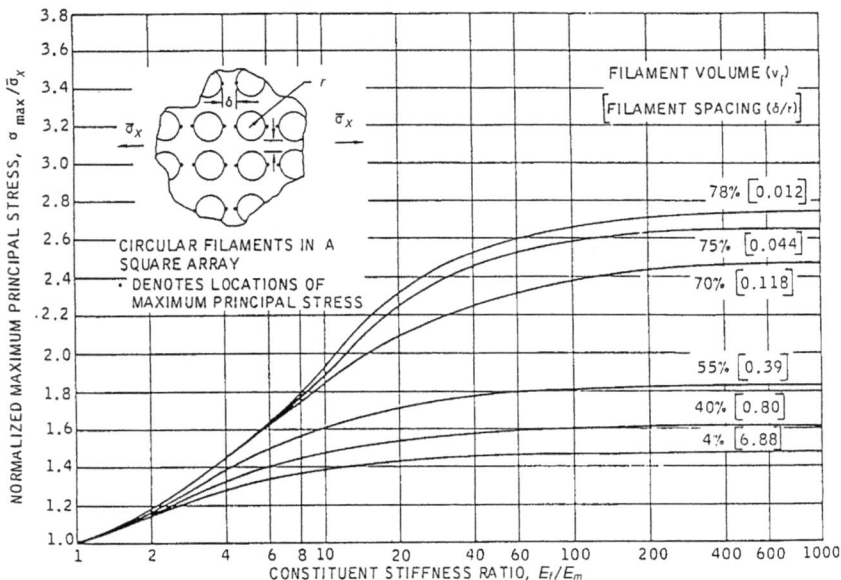

Figure 2.22 Stress concentration factor versus ratio of fiber to matrix elastic modulus for a number of fiber volume ratios in a composite with square array of fibers subjected to transverse normal tensile stress, according to Adams and Doner.[21]

loading is nonlinear due to the nonlinear behavior of the matrix. The strain concentration factor is

$$k_\varepsilon = \frac{\varepsilon_{max}}{\varepsilon_2} = k_\sigma \left(\frac{E_2}{E_m}\right) \frac{(1 + v_m)(1 - 2v_m)}{1 - v_m} \tag{2.69}$$

where ε_{max} = maximum strain in the composite
$\quad\quad \varepsilon_2$ = average applied strain
$\quad\quad v_m$ = matrix Poisson's ratio
$\quad\quad E_2$ = transverse elastic modulus of the composite
$\quad\quad E_m$ = matrix elastic modulus

Results for k_ε were obtained analytically by elasticity, finite element, finite difference, or boundary element methods and experimentally by photoelastic methods.[22]

Assuming a maximum tensile stress or maximum strain failure criterion and linear elastic behavior up to failure for the matrix, the transverse tensile strength of the composite, F_{2t}, is given by

$$F_{2t} = \frac{1}{k_\sigma}(F_{mt} - \sigma_{rm}) \tag{2.70}$$

for the maximum tensile stress criterion, and

$$F_{2t} = \frac{1 - v_m}{k_\sigma(1 + v_m)(1 - 2v_m)}(F_{mT} - \varepsilon_{rm}E_m) \tag{2.71}$$

for the maximum tensile strain criterion with very stiff fibers perfectly bonded to the matrix, where

$$F_{mT} = \text{matrix strength in tension}$$
$$\sigma_{rm} = \text{radial residual stress}$$
$$\varepsilon_{rm} = \text{radial residual strain}$$

Kies,[23] using a mechanics of materials analysis, gave for the strain concentration factor

$$k_\varepsilon = \frac{\varepsilon_{max}}{\varepsilon_2} = \frac{1}{\dfrac{2r}{s}\left[\dfrac{E_m}{E_f} - 1\right] + 1} \tag{2.72}$$

where s is the distance between the centers of the circular cross sections of adjacent fibers.

Equations (2.70) or (2.71) indicate that the transverse strength of the composite is inversely proportional to the stress concentration factor, k_σ. Based on the results of Fig. 2.22, k_σ increases with increasing fiber stiffness and increasing fiber volume ratio. Thus, while increasing fiber stiffness and increasing fiber volume ratio have a beneficial effect on the longitudinal strength of the composite, they have an adverse effect on the transverse strength.

The above prediction is based on the application of the maximum stress or strain criterion at a point in the composite. As in the case of longitudinal tensile loading, failure initiates from interfacial or matrix micro-cracks which increase with increasing loading and coalesce to form a macro-crack. For example, in the case of ceramic composites, the strength of the matrix and the interface varies with a statistical distribution in the composite. The effect of the stiffness of the interface on the failure micromechanisms of a ceramic composite was studied by Anastassopoulos and Daniel.[24] The analytical model consists of an isolated fiber in an infinite medium with a third phase (interphase) between the fiber and the matrix. An elasticity solution for stresses and displacements was obtained. Figure 2.23 presents the radial and circumferential stresses in the matrix around the fiber for various values of the interphase stiffness modulus for a silicon carbide/glass

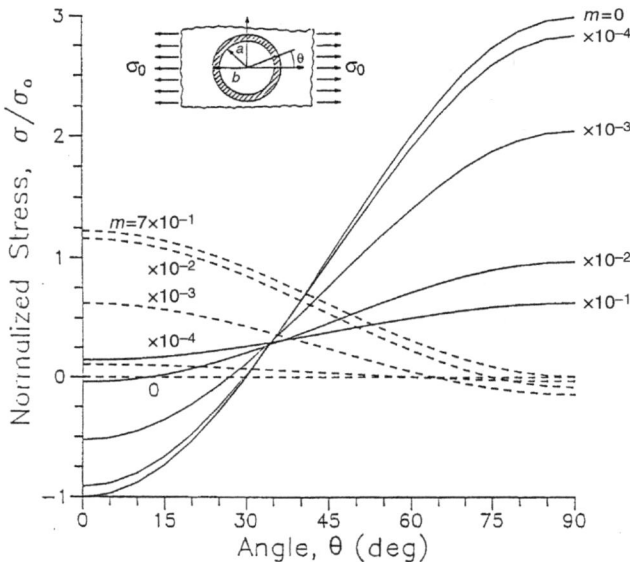

Figure 2.23 Radial and circumferential stress distributions around fiber inclusion for various values of the interphase stiffness modulus for a SiC/CAS composite, according to Anastassopoulos and Daniel.[24] —— $\sigma_{\vartheta\vartheta}$; ----- σ_{rr}; $b/a = 1.02$; $E_f = 206.8\,\text{GPa}\ (30\,\text{Msi})$; $E_m = 97.9\,\text{GPa}\ (14.2\,\text{Msi})$; $v_f = 0.2$; $v_m = 0.2$; $m = E_{int}/E_m$

ceramic composite with $E_f = 206.8\,\text{GPa}$, $E_m = 97.9\,\text{GPa}$, $v_f = v_m = 0.2$, and $b/a = 1.02$. It is shown from figure that for high interphase moduli, the radial stress at the $0°$ location is the most critical one, leading to interfacial cracking. However, for lower interphase moduli, the circumferential stress at the $90°$ location becomes critical resulting to radial cracks.

Microscopic failure mechanisms for the above silicon carbide/glass ceramic composite were studied in Reference 24 by testing the specimens under the microscope. The first micro-cracks originating at the fiber–matrix interface are nearly normal to the interface, which indicates that failure is caused by the circumferential tensile stress in the matrix. For closely packed fibers, radial cracks initiate at approximately $45°$ from the loading axis. However, when fibers are farther apart and are surrounded by a relatively large volume of matrix, radial cracks occur at approximately $90°$ from the loading axis. This agrees with the elasticity solution of the three-phase system for low interphase moduli. As the load increases, interface cracks are formed in an area along the loading axis over an arc $20°$ about the $0°$ points. These interface cracks are not immediately connected to the radial cracks developed earlier. Additional radial cracks may develop in this second stage of damage development. In the third phase of damage development radial and interfacial cracks are connected to form a long continuous crack. The various failure mechanisms and stages of damage development are illustrated schematically in Fig. 2.24.

The effect of the interfacial shear strength (ISS) on the transverse tensile strength of graphite/epoxy composites was studied by Madhukar and Drzal.[11] They found that the transverse tensile strength increases approximately in the same ratio as the increase of the ISS. The fracture surfaces showed the presence of interfacial failure for all values of ISS examined. For higher values of ISS the matrix failure becomes more dominant.

The analysis of transverse failure of unidirectional composites under longitudinal compression is analogous to that of uniaxial tension on replacement of the tensile stresses, strains, and strengths with the corresponding compressive stresses, strains, and strengths. For example, Eq. (2.70) for compression is written as

$$F_{2c} = \frac{1}{k_\sigma}(F_{mc} + \sigma_{rm}) \tag{2.73}$$

where F_{mc} is the compressive strength of the matrix and σ_{rm} is the maximum compressive residual radial stress at the interface.

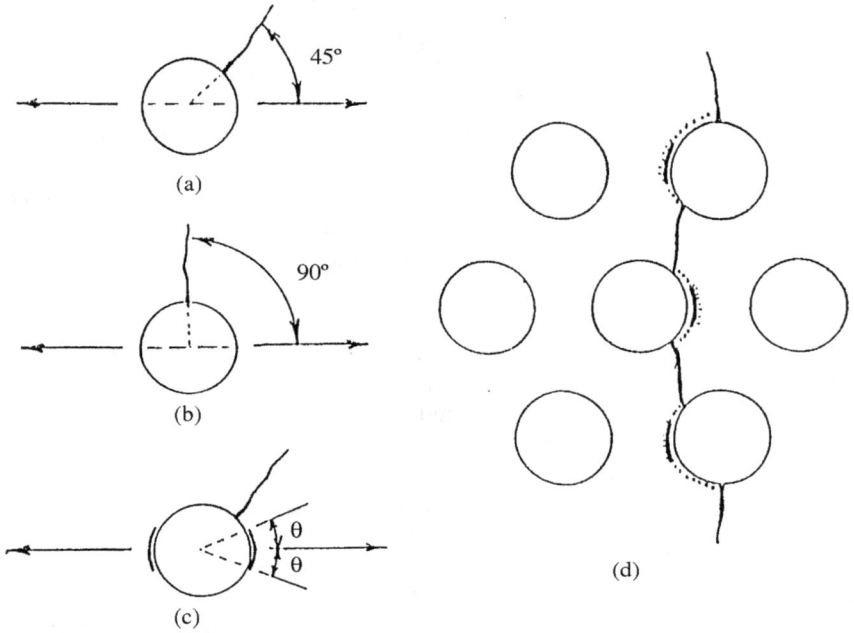

Figure 2.24 Failure mechanisms and stages of damage development in a transversely loaded ceramic composite. (a) Initial radial cracks around closely packed fibers; (b) critical radial cracks around isolated fibers; (c) interfacial cracks; and (d) interconnection of radial and interfacial cracks, according to Anastassopoulos and Daniel.[24]

2.5   Micromechanics of Failure under In-plane Shear

In-plane shear results in high stress concentrations in the composite. The in-plane shear strength can be predicted using a local deterministic failure criterion at a point, in the same way as in the case of transverse loading. Kies,[23] using a mechanics of materials analysis, gave an expression for the in-plane shear strain concentration factor analogous to Eq. (2.72):

$$k_\tau = \frac{\gamma_m}{\gamma_c} = \frac{1}{\dfrac{2r}{s}\left[\dfrac{G_m}{G_f} - 1\right] + 1} \tag{2.74}$$

where γ_m = average matrix in-plane shear strain
           γ_c = maximum composite in-plane shear strain
G_m, G_f = matrix, fiber in-plane shear modulus

From Eq. (2.74) it can be seen that k_τ increases with increasing fiber stiffness and increasing fiber volume ratio. Since the composite in-plane shear strength is inversely proportional to k_τ, it is concluded that increasing fiber stiffness and fiber volume ratio have an adverse effect on the in-plane shear strength. The composite shear strength is always lower than the matrix shear strength, which can be used as an upper limit for the composite strength. Due to the high stress concentration at the fiber–matrix interface, the basic failure mechanisms of the composite are shear failure of the matrix and fiber–matrix debonding.

A method for predicting the composite in-plane shear strength using a finite element micromechanics model was presented by King et al.[25] The results of this method can be used to predict the effect of the matrix strength, the fiber surface treatment and sizing, and fiber volume on the composite shear strength. An experimental study to determine the effect of the interfacial shear strength (ISS) on the composite in-plane shear strength was conducted by Madhukar and Drzal.[26] They found that the in-plane shear strength increases approximately in the same ratio as the interfacial shear strength. From the fracture surface analysis, the major failure modes were revealed. For low values of ISS the failure surface showed extensive fiber–matrix interfacial failure. For intermediate values of ISS, the failure was characterized by a combination of interfacial and matrix failure, whereas high ISS resulted in primarily matrix failure or brittle failure of the interphase.

2.6 Macromechanical Failure Criteria

2.6.1 Introduction

In the previous sections the micromechanical analysis of failure of a unidirectional lamina under axial or transverse tension or compression or in-plane shear loading was presented. The basic failure mechanisms such as fiber breakage, fiber buckling, matrix cracking, and fiber–matrix debonding vary greatly with material properties and type of loading. The prediction of failure initiation was based on micromechanical analysis and point failure criteria. The micromechanical analysis of the failure process is complicated by the fact that the failure mechanisms occur in various sequences and interact in many ways to form macroscopic failures leading to catastrophic fracture. Complete description of the initiation and growth of the microfailure modes under a general type of loading is a formidable task. Studies of micromechanical failure mechanisms generally require sophisticated theoretical and experimental analyses and have mostly been performed under simple types of loading.

On the other hand, for design purposes it is important to be able to reliably and quickly predict the composite strength for various combinations of normal and shear loading in terms of a number of basic strength parameters referred to the principal material directions. The basic strength parameters are determined by the experiment. For these reasons the problem of macromechanical or phenomenological approach to composite strength analysis appears to be preferable. A number of macromechanical strength theories or failure criteria have appeared in the literature. The strength theories primarily attempt to predict the onset, not the mode, of failure from a macroscopic viewpoint. Most failure criteria for the unidirectional lamina are based on generalization of isotropic yield criteria that have been developed to predict the transi-tion from elastic to plastic behavior of isotropic metals.

In this section the most representative and widely used failure criteria will briefly be discussed. For an in-depth analysis, refer to References 27 to 32.

2.6.2 Macroscopic strength parameters

As was discussed previously, the strength of a lamina is an anisotropic property, that is, it depends highly on direction. Despite the underlying microscopic failure mechanisms, the macroscopic strength of a lamina may be defined as the ultimate value of the applied stress that causes lamina failure under a simple state of stress. Figure 2.25 shows the uniaxial stress–strain response of a lamina under longitudinal tensile or compressive (a), transverse tensile or compressive (b), and in-plane shear (c) loading. The longitudinal tensile or compressive strength of a lamina, F_{1t} or F_{1c}, is much greater than the corresponding transverse strength, F_{2t} or F_{2c}. Furthermore, the longitudinal or transverse compressive strengths, F_{1c} or F_{2c}, are different from the corresponding tensile strengths, F_{1t} or F_{2t}. The in-plane shear strength, F_6, referred to the principal material axes, is an independent material property. Table 2.3 gives experimental values of the macroscopic lamina strength for some composites. Note that the transverse tensile strength is the lowest of all strengths. The longitudinal or transverse compressive strengths are not equal to the corresponding tensile strengths. The transverse tensile strengths are lower than the transverse compressive strengths, while an analogous conclusion for the longitudinal tensile or compressive strengths cannot be drawn.

For in-plane load, a lamina may be characterized by five strength parameters, F_{1t}, F_{1c}, F_{2t}, F_{2c}, F_6, as shown in Fig. 2.25. When a linear

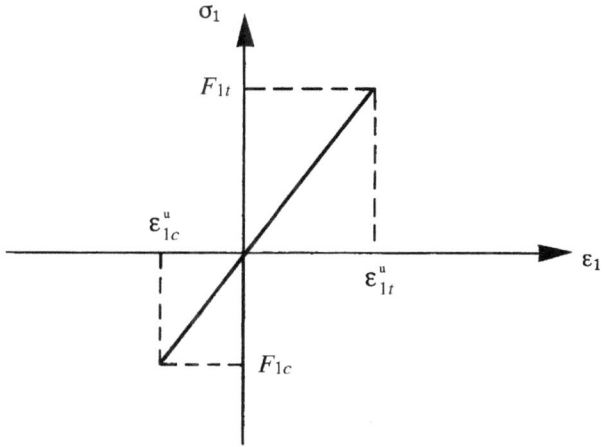

(a) Longitudinal loading

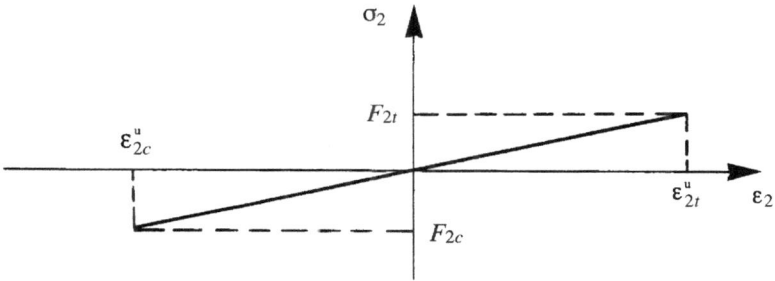

(b) Transverse loading

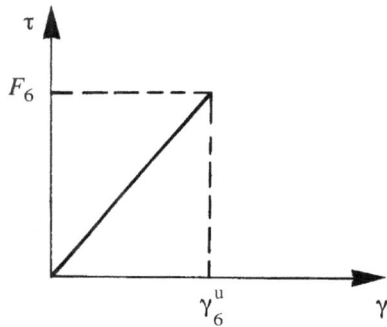

(c) Shear loading

Figure 2.25 Uniaxial stress–strain response of a unidirectional lamina showing ultimate stresses and strains.

TABLE 2.3

Property	E-glass/epoxy	S-glass/epoxy	Woven-glass/epoxy (7781/5245C)	Kevlar/epoxy (Aramid 149/epoxy)	Carbon/epoxy (AS4/3501-6)	Carbon/PEEK (AS4/APC2)
Longitudinal tensile strength (F_{1t}, MPa [ksi])	1080 [157]	1280 [185]	367 [53]	1280 [185]	2280 [330]	2060 [299]
Transverse tensile strength (F_{2t}, MPa [ksi])	39 [5.7]	49 [7.1]	367 [53]	30 [4.2]	57 [8.3]	78 [11.4]
In-plane shear strength (F_6, MPa [ksi])	89 [12.9]	69 [10.0]	97.1 [14.1]	49 [7.1]	71 [10.3]	157 [22.8]
Ultimate longitudinal tensile strain (ϵ_{1t}^u)	0.028	0.029	0.025	0.015	0.015	0.016
Ultimate transverse tensile strain (ϵ_{2t}^u)	0.005	0.006	0.025	0.005	0.006	0.009
Longitudinal compressive strength (F_{1c}, MPa [ksi])	620 [90]	690 [100]	549 [80]	335 [49]	1440 [209]	1080 [156]
Transverse compressive strength (F_{2c}, MPa [ksi])	128 [18.6]	158 [22.9]	549 [80]	158 [22.9]	228 [33]	196 [28.4]

stress–strain response is assumed, the corresponding ultimate strain parameters are obtained as

$$\varepsilon_{1t}^{u} = \frac{F_{1t}}{E_1} \qquad \varepsilon_{1c}^{u} = \frac{F_{1c}}{E_1}$$

$$\varepsilon_{2t}^{u} = \frac{F_{2t}}{E_2} \qquad \varepsilon_{2c}^{u} = \frac{F_{2c}}{E_2} \qquad (2.75)$$

$$\gamma_{6}^{u} = \frac{F_6}{G_{12}}$$

For the case of in-plane shear it is important to examine the effect of the sign of the shear stress on the lamina in-plane shear strength. For a unidirectional lamina subjected to positive and negative shear stress, as can be seen from the stress transformation equations, both cases correspond to equal tensile and compressive normal loading at 45° with the fiber direction (Fig. 2.26). Thus positive and negative shear strength referred to the principal materials directions are equivalent.

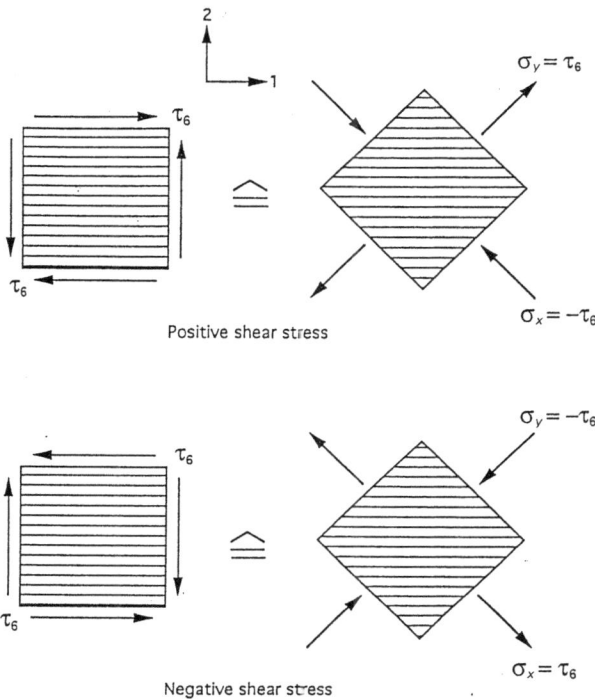

Figure 2.26 Positive and negative shear stress acting along principal material directions.

On the other hand, this is not the case when the shear stress is applied at an angle to the principal material directions. According to the stress transformation equations, it can be seen from Fig. 2.27, that a positive applied shear stress produces longitudinal tension and transverse compression along the principal material axes, whereas a negative applied shear produces longitudinal compression and transverse tension. Since the longitudinal and transverse strengths of a lamina are different in tension and compression, the in-plane shear strength is different for positive or negative in-plane shear stress. Since most composites are weaker in transverse tension than in transverse compression, the lamina will be stronger under positive shear.

Thus, the sign of the shear stress makes no difference for shear strength along the principal material axes, while it affects the shear strength for shear applied at an angle with the principal materials axes.

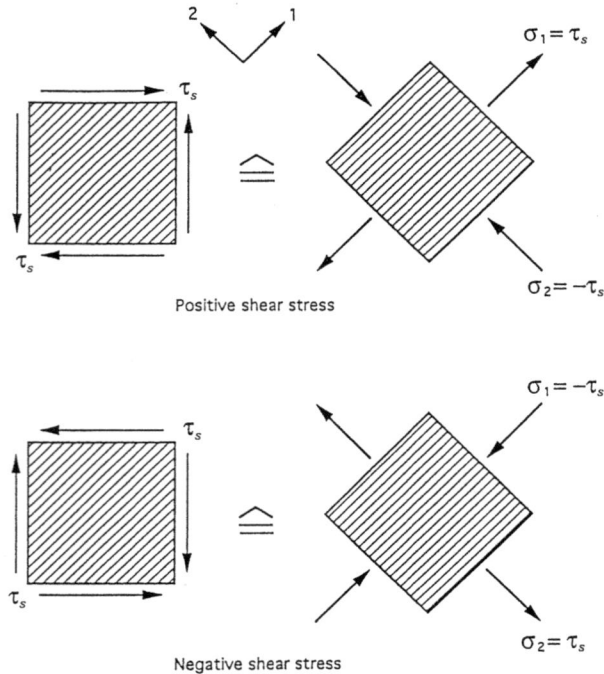

Positive shear stress

Negative shear stress

Figure 2.27 Positive and negative shear stress acting at 45° with principal material directions.

2.6.3 Isotropic failure criteria

Most failure criteria for composites are based on extensions of concepts employed with isotropic materials. The formulation of failure criteria for homogeneous isotropic materials is mostly based on limiting stress, strain, or energy and they are used to predict the transition from elastic to plastic behavior. They use the concept of the failure surface or failure envelope created by plotting the limiting states of stress in a stress space. Any stress state that falls inside the failure surface does not cause failure, whereas those stress combinations that fall on or outside the failure surface cause failure. The most widely used failure criteria for isotropic materials that are relevant to the failure of composite materials are the following.

(a) *Maximum normal stress criterion (Rankine, Lamé, Clapeyron).* According to this criterion, failure occurs when the principal stresses, σ_1, σ_2, σ_3 satisfy separately the expressions

$$\sigma_1 = \sigma_y \quad \text{or} \quad \sigma_2 = \sigma_y \quad \text{or} \quad \sigma_3 = \sigma_y \quad (2.76)$$

where σ_y is a material parameter obtained from a uniaxial tensile test. The criterion is used to predict fracture of brittle materials. It is simple but does not account for the effect of the other two principal stresses. A number of experimental results contradict Eq. (2.76).

(b) *Maximum principal strain criterion (Saint Venant).* According to this criterion, failure occurs when the maximum strain reaches some critical value for the material obtained from a uniaxial tensile test. In the application of a criterion, a linear elastic response is usually assumed.

(c) *Total strain energy criterion (Beltrami).* Failure by yielding occurs when the total strain energy under multiaxial loading reaches its critical value for uniaxial tension, i.e.,

$$(1 + v)[(\sigma_1 - \sigma_2)^2 + (\sigma_2 - \sigma_3)^2 + (\sigma_3 - \sigma_1)^2]$$
$$+ (1 - 2v)(\sigma_1 + \sigma_2 + \sigma_3)^2 = 3\sigma_y^2 \quad (2.77)$$

where v is Poisson's ratio and σ_y is the yield stress under uniaxial tension.

(d) *Distortional strain energy criterion (Von Mises).* Failure by yielding occurs when the distortional strain energy under multiaxial loading reaches its critical value for uniaxial tension, i.e.,

$$(\sigma_1 - \sigma_2)^2 + (\sigma_2 - \sigma_3)^2 + (\sigma_3 - \sigma_1)^2 = 2\sigma_y^2 \quad (2.78)$$

This criterion comes from the total strain energy criterion by dropping the contribution of the hydrostatic pressure for the case of failure by yielding.

Under conditions of plane stress ($\sigma_3 = 0$), Eq. (2.78) becomes

$$\sigma_1^2 + \sigma_2^2 - \sigma_1\sigma_2 = \sigma_y^2 \tag{2.79}$$

Since the state of stress in a lamina is plane stress, the failure surface will be two-dimensional. In the σ_1–σ_2 plane the failure surface for the maximum stress criterion is a rectangle; for the maximum strain criterion it is a skewed parallelogram; and for the distortional strain energy criterion it is an ellipse (Fig. 2.28). Because of the assumption of equal strength in tension and compression, the rectangle, the parallelogram, and the ellipse are symmetric about the origin. All three yield surfaces intersect the σ_1 and σ_2 axes at the same points ($\sigma_1, 0$) and ($0, \sigma_2$), respectively.

2.6.4   Anisotropic failure criteria

A failure criterion for a lamina expresses the lamina strength under off-axis or multiaxial loading in terms of the five lamina strength parameters: F_{1t}, F_{1c}, F_{2t}, F_{2c}, F_6. These parameters are referred to the principal material axes. For the application of a failure criterion, all applied stresses should be transformed to the principal material axes. With a homogeneous but orthotropic lamina, development of a failure criterion has involved extending an isotropic criterion to account for anisotropy. Failure criteria for isotropic materials predict initial failure by yielding, which does not necessarily cause catastrophic fracture. In fiber reinforced composites yielding does not occur in the same sense as in metals. Failure criteria for composites constitute analytical expressions to describe experimental results of failure under multiaxial loading. In this respect they are phenomenological and empiri-

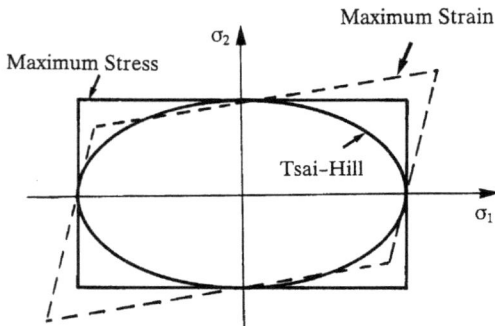

Figure 2.28   Failure surfaces in σ_1–σ_2 plane according to maximum stress, maximum strain and distortional strain energy criteria for an isotropic material.

cal, and constitute mathematical representations of strength for a given composite.

In the following the maximum stress, the maximum strain, the deviatoric strain energy and the interactive tensor polynomial criterion for orthotropic materials will be briefly presented.

2.6.5 Maximum stress criterion

This criterion predicts failure when any stress component along a principal material axis exceeds the corresponding strength in that direction. When σ_1, σ_2, τ_6 are the stresses in a lamina along the principal material axes the criterion is expressed as

$$\sigma_1 = \begin{cases} F_{1t} & \text{when } \sigma_1 > 0 \\ -F_{1c} & \text{when } \sigma_1 < 0 \end{cases} \tag{2.80a}$$

$$\sigma_2 = \begin{cases} F_{2t} & \text{when } \sigma_2 > 0 \\ -F_{2c} & \text{when } \sigma_2 < 0 \end{cases} \tag{2.80b}$$

$$|\tau_6| = F_6 \tag{2.80c}$$

The failure surface for $\tau_6 = 0$ takes the form of a rectangle as shown in Fig. 2.29. Due to the different values of longitudinal and transverse tensile and compressive strengths, the rectangle is not symmetrical with respect to both σ_1 and σ_2 axes. Note that the criterion does not account for interaction between the stress components, that is, the critical value of a stress component is independent of the values of the other stress components. Failure is predicted by the least of the three expressions of Eq. (2.80).

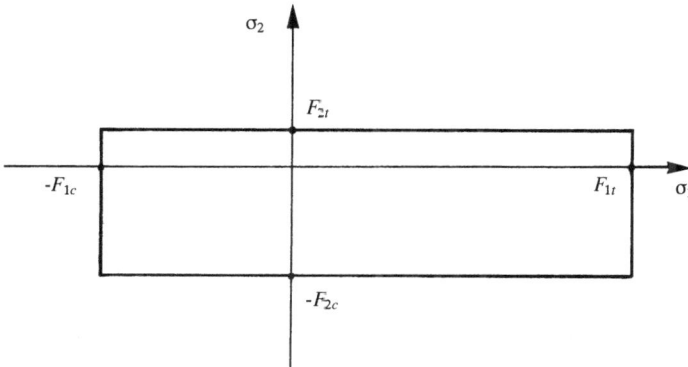

Figure 2.29   Failure surface in σ_1–σ_2 plane for an unidirectional lamina according to the maximum stress criterion.

As an example, consider the case of a unidirectional lamina subjected to off-axis tensile stress σ_x (Fig. 2.30). This stress produces the following biaxial state of stress along the principal material axes:

$$\sigma_1 = \sigma_x \cos^2 \theta, \qquad \sigma_2 = \sigma_x \sin^2 \theta, \qquad \tau_6 = -\sigma_x \sin \theta \cos \theta \qquad (2.81)$$

From Eq. (2.80) we obtain that the maximum allowable stress σ_{xt}^{cr} is the smallest of the following:

$$\sigma_{xt}^{cr} = \frac{F_{1t}}{\cos^2 \theta} \qquad \sigma_{xt}^{cr} = \frac{F_{2t}}{\sin^2 \theta} \qquad \sigma_{xt}^{cr} = \frac{F_6}{\sin \theta \cos \theta} \qquad (2.82)$$

For a compressive stress σ_x, the maximum allowable stress σ_{xc}^{cr} is the smallest of the following:

$$\sigma_{xc}^{cr} = \frac{F_{1c}}{\cos^2 \theta} \qquad \sigma_{xc}^{cr} = \frac{F_{2c}}{\sin^2 \theta} \qquad \sigma_{xc}^{cr} = \frac{F_6}{\sin \theta \cos \theta} \qquad (2.83)$$

The maximum stress criterion is expected to be in good agreement with experimental data when the applied stress is uniaxial along or close to the principal material directions. For biaxial states of stress, however, due to the fact that it does not take into account any stress interaction, no good agreement with experimental data is expected.

2.6.6 Maximum strain criterion

This criterion predicts failure when any strain component along a principal material axis exceeds the corresponding ultimate strain in that direction. It is expressed as

$$\varepsilon_1 = \begin{cases} \varepsilon_{1t}^u & \text{when } \varepsilon_1 > 0 \\ \varepsilon_{1c}^u & \text{when } \varepsilon_1 < 0 \end{cases} \qquad (2.84\text{a})$$

$$\varepsilon_2 = \begin{cases} \varepsilon_{2t}^u & \text{when } \varepsilon_2 > 0 \\ \varepsilon_{2c}^u & \text{when } \varepsilon_2 < 0 \end{cases} \qquad (2.84\text{b})$$

$$|\gamma_6| = 2|\varepsilon_{12}| = \gamma_6^u \qquad (2.84\text{c})$$

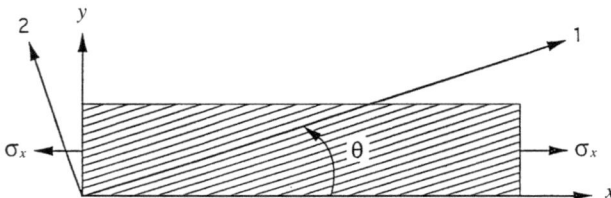

Figure 2.30 Off-axis tensile stress of a unidirectional lamina.

where ε_1, ε_2, γ_6 are the strain components referred to the principal material axes, and where

$$\varepsilon_{1t}^u = \text{ultimate longitudinal tensile strain}$$
$$\varepsilon_{1c}^u = \text{ultimate longitudinal compressive strain}$$
$$\varepsilon_{2t}^u = \text{ultimate transverse tensile strain}$$
$$\varepsilon_{2c}^u = \text{ultimate transverse compressive strain}$$
$$\gamma_6^u = \text{ultimate in-plane shear strain}$$

Let the criterion apply to a two-dimensional state of stress with stress components along the principal material axes σ_1, σ_2, and τ_6. The strain components along the principal material axes, ε_1, ε_2, γ_6, are calculated as

$$\varepsilon_1 = \frac{1}{E_1}(\sigma_1 - \nu_{12}\sigma_2) \qquad \varepsilon_2 = \frac{1}{E_2}(\sigma_2 - \nu_{21}\sigma_1) \qquad \gamma_6 = \frac{\tau_6}{G_{12}} \qquad (2.85)$$

For linear elastic response the ultimate strains of the lamina are obtained from the basic strength parameters of the lamina from Eq. (2.75).

Using Eqs. (2.75) and (2.85), the failure criterion in Eq. (2.84) is expressed as

$$\sigma_1 - \nu_{12}\sigma_2 = \begin{cases} F_{1t} & \text{when } \varepsilon_1 > 0 \\ -F_{1c} & \text{when } \varepsilon_1 < 0 \end{cases} \qquad (2.86a)$$

$$\sigma_2 - \nu_{21}\sigma_1 = \begin{cases} F_{2t} & \text{when } \varepsilon_2 > 0 \\ -F_{2c} & \text{when } \varepsilon_2 < 0 \end{cases} \qquad (2.86b)$$

$$|\tau_6| = F_6 \qquad (2.86c)$$

For a biaxial stress field with $\tau_6 = 0$, Eq. (2.86a) represents in the σ_1, σ_2 coordinate system two parallel lines that intercept the σ_1-axis at points $(F_{1t}, 0)$ and $(-F_{1c}, 0)$ and have slopes $-\nu_{12}$. Similarly, Eq. (2.86b) represents two parallel lines that intercept σ_2-axis at points $(0, F_{2t})$ and $(0, -F_{2c})$ and have slopes $-\nu_{21}$. Thus Eqs. (2.86a) and (2.86b) represent in the σ_1, σ_2 coordinate system a parallelogram that is not symmetric with respect to the origin (Fig. 2.31). The intercepts of this parallelogram for the maximum strain criterion are the same as the intercepts of the rectangle for the maximum stress criterion (Fig. 2.29).

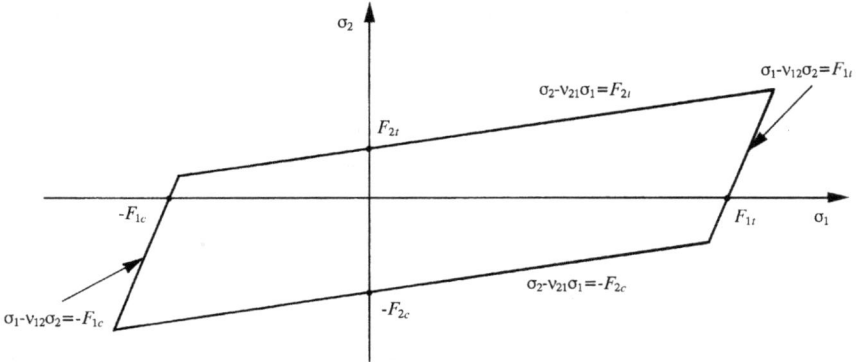

Figure 2.31 Failure surface in σ_1–σ_2 plane for a unidirectional lamina according to the maximum strain criterion.

For a unidirectional lamina subjected to an off-axis tensile stress σ_x, we obtain from Eqs. (2.81) and (2.86) that the maximum allowable stress σ_{xt}^{cr} is the smallest of the following:

$$\sigma_{xt}^{cr} = \frac{F_{1t}}{\cos^2 \theta - v_{12} \sin^2 \theta}$$

$$\sigma_{xt}^{cr} = \frac{F_{2t}}{\sin^2 \theta - v_{21} \cos^2 \theta} \qquad (2.87)$$

$$\sigma_{xt}^{cr} = \frac{F_6}{\sin \theta \cos \theta}$$

For a compressive stress σ_x, the maximum allowable stress σ_{xc}^{cr} is the smallest of the following:

$$\sigma_{xc}^{cr} = \frac{F_{1c}}{\cos^2 \theta - v_{12} \sin^2 \theta}$$

$$\sigma_{xc}^{cr} = \frac{F_{2c}}{\sin^2 \theta - v_{21} \cos^2 \theta} \qquad (2.88)$$

$$\sigma_{xc}^{cr} = \frac{F_6}{\sin \theta \cos \theta}$$

Contrary to the maximum stress criterion, the maximum strain criterion allows for some interaction of stress components due to Poisson's effect.

2.6.7 Distortional strain energy criterion (Tsai–Hill)

This criterion is an extension of the distortional strain energy or Von Mises yield criterion to account for anisotropy. It was proposed by

Hill[33] to include the effect of induced anisotropy of initially isotropic metals under large plastic deformation. For a biaxial state of stress with stresses σ_1, σ_2, τ_6, Eq. (2.79) is modified as

$$A\sigma_1^2 + B\sigma_2^2 + C\sigma_1\sigma_2 + D\tau_6^2 = 1 \qquad (2.89)$$

where A, B, C, and D are material parameters. Azzi and Tsai[34] suggested the extension of Eq. (2.89) to predict failure of a unidirectional lamina with transverse isotropy. The stresses σ_1, σ_2, τ_6 refer to the principal material axes. The parameters of Eq. (2.89) can be related to the basic strength parameters of the lamina by considering the following states of stress.

(i) *Uniaxial longitudinal stress* ($\sigma_1 \neq 0$, $\sigma_2 = \tau_6 = 0$).   At failure we have $\sigma_1 = F_1$ and Eq. (2.89) yields

$$A = \frac{1}{F_1^2} \qquad (2.90)$$

(ii) *Uniaxial transverse stress* ($\sigma_2 \neq 0$, $\sigma_1 = \tau_6 = 0$).   At failure we have $\sigma_2 = F_2$ and Eq. (2.89) yields

$$B = \frac{1}{F_2^2} \qquad (2.91)$$

(iii) *In-plane shear stress* ($\tau_6 \neq 0$, $\sigma_1 = \sigma_2 = 0$).   At failure we have $\tau_6 = F_6$ and Eq. (2.89) yields

$$D = \frac{1}{F_6^2} \qquad (2.92)$$

(iv) *Equal biaxial stress* ($\sigma_1 = \sigma_2 \neq 0$, $\tau_6 = 0$).   Failure can be assumed to occur according to the maximum stress criterion when the transverse stress σ_2 reaches the value of the transverse strength F_2, which is smaller than the value of the longitudinal strength F_1. Equation (2.89) yields

$$C = -\frac{1}{F_1^2} \qquad (2.93)$$

Introducing the values of the parameters A, B, C, and D from Eqs. (2.90)–(2.93) into Eq. (2.89), we obtain the Tsai–Hill criterion:

$$\frac{\sigma_1^2}{F_1^2} - \frac{\sigma_1\sigma_2}{F_1^2} + \frac{\sigma_2^2}{F_2^2} + \frac{\tau_6^2}{F_2^2} = 1 \qquad (2.94)$$

Equation (2.94) represents the yield surface or yield envelope of the Tsai–Hill failure criterion. Failure occurs when the left-hand side of Eq. (2.94) is ≥ 1, whereas failure is avoided when the left-hand side of Eq. (2.94) is <1. For constant values of $k = \tau_6/F_6$, Eq. (2.94) takes the form

$$\frac{\sigma_1^2}{F_1^2} - \frac{\sigma_1\sigma_2}{F_1^2} + \frac{\sigma_2^2}{F_2^2} = 1 - k^2 \tag{2.95}$$

and represents an ellipse in the coordinate system of axes σ_1 and σ_2.

Equation (2.95) does not distinguish between tensile and compressive strengths. However, it is suggested that the strengths F_1 and F_2 take the values F_{1t} or F_{1c} and F_{2t} or F_{2c} when the stresses σ_1 and σ_2 are tensile or compressive. For example, for $\sigma_1 > 0$ (tensile) and $\sigma_2 < 0$ (compressive) $F_1 = F_{1t}$ and $F_2 = F_{2c}$. Thus, Eq. (2.95) represents four different elliptical arcs in each quadrant joined at the σ_1, σ_2 axes.

The Tsai–Hill criterion is expressed by a single equation instead of three as in the case of the maximum stress or maximum strain criteria. Furthermore, this criterion accounts for some stress interaction which, however, is entered into the criterion by specifying the strength parameters according to the given state of stress.

2.6.8 Interactive tensor polynomial criterion (Tsai–Wu)

A general tensor polynomial failure criterion for anisotropic materials was proposed by Gol'denblat and Kopnov.[35] The criterion is based on the concept of strength tensors and has the form of invariants of stress and strength tensor quantities. The criterion accounts for different strengths in tension and compression. It has the form

$$(f_i\sigma_i)^\alpha + (f_{ij}\sigma_i\sigma_j)^\beta + (f_{ijk}\sigma_i\sigma_j\sigma_k)^\gamma + \cdots = 1 \tag{2.96}$$

where $f_i, f_{ij}, f_{ijk}, \ldots$ are strength tensors of second, fourth, sixth, and higher orders that can be obtained from the basic strength parameters of the material.

Tsai and Wu[36] proposed a failure criterion based on a simplified form of Eq. (2.96) as

$$f_i\sigma_i + f_{ij}\sigma_i\sigma_j = 1, \qquad i = 1, 2, \ldots, 6 \tag{2.97}$$

For a state of plane stress, Eq. (2.97) in expanded form is expressed as

$$(f_1\sigma_1 + f_2\sigma_2 + f_6\tau_6) + (f_{11}\sigma_1^2 + f_{22}\sigma_2^2 + f_{66}\tau_6^2)$$
$$+ (2f_{12}\sigma_1\sigma_2 + 2f_{16}\sigma_1\tau_6 + 2f_{26}\sigma_2\tau_6) = 1 \qquad (2.98)$$

The linear terms in Eq. (2.98) account for different strengths in tension and compression. All linear terms in the shear stress τ_6 are equal to zero since the strength of a lamina loaded in shear along the principal material directions is independent of the sign of the shear stress, i.e.,

$$f_6 = f_{16} = f_{26} = 0$$

As in the case of the distortional strain energy criterion, the parameters in Eq. (2.98) can be related to the basic strength parameters of the lamina by considering the following states of stress.

(i) *Uniaxial longitudinal tensile stress* $(\sigma_1 > 0, \; \sigma_2 = \tau_6 = 0)$. At failure $\sigma_1 = F_{1t}$ and Eq. (2.98) yields

$$f_1 F_{1t} + f_{11} F_{1t}^2 = 1 \qquad (2.99)$$

(ii) *Uniaxial longitudinal compressive stress* $(\sigma_1 < 0, \; \sigma_2 = \tau_6 = 0)$. At failure $\sigma_1 = -F_{1c}$ and Eq. (2.98) yields

$$-f_1 F_{1c} + f_{11} F_{1c}^2 = 1 \qquad (2.100)$$

From Eqs. (2.99) and (2.100) we obtain for the coefficients f_1 and f_{11}

$$f_1 = \frac{1}{F_{1t}} - \frac{1}{F_{1c}} \qquad (2.101\text{a})$$

$$f_{11} = \frac{1}{F_{1t}F_{1c}} \qquad (2.101\text{b})$$

(iii) *Uniaxial transverse tensile and compressive stress* $(\sigma_2 \neq 0, \; \sigma_1 = \tau_6 = 0)$. We obtain

$$f_2 = \frac{1}{F_{2t}} - \frac{1}{F_{2c}} \qquad (2.102\text{a})$$

$$f_{22} = \frac{1}{F_{2t}F_{2c}} \qquad (2.102\text{b})$$

(iv) *Pure shear stress* $(\tau_2 \neq 0, \sigma_1 = \sigma_2 = 0)$. We obtain

$$f_{66} = \frac{1}{F_6^2} \tag{2.103}$$

(v) *Equal biaxial normal stress* $(\sigma_1 = \sigma_2 \neq 0, \tau_6 = 0)$. We obtain

$$(f_1 + f_2)F_{(12)} + (f_{11} + f_{22} + 2f_{12})F_{(12)}^2 = 1$$

where $\sigma_1 = \sigma_2 = F_{(12)}$ is the strength under equal biaxial tensile loading. Introducing the values of f_1, f_2, f_{11}, and f_{12} from Eqs. (2.101a), (2.102a), (2.101b), and (2.102b), we obtain for f_{12} as a function of the basic strength parameters of the lamina and the biaxial strength $F_{(12)}$

$$f_{12} = \frac{1}{2F_{(12)}^2}\left[1 - F_{(12)}\left(\frac{1}{F_{1t}} - \frac{1}{F_{1c}} + \frac{1}{F_{2t}} - \frac{1}{F_{2c}}\right) - F_{(12)}^2\left(\frac{1}{F_{1t}F_{1c}} + \frac{1}{F_{2t}F_{2c}}\right)\right] \tag{2.104}$$

Tsai and Hahn[37] have proposed an approximation for f_{12} as

$$f_{12} = -\tfrac{1}{2}(f_{11}f_{22})^{1/2} \tag{2.105}$$

The Tsai–Wu criterion takes the form

$$f_1\sigma_1 + f_2\sigma_2 + f_{11}\sigma_1^2 + f_{22}\sigma_2^2 + f_{66}\tau_6^2 + 2f_{12}\sigma_1\sigma_2 = 1 \tag{2.106}$$

where the coefficients f_1, f_2, f_{11}, f_{22}, f_{66}, and f_{12} were defined above in terms of the basic strength parameters.

A failure envelope in $\sigma_1 - \sigma_2$ plane for a unidirectional lamina according to Tsai–Wu criterion is shown in Fig. 2.32.

It is interesting to note that when it is assumed that the tensile and compressive strengths are equal, Eq. (2.106) which expresses the Tsai–Wu criterion reduces to Eq. (2.94) which expresses the Tsai–Hill criterion.

The Tsai–Wu failure criterion satisfies the invariant requirements under rotation of coordinates, transforms according to the laws of tensors, provides independent interaction among stress components, and accounts for the difference in tensile and compressive strengths. The strength tensors have symmetrical properties similar to those of stiffnesses and compliances. The criterion provides a better fit to the experimental data than the other failure criteria. It is operationally simple and amenable to computational procedures. The criterion is more applicable when ductile behavior is predominant.

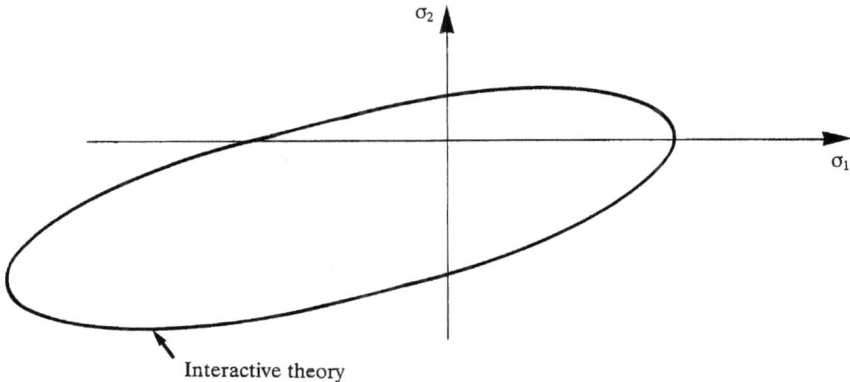

Figure 2.32   A failure envelope in $\sigma_1 - \sigma_2$ plane for a unidirectional lamina according to Tsai–Wu criterion.

References

1. H. L. Cox (1952) The elasticity and strength of paper and other fibrous materials, *Br. J. Appl. Phys.*, **3**, 72–79.
2. N. F. Dow (1961) *Study of Stresses Near a Discontinuity in a Filament-Reinforced Composite Metal*, Space Mechanics Memo 102, General Electric Space Sciences Lab, Report No. 63 SD 61.
3. B. W. Rosen (1965) Mechanics of composite strengthening, in *Fiber Composite Materials*, American Society for Metals, Metals Park, OH, 37–75.
4. A. Kelly and W. R. Tyson (1965) Tensile properties of fibre reinforced metals: Copper/tungsten and copper/molybdenum, *J. Mech. Phys. Solids*, **13**, 329–350.
5. J. M. Whitney and L. T. Drzal (1987) Axisymmetric stress distribution around an isolated fiber fragment, in *Toughened Composites* (N. J. Johnson, ed.), ASTM STP 937, American Society for Testing and Materials, Philadelphia, 179–196.
6. L. N. McCartney (1989) New theoretical model of stress transfer between fibre and matrix in a uniaxially fibre-reinforced composite, *Proc. Royal Soc. London*, **A425**, 215–244.
7. J. A. Nairn (1992) A variational mechanics analysis of the stresses around breaks in embedded fiber, *Mech. Mater.*, **13**, 131–154.
8. B. W. Rosen (1964) Tensile failure of fibrous composites, *AIAA J.*, **2**, 1985–1991.
9. W. R. Tyson and G. I. Davies (1965) A photoelastic study of the shear stresses associated with the transfer of stress during fibre reinforcement, *Br. J. Appl. Phys.*, **16**, 199–205.
10. J. Mullin, J. M. Berry, and A. Gatti (1968) Some fundamental fracture mechanisms applicable to advanced filament reinforced composites, *J. Compos. Mater.*, **2**, 82–103.
11. M. S. Madhukar and L. T. Drzal (1991) Fiber–matrix adhesion and its effect on composite mechanical properties: II. Longitudinal (0°) and transverse (90°) tensile and flexure behavior of graphite/epoxy composites, *J. Compos. Mater.*, **25**, 985–991.
12. C.-H. Hsueh (1988) Analytical evaluation of interfacial shear strength for fiber-reinforced ceramic composites, *J. Am. Ceram. Soc.*, **71**, 490–493.
13. J.-W. Lee and I. M. Daniel (1992) Deformation and failure of longitudinally loaded brittle-matrix composites, in *Composite Materials: Testing and Design*, vol. 10 (G. C. Grimes, ed.), ASTM STP 1120, American Society for Testing and Materials, Philadelphia, 204–221.
14. I. M. Daniel and G. Anastassopoulos (1993) The behavior of ceramic matrix fiber composites under longitudinal loading, *Compos. Sci. Technol.*, **46**, 105–113.
15. E. E. Gdoutos and A. Giannakopoulou (1999) Stress and failure analysis of brittle matrix composites. Part I: Stress analysis, *Int. J. Fracture*, **98**, 263–276.

16. E. E. Gdoutos, A. Giannakopoulou, and D. A. Zacharopoulos (1999) Stress and failure analysis of brittle matrix composites. Part II: Failure analysis, *Int. J. Fracture*, **98**, 277–289.
17. J. M.Whitney, I. M. Daniel, and R. B. Pipes (1985) *Experimental Mechanics of Fiber-Reinforced Composite Materials*, Society for Experimental Mechanics, Monograph No. 4, Prentice-Hall, Englewood Cliffs, N J.
18. H. Schuerch (1966) Prediction of compressive strength in uniaxial boron fiber metal matrix composites, *AIAA J.*, **4**, 102–106.
19. L. B. Creszczuk (1974) Microbuckling of lamina-reinforced composites, in *Composite Materials: Testing and Design (Third Conference)*, (C. A. Berg et al., eds.), ASTM STP 546, American Society for Testing and Materials, Philadelphia, 5–29.
20. B. D. Agarwal and L. J. Broutman (1990) *Analysis and Performance of Fiber Composites*, Wiley, New York.
21. D. F. Adams and D. R. Doner (1967) Transverse normal loading of a unidirectional composite, *J. Compos. Mater.*, **1**, 152–164.
22. I. M. Daniel (1974) Photoelastic investigation of composites, in *Composite Materials, Vol. 2, Mechanics of Composite Materials* (G. P. Sendeckyj, Vol. ed.; L. J. Broutman and R. H. Krock, Series eds.), Academic Press, New York, 433–489.
23. J. A. Kies (1962) Maximum strains in the resin of fiber glass composites, *US Naval Research Laboratory Report No. 5752*.
24. G. J. Anastassopoulos and I. M. Daniel (1991) Investigation of interphase stiffness in a ceramic matrix composite, *Proceedings of Conference on Advanced Composites in Emerging Technologies* (S. A. Paipetis and T. P. Philippidis, eds.), Amatec, Greece, 448–459.
25. T. R. King, D. M. Blackketter, D. E. Walrath, and D. F. Adams (1992) Micromechanics prediction of the shear strength of carbon fiber/epoxy matrix composites: The influence of the matrix and interfacial strengths, *J. Compos. Mater.*, **26**, 558–573.
26. M. S. Madhukar and L. T. Drzal (1991) Fiber–matrix adhesion and its effect on composite mechanical properties: I. In plane and interlaminar shear behavior of graphite/epoxy composites, *J. Compos. Mater.*, **25**, 932–957.
27. Z. Hashin (1980) Failure criteria for unidirectional fiber composites, *J. Appl. Mech.*, **47**, 329–334.
28. C. C. Chamis (1969) Failure criteria for filamentary composites, in *Composite Materials: Testing and Design*, ASTM STP 460, American Society for Testing and Materials, Philadelphia, 336–351.
29. G. P. Sendeckyj (1972) A brief survey of empirical multiaxial strength criteria for composites, in *Composite Materials: Testing and Design (Second Conference)*, ASTM STP 497, American Society for Testing and Materials, Philadelphia, 41–51.
30. R. E. Rowlands (1985) Strength (failure) theories and their experimental correlation, in *Failure Mechanics of Composites* (G. C. Sih and A. M. Skudra, eds.), North-Holland, Amsterdam, 71–125.
31. P. S. Theocaris (1994) Failure criteria for anisotropic bodies, in *Handbook of Fatigue Crack Propagation in Metallic Structures* (A. Carpinteri, ed.), Elsevier, Amsterdam, 3–45.
32. P. S. Theocaris (1989) The paraboloid failure surface for the general orthotropic material, *Acta Mech.*, **79**, 55–79.
33. R. Hill (1948) A theory of the yielding and plastic flow of anisotropic metals, *Proc. Royal Soc. London, Series A*, **193**, 281–297.
34. V. D. Azzi and S. W. Tsai (1965) Anisotropic strength of composites, *Exp. Mech.*, **5**, 283–288.
35. I. I. Gol'denblat and V. A. Kopnov (1965) Strength of glass- reinforced plastics in complex stress state, *Mekhanika Polimerov*, 1, 70–78 (in Russian) [English translation (1966) *Polymer Mechanics*, 1, Faraday Press, 54.]
36. S. W. Tsai and E. M. Wu (1971) A general theory of strength for anisotropic materials, *J. Compos. Mater.*, **5**, 58–80.
37. S. W. Tsai and H. T. Hahn (1980) *Introduction to Composite Materials*, Technomic, Lancaster PA.

3

Delamination and Debonding

G. C. Papanicolaou

Department of Mechanical and Aeronautical
Engineering, University of Patras, Patras, Greece

3.1 Delamination

As fiber reinforced polymer matrix composites (PMCs) are widely applied in aircraft structures, increased attention is being devoted to the understanding and characterizing the significant failure modes of composites. Perhaps the most commonly observed failure mode in composites is **delamination** which is *a separation of the individual plies.* Delamination failure of bonded structures and laminated composite materials has been the subject of extensive study for many years. This mode of failure may start near free edges, around holes at the end of bonded components, or due to impact that causes separation between two layers. Also, compression loading drives a mixed buckling-delamination failure.[1]

The free-edge region of a fibrous composite has been referred to as the "Achilles Heel" of fiber reinforced composite materials. This is the region where the load is transferred from the fibers to the relatively weak matrix, causing failure to occur.[2]

The interlaminar-shear deformation in a multi-ply laminate can be understood by considering the in-plane extensional behavior of a single ply. Figure 3.1 illustrates single-ply shear-coupling behavior. Note the square-shaped undeformed reference diagram and the direction of the fibers. The two deformed single-ply strips, one with fibers at $+\alpha$ and the other with fibers at $-\alpha$ demonstrate the dependence of the direction of shear deformation on fiber orientation.[2]

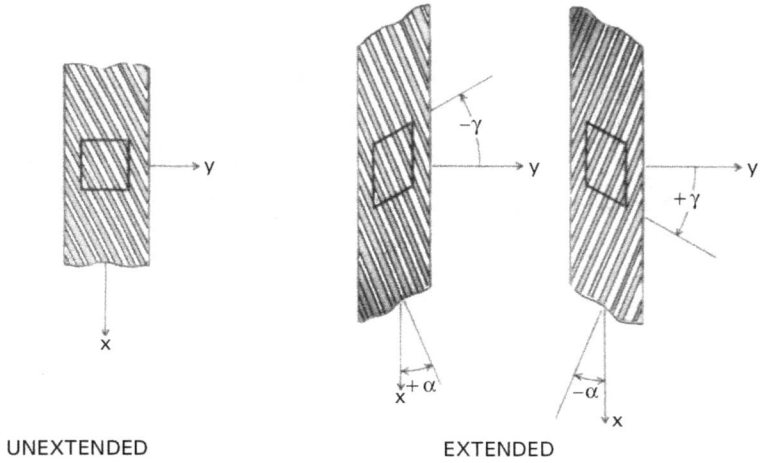

Figure 3.1 Single-ply shear-coupling behavior.[2]

Bonding these two single plies together to form a two-ply, $\pm\alpha$ laminate causes each ply to interact with the other. Extension of such a two-ply laminate produces shear deformations of opposite direction in each ply. This results in interply shear. This action is demonstrated in Fig. 3.2.

Figure 3.3 is an edge view of this two-ply, $\pm\alpha$ laminate. A series of lines that form a reference grid has been scribed on the edge of the specimen. These lines are originally perpendicular to the edges of the strip before deformation. Extension of this strip causes shear deforma-

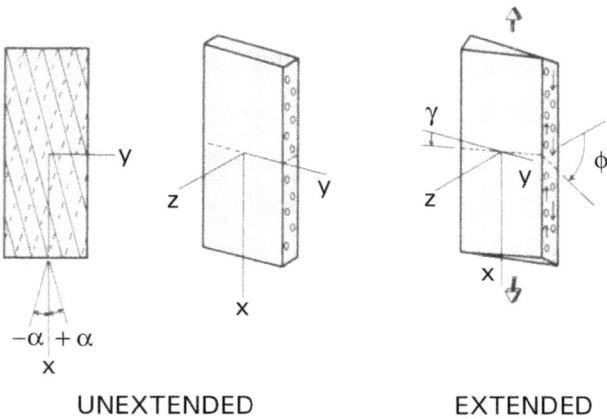

Figure 3.2 Behavior of a two-ply $\pm\alpha$ laminate under uniaxial loading.[2]

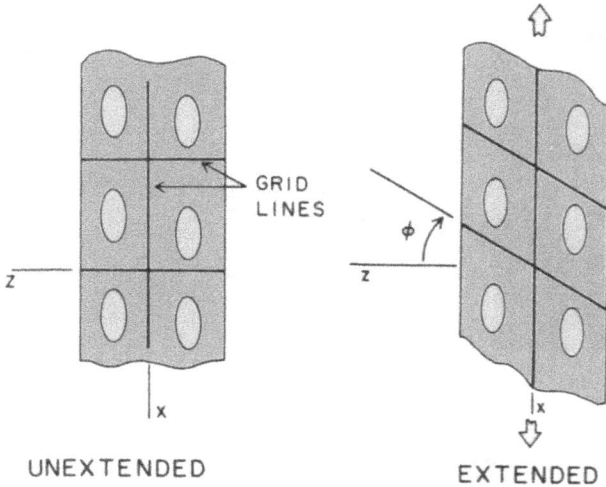

Figure 3.3   Edge view of a two-ply ±α laminate showing rotation of the grid lines due to uniaxial loading.[2]

tion of opposite direction in each ply. Such a deformation is evident from the rotation of these lines through an angle φ. This angle of rotation is a measure of interply (interlaminar) shear occurring between the plies.[2]

Since composite laminates are used as skins and in supporting structures such as beams and wing spars, they are likely to include holes and cutouts made for riveted or bolted attachements, and access holes of all kinds. At the straight edges and at the hole and cutout edges, high values of stresses are obtained in addition to the fact that these locations are natural stress raisers. That is because of the ply deformation mismatching caused by different fiber orientations and possibly different materials at the interface between adjacent plies.

Figure 3.4 shows isochromatic-fringe patterns in photoelastic coating of [0/±45/90]s graphite/epoxy specimen with 3.54 cm diameter hole under equal biaxial loading.[3] The edge effect is shown in Fig. 3.5 where dark field isochromatic fringes for an elliptical hole near the edge of a plate in uniform tension are illustrated.[4]

In many applications low velocity impacts are common, for example, stones thrown up from the runway hitting the wing of an aircraft, or damage occurring during maintenance. The effect of this damage can be greater than that of a high velocity impact that creates a neat puncture of the component, especially if the damage goes undetected and grows under subsequent loading. Impacts result in complicated damage in the form of varying amounts of delaminations between the

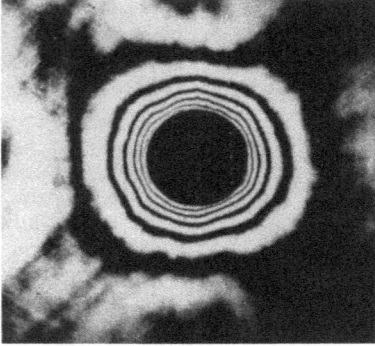

Figure 3.4 Isochromatic-fringe patterns in photoelastic coating of [0/±45/90]s graphite/epoxy specimen with 2.54 cm diameter hole under equal biaxial loading.[3]

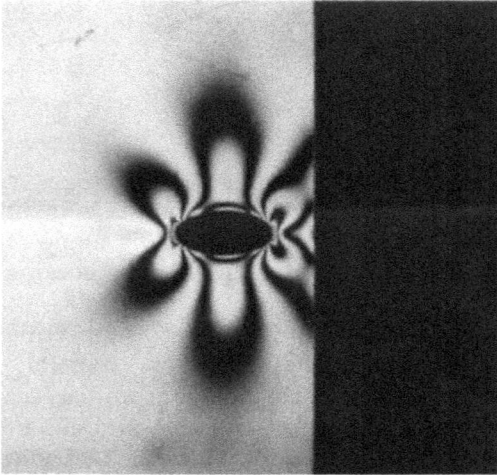

Figure 3.5 Dark field isochromatic fringes for an elliptical hole near the edge of a plate in uniform tension.[4]

plies, matrix cracks between the fibers, and fiber breaks.[5] Figure 3.6 illustrates the delamination mechanism in an impacted laminated composite. The shape of delamination is in general irregular. However, the major axis of the delamination is, in general, parallel to the fiber direction of the lower ply. Therefore, the delamination between any two plies is characterized by the length and width, which are taken to be the dimensions of the delamination in the directions parallel and perpendicular to the fiber direction in the lower ply, i.e., in the ply farther from the impactor (Fig. 3.6). The length of the delamination l_D depends on many parameters, of which the most important ones are:[5]

- the stresses at the location of the damage, σ
- the rate at which the stresses change, R

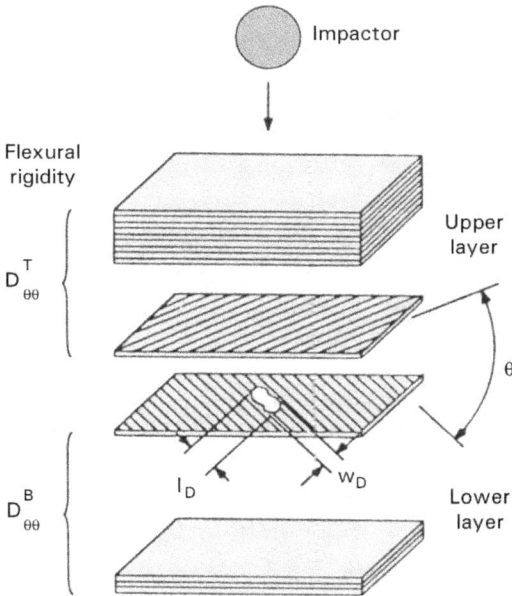

Figure 3.6 Delamination mechanism in an impacted laminated composite.[5]

- the duration of the stresses, t_f
- the difference in reduced stiffnesses of the two plies adjoining the delamination, ΔQ
- the flexural rigidities of the layers above, D^T, and below D^B, the interface where delamination occurs
- the initial size of the flaw which is present prior to the impact, l_0
- the resistance of the material to separation, K_c

Composite delamination in structures subjected to in-plane loading is a subcritical failure mode whose effect may be (1) a stiffness loss that is benign in terms of the structural failure, (2) a local tensile strain concentration in the load-bearing plies that causes tensile failure, or (3) a local instability that causes further growth that ends in compressive failure. In the last two cases the delamination leads to redistribution of structural load paths, which, in turn, precipitates structural failure.

Because delamination occurs so frequently, interest in analyzing and characterizing delamination behavior has increased. In spite of the large variety of loading, geometry, and material that has been studied experimentally and analytically, a unique criterion for the initiation and extension of the delamination failure has not yet been established. This is due to the multiparameter features of the phenomena.

3.2 Mechanics of Delamination

3.2.1 Strain energy release rate analysis

Strain energy release rate, G, is a quantitative measure of the interlaminar fracture toughness of a material. The precise evaluation of the fracture toughness in a composite system seems to be a difficult task since, apart from the complicated failure modes, a certain number of experimental parameters are seriously involved, such as the strain rate, temperature, and the shape of the specimen. These parameters have always to be taken into account and further investigation is essential on the effect of strain and/or loading rate on the G_{Ic} values derived from DCB (double cantilever beam) Mode I experiments (Fig. 3.7), as a great number of contradictory results have emerged in literature.

The strain energy release rate, G, is defined as

$$G_{Ic} = -\frac{\delta U}{B\,\delta a} = -\frac{\delta U}{\delta A} \tag{3.1}$$

where U is the total energy of the system, B is the uniform thickness of the body, δa is the small change in crack length that results in the energy change δU, and δA is the corresponding variation of the crack area for the respective infinitesimal crack propagation ($\delta A = B\,\delta a$). The negative sign is used since reduction in system energy results in an increase in energy available for external work.

The threshold condition for crack propagation is characterized by the *critical value of G_I*, termed G_{Ic}, which can be determined experimentally. More precisely: when $G_I > G_{Ic}$, crack propagation takes

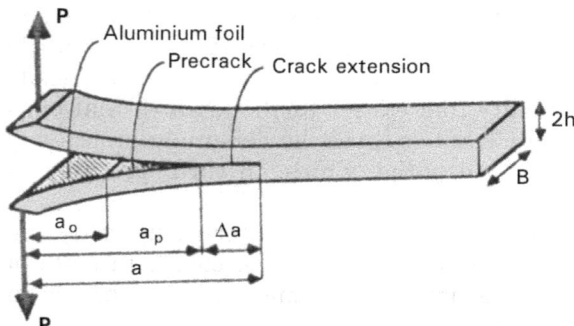

Figure 3.7 A DCB Mode I experiment.[15]

place. Then, the type of crack propagation observed depends on the following subconditions:

- If $\delta G_I/\delta A < 0$, propagation is stable.
- If $\delta G_I/\delta A > 0$, propagation is unstable.

Finally, if $G_I < G_{Ic}$, there is no propagation.[6]
Several investigators have analyzed the strain energy release rate, G, associated with edge delamination growth. From finite element analysis they have found that, once the delamination progresses beyond a distance equal to a few ply thicknesses from the edge, G reaches a constant plateau given by

$$G = \frac{\varepsilon^2 t}{2}(E_{lam} - E^*) \quad \text{or} \quad \varepsilon_c = \sqrt{\frac{2G_c}{t[E_{lam} - E^*]}} \qquad (3.2)$$

This equation, derived from a rule of mixtures and laminated plate theory, shows that G is independent of delamination size and depends only on the remote strain, ε, the laminate thickness, t, and two modulus terms, and E_{lam} and E^*, that correspond to the laminate modulus before and after delamination. It was found that the predictions agreed well with experimental data and captured the trend of decreasing delamination onset strain with increasing thickness. An analysis based on critical values of interlaminar stresses at the straight edge would not account for this thickness dependence because the interlaminar stresses are independent of the number of plies, n.[7,8]
No rigorous analytical explanation has been presented to explain the ability of the strain energy release rate (which is defined only when a crack of finite size exists) to predict delamination onset when a stress criterion cannot. Although the interlaminar stresses in the interfaces that delaminate are singular at the straight edge, these singularities are not the same as those that exist when the delamination has formed. Perhaps the most logical explanation is the presumption that inherent interlaminar flaws of microscopic dimension are initially present in the interface near the edge, but they do not form delaminations of a finite detectable size until a critical G value is reached for delamination extension.
A similar analysis for a local delamination growing from a matrix ply crack in an off-axis ply yielded the following equation

$$G = \frac{P^2}{4w^2}\left[\frac{1}{E_{LD}t_{LD}} - \frac{1}{E_{lam}t}\right] \quad \text{or} \quad \varepsilon_c = \frac{1}{E_{lam}t}\sqrt{\frac{4G_c}{\dfrac{1}{t_{LD}E_{LD}} - \dfrac{1}{tE_{lam}}}} \qquad (3.3)$$

Equation (3.3) also shows that G is independent of delamination size and depends only on the remote load, P, the laminate width, w, the laminate thickness, t, the laminate modulus, E_{lam}, and the thickness and modulus of the uncracked, locally delaminated regions, t_{LD} and E_{LD}, respectively. Again, just as the dependence of ε_c on thickness for edge delamination was predicted by the G analysis, so the trend of decreasing ε_c with increasing thickness of the cracked ply, and hence increasing n, was predicted for local delaminations.

Because delamination is constrained to grow between individual plies, both interlaminar tension and shear stresses are commonly present at the delamination front. Therefore, delamination is often a mixed-mode fracture process. A boundary value problem must be formulated and solved to determine the G_I interlaminar tension, G_{II} sliding shear, and G_{III} scissoring shear components of the total strain energy release rate. It was found that the total G is equal to the sum of G_I and G_{II} components, while G_{III} component was negligible. Furthermore, like the total G, the G_I and G_{II} components are also independent of delamination size for edge delamination.

Because of the complex mixed-mode nature of composite delamination, no closed-form solutions have been developed to lay a strong theoretical foundation for understanding the governing parameters that control delamination behavior.

Delamination in composite laminates may interact with other damage mechanisms and result in different growth behavior from that which may be anticipated from an elastic analysis. For example, under quasi-static loading, edge delamination formation is followed by matrix cracking in the 90° plies. These matrix cracks may extend throughout the laminate width, ahead of the delamination front, causing a perturbation in the three-dimensional stress field at the delamination front. This interaction between matrix crack and the delamination causes stable delamination growth.

If the composite has a tough matrix, however, the matrix cracking may be suppressed. Delaminations will form at higher G values, but they will immediately extend through the laminate width in an unstable manner. This edge delamination behavior in a toughened-matrix composite is the opposite of the stable crack growth observed in a ductile metal in which large plastic zones at the crack tip result in more stable growth.

In the case of laminated composites, G_{Ic} depends on a great number of parameters. Among these parameters one may include matrix structure, reinforcement volume fraction, existence of matrix-rich and matrix-poor regions, fiber bridging, specimen lay-up, loading rate, environmental conditions, time and temperature. The effect of each of the already mentioned parameters is as described in the following.

3.3   Effect of Matrix Material on Delamination

Matrix structure depends on, among others: (a) the degree of cross-linking, (b) the cure and post-cure conditions, and (c) the void content.

(a) Starting from the first of these parameters, it is well known that highly cross-linked epoxy resins suffer from low crack resistance. In cases where epoxy resins are used as matrices for composite materials, the enhancement of matrix toughness is one of the main goals in order to improve the performance of fiber reinforced polymers. As has been demonstrated by Murakami et al.,[9] the addition of thermoplastic resin (PEI) can significantly increase the toughness of highly cross-linked epoxy resins by a factor of 8 in terms of G_{Ic}, without a substantial deterioration in mechanical and thermal properties.

(b) Cure and post-cure conditions also influence G_{Ic} values. Cooling rate and annealing constitute parameters of cardinal importance when dealing with thermoplastic composites since they affect the degree of crystallinity.[10,11]

(c) Composite void content also has a significant effect on fracture toughness. More precisely, it has been found that Mode I interlaminar fracture toughness increased with void content because of crack tortuousity.[12] Finally, mode I and mode II fracture energies were found to increase with the extent of interlayering in BMI systems.[13]

One area of particular interest has been towards producing tougher, or more ductile, matrix phase resins for the formulation of advanced polymeric materials. While the use of tougher matrix resins has resulted in an improved fracture toughness for composite materials, it is restricted to an upper limit. Once the resin toughness has been increased beyond a certain critical value, no further increases in the toughness of the resulting composite materials are usually observed. Two factors that have been considered important causes of the unusual behaviour of these composites involve the propagation of cracks in the resin phases. These factors are the limitation of the plastic zone size adjacent to the crack tip by the fabric layers and the interpenetration of the monofilaments into the resin lamellae. Both factors are directly related to the microstructure of the manufactured laminate. It has been argued that the necessary plastic deformation zones for the generation of the maximum resin phase toughness cannot be fully developed since the fibres constrain and hence limit the volume of the resin deformation zone. The size of this deformation, or process zone, will depend on several factors such as the volume fraction of fibres in the composite and the microscopic resin distribution. It is supposed that this zone size constraint causes a decreasing matrix contribution to the composite toughness as the intrinsic matrix toughness is increased.

The enhancing of matrix strength and ductility obviously increases the critical loads for delamination onset. However, the enhancing rates for critical loads are not proportional to the matrix strength and ductility. They are closely related to the transverse tensile strength of the composite materials. It was also found that the enhancing of matrix ductility obviously increases the resistance to delamination growth in composite laminates under static loading, and significantly reduces the delamination growth rate under cyclic loading. Finally, it was shown that it is possible to choose a matrix resin that can reduce the influence of delamination on the behavior of structural composite laminates.

3.4   Effect of Fiber–Matrix Adhesion

Another commonly adopted approach in the attempt to improve the performance of composite materials is to change the adhesive strength of the fiber–matrix interface by modifying the surface of the reinforced fibers to increase their chemical activity. It has been observed that laminates that possessed the expected weakest bonding between the fiber and the matrix proved to be the toughest composites. It was concluded that enhanced fiber bridging, which led to a significant increase in fracture toughness, was related to surface treatment-induced changes in the fiber–fiber friction. These interfiber frictional effects in turn were postulated as an important factor in controlling fiber migration into the interply zones during fabrication. The migration/intermingling of individual fibers from the plies into the resin-rich interply zones has previously been postulated to be a major factor in the occurrence of fiber bridging.

3.5   Effect of Reinforcement Volume Fraction

Fiber volume fraction has a complex effect on fracture toughness and no general rule can be deduced. This is because fiber volume fraction in combination with fiber distribution is a main parameter affecting G_{Ic} values, this effect being related to fiber bridging and the existence of distinct matrix-rich interlaminar regions.[10]

Irregularities in fabrication, such as fiber misalignment are beneficial for raising the fracture toughness of composites with a minimal loss of stiffness. There exists an optimal thickness for the resin-rich zones in graphite epoxy laminates which in turn depends on the type of the matrix phase employed. The extent of fiber bridging increases as the plies are brought closer together through changes in the fibre volume for graphite fiber-epoxy laminates.

In the case of particulate composites, the fracture toughness of both rubber- modified and glass-filled epoxies reaches a maximum at a characteristic volume fraction of particles and an additional increase in volume fraction does not result in significant improvement in fracture toughness. However, hybrid epoxy composites containing hollow glass spheres exhibited synergistic toughening, i.e., the toughness of such systems can be greater than the toughness of systems modified with either the rubber particles or the hollow spheres alone.[14]

3.6   Effect of Fiber Bridging and Crack-tip Splitting

Fiber bridging and crack-tip splitting are two important mechanisms involved in Mode I interlaminar failure of fiber reinforced composite materials. Fiber bridging takes place in the area behind the crack tip where a large amount of fibers are linked across the crack from one half-beam to the other, so forming a bridge. In contrast, by the term "crack-tip splitting" is meant the formation of disconnected microscopic subcritical cracks in front of the crack tip, leading to further fiber bridging.[15]

The observed mode of variation of G_{Ic} with crack length can be explained by means of both mechanisms.[15] Also, with increasing intralaminar DCB specimen thickness a decrease in the bridging contribution has been observed.[16] A starting attempt to evaluate fiber bridging on the interlaminar fracture of composites is presented.[10]

Fiber bridging that occurs when a delamination is forced to grow between plies of similar orientation often occurs in the double cantilever beam (DCB) test used to measure Mode I interlaminar fracture toughness.

In many material systems it has been observed that the deduced fracture toughness increases with crack length. Materials that exhibit this phenomenon are classed as having a "R" or "resistance" curve behavior. One of the most common mechanisms attributed to the occurrence of this behavior is the formation of a fiber bridging zone behind the propagating crack tip. Here unbroken fibers can bridge from one side of the crack to the other, thus effectively constraining the crack growth. Enhanced levels of fiber debonding, fiber pull-out and fiber fracture will occur, during fracture, for systems in which fiber bridging is present. Enhanced fracture toughness would thus be expected, as is observed.

3.7   The Effect of Anisotropy and Fiber Orientation[17]

Since the angle-ply laminate is a simple and basic component of many composite laminates, the comprehensive understanding of its failure behavior is of fundamental importance to the structural design and integrity prediction of advanced fibrous composites. The failure theory for the angle-ply laminate predicts that the stress at the first ply failure occurs within the laminate. When the angle-ply laminate is loaded in the direction of bisectors of fiber angle, all laminae have the same direct stress but shear stress of opposite sign. Therefore, failure of the entire laminate is assumed to occur immediately after the initial failure of either ply. This fact may lead to the belief that strength of angle-ply laminate can be predicted from the first ply failure criterion.

Experimental data on the strength agree reasonably well with the failure theory except for the small fiber orientations under tension. The early failure may be due to the interlaminar stresses occurring in the edge region, because considerable interlaminar separation and cracking were present in failed specimens.

Analytical studies have shown that the state of stress in the vicinity of the free edge of such a laminate is fully three-dimensional and may not be predicted by laminated plate theory. These results show that interlaminar shear stress, τ_{xz}, may dominate free edge delamination in $[\pm\theta]$ laminate. The free edge delamination is a unique failure mode in composite laminate, and it has been observed as a matrix-dominated mechanism occurring in the resin-rich interlaminar region. The presence of delamination crack is known to cause a progressive stiffness reduction, material degradation, and final fracture of the composites. Also, the tensile strength of quasi-isotropic laminates was predicted by using the rule of mixtures analysis for stiffness. More recent results indicated the fundamental nature of free edge delamination in angle-ply laminate subjected to uniform axial extension. Fracture mechanics parameters such as stress intensity factors and associated strain energy release rate were defined, and the effect of delamination length, fiber orientation, lamination, and geometric variables were considered. However, a systematic discussion of the relationship between the tensile strength of angle-ply laminate for small fiber angle and free edge delamination has never been presented.

Since composite laminates undergo uniform thermal loading associated with cooling the laminate from processing temperature, significant residual thermal interlaminar stresses sufficient to cause failure of layers within laminate can be developed.

The effect of delamination on the tensile strength for small fiber angle was also evaluated using the rule of mixtures and energy release rate concept in classical fracture mechanics. Due to the complexities of the problem, a quasi-three-dimensional finite element method was used to obtain energy release rate associated with delamination growth. Finally, predicted failure modes and tensile strengths were compared with known experimental data to asses the prediction accuracy of the delamination growth model. It was found that Mode III is dominant fracture mode of free edge delamination and the stable growth region occurs after unstable growth. The behavior of delamination is influenced by the fiber orientation, ply thickness, thermal residual stress, and loading conditions (tension/compression). The presence of thermal residual stress plays a significant role in the initiation and propagation of delamination according to the loading conditions. It was also found that thicker laminates will delaminate at a lower strain level than thinner laminates.

Most of the existing experimental results refer to unidirectional laminated composites. However, in general, G_{Ic} values obtained for lay-ups other than UD are considerably higher than respective values obtained on UD materials.[9,10] It was also found that for laminates with a given lay-up but different stacking sequences, the total G was nearly identical for all stacking sequences but the G_I component varied significantly.

3.8 Effect of Loading Rate

A large number of contradictory results concerning loading rate effect on G_{Ic} values can be found in literature. Thus, this effect is quite obscure and needs further study.[10]

3.9 Effect of Environmental Conditions[18]

It has been recognized for quite some time that the static mechanical properties of Polymeric Matrix Composite (PMC) are affected by various environments. Of particular interest is the influence of humid environment (moisture) on mechanical properties, particularly transverse static properties, and fracture behavior that is dominated by the matrix and interfacial properties. Depending on matrix phase chemistry, other fluids that are absorbed by the bulk matrix and by the interface phase can induce degradation. For most thermoset (epoxy) composites, the absorption of moisture modifies the strength, stiffness, and delamination fracture behavior. In contrast, for thermoplastic composites, experimental results show essentially no effect of moisture on longitudinal or transverse mechanical properties. Moisture-induced property degradation manifests a strong correlation with the degree of sorbed moisture in the matrix. The rate of moisture sorption and solubility limits are governed by the matrix phase chemistry and by the fiber/matrix interface phase composition. Moisture sorption can be up to 10 times greater in thermoset composites than in thermoplastic composites. Moisture-related degradation is associated with chemisorption and reactivity characteristics of the permeating environment and matrix materials. Moreover, defects such as void, cracks, and other microstructural inhomogeneities associated with the fiber/matrix interface contribute to severity of moisture effects in resin matrix composites. The deleterious effects of environments, such as moisture, on static (strength and stiffness) properties of PMCs have been well documented. However, research emphasizing the effects of moisture on fracture mechanisms and crack propagation behavior is inadequate. However, persistent demands for use of PMCs in critical applications dictate that extensive characterization should be performed regarding external environmental effects on fracture aspects.

Plasticization of the bulk matrix and the interface phase by sorbed moisture influences the reduction in fracture toughness. Moisture-induced plasticization causes softening of the resin and loss of strength. Moreover, moisture tends to weaken resin in vicinity of the fiber/matrix interface. The degrading effect of moisture in epoxy composites is manifested by preferential cracking along the weakened fiber/matrix interface. Persistent fracture mechanisms appear to be interfacial cracking and debonding and microcrack growth and coalescence in the plasticized matrix material.

There has been a growing interest in using fiber reinforced polymeric composites in offshore applications.[19] The long-term durability of composites, however, must be characterized before their full potential can be realized in the aggressive environment of seawater, in which a composite structure is subjected to fatigue wave loading and moisture absorption. Fatigue and damage development in composites have been the subjects of many investigations in recent years. Of the several forms of damage, delamination has been identified as the most dominant in the fatigue life of fiber reinforced composites.

As mentioned above, the effect of moisture on the delamination is quite complex. It can be beneficial due to the relief of residual thermal curing stresses and matrix plasticization. It can also be detrimental because of the induced chemical and/or physical degradation. Several investigations have examined the fatigue behavior of composites in seawater environment, but few have investigated the combined effect of fatigue and seawater exposure on the damage development and rate of damage growth.

As the amount of seawater in the sample increases, the dominant crack path changes from interply to intraply cracking, but the overall growth of the cracking remains approximately the same. It has been observed that moisture tends to decrease the available strain energy release rate. Exposure to seawater does cause degradation of the interfacial strength, but this degradation leads to delamination mode switch. The overall damage growth resistance of preconditioned samples is enhanced by the seawater exposure due to a combination of moisture-induced stress relief and damage mode switch.

3.10   Effect of Time and Temperature[20]

The time-dependent process in which the polymer in the glassy state undergoes a structural recovery with volume and enthalpy relaxation below its glass transition temperature is called *physical aging*.[20] Upon aging, the polymer becomes rigid and brittle and shows a lower damping and a slower stress-relaxation rate. Energy absorbed during impact decreases with increase of aging temperature and period.

To characterize the intralaminar fracture behavior of a unidirectional carbon fiber/epoxy resin composite material, two distinct sets of experiments were performed.[21] First, the tensile mechanical properties of the laminae were determined by testing specimens with different fiber orientation at varying rates of deformation. Then, the double torsion (DT) technique was applied to characterize the fracture resistance of the material. The strain energy release rate vs. crack speed curve so derived is in agreement with measurements conducted using the double cantilever beam and compact tension techniques. The results of the research indicate that double torsion is an attractive technique for studying the time-dependent fracture of composite materials.

Double torsion fracture testing was also applied[22] to investigate the rate and temperature dependence of intralaminar fracture toughness in both unidirectional and woven carbon fiber/epoxy resin laminates. A noteworthy variation was found in the fracture toughness and surface morphology of unidirectional laminates when the testing temperature was varied from below that of the glass transition of the epoxy matrix. At the higher temperature, the slope of the fracture toughness vs. crack speed curve is slightly positive, indicating that fracture is dominated by the far-field viscoelastic behavior of the matrix. At room temperature, however, the slope becomes negative, which can be attributed to a time-dependent fracture mechanism taking place inside the process zone. At this temperature, the variation in fracture toughness observed by varying crack speed is comparable to that observed by changing the testing temperature. Finally, at the lower temperature, fracture toughness is lower and remains constant with varying crack speed, the material behaving as if it were effectively elastic.

The rate dependence of the yield stress of the laminate in plain strain compression is consistent with the explanation given here for the fracture behavior observed.

A general increase in G_{Ic} values for both thermosetting and thermoplastic composites with increasing test temperature is usually observed.[10]

Unlike the stable delamination growth that occurs under quasi-static loading at G levels above G_c, cyclic delamination growth may occur at levels well below the static interlaminar fracture toughness.

From the above presentation of the parameters affecting interlaminar fracture resistance of laminated composites, it becomes clear that prediction of G_{Ic} values is a difficult task. Thus, a number of data-reduction methods have been developed for the evaluation of the interlaminar fracture toughness.

3.11   Data-reduction Methods

The experimental method usually applied for the determination of the critical strain energy release rate G_{Ic} is the double cantilever beam (DCB) method. The following data reduction methods are used to calculate G_{Ic} values.

(i) *The simple beam theory.*   According to this method, the expression for G_{Ic} of a perfectly built-in DCB specimen is

$$G_{Ic} = \frac{3P\,\delta}{2Ba} \tag{3.4}$$

where P = load, δ = displacement, B = specimen width, a = crack length.

(ii) *The corrected beam theory.*   As the beam is not perfectly built-in, the above expression underestimate the compliance. Thus, a correcting factor Δ is introduced and the expression for G_{Ic} becomes

$$G_{Ic} = \frac{3P\,\delta}{2\varepsilon(a+\Delta)} \tag{3.5}$$

According to this method, the beam is treated as containing a slightly longer crack $a + \Delta$, and Δ may be found experimentally by plotting the cube root of compliance, $C^{1/3}$, as a function of crack length.

(iii) *The compliance method.*   This method is based on the basic concept according which the value of the fracture energy from an

Linear Elastic Fracture Mechanics (LEFM) test is given by

$$G_{Ic} = \frac{P^2}{2B} \frac{dC}{da} \tag{3.6}$$

where P is the load and C is the compliance given by

$$C = \frac{\delta}{P} \tag{3.7}$$

and where δ is the displacement corresponding to a load P.

To evaluate G_{Ic} via this method, we first plot C against crack length, a, and the curve is fitted using an appropriate polynomial function. Next, knowing the values of the load, P, and the differential, dC/da, at a given crack length, the G_{Ic} value at any crack length is then evaluated using Eq. (3.6)

(iv) *The load method.* This method is based on simple beam method considerations according which the compliance, C, is given by

$$C = \frac{\delta}{P} = \frac{2a^3}{3EI} \tag{3.8}$$

and, since $I = Bh^3/12$,

$$C = \frac{8a^3}{BEh^3} \tag{3.9}$$

where E is the flexural modulus, I is the second moment of area, and $2h$ is the total thickness of the DCB specimen. Thus, from Eqs. (3.6) and (3.8),

$$G_{Ic} = \frac{P^2 a^2}{BEI} \tag{3.10}$$

(v) *The experimental compliance method.* This method involves plotting compliance against crack length on a log-log plot, and then to using the slope of this plot, n, to give G_{Ic} as follows

$$C = ka^n \quad \text{and} \quad G_{Ic} = \frac{nP\delta}{2Ba} \tag{3.11}$$

(vi) *The area method.* The crack extension is related directly to the area enclosed between loading and unloading curves. Thus, G_{Ic} is defined as

$$G_{Ic} = \frac{\Delta a}{B \, \delta a} \tag{3.12}$$

where Δa is the area enclosed by the loading–unloading path.

All the theories described above refer to specimens with one initial crack, and they may be modified in order to apply to specimens having two parallel initial cracks. Also, they are based on certain assumptions, such as: (1) homogeneity and isotropy; (2) certain geometric conditions are respected; (3) St'Vainant's principle applies; and (4) plane sections remain plane after loading.

However, the above assumptions are hardly fulfilled in practice. Moreover, the correction factor, Δ, introduced in the corrected beam theory has no physical meaning since practically Δ cannot remain constant during the whole crack propagation procedure and a variable correcting factor would be a better approach. On the other hand, the compliance method is a two-parameter method, where these two parameters are linked and a change in one results in the other also changing. Finally, although the area method is still valid if a nonlinear elastic load–displacement response is observed, in contrast to the compliance method, for which only linear response can be treated, this method is only applicable to stable propagation.

In addition to the methods described, some other interesting methods have also been developed. Among them one may include the method developed by Devitt et al.,[23] the method developed by Williams,[24] and finally the method developed by Hashemi et al.[10] A general characteristic of all these methods is that large displacements as well as correction factors for nonlinear behavior are considered. A more detailed description of these methods is given.[25]

However, differences of the order of 40% can be observed between the predictions derived from the above theories when applied to the same type of materials. Thus, there is a need for a new data-reduction method that can be applied for a more accurate calculation of G_{Ic}.

3.12   Delamination in Sandwich Structures

As the demand for higher performance aircraft continues to increase, so does the need for developing advanced materials and processing methods to meet their requirements. The most important requirements are both reduced structural weight fractions and lower costs. Such reductions are attainable through improved materials, low-cost fabrication methods, innovative design concepts, and improved analytical methods.[26]

One area of advancement is that of sandwich beams. A typical sandwich beam consists of two skins made of metal or laminated composite and a core (Fig. 3.8). The core material can be honeycomb, metallic or non-metallic, plastic foam,[27] or laminated composite. Sandwich structures combine high strength, low density, and high

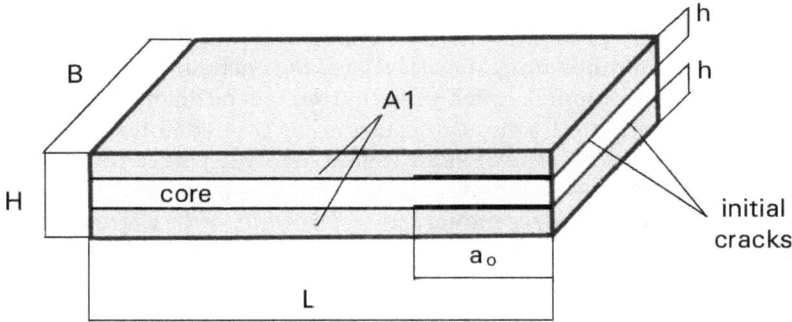

Figure 3.8 A typical sandwich beam.[33]

modulus of elasticity. They are also characterized by improved toughness and corrosion resistance, and good fatigue properties.

The function of the core material is to transfer shear and normal stresses to the skin through the skin–core interface. Thus, one of the parameters of cardinal importance for the final overall performance for good performance of sandwich structures is the degree of bonding between core and skin.[28] The most sensitive site in a composite sandwich structure is the skin–core interface, where delamination may occur.

As in the case of composite laminates, fracture mechanics tests can be used to characterize delamination behavior of composite sandwich structures. Critical strain energy release rate is a measure of the interlaminar fracture toughness of these materials, while for the development of interfacial delamination, both interlaminar tensile stresses and interlaminar shear stresses are responsible. Thus, three different modes of fracture are often present in most delamination phenomena, namely, opening Mode I, a sliding shear Mode II, and a scissoring Mode III.

The most common test used for the evaluation of the Mode I interlaminar fracture toughness (G_{Ic}), is the double cantilever beam (DCB) test.[29] Phenomena such as fiber nesting, fiber bridging behind the crack tip, and crack-tip splitting in the area ahead of the crack[29,30] are included in the complex mechanism of crack propagation in a composite DCB specimen. Under the condition that the insert is sufficiently small, one may assume that the onset value of G_{Ic} measured from the insert may be representative of a naturally occurring delamination.[31]

From the definition of G_{Ic} as the lowest fracture energy it becomes clear that for the Mode I fracture and in the case of isotropic materials, the crack will always propagate along a path normal to the direction of maximum local principal stress regardless of the orientation of the initial crack with respect to the applied load.[15] The failure of isotropic materials is usually precipitated by a dominant crack that propagates in a self-similar manner until a critical size is reached.[32]

However, a more complex crack propagation mechanism is observed in the case of laminated composite materials, this complexity being more pronounced in the case of sandwich structures where laminated composites are used as core material. In this case there will be three dominant modes of fracture, namely, *interlaminar matrix cracking or delamination, intralaminar matrix cracking*, and *fiber fracture*. Both fiber orientation with respect to the loading direction and stacking sequence affect the non-self-similar propagation of matrix cracks. Moreover, fibers act as crack stoppers, resulting in a crack-arresting procedure, while interaction between interlaminar and intralaminar cracks leads to a structural and mechanical degradation of the composite material. Stiffness reduction and local stress concentrations and redistributions are the main effects observed due to crack interaction.

Based on different assumptions, several data-reduction methods for G_{Ic} evaluation have been developed. However, due to their different assumptions, considerable discrepancy between G_{Ic} values determined for the same material by means of these methods is observed in the literature. As an example of this kind of difficulty, we may report the precracking procedure followed by many researchers. More precisely, since the initiation fracture energy, $G_{Ic(init)}$, is always a difficult parameter to define precisely, as reported,[15] it has been defined as the value of the fracture energy after 2 mm increment of crack growth ahead of the starter foil. A precracking procedure was also followed.[32] However, in this case 1 mm precracking was used.

3.13 The Variable Correction Factor Model VCFM [33]

One of the most important factors affecting the calculated G_{Ic} values, irrespective of the specific method used, is the error made when measuring crack length. Also, this length is almost always measured from the front edge of the specimen, while a different crack length may exist in the rear edge of the specimen.

The VCFM-model refers to the specimen geometry shown in Figs. 3.9 and 3.10. According to this geometry, two parallel interfacial cracks have been considered. Thus, taking into account the simple beam considerations

$$f = \frac{\delta}{2} = \frac{4Pa_{th}^3}{EBh^3} \tag{3.13}$$

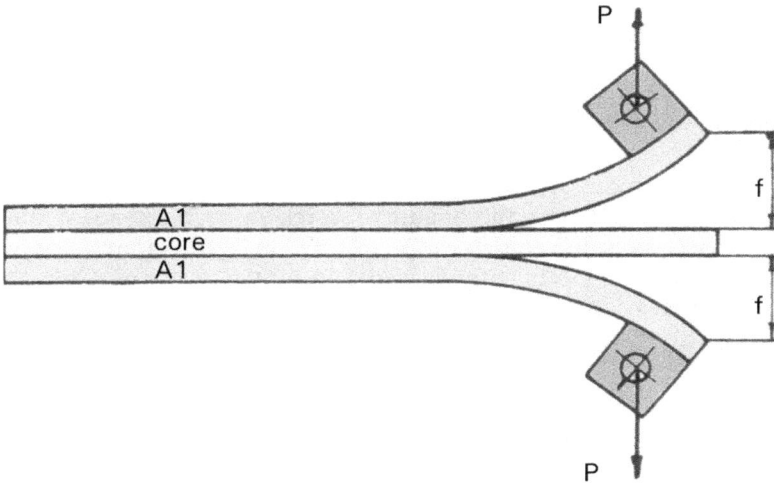

Figure 3.9   Basic concept for computation of G_1.[33]

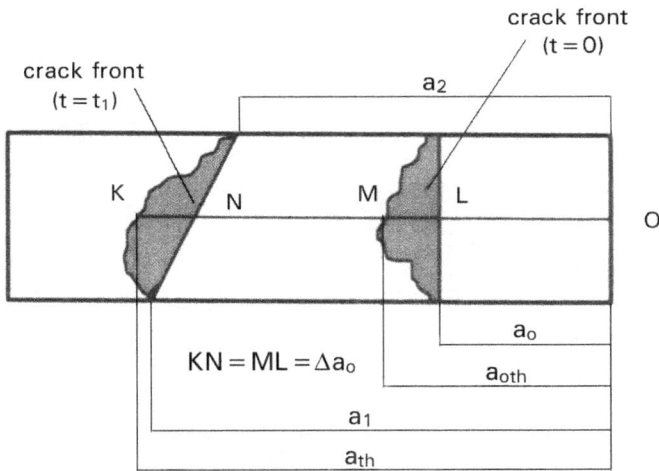

Figure 3.10   Top view of a specimen showing a general mode of crack-front propagation.[33]

and

$$C = \frac{\delta}{P} = \frac{8a_{th}^3}{EBh^3} \qquad (3.14)$$

where δ = displacement, P = load, E = bending modulus, B = specimen width, h = thickness, C = compliance and a_{th} is the theoretical crack length value as derived from Eq. (3.15)

$$a_{th} = \frac{h}{2}\sqrt[3]{EBC} \qquad (3.15)$$

This value is not always equal to the experimentally measured one. Next, from the definition of G_{Ic}, it follows that

$$G_{Ic} = \frac{P^2}{2B} \frac{dC}{d(2a_{th})} = \frac{6P^2 a_{th}^2}{EB^2 h^3} = \frac{3P \delta}{4Ba_{th}} \tag{3.16}$$

Since one of the main problems in defining the value of the initiation fracture energy, G_{Ic}, is the increment of crack growth ahead of the starter foil corresponding to this value, the following assumption has been made. Instead of having a precracked specimen, as many authors do, an initial crack length correcting factor Δa_0 is defined as follows

$$\Delta a_0 = a_{0th} - a_0 \tag{3.17}$$

where a_{0th} is the initial nominal value of a, as evaluated from Eq. (3.15) with P and δ values corresponding to the final point of the linear part of the P–δ curve and a_0 is equal to the length of the starter foil used reduced by the half-length of the end block. This arises from the fact that with introduction of the starter foil an abrupt initial crack-front propagation is observed (Fig. 3.10).

The only basic assumption made in the present model is that the Δa_0 value remains constant during the crack propagation, when measured in the direction of the central axis of the specimen (Fig. 3.10). From the specimen geometry (Fig. 3.10) it can easily be derived that

$$a_{th} = \frac{a_1 + a_2}{2} + \Delta a_0 \tag{3.18}$$

where a_2 is the length of the crack propagating in the rear edge of the specimen.

A better estimation of the equivalent crack length (i.e., the equivalent length of the bending beam) will be achieved if a mean crack length, a'_{th} is considered instead of a_{th}

$$a'_{th} = \frac{a_1 + a_2 + a_{th}}{3} = a_{th} - \frac{2 \Delta a_0}{3} \tag{3.19}$$

By introducing this value into Eq. (3.16), the following simple expression for G_{Ic} is derived

$$G_{Ic} = \frac{3P \delta}{4B'_{th}} \tag{3.20}$$

It is clear that by means of the above procedure, $G_{I(prop)}$ values may also be determined. Therefore, the proposed model offers a physical meaning for all the parameters introduced, and gives a continuous description of the crack front propagation history. Moreover, as shown,[25] the proposed model may be applied to all types of mechanical behavior, i.e., linear elastic, nonlinear elastic, and inelastic.

The model has been applied to the DCB experimental results described.[33] According to this work, specimens had the geometry shown in Fig. 3.8. According to this figure, the dimensions are $L = 125$ mm, $B = 20$ mm, $H = 2.4$ mm, $h = 0.8$ mm, $a_0 = 50$ mm. The starter film was made of PTFE having a thickness of 20 µm. The specimens had two initial cracks as shown in Fig. 3.8. The load was introduced via machined loading blocks allowing free rotation and minimal stiffening of the specimen. Sandwich beams had a skin made of an aluminum alloy 2024-T6 and a core of epoxy Ciba Geigy resin type XB 3052 A, along with a hardener of the type XB 3052 B, reinforced with unidirectional E-glass fibers placed parallel to the specimen's axis. The measured weight fraction of the fibers in the core was approximately 50%.

Depending on the surface treatment applied to the aluminum skins, three different types of specimens were prepared: (a) without any treatment, (b) roughened with emery paper (No. 80), and (c) roughened by means of sandblasting.

The DCB specimens were loaded in an Instron machine operated in a displacement control mode with a constant rate of 2 mm/min. The tensile load was applied perpendicular to the plane of delamination growth. Five specimens from each type of skin surface treatment were tested.

Figures 3.11–3.13 show characteristic load–displacement traces for the three cases studied. From these figures it becomes clear that a linear elastic behavior in all cases was observed initially. Thus, there is no need to apply the theories modified for nonlinear behavior.

G_{Ic} values as predicted by the present theoretical model along with respective values obtained by means of the simple beam theory, (SBT), the corrected beam theory, (CBT), the compliance method (CM) and Berry's method are tabulated in Table 3.1 for comparison. Figure 3.14 shows G_{Ic} values for different aluminum surface treatments as derived by the model. As expected, G_{Ic} values derived from the different data-reduction methods depend on the specific method considered. A comparison between different predictions is shown in Fig. 3.15.

It must be noted that the elastic modulus, E, used in the proposed model was kept constant and equal to that determined from the three-point flexure test. This is in accordance with the observations made by Hashemi et al.,[15] where correction factors for the crack-tip rotation, the stiffening effect, and large displacements were been introduced in their model so that more complex expressions in form for G_{Ic} and C were derived, and where finally it was shown that the calculated modulus values (termed $E_{11(corr)}$) exhibited no significant variation

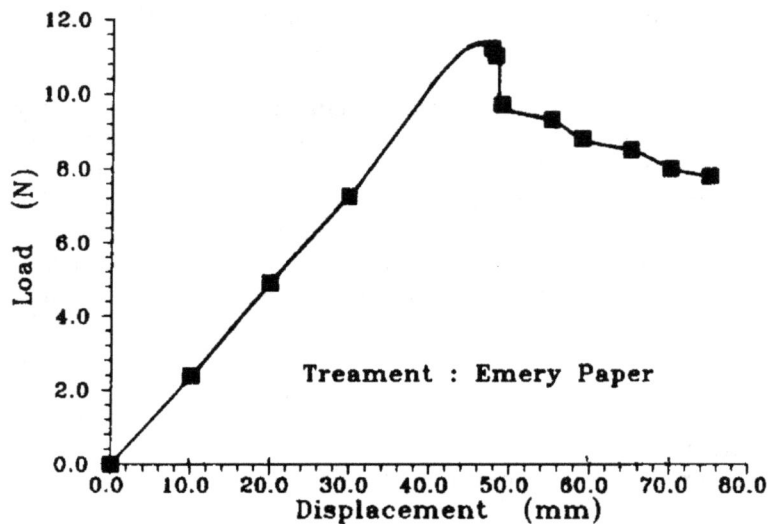

Figure 3.11 Characteristic load–displacement curve for a DCB-sandwich specimen treated with emergy paper.[33]

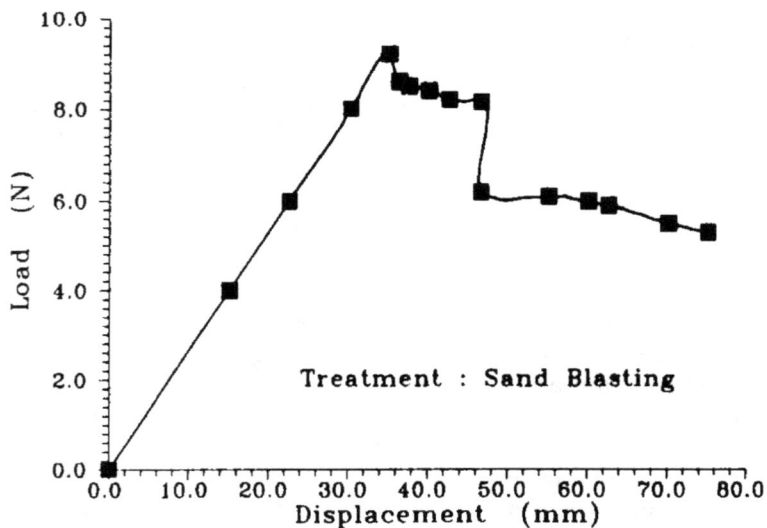

Figure 3.12 Characteristic load–displacement curve for a DCB-sandwich specimen treated by sandblasting.[33]

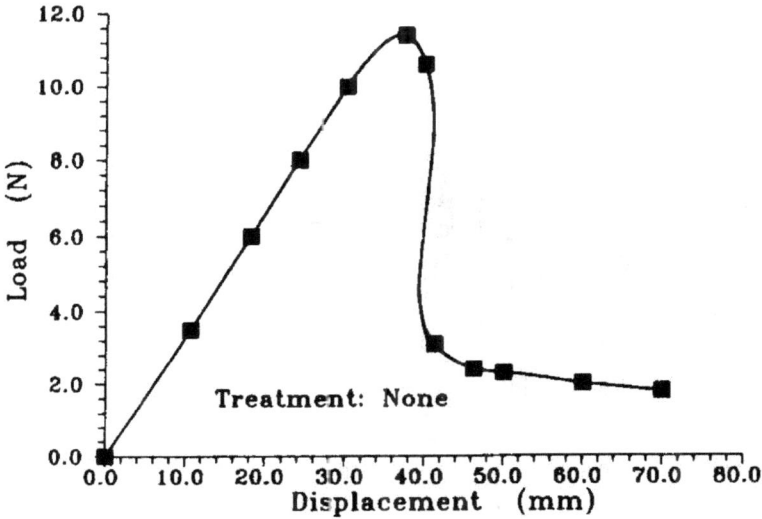

Figure 3.13 Characteristic load–displacement curve for a DCB-sandwich specimen without any treatment [33]

with crack length (i.e., E_{11} = const.) and were in good agreement with the value determined from the three-point flexure test.

From the same diagram it can be seen that maximum values were obtained for the case of the emery paper-treated skins while, in most cases, minimum values were obtained in the case of the sandblasting

TABLE 3.1 G_{Ic} **values for the DCB sandwich beams as derived by the proposed method.**[33]

Treatment	Spec. No.	Δa_0 (mm)	a'_{th} (mm)	G_{Ic} (KJ/m^2)	Mean Value	Value Deviation
None	1	24.57	50.34	0.319		
	2	23.62	50.87	0.326		
	3	25.19	51.40	0.316	0.318	0.0074
	4	24.84	51.27	0.308		
	5	23.38	50.79	0.318		
Sandblasting	1	28.89	52.62	0.233		
	2	29.42	52.81	0.245		
	3	27.31	52.10	0.240	0.244	0.0292
	4	29.13	52.70	0.213		
	5	27.69	52.22	0.298		
Emery paper	1	13.62	47.53	0.355		
	2	29.58	52.86	0.385		
	3	25.98	51.66	0.408	0.384	0.0181
	4	25.90	51.63	0.375		
	5	26.04	51.68	0.396		

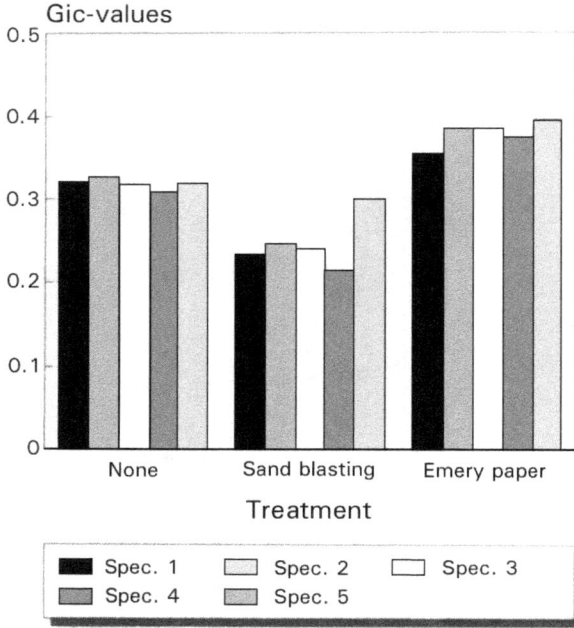

Figure 3.14 G_{Ic} values for different aluminum surface treatments as derived by the proposed method.[33]

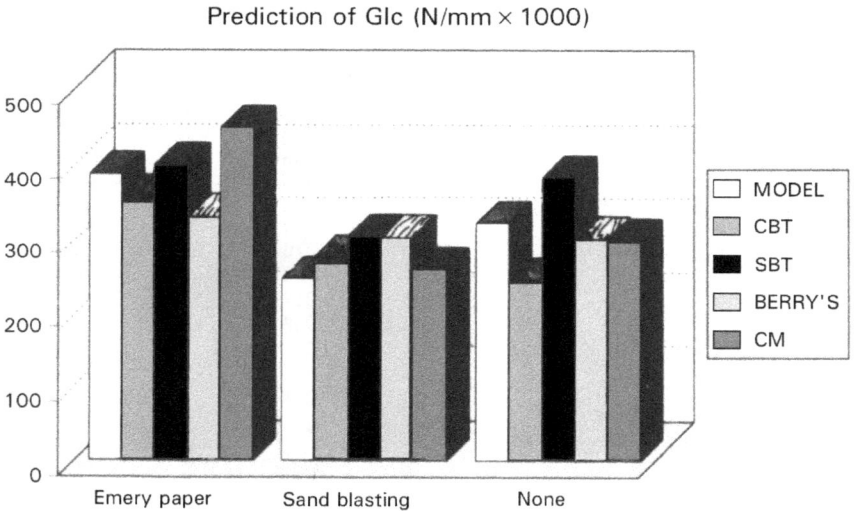

Figure 3.15 Comparison between G_{Ic} mean values as they are predicted by the different methods.[33]

treatment. This kind of behavior was attributed to the fact that, in the case of sandblasting, sand is trapped into the skin, which in turn facilitates the skin–core debonding. Moreover, sandblasting could transfer grease or humidity to the aluminum surface during the whole procedure. In contrast, with the emery paper-treated specimens a better surface roughness was achieved that resulted in better mechanical bonding between skin and core. These results are in good agreement with those derived by the authors in a previous publication[32] where the effect of skin treatment on the bending behavior of sandwich structures was investigated. In that case it was observed that specimens treated with emery paper showed the highest bending moduli.

Another important parameter describing interlaminar crack propagation behavior in sandwich structures is the crack propagation speed. Figures 3.16–3.18 show the variation of crack lengths a_1 and a_2 with time for the three types of skin treatment, while Figs. 3.19–3.21 show the respective variation of crack propagation speed, da_1/dt and da_2/dt with time. From these figures it becomes clear that in all cases, an increase of the crack speed is observed, followed by a subsequent decrease. In the case of emery paper-treated specimens, quite low values of maximum crack speed are observed, showing the good bonding conditions achieved by this treatment (Figs. 3.19 and 3.22). Moreover, due to the effects of sand, an initial increase of the speed was observed, while a subsequent decrease was not very obvious, even though crack speed values were kept very low (Figs. 3.20 and 3.23). Finally, specimens without any skin treatment, although they are

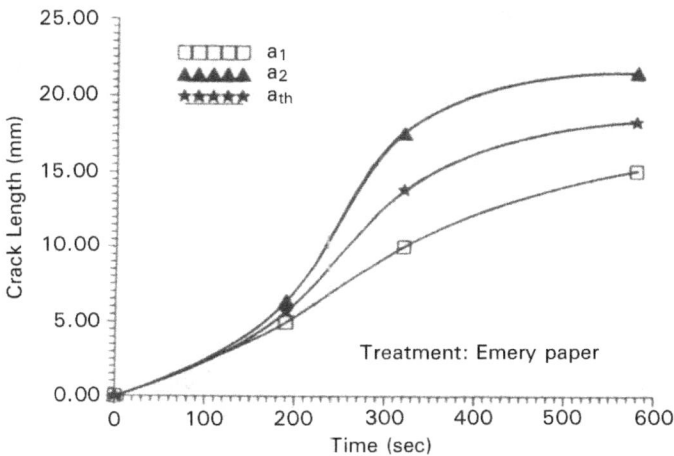

Figure 3.16 Variation of crack lengths a_1, a_2, and a_{th} with time in DCB-sandwich specimens treated with emery paper.[33]

characterized by a relatively high initial G_{Ic} value, in the sequence showed an abrupt crack propagation in the rear edge of the specimen, where very high values of crack speeds were apparent (Figs. 3.21 and 3.24).

It should be noted that although the final equation of the proposed model is of the same form as that given by the corrected beam method, a completely different physical meaning is introduced. The correction factor included in the expression of a'_{th} has a variable value as the crack propagates. Thus, the proposed model determines at any instant the location of the crack front, taking into account the initial effects produced by the introduction of the starter foil.

3.14 Debonding

It is well known that the fiber/matrix interface is of cardinal importance when dealing with the mechanical properties of fiber reinforced composites. Hence appropriate measurement techniques are needed for quantifying the degree of interfacial adhesion between the fibers and the matrix. Among the different techniques developed in this area, three micromechanical test methods are the most commonly applied. Unfortunately, they often give different results even with the same fiber/matrix systems prepared under identical conditions. The reasons for these discrepancies are based on the different experimental errors appearing in each of them as well as on the different theoretical assumptions made for the evaluation of stresses.

Different analytical models have been developed for the prediction and evaluation of stress transfer from a matrix to a fiber.[35,36] Figure 3.25a shows schematically a fiber of length l embedded in a continuous matrix subjected to stress. It is found that following the shear-lag analysis of Cox[36] the variation of tensile stress, σ, in the fiber with distance along the fiber, x, is given by

$$\sigma = E_f\left[1 - \frac{\cosh\left[\beta\left(\frac{l}{2} - x\right)\right]}{\cosh(\beta l/2)}\right] \qquad \text{where} \qquad \beta = \sqrt{\left(\frac{G_m}{E_f}\right)\left(\frac{2\pi}{A_f \ln(R/r_0)}\right)}$$

$$(3.21)$$

where E_f is the Young's modulus of the fiber, e is the matrix strain, G_m is the shear modulus of the matrix, A_f the cross-sectional area of the fiber, r_0 is the fiber radius, and R is the radius of the cylinder of resin around the fiber. The variation of σ with x is shown schematically in Fig. 3.25b and it can be seen that s rises from zero at the fiber ends to a constant plateau value along the center of the fiber and then falls

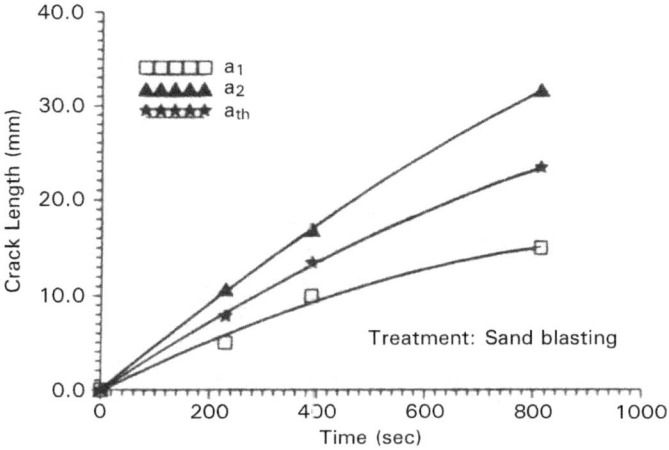

Figure 3.17 Variation of crack lengths a_1, a_2, and a_{th} with time in DCB-sandwich specimens treated by sandblasting.[33]

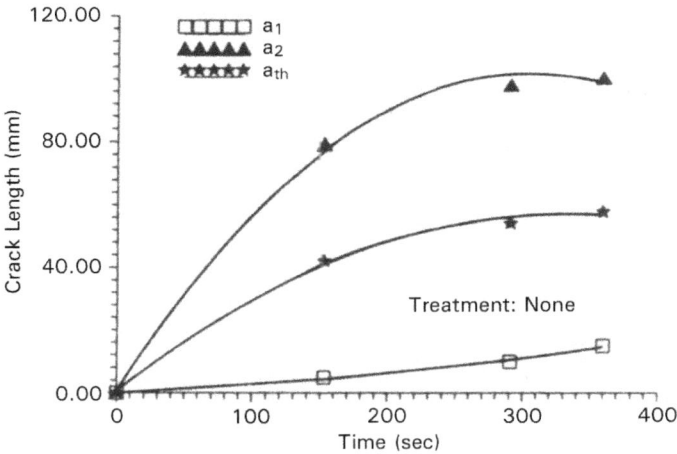

Figure 3.18 Variation of crack lengths a_1, a_2, and a_{th} with time in DCB-sandwich specimens without any treatment.[33]

to zero at the other end. The shear stress in the matrix at the fiber/matrix interface, τ, can also be determined and is given by

$$\tau = E_f \sqrt{\left(\frac{G_m}{E_f 2 \ln(R/r_0)}\right)} \frac{\sinh[\beta(l/2 - x)]}{\cosh(\beta l/2)} \qquad (3.22)$$

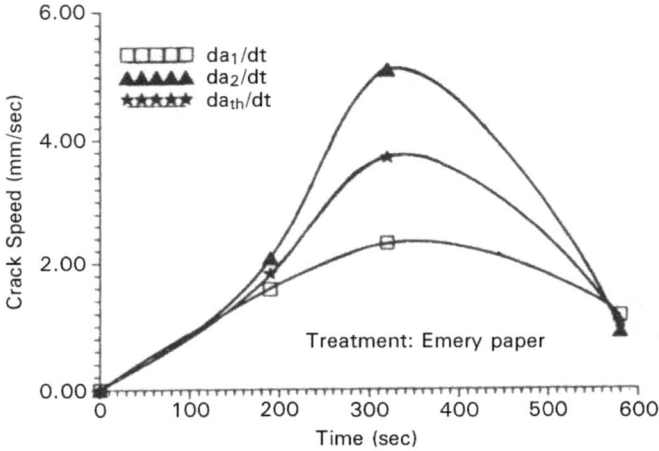

Figure 3.19 Variation of crack tip speeds da_1/dt, da_2/dt, and da_{th}/dt with time in DCB-sandwich specimens treated with emery paper.[33]

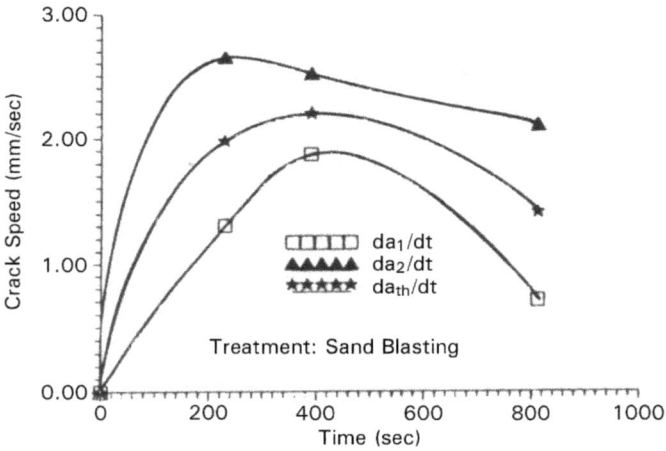

Figure 3.20 Variation of crack tip speeds da_1/dt, da_2/dt, and da_{th}/dt with time in DCB-sandwich specimens treated by sandblasting.[33]

This function is also plotted schematically in Fig. 3.25b, where it can be seen that it is a maximum at the fiber ends and falls to zero along the middle of the fiber.

The main assumptions made in the shear lag analysis are as follows:

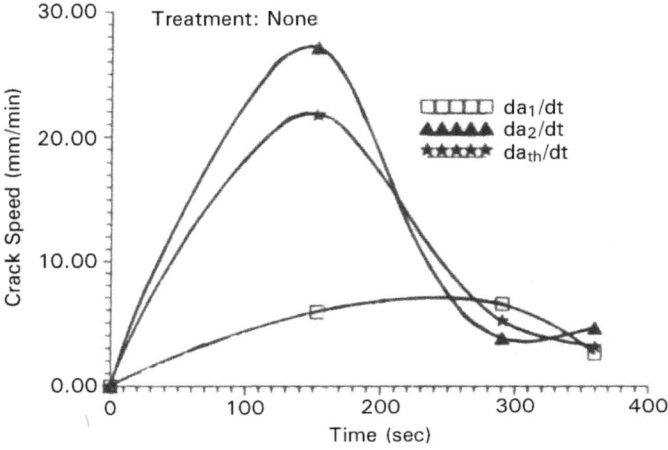

Figure 3.21 Variation of crack tip speeds da_1/dt, da_2/dt, and da_{th}/dt with time in DCB-sandwich specimens without any treatment.[33]

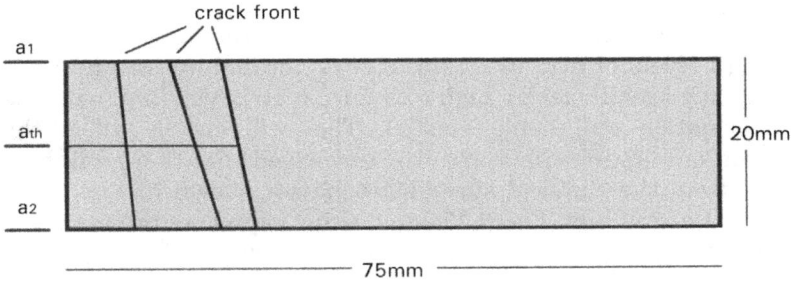

Figure 3.22 Top view of a specimen showing the crack-front propagation history in an emery paper-treated DCB-sandwich beam.[33]

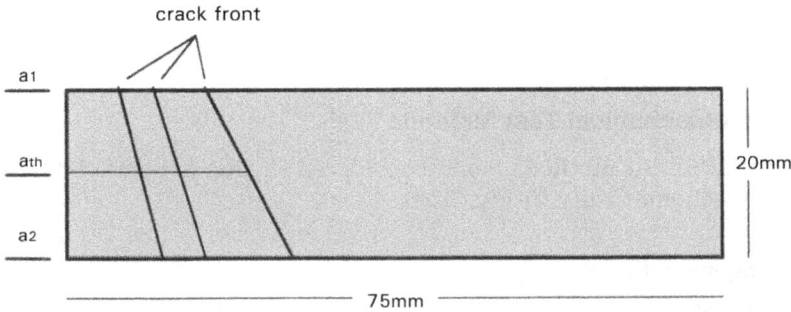

Figure 3.23 Top view of a specimen showing the crack-front propagation history in a sandblasting-treated DCB-sandwich beam.[33]

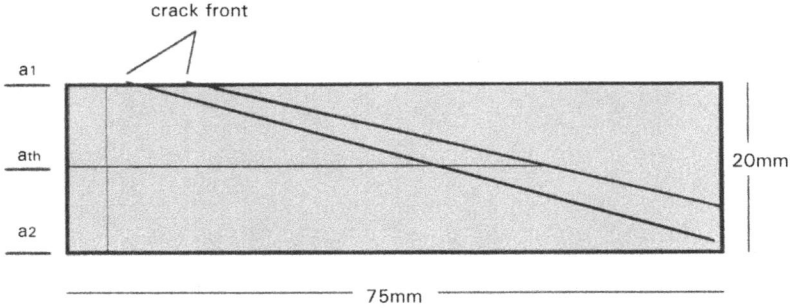

Figure 3.24 Top view of a specimen showing the crack-front propagation history in a DCB-sandwich beam without any treatment.[33]

1. It is assumed that there is no adhesion across the ends of the fibers.

2. It is assumed that all of the deformation is linear elastic. Hence, at best, it will only be applicable at low strains.

3. It is assumed there is no debonding.

There will be a concentration of shear stress at the ends of the fiber, and at high levels of overall strain, e ($<e_f$, the failure strain of the fiber) this may be sufficiently high to induce matrix yielding (with the fiber deformation still being elastic). This will clearly affect the distribution of σ and τ with x and it is envisaged that there will be a transition from the classical shear-lag behavior shown in Fig. 3.25b through partial yielding (Fig. 3.25c) to a situation where the matrix is yielded at the fiber ends but the fiber deformation is still elastic (Fig. 3.25d).[34] The tensile stress in the fiber can be determined from a simple balance of forces argument whereby this stress is balanced by the shear stress at the fiber/matrix interface. This leads to

$$\frac{d\sigma}{dx} = \frac{2\tau}{r_0} \quad \text{so for} \quad 0 \le x < \alpha \quad \sigma = \frac{2\tau_y x}{r_0} \quad (3.23)$$

3.15 Micromechanical Test Methods

The three different methods being considered in this present chapter are shown schematically in Fig. 3.26.

3.15.1 Fragmentation test

The fragmentation method proposed by Fraser et al.[37] is widely used for measuring the interfacial shear strength in polymeric matrix composites. In the fragmentation test, a tensile dumbbell specimen is fabricated in which a single fiber is embedded in the resin and

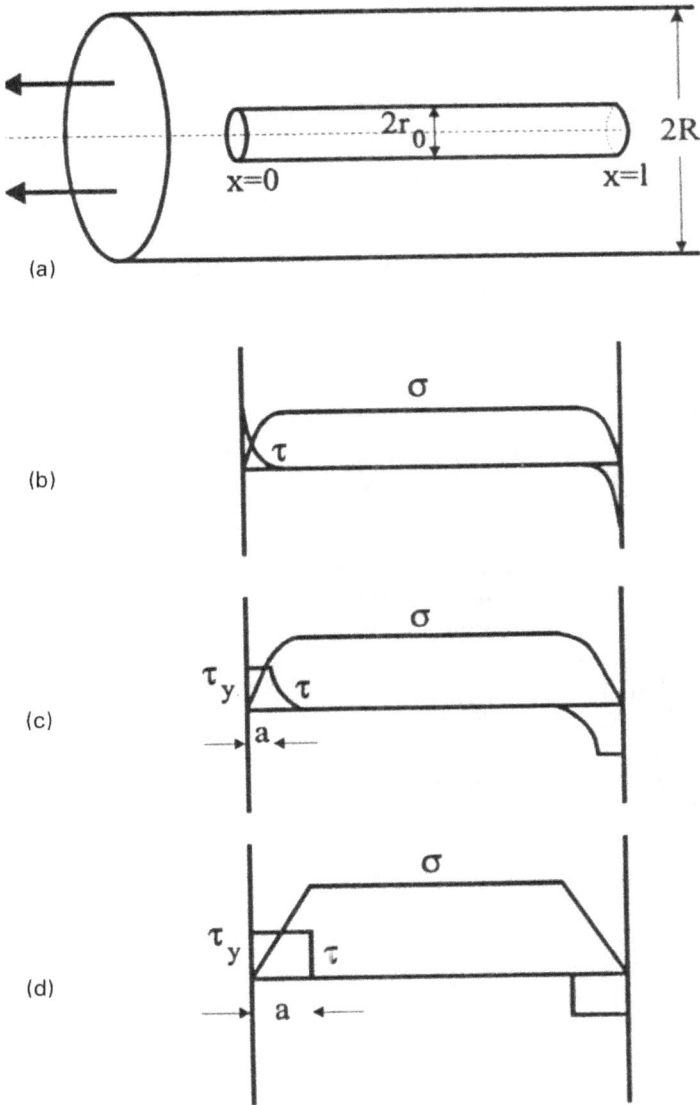

Figure 3.25 Schematic diagram of the micromechanics of stress transfer between matrix and fibre.[34]

aligned parallel to the tensile axis. The length of the fiber can vary from a few millimeters to the entire length of the specimen. The conventional way of using the test is to increase the stress on the specimen until fiber fracture occurs and, in order to do this, the fracture strain of the matrix needs to be higher than that of the

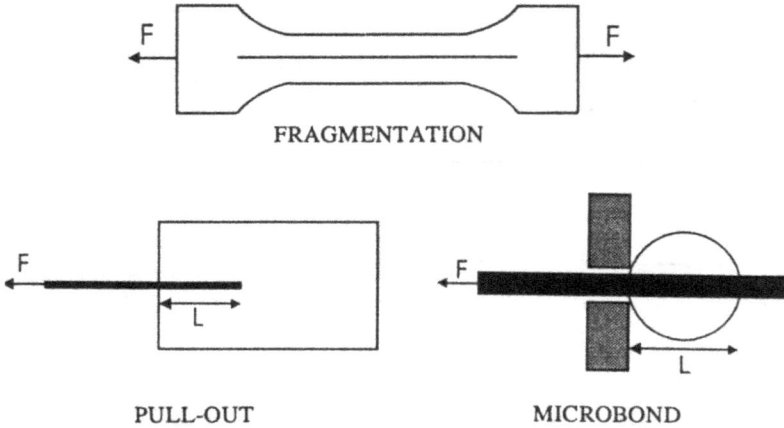

Figure 3.26 Schematic diagrams of the three main micromechanical test methods.[34]

fiber. As the applied strain increases, the fiber breaks at any point where the stress equals the fiber strength. A continuous increase of the strain induces further random breaks until the fragments are too short to allow the transfer of stress equal to or greater than the fiber tensile strength. Such a critical point is defined as the saturation of the fragmentation process. The limiting final fragment length is normally referred to as *critical length* l_c. It is assumed that the distribution of stress in the fragment is initially as illustrated in Fig. 3.25d and the fragmentation process is then though to proceed as shown in Fig. 3.27.

Fiber fracture can be monitored either directly using optical microscopy with polarised light or indirectly by the use of acoustic emission.

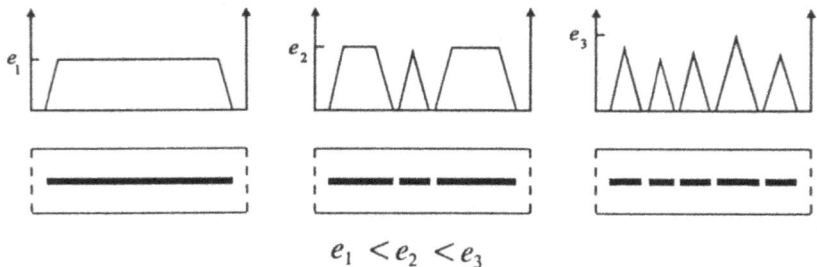

$$e_1 < e_2 < e_3$$

Figure 3.27 Schematic illustration of the distribution of fibre strain during fragmentation.[34]

The mean length of the fragments at saturation is related to the interfacial shear strength (τ) by the relation

$$\tau = \frac{\sigma_f r_0}{l_c} \qquad (3.24)$$

where σ_f is the mean fiber strength. This equation, however, simplifies the analysis since, towards the end of the fragmentation process, fragments with lengths in the range $l_c < l < 2l_c$ will continue to break and will lead to fragments in the size range $l_c/2 < l < l_c$ as illustrated in Fig. 3.27. Hence a better estimate of τ can be given by a modified equation,

$$\tau \approx \frac{3\sigma_f r_0}{4\langle l_c \rangle} \qquad (3.25)$$

where $\langle l_c \rangle$ is the mean value of the fragment length.

In order to apply Eq. (3.25) it is necessary to know the value of σ_f for the appropriate fragment length. In practice this is difficult to determine since fragments are generally less than 1 mm long and it is difficult to undertake tensile tests on fibers with such short gauge lengths (say 5–100 mm) and extrapolate to the value of σ_f at l_c using Weibull methods or other statistical techniques.

3.15.2 Pull-out

Fundamental information on the nature and strength of the interfacial bond between the fiber and the matrix in fiber reinforced composite material can be obtained using the single-fiber concentric pull-out test specimen and configuration shown in Fig. 3.26.

Stress distributions in this specimen are the same as in the common failure mode for discontinuous fiber composites with a brittle or elastic matrix.

Resistance to the debonding and pull-out processes is principally a function of the fiber/matrix interfacial bond shear strength and the interfacial area. It can be used to determine both the forces required to debond the fiber and also a measure of the frictional forces during pull-out. Specimen preparation can be quite difficult, especially if only small embedded lengths are required and during pull-out the axial tensile stresses in the fiber can lead to Poisson's contraction, which makes the situation at the interface different from that of a real composite.

If deformation of the fiber and matrix are assumed to be elastic then a Cox-type shear-lag analysis analogous to that described earlier can be used. If it is assumed that the interface fails at a shear stress of τ, then the analysis gives a value of the debonding force F_d for a fiber with an embedded length of L as

$$F_d = \left(2\pi r_0^2 \frac{\tau}{\eta}\right)\tanh\left(\frac{\eta L}{r_0}\right) \qquad \text{where} \qquad \eta^2 = \frac{E_m}{E_f(1 + v_m)\ln(R/r_0)} \qquad (3.26)$$

where E_m and v_m are the Young's modulus and Poisson's ratio of the matrix, respectively. As before, at high stresses yielding of the interface is also possible and in this case it is assumed that in the case of no work hardening the interfacial shear stress is the shear yield stress τ_y and the debonding force is a linear function of the embedded length such that

$$F_d = 2\pi r_0 \tau_y L \qquad (3.27)$$

Other analyses of the pull-out test have considered failure being controlled by an energy criterion rather than a stress criterion, but it has been shown that this leads to a similar dependence of F_d upon L as found in Eq. (3.26).

A potential breakthrough in this area has been obtained using Raman spectroscopy. The use of Raman spectroscopy to determine the variation of fiber strain with position along a fiber in a resin is described.[34]

3.15.3 Microbond test

The geometry of the microbond test is shown schematically in Fig. 3.26. A small droplet of resin is applied to a fiber. The force required to remove the droplet, F_d, by pulling the fiber and droplet between two knife edges is measured and the interfacial shear strength is conventionally determined by assuming that it is uniform along the fiber/matrix interface. Hence τ is given by

$$\tau = \frac{F_d}{\pi DL} \qquad (3.28)$$

where D is the fiber diameter and L is the embedded length.

In spite of the different experimental techniques as well as the analytical models developed by a large number of investigators, the problem of a clear definition of bonding strength remains open.

References

1. T. K. O'Brien (1990), Delamination of composite materials, in *Fatigue of Composite Materials* (K. L. Reifsnider, ed.) Elsevier Science, 181–198.
2. D. O. Stalnaker, R. H. Kennedy, and J. L. Ford (1980) Interlaminar-shear strain in a two-ply balanced cord-rubber composite, *Exp. Mech.*, 87–94.

3. Isaac M. Daniel (1980) Behavior of graphite/epoxy plates with holes under biaxial loading, *Exp. Mech.*, **20**, 1–8.

4. J. B. Hanus and C. P. Burger (1981) Stress-concentration factors for elliptical holes near an edge, *Exp. Mech.*, **21**, 336–340.

5. His-Yung T. Wu and George S. Springer (1988) Impact induced stresses, strains, and delaminations in composite plates, *J. Compos. Mater.*, **22**, 533–560.

6. D. Guedra-Degeorges, S. Maison, D. Trallero, and J. L. Petitnio (1992) Buckling and post-buckling behaviour of a delamination in a carbon-epoxy laminated structure: experiments and modelling, *Proceedings of the 2nd International Conference on "Deformation and Fracture of Composites"*, UMIST, Manchester, UK, The Institute of Materials, Wandsworth, London, 7/1–7/11.

7. Doron Shalev and K. L. Reifsnider (1990) Study of the onset of delamination at holes in composite laminates, *J. Compos. Mater.*, **24**, 42–71.

8. T. K. O'Brien (1982) Characterization of delamination onset and growth in a composite laminate, *Damage in Composite Materials*, ASTM STP 775 (K. L. Reifsnider, ed.), 140–167.

9. A. Murakami, D. Saunders, K. Ooishi, M. Murakami, T. Yoshiki, O. Watanabe, and M. Takezawa (1993) Delamination toughness and compression strength after impact in improved composite laminates, *Proceedings of the 2nd International Conference on "Deformation and Fracture of Composites"*, UMIST, Manchester, UK, The Institute of Materials, Wandsworth, London, 8/1–8/9.

10. P. Davies and M. L. Benzeggagh (1989) *Application of Fracture Mechanics to Composite Materials*, (K. Friedrich, ed.), Elsevier Science, 81.

11. G. C. Christopoulos, G. C. Papanicolaou, and K. Friedrich (1992) Mode I interlaminar fracture of a continuous ECR class fibre-polyamide12-composite as a function of thermal treatment, *Debonding/Delamination of Composites*, 74th Meeting of AGARD Structures and Materials Panel, Patras, Greece AGARD-CP-530, 30/1–30/4.

12. J. W. Putnam, J. C. Seferis, and T. Pelton (1993) Influence of material characteristics on the fracture behaviour of polymeric composites, *Proceedings of the 2nd International Conference on "Deformation and Fracture of Composites"*, UMIST, Manchester, UK, The Institute of Materials, Wandsworth, London, 10/1–10/11.

13. Y. B. Shi and A. F. Yee (1993) Delamination fracture of interlayered BMI composites, *Proceedings of the 2nd International Conference on "Deformation and Fracture of Composites"*, UMIST, Manchester, UK, The Institute of Materials, Wandsworth, London, 31/1–31/10.

14. R. A. Pearson, A. K. Smith, and A. F. Yee (1993) Synergistic toughening in hybrid epoxy composites, *Proceedings of the 2nd International Conference on "Deformation and Fracture of Composites*, UMIST, Manchester, UK, The Institute of Materials, Wandsworth, London, 9/1–9/10.

15. S. Hashemi, A. J. Kinloch, and J. G. Williams (1990) Mechanics and mechanisms of delamination in a poly(ether sulphone)–fibre composite, *Comp. Sci. Technol.*, **37**, 429–462.

16. W. Hwang and K. S. Han (1989) Interlaminar fracture behavior and fiber bridging of glass–epoxy composite under mode I static and cyclic loadings, *J. Compos. Mater.*, **23**, 396–430.

17. K. S. Kim and C. S. Hong, Delamination in angle-ply laminated composites, *J. Compos. Mater.*, **20**, 423–438.

18. J. P. Lucas and J. Zhou (1993) Moisture absorption effects on delamination mechanisms of carbon fiber polymeric matrix composites, *Proceedings, Ninth International Conference on Composite Materials (ICCM/9)*, Madrid, Spain, 633–641.

19. P. Chiou and W. L. Bradley (1993) The effect of seawater exposure on the fatigue edge delamination growth of a carbon/epoxy composite, *Proceedings, Ninth International Conference on Composite Materials (ICCM/9)*, Madrid, Spain, 516–523.

20. Chen-Chi M. Ma, Chang-Lun Lee, Min-Jong Chang, and Nyan-Hwa Tai (1992) Effect of physical aging on the toughness of carbon fiber-reinforced poly(ether ether ketone) and poly(phenylene sulfide) composites I, *Polym. Compos.*, **13**, 441–447.

21. R. Frassine, M. Rink, and A. Pavan (1993) Viscoelastic effects on interlaminar fracture toughness of epoxy/carbon-fibre laminates, *J. Compos. Mater.*, **27**, No. 9, 921–933.

22. R. Frassine (1992) The application of double torsion testing to unidirectionally reinforced composite materials, *J. Compos. Mater.*, **26**, No. 9, 1339–1350.
23. D. F. Devitt, R. A. Schapery, and W. L. Bradley (1980) A method for determining the mode I delamination fracture toughness of elastic and viscoelastic composite materials, *J. Compos. Mater.*, **14**, 270–285.
24. J. G. Williams (1987) Large displacement and end block effects in the "DCB" interlaminar test in modes I and II, *J. Compos. Mater.*, **21**, No. 4, 330–361.
25. G. C. Papanicolaou, D. Phalippon, and D. Bakos (1994) Effect of strain-rate on the mode-I interlaminar fracture toughness of sandwich double cantilever beams, *Proceedings of the International Conference on Composite Materials on "Advancing with Composites '94"*, Milan, Italy, 143–157.
26. D. C. Ruhmann, W. F. Bates Jr., H. B. Dexter, and R. R. June (1992) New materials drive high-performance aircraft, *Aerospace America*, **6**, 46–49.
27. Y. Frostig and M. Baruch (1990) Bending of sandwich beams with transversely flexible core, *AIAA J.*, **28**, No. 3, 523–531.
28. D. Bakos and G. C. Papanicolaou (1993) Effect of the skin treatment and core material on the bending behaviour of sandwich beams, *Compos. Sci. Technol.*, **49**, 35–43.
29. J. M. Whitney, C. E. Browning, and W. Hoogsteden (1982) Double cantilever beam tests for characterising mode I delamination of composite materials, *J. Reinforced Plast. Compos.*, **1**, No. 4, 297–313.
30. W. S. Johnson and P. D. Mangalgiri (1987) Influence of the resin on interlaminar mixed-mode fracture, *Toughened Plast.*, ASTM STP 937, 295–315.
31. T. K. O'Brien and W. Elber (1992) Delamination and fatigue of composite materials: a review, *Debonding/Delamination of Composites*, 74th Meeting of the AGARD Structures and Materials Panel, Patras, Greece, AGARD-CP-530, 2/1–2/5.
32. C. Poon, N. C. Bellinger, Y. Xiong, and R. W. Gould (1992) Edge delamination of composite materials, *Debonding/Delamination of Composites*, 74th Meeting of the AGARD Structures and Materials Panel, Patras, Greece, AGARD-CP-530, 12/1–12/13.
33. G. C. Papanicolaou and D. Bakos (1995) Effect of treatment conditions on the mode I delamination fracture toughness of sandwich structures, *J. Compos. Mater.*, **29**, 2295–2316.
34. R. J. Young (1993) Composite micromechanical test methods and their interpretation, *Proceedings of the 2nd International Conference on Deformation and Fracture of Composites*, UMIST, Manchester, UK, The Institute of Materials, Wandsworth, London, K-1–K-10.
35. A. Kelly and N. H. Macmillan (1986) *Strong Solids*, 3rd ed., Clarendon Press, Oxford.
36. H. L. Cox (1952) *Br. J. Appl. Phys.*, **3**, 72.
37. W. A. Fraser, F. H. Ancker and A. T. Dibenedetto (1975) in *Proceedings of the 30th Ann. Techn. Conf. SPI Reinf. Plastics Division-Composite Inst.*, Washington D.C. USA, 22-A, p.1.

Fracture Toughness and Fatigue Crack Growth

V. Kostopoulos

Applied Mechanics Laboratory, University of Patras,
Patras, Greece

Nomenclature

α, α_0	Crack length, crack correction term
α_n, α_{n-1}	Crack length at two successive cycles n and $n-1$, respectively
A	Crack area
b, b_1, b_2	Material constants
B	Width of test specimen
C	Compliance
C_P, C_S, C_R	Wave propagation velocities of P, S, and R waves
ε	Strain
E_x, E_y	Flexural modulus in x and y direction respectively
f	Function of θ
F	Function of θ; F_1 calibration factor for large displacements
g	Function of θ
G_{xz}	Shear modulus in xz plane
G	Energy release rate, separated in G_{I} for mode I, G_{II} for mode II
t	Thickness of laminate
I	Second moment of area
J	Path-independent J integral
K	Stress intensity factor

L	Distance in beam test; variable ratio
M	Bending moment
N	Compliance correction term due to end block stiffening
p	Pressure
P, P_c	Load, critical load
Q	Shear force
r	Radius; distance from crack tip
R	Crack resistance
u	Displacement in x direction
v	Displacement in y direction
U	Elastic strain energy
W	Work done by external forces
T	Kinetic energy
$d\alpha/dN$	Crack propagation rate
F	Energy available for crack formation
x, y, z	Cartesian coordinates
Y	Finite-width correction factor
γ	Shear strain
Γ	Compliance function
δ	Displacement at crack tip
Δ	Compliance function; length of crack extension
θ	Angle from crack line
κ	Timoshenko's shear coefficient
λ	Fraction of the total mechanical energy corresponding to irreversible fracture mechanisms
μ	Shear modulus
ν	Poisson's ratio
ξ	Dimensionless length
σ	Normal stress
τ	Shear stress
Φ_{ir}	Energy rate related to irreversible/plastic deformation
φ	Slope of beam

4.1 Introduction

Fracture in continuous fiber composites may occur in many complex and different ways due to their anisotropic nature. Although a significant effort has been accumulated during the last 40 years in understanding and describing the damage, failure, and fracture of fiber composites, knowledge in the field is much less complete than that for traditional metallic materials.

As a consequence of the nature of the materials and the fabrication processes, a number of manufacturing defects already exist within the structure of the fabricated composite components before they are put into service. Under service conditions the initial flaws grow and new defects are accumulated. At the microscopical level this damage mainly consists of cracks in matrix material, debonding at the interface between fiber and matrix, and breakage of fibers. The evolution of damage under critical loading conditions forms cracks on a macroscopic scale. Crack advance can cause serious reduction in stiffness and strength and finally lead to interlaminar, intralaminar, or translaminar catastrophic fracture.

Interlaminar fracture describes fracture parallel to the constituent plies of a composite laminate; intralaminar fracture refers to fracture that occurs within the body of a ply parallel to the fiber direction; while translaminar fracture is defined as that oriented normal to the laminated plane, as shown in Fig. 4.1.

Parameters that control damage development, and failure of a composite component are among others, the strength of the constituent materials of the composite lamina, the fiber–matrix interfacial strength, and the ability of the composite to absorb energy under critical loading.

In several loading cases, such as slight overloading in service, low-velocity impacts, and strikes by projectiles of low mass, only a finite amount of energy is available to produce damage. Then, the ability of the composite to absorb energy minimizes the developed damage and the related material toughness secures its resistance to crack propagation. Thus, in a number of loading cases, toughness rather than strength is the key property of the composite material.

Based on the nature of the composite material and the manufacturing process of the composite components, it is clear that their weak point is the poor interlaminar and intralaminar strength, which is mainly controlled by the properties of the matrix material. On the other hand, design with composites very often leads to design details that produce stress concentration sites and promote high interlaminar stresses known as edge effects. Then, the ability of the material to withstand the formation and propagation of delamination is again a toughness-controlled parameter.

(a)

(b)

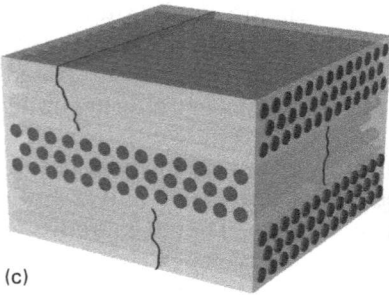

Figure 4.1 Schematical representations of typical fracture
modes in composite laminates.
(a) Interlaminar fracture; (b)
intralaminar fracture; and (c)
transalaminar fracture.

(c)

For design purposes, it is very important for the engineer to know
the fracture toughness of composite materials under the different
loading modes. Figure 4.2 presents the typical remote loading conditions applied to a cracked structure.

All the fracture types (interlaminar, intralaminar, and translaminar) mentioned above may occur under mode I (tensile or opening
mode), mode II (in-plane shear or sliding mode), mode III (antiplane
shear or tearing mode), or any combination of these loading configurations.

(a)

(b)

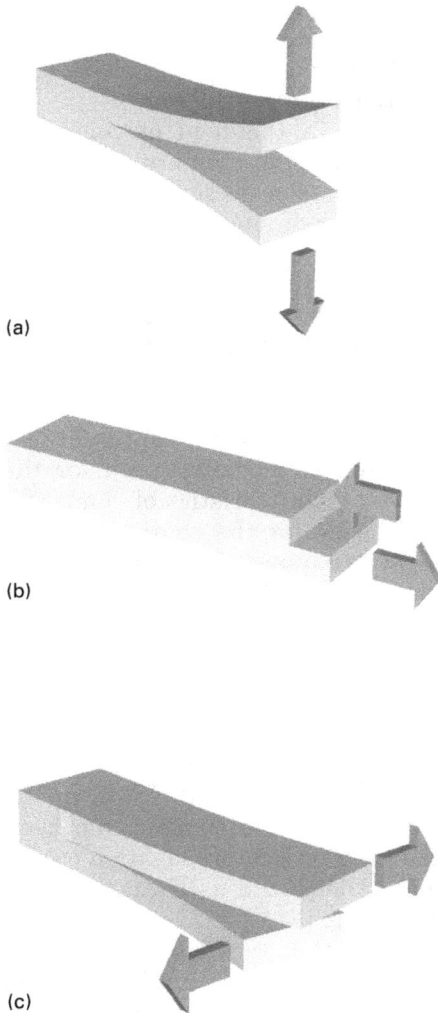

Figure 4.2 Schematical representations of the typical remote loading conditions. (a) Mode I (opening mode); (b) mode II (shear mode); and (c) mode III (tearing mode).

(c)

It is evident that since interlaminar and intralaminar fracture occurs in the direction parallel to the fiber reinforcement, the behavior of the composite laminate is mainly controlled by the matrix fracture properties and the fiber–matrix interface. In contrast, translaminar fracture includes basically fiber fractures and thus is a fiber-dominated fracture type, where again the fiber–matrix interface plays an important role.

The main part of this chapter will focus on the study of the interlaminar and the intralaminar fracture of organic matrix composites under mode I and mode II remote loading conditions and the study of translaminar fracture of both organic and ceramic matrix composites under mode I loading configuration. In all the cases presented, both quasi-static and fatigue loading conditions are considered.

It is important to mention here that in the field of composite materials the three typical remote loading conditions are still under investigation. However, for the cases chosen to be presented within the framework of the present chapter, we provide enough experimental data to permit a more or less complete description of the microscopical damage modes and the resulting failure characteristics.

The purpose of this chapter is to present a clear, concise, and thorough understanding of the fracture behavior of composite laminates. Since this book is primarily aimed at multidisciplinary engineering students in the final year of their undergraduate studies or in the first years of their postgraduate studies, this chapter focuses on the detail presentation of the analytical modeling used for the determination of the fracture toughness characteristics of composite materials. The experimental methods used for the characterization of the fracture toughness in composite laminates are also presented and the data-reduction methodologies are given.

4.2   Basics of Fracture Mechanics

Although significant criticism has been directed against the applicability of classical linear elastic fracture mechanics (LEFM), in the case of composite materials and structures such an analysis has been proven effective in a number of cases where crack propagation occurs in a plane perpendicular to the applied load and a relatively small damage zone is formed ahead the crack tip.

The material anisotropy does not affect the crack tip stress singularity [$\sigma \sim (1/r)$]; however, the angular stress distribution is disturbed by the material constants.[1] According to the Griffith energy concept, an elastic body subjected to externally applied loads must satisfy the following energy balance[2,3]:

$$W = U + T + F \qquad (4.1)$$

where W is the work done by the external loads, U is the elastic strain energy stored in the system, T is the kinetic energy and F is the energy dissipation due to the crack formation.

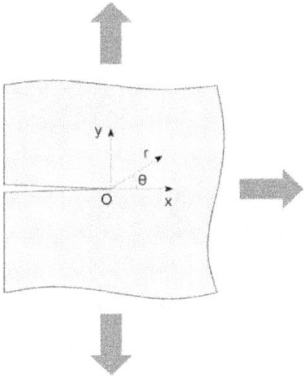

Figure 4.3 Typical cracked body configuration in the presence of remote loading.

A typical cracked body configuration is given in Fig. 4.3. Crack advance of $d\alpha$ causes a rearrangement of the terms of Eq. (4.1) in the sense that energy balance remains valid. Thus,

$$dW = dU + dT + dF \qquad (4.2)$$

The available elastic energy in the system $(dW - dU)$ will be delivered for the creation of new crack surface and possibly the increase of its kinetic energy. In order to increase the crack size, the available energy must be sufficient to overcome the surface energy of the material. Under quasi-static loading conditions, the variation of the kinetic energy of the system is negligible.

Expressing the energy differences per unit crack area,

$$\frac{dW}{dA} - \frac{dU}{dA} = \frac{dF}{dA} \qquad (4.3)$$

which is the expression of the Griffith energy balance for an increase of the crack area under equilibrium conditions. The left part of Eq. (4.3) is the so called "energy release rate" and plays the role of the crack driving force. It is denoted G. The right part of Eq. (4.3) represents the energy consumed during crack propagation and is called "crack resistance," denoted by R.

The energy balance given in Eq. (4.3) states that G must be equal to or greater than R for crack advance to occur, assuming in a first approximation that energy required for crack propagation is constant for any $d\alpha$. Hence the crack advances when G exceeds a critical value $G_c(=R)$, which is called "critical energy release rate."

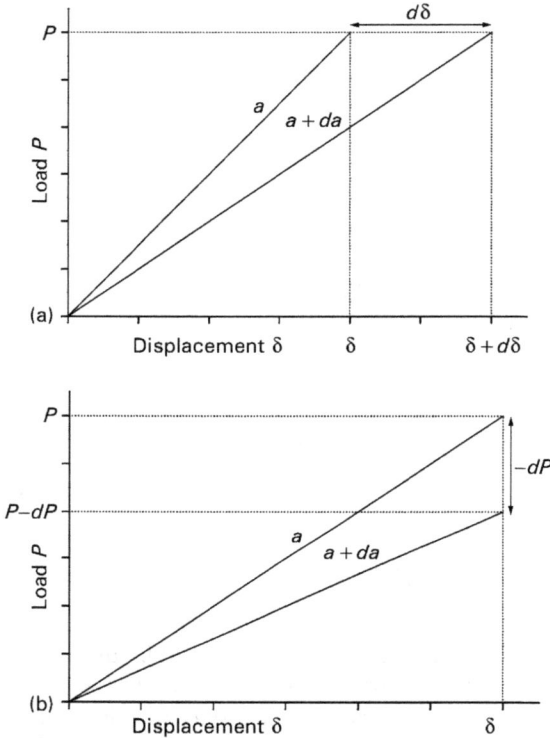

Figure 4.4 Load-displacement curve of a cracked body. (a) Loaded under constant force P; (b) loaded under constant displacement δ.

Next consider a cracked plate loaded with a constant force P. Then, as shown in Fig. 4.4a, the work done by the externally applied load P is given by

$$W = P\delta \tag{4.4}$$

where δ is the displacement produced at the loading point by the applied load P, and the elastic energy stored in the body is

$$U = \int_0^\delta P \, d\delta = \frac{P\delta}{2} \tag{4.5}$$

According to Eq. (4.3),

$$G = \frac{1}{B}\left[\frac{d(W - U)}{d\alpha}\right]_P = \frac{P}{2B}\left(\frac{d\delta}{d\alpha}\right)_P \tag{4.6}$$

On the other hand, if the cracked plate is assumed to be under displacement control as shown in Fig. 4.4b, then $W = 0$ and U is given again by Eq. (4.6). In this case G is given by

$$G = -\frac{1}{B}\left(\frac{dU}{d\alpha}\right)_\delta = -\frac{\delta}{2B}\left(\frac{dP}{d\alpha}\right)_\delta \qquad (4.7)$$

Introducing the definition of compliance C for the cracked structure as the inverse of plate stiffness, i.e.,

$$C = \frac{\delta}{P} \qquad (4.8)$$

and substituting (4.8) into (4.6), and (4.7), one obtains

$$G = \frac{P^2}{2B}\frac{dC}{d\alpha} \qquad (4.9)$$

which holds true for both load control and displacement control conditions.

As has been mentioned earlier, crack extension occurs when G equals R. Then stable crack propagation is attained whenever

$$\frac{dG}{da} \le \frac{dR}{da}$$

and unstable crack growth occurs when

$$\frac{dG}{da} > \frac{dR}{da}$$

A crack may propagate in any combination of the three fracture modes introduced in Fig. 4.2. Experimental evidences support the idea that resistance in crack propagation of most materials depends upon the mode of loading at the crack tip area. Thus, in order to characterize completely the fracture behavior of a particular material, the critical energy release rates G_{IC}, G_{IIC} and G_{IIIC}, resulting under different loading configuration of the crack tip are needed.

Some basic stress analysis results referring to certain cracked bodies subjected to external loading under the assumption of linear elastic and isotropic material behavior, are given next. Such results have been proposed by pioneers in the field of LEFM such as like Westergaard,[4] Irwin,[5] Williams,[6] etc.

In general, the stress field within the crack-dominated zone is given by

$$\sigma_{ij} = \frac{k}{\sqrt{r}}f_{ij}(\theta) \qquad (4.10)$$

where σ_{ij} is the stress tensor at a given point and r, θ are polar coordinates, k is a constant, and $f_{ij}(\theta)$ is a dimensionless function of θ.

The polar coordinate system has its origin at crack tip as shown in Fig. 4.3. As is shown, the behavior of the stress field in the crack-dominated area is proportional to $1/\sqrt{r}$. As $r \rightarrow 0$ the stresses approach infinity regardless of the cracked body.

Each mode of loading causes the $1/\sqrt{r}$ singularity at the crack tip; however, the constant k and the dimensionless function $f_{ij}(\theta)$ depend upon the applied mode.

Introducing the stress intensity factor $K = k\sqrt{2\pi}$ and denoting by K_{I}, K_{II} and K_{III} the respective stress intensity factors under mode I, II, and III loading conditions, then the stresses in the crack dominated zone are given by

$$\sigma_{ij}^{\mathrm{I}} = \frac{K_{\mathrm{I}}}{\sqrt{2\pi r}} f_{ij}^{\mathrm{I}}(\theta)$$

$$\sigma_{ij}^{\mathrm{II}} = \frac{K_{\mathrm{II}}}{\sqrt{2\pi r}} f_{ij}^{\mathrm{II}}(\theta)$$

$$\sigma_{ij}^{\mathrm{III}} = \frac{K_{\mathrm{III}}}{\sqrt{2\pi r}} f_{ij}^{\mathrm{III}}(\theta)$$

Under mixed mode loading conditions, the application of the linear superposition principle leads to

$$\sigma_{ij}^{\mathrm{mixed}} = \sigma_{ij}^{\mathrm{I}} + \sigma_{ij}^{\mathrm{II}} + \sigma_{ij}^{\mathrm{III}}$$

The complete expressions for the elements of the stress tensor σ_{ij} under mode I, mode II, and mode III loading conditions are given in the following.[7]

Stress field ahead of a crack tip for mode I and mode II

MODE I LOADING:

$$\sigma_{xx} = \frac{K_{\mathrm{I}}}{\sqrt{2\pi r}} \cos\left(\frac{\theta}{2}\right)\left[1 - \sin\left(\frac{\theta}{2}\right)\sin\left(\frac{3\theta}{2}\right)\right]$$

$$\sigma_{yy} = \frac{K_{\mathrm{I}}}{\sqrt{2\pi r}} \cos\left(\frac{\theta}{2}\right)\left[1 + \sin\left(\frac{\theta}{2}\right)\sin\left(\frac{3\theta}{2}\right)\right]$$

$$\tau_{xy} = \frac{K_{\mathrm{I}}}{\sqrt{2\pi r}} \cos\left(\frac{\theta}{2}\right)\sin\left(\frac{\theta}{2}\right)\cos\left(\frac{3\theta}{2}\right)$$

$\sigma_{zz} = 0$ (plane stress)   or $\sigma_{zz} = v(\sigma_{xx} + \sigma_{yy})$ (plane strain)

$\tau_{xz} = 0$ $\tau_{yz} = 0$

where v is the Poisson's ratio.

MODE II LOADING:

$$\sigma_{xx} = -\frac{K_{II}}{\sqrt{2\pi r}} \sin\left(\frac{\theta}{2}\right)\left[2 + \cos\left(\frac{\theta}{2}\right) \cos\left(\frac{3\theta}{2}\right)\right]$$

$$\sigma_{yy} = \frac{K_{II}}{\sqrt{2\pi r}} \sin\left(\frac{\theta}{2}\right) \cos\left(\frac{\theta}{2}\right) \cos\left(\frac{3\theta}{2}\right)$$

$$\sigma_{xy} = \frac{K_{II}}{\sqrt{2\pi r}} \cos\left(\frac{\theta}{2}\right)\left[1 - \sin\left(\frac{\theta}{2}\right) \sin\left(\frac{3\theta}{2}\right)\right]$$

$$\sigma_{zz} = 0 \text{ (plane stress)} \qquad \text{or} \qquad \sigma_{zz} = \nu(\sigma_{xx} + \sigma_{yy}) \text{ (plane strain)}$$

$$\tau_{xz} = 0 \qquad \tau_{yz} = 0$$

Nonzero stress components in mode III loading

$$\tau_{xz} = -\frac{K_{III}}{\sqrt{2\pi r}} \sin\left(\frac{\theta}{2}\right)$$

$$\tau_{yz} = \frac{K_{III}}{\sqrt{2\pi r}} \cos\left(\frac{\theta}{2}\right)$$

4.3 Mode I Interlaminar Fracture

The use of double cantilever beam (DCB) test specimen is the most popular way of determining mode I interlaminar fracture toughness of composite laminates. The specimen was first introduced by Ripling et al.[8] for the fracture toughness measurements of adhesively bonded metals (single lap adhesive joints). Its modification and application in composites for the measurement of their mode I interlaminar fracture toughness was the natural next step.

The straight strip DCB test specimen developed by Wilkins et al.[9] has been used more extensively compared to the tapered width specimen proposed by Bascom et al.[10] The typical configuration of a straight strip DCB specimen is shown in Fig. 4.5.

The load is applied on the DCB specimen either using piano hinges or through special aluminum end blocks bonded onto the surface of the specimen, in order to secure load application across the same line. This is absolutely necessary for pure mode I loading conditions at the crack tip. The initial delamination is introduced in the composite test specimen by the positioning of a starter thin film (Teflon, Kapton, or Upilex of 13 μm thickness, sprayed with a liquid release agent) at the mid-plane of composite laminate during the lay-up stage of composite manufacturing. In this way, thermal mismatch between the starter

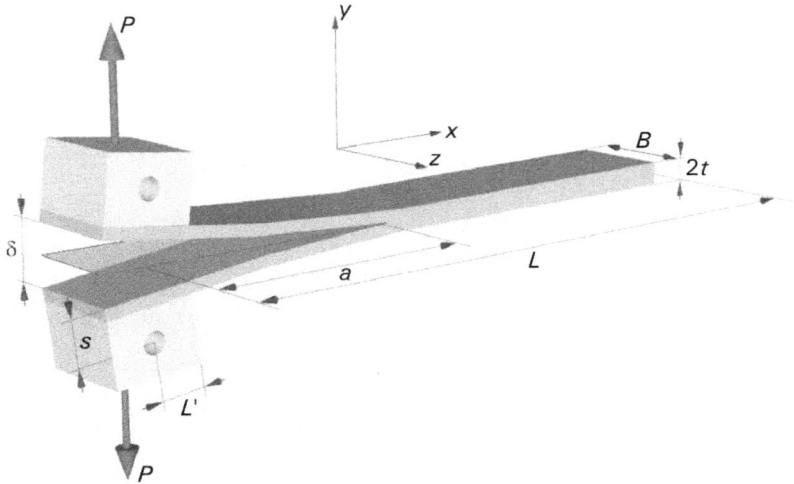

Figure 4.5 Schematical representation of the straight strip DCB specimen.

film and the composite is reduced and residual thermal stresses at the resin-rich crack tip area are minimized.

Typical specimen dimensions are given in Table 4.1

For the analysis, the two arms of the DCB specimen are considered to be fixed at the crack tip. Then, according to the linear beam theory,

$$\frac{\delta''(x)}{2} = \frac{M}{E_x I_z} = \frac{P(\alpha - x)}{EI_z} \tag{4.11}$$

where E_x is the flexural modulus of elasticity of the two cantilever beams and I_z is the moment of inertia along the z-axis. After integrating Eq. (4.11) twice and applying boundary conditions for $x = \alpha$ [$\delta(\alpha) = 0$ and $d\delta(\alpha)/dx = 0$], the following formula is reached:

$$\frac{\delta(x)}{2} = \frac{P}{E_x I_z} \left(\frac{\alpha x^2}{2} - \frac{x^3}{6} \right) \tag{4.12}$$

TABLE 4.1   Straight Strip DCB Specimen Dimensions

L (mm)	B (mm)	$2t$ (mm)	α (mm)
> 200 (for permitting 150 mm crack propagation)	≥ 20 (for avoiding significant contribution of edge effects)	2 (in case where aluminum end blocks are used)	40 ($\alpha/t \sim 40$ for avoiding shear correction)

Then substituting I_z $(= \frac{1}{12} Bt^3)$ and calculating δ at $x = \alpha$, one obtains

$$\frac{\delta(\alpha)}{2} = \frac{4P\alpha^3}{E_x Bt^3} \tag{4.13}$$

and the compliance of the system, which is defined in Eq. (4.9), is given by

$$C(\alpha) = \frac{\delta}{P} = \frac{8\alpha^3}{E_x Bt^3} \tag{4.14}$$

Substituting the expression for the compliance of the system, C, into Eq. (4.9), the energy release rate for the DCB specimen is obtained as

$$G_{\mathrm{I}} = \frac{12P^2\alpha^2}{E_x B^2 t^3} \tag{4.15}$$

which constitutes the beam model expression for the DCB test specimen.[11–14] When the applied load P is equal to the critical load P_c, which corresponds to crack advance, Eq. (4.15) provides the critical energy release rate G_{Ic}, which according to the LEFM is a material parameter,

$$G_{\mathrm{Ic}} = \frac{12P_c^2\alpha^2}{E_x B^2 t^2} \tag{4.16}$$

In order for Eq. (4.16) to hold true, a number of assumptions must be fulfilled:

- The composite under consideration behaves macroscopically as homogeneous and orthotropic material in the general sense.

- The problem under consideration does not involve either material or geometrical nonlinearities.

- Load application and boundary conditions as well edge effects do not affect the crack delamination advance at the crack tip area (St. Venant's principle).

However, these idealized conditions are far from applying in the case of composite laminates. As a result, a number of modified expressions have been proposed in the literature and the most interesting of them will be listed below.

Modifications due to crack tip compliance. As has been stated above, according to the beam model the arms of the DCB specimen are considered to be fixed at the crack tip. However, this is not the case. The arms of the DCB specimen are not rigidly clamped at the crack tip. Thus, the beams rotate about the z-axis and also a displacement parallel to the y-axis appears. Under these conditions, the argument of fixed support of the DCB arms is more likely satisfied at a point ahead the crack tip, where both displacement and rotation are negligible.

Then an effective crack length[15] may introduced in Eqs. (4.14) and (4.17) given by

$$\alpha_{\text{eff}} = \alpha + \alpha_0 \tag{4.17}$$

where α_0 is a correction term that takes into account the crack tip compliance. According to the literature, α_0 is positive and is given by

$$\alpha_0 = bt_0 \tag{4.18}$$

where t_0 is the ply thickness in composite laminates or the thickness of the adhesive layer in adhesively bonded metals. The proportionality coefficient b is determined from the compliance data and for the case of bonded metals has been found to be $b = 0.37$.[16] In the case of composite materials various values of α_0 have been reported.[17-20] It is important to mention here that a_0 is not always a positive correction as mentioned earlier. In some cases of thermoplastic composites made of low-modulus fibers coupled with a tough matrix, negative a_0 values have been found.[21,22]

The effects of crack tip compliance have been also treated in a different way. Kanninen has introduced an analytical model that assumes that the arms of the DCB specimen are supported on an elastic foundation.[23] He proposes the following expression for the compliance:

$$C(\alpha) = \frac{4\alpha^3}{E_x B t^3} \left[1 + 1.92\left(\frac{t}{\alpha}\right) + 1.22\left(\frac{t}{\alpha}\right)^2 + 0.39\left(\frac{t}{\alpha}\right)^3 \right] \tag{4.19}$$

Equation (4.19) provides good agreement with experimental data for both thin and thick laminates. However, Kanninen's model is mainly applicable in the case of isotropic materials.

A modification of Eq. (4.19) has been proposed by Ashizawa[24] in order to take into account the anisotropic nature of composites laminates. The modified compliance exhibits the form

$$C(\alpha) = \frac{8\alpha^3}{E_x B t^3} \left[1 + 1.92\left(\frac{t}{\alpha}\right)\left(\frac{E_x}{E_y}\right)^{0.25} + 1.22\left(\frac{t}{\alpha}\right)^2\left(\frac{E_x}{E_y}\right)^{0.5} \right.$$
$$\left. + 0.39\left(\frac{t}{\alpha}\right)^3\left(\frac{E_x}{E_y}\right)^{0.75} \right] \tag{4.20}$$

where E_x and E_y are the tensile moduli of the composite laminate along the x and y directions.

Modifications due to transverse shear. In a number of practical cases the ratio of thickness to crack length is not very small and/or, since the material is transversely isotropic, the shear modulus G_{xz} is much lower compared to the flexural modulus E_x.

Then the developed shear deformation significantly affects the compliance of the arms of the DCB specimen. The overall compliance of the system under these conditions is given according to Whitney et al.[11] by the relation

$$C(\alpha) = \frac{8\alpha^3}{E_x Bt^3}\left[1 + \frac{3}{10}\frac{E_x}{G_{xz}}\left(\frac{t}{\alpha}\right)^2\right] \tag{4.21}$$

The second term in the bracket of the right-hand side of Eq. (4.21) is called shear correction factor.

In this case the critical energy release rate G_{Ic} is given by

$$G_{Ic} = \frac{12P^2}{E_x B^2 t}\left[\left(\frac{\alpha}{t}\right)^2 + \frac{1}{10}\frac{E_x}{G_{xy}}\right] \tag{4.22}$$

Taking into account the shear correction factor, the critical energy release rate may increase under realistic conditions by up to 2–3%.

Modification due to end blocks. It is obvious that the arms of the DCB specimen are stiffened by the presence of the bonded end blocks that are used for the application of the load. The increase of the bending stiffness leads to a decrease in energy release rate.

However, the stiffening effect due to the end blocks is reduced as the crack extends. Under these conditions a modified compliance must be used for the calculation of G_{Ic} which is given by $C' = C/N$,[25] where

$$N = 1 - \left(\frac{L'}{\alpha}\right)^3 - \left(\frac{9}{8}\right)\left[1 - \left(\frac{L'}{\alpha}\right)^2\right]\frac{\delta t_0}{\alpha^2} - \frac{9}{35}\left(\frac{\delta}{\alpha}\right)^2 \tag{4.23}$$

where L' and t_0 are shown in Fig. 4.5 and δ is the crack opening displacement (the loading point displacement).

Modification due to large displacement. There are some cases where the compliance of the DCB specimen is very high and this leads to large deflections of the arms of the DCB specimen. Therefore a nonlinear, elastic approach to the problem of the bending of a cantilever composite beam seems more adequate for such problems.

As has been shown in,[26] the large deflection effect has to be taken into consideration whenever $\delta/\alpha > 0.3$. In these cases a parameter F is introduced by the relation[25]

$$F = 1 - \frac{3}{10}\left(\frac{\delta}{\alpha}\right)^2 - \frac{3}{2}\left(\frac{\delta t}{\alpha^2}\right) \qquad (4.24)$$

In order to take into account the large displacement effect, one multiplies the calculated G_{Ic} by F to obtain the corrected critical energy release rate.

4.3.1 Data-reduction schemes

According to the literature there are two different ways of producing load–displacement data in the case of DCB tests specimens. The first assumes monotonically increasing loading conditions, while the second applies a cyclic loading profile.

During the monotonically increasing loading, the test specimen is loaded under displacement control conditions. A crosshead velocity of 0.5 mm/min is used. The loading rate may be increased after the first 5 mm of delamination growth to 1 mm/min. The delamination front is observed visually at discrete time intervals and the crack advance is measured from the end of the insert to the crack tip, on both edges of the test specimen. During the first 5 mm of crack advance the density of the crack growth measurement must be made at 1 mm intervals, or the closest possible interval. After the first 5 mm of delamination growth, measurements must be obtained at least every 5 mm of crack advance.

All the crack growth measurements correspond to the points of the load–displacement curve for which they have been measured. When the delamination has extended 25 mm beyond the insert edge, the specimen is unloaded and the test is finished.

When a cyclic loading is applied on the DCB specimen, then during the initial cycle the specimen must be loaded until the crack extends at about 10 mm. The machine is stopped and the actual crack length is measured precisely. In this sequence the specimen is unloaded. The procedure is repeated until the crack advances for approximately 150 mm. Crack length measurements are made at intervals of 10 mm. During the loading phase of the cycles a crosshead velocity of 1 mm/min must be used. The method assumes that the material behaves elastically, which means that during the unloading phase the displacement returns to zero when the load is removed. Permanent displacement indicates that other mechanisms in addition to the delamination growth contribute to the energy dissipation. Therefore, increased values of the critical energy release rate will be calculated. This effect will be discussed later in the section on fiber bridging.

No matter what the type of load–displacement, the data-reduction schemes follow the same philosophy. Next a complete presentation of the data reduction principles will be given.

Beam method. Using the analysis of the beam model presented earlier, and assuming the effects of crack tip compliance, transverse shear, and large deformation to be negligible, one arrives at the following expression for the critical energy release rate under mode I loading conditions:

$$G_{\text{Ic}} = \frac{3}{2B} \frac{P_c \delta_c}{\alpha_c} \tag{4.25}$$

where P_c, δ_c and α_c are the critical applied load, displacement and crack length, respectively.

Expression (4.25) is much more useful for the calculation of the critical energy release rate using experimental data, because

1. It does not include the flexural modulus E_x of the composite under investigation and knowledge of it is not necessary.

2. It does not include the thickness of the test specimen.

3. It minimizes errors produced by inaccurate measurements of crack length, since G_{Ic} is proportional to (α) and not to (α^2) as in Eq. (4.15).

Empirical compliance method. The empirical compliance method is based upon the curve fitting of the compliance measurements $C = C(\alpha)$ recorded during the experiment. It was first introduced by Wilkins et al.[9] This method proposes a general relation of the compliance versus the crack length of the form

$$C(\alpha) = \frac{\alpha^b}{A} \tag{4.26}$$

where A and b are constants determined by curve fitting of a $\log(C)$–$\log(\alpha)$ plot. Then the critical energy release rate is given by

$$G_{\text{Ic}} = \frac{bP_c \delta_c}{2B\alpha_c} \tag{4.27}$$

where again P_c, δ_c, and α_c are the same quantities as defined earlier.

It is important to notice that the empirical compliance method inherently involves all the necessary corrections coming from either material or geometrical nonlinearities in the calculation of the parameters A and b. In the case where the contribution of the various effects is negligible, then $b = 1/3$ and Eq. (4.27) obtains the same form as Eq. (4.25).

Areas method. A very popular alternative to the methods based on compliance for the measurement of the mode I delamination critical energy release rate is a direct energy method known as the areas method.[11,27]

Figure 4.6 shows a series of loading–unloading cycles during mode I interlaminar fracture toughness test, performed on DCB test specimen. During a typical loading cycle the load increases up to a certain level P, where it causes the propagation of a preexisting crack. When the crack extends to a new crack length $\alpha + \Delta\alpha$, the applied load drops to $P - \Delta P$ and the crack opening displacement increases to $\delta + \Delta\delta$. Then the load is removed.

The area ΔA enclosed within the loading–unloading curve represents the decrease of the stored elastic energy due to the crack advance. The critical energy release rate is given by

$$G_{Ic} = \frac{\Delta A}{B\,\Delta\alpha} \tag{4.28}$$

The product $B\,\Delta\alpha$ represents the new crack surface created by the crack advance.

The method is also valid in the case of nonlinear elastic behavior of the material and the effects of the nonlinearities are contained in the calculation of the areas ΔA. It is not valid whenever unstable crack propagation occurs, because the contribution of the kinetic energy term is not included in Eq. (4.28).

J-integral method. The analytical expressions provided so far by the beam method have been limited to the case of linear elastic behavior of

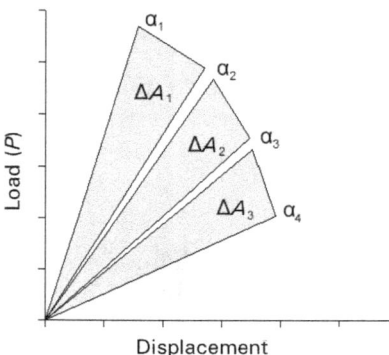

Figure 4.6 Schematical representation of the load–displacement curve showing the application principles of the area method for the determination of G_{IC}.

the material with very narrow fracture process zone and plasticity effects. The presence of nonlinearity was treated using correction factors.

Another elegant way of determining the energy release rate in the case of a nonlinear elastic load–displacement curve is the use of the J-integral method. By definition, the J-integral in two-dimensional form is given by[2]

$$J = \int_\Gamma W \, dy - T \frac{\partial u}{\partial x} \, ds \tag{4.29}$$

where Γ is a curve that surrounds the crack tip, T is the traction vector $(T = \sigma_{ij} n_j)$, u is the displacement along the x-axis, and W is the strain energy density and defined by

$$W(x, y) = \int_0^\varepsilon \sigma_{ij} \, d\varepsilon_{ij} \tag{4.30}$$

The y direction is taken normal to the crack line and ds is an element on curve Γ.

It can be shown that along any closed contour, $J = 0$. Rice[28] has shown that the J-integral as defined along a contour around crack tip is the change in potential energy for a virtual crack exclusion $d\alpha$,

$$J = -\frac{1}{B} \frac{\partial U}{\partial \alpha} \tag{4.31}$$

where U is the strain energy. For a linear elastic material,

$$J = G \tag{4.32}$$

Figure 4.7 shows the load–displacement curves in the case of nonlinear elastic response of the DCB specimen for the delamination length (α) and ($\alpha + d\alpha$). The area between the two load–displacement curves gives the energy release rate.

As is clearly stated[29] the material unloads along the same curve along which it was loaded. The stress–strain behavior is nonlinear but elastic.

At this point it is important to focus on a remark given[2]: "Many problems in plasticity can be dealt with by treating the material as nonlinear elastic through the deformation theory of plasticity. In general however, the deformation theory of plasticity cannot be used for problems in which unloading occurs, for the obvious reason that unloading of a real material follows a different stress–strain curve. In that case one has to use the incremental theory of plasticity." Keeping

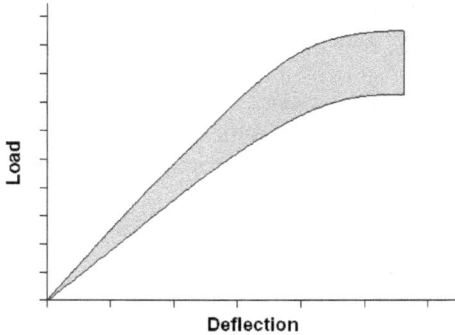

Figure 4.7 Schematical representation of the load–displacement curve in the case of non-linear elastic response of a DCB test specimen.

this in mind one may calculate the strain energy $U(\delta)$ by integrating the load–displacement curve:

$$U(\delta) = \int_0^{\delta} P(\delta') \, d\delta' \qquad (4.33)$$

and the J-integral is given by the application of Eq. (4.31):

$$J = -\frac{1}{B}\frac{dU}{d\alpha} \qquad (4.34)$$

The critical value of the J-integral, J_{Ic}, is then taken at the critical displacement δ_c, just before crack advances.

A detailed methodology for the calculation of J-integrals is given.[29–31]

4.3.2 Fiber bridging

The fiber cross-over of the delaminated surface during the mode I delamination test in DCB specimens increases the demanded load for crack advance, and this phenomenon is known as the fiber bridging effect. This effect enhances the crack growth resistance of the material, giving rise to an increase in the crack growth resistance curve described as a function of crack extension. This behavior is called R-curve behavior of the material and it was first reported in the case of composites as long ago as 1971.[32] Since then, further studies have been performed[33,34] and these have been summarized.[35]

Whenever the bridging is restricted to a zone much smaller than the characteristic specimen dimension (the smaller specimen dimension), it is known as small-scale bridging and the resulting R-curve can be regarded as a material property. However, in many cases the bridging zone is of comparable size to the characteristic specimen dimension. Then it is called large-scale bridging,[36] and the shape of the R-curve depends on the geometry of the testing sample, no longer being a material parameter.

Under these conditions, a damage process zone is formed ahead the crack tip, which is first fully developed and then propagates in a self-similar way. The reason, which promotes fiber bridging in the test coupons used for the determination of mode I interlaminar critical energy release rate, is that the DCB specimen is mainly constructed as unidirectional 0° laminates. There are serious arguments for this choice:

- High bending stiffness minimizes the large deflection effect.

- Delamination propagates at the interface between two adjacent plies in a self-similar way, minimizing the diffusion of matrix cracking in neighboring plies.

- It offers completely symmetric laminate and symmetric sublaminate arms in the DCB specimen, eliminating any twisting of the system.

In contrast, the 0° laminate configuration permits fiber nesting and migration into adjacent plies, so that no distinct interlaminar plane is available for crack propagation, and also the formation of additional cracks above and below the main crack plane that propagate together. This may be caused by the preexisting defects, particularly in tough materials.

Thus the fiber bridging effect is present during mode I interlaminar fracture toughness tests. Although fiber bridging is a beneficial effect in practical applications of composites since it increases interlaminar fracture toughness, it is undesirable during the testing procedure because it leads to the determination of apparent fracture toughness.

As has already been stated, the problem appears particularly in the case of large-scale bridging, which cannot be accounted as a material parameter. In practice, this is the case for thermoplastic and ceramic matrix composites where, for different reasons in each case, the composite exhibits a tougher behavior.

In the following, a relatively new methodology is proposed for the characterization of the interlaminar energy release rate under mode I loading conditions in the case where large-scale bridging exists. Following a loading–unloading testing procedure; both the energy release rate corresponding to the formation of new crack area and the energy dissipation, which comes mainly from the bridging effect and the crack tip plasticity, are calculated. Then, the sum of these two contributions establishes the R-curve behavior of the composite. The proposed methodology has been found particularly useful in the case of thermoplastic composites, for example for the bridging effect and the crystallinity effect of the matrix material, which varies depending on the thermal history of the material.

The analysis aims to determine the crack growth resistance R as the sum of two energy rate contributions:

- The nonlinear strain energy release rate G_R^*, which can be directly correlated to the G_{Ic} fracture parameter, calculated by LEFM.
- The plastic energy dissipation rate Φ_{ir} which is the energy consumed due to the process of the formation of the damage zone that surrounds the crack tip region.

The crack growth resistance R, the nonlinear strain energy release rate G_R^*, and the plastic energy dissipation rate Φ_{ir} were calculated using the loading–unloading procedure applied for the DCB specimen configuration. The establishment of this nonlinear semi-empirical approach, based on the form of the P–δ curve at each stage of crack increment, is a necessary requirement for understanding the fracture behavior, tracking down the inherent irreversible mechanisms, and calculating the rate of energy consumption needed for the formation of the crack area.

In Fig. 4.8 a representative curve in case of the linear elastic fracture behavior is shown. At the instant a critical load is reached, a crack increment occurs. As a result, the curve exhibits a nonlinear form. Successive unloading at different stages of crack propagation shows that no plastic phenomena are present. In this case the characteristic parameters are the load P, the crack opening displacement (COD) δ, and the compliance from the origin, $C = \delta/P$. Therefore, the effective crack length may easily be calculated. Additionally, it is obvious that the nonlinearity of the P–δ curve is attributable only

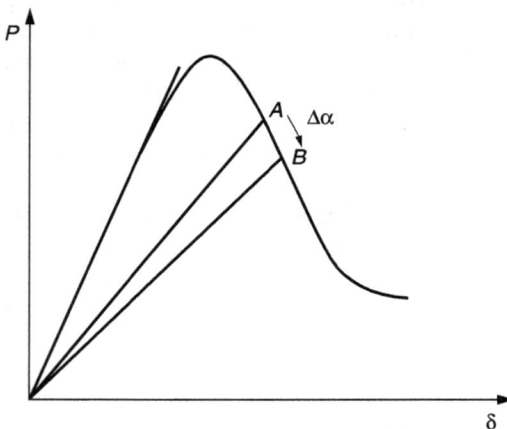

Figure 4.8 Schematical representation of the Load-Displacement curve in the case of linear elastic fracture behaviour.

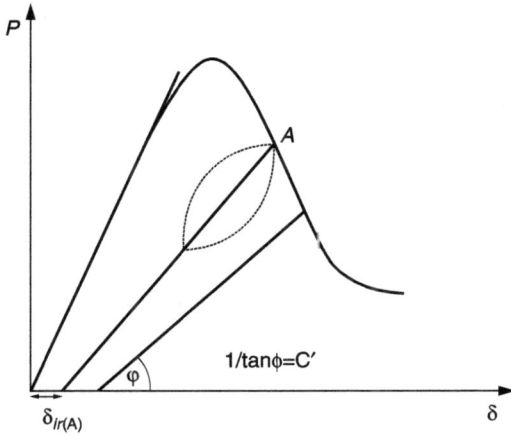

Figure 4.9 Schematical representation of the Load-Displacement curve in the case of linear non-elastic fracture behaviour.

to the irreversible loss of energy consumed in creating a new crack surface.

Figure 4.9 shows a schematic representation for the case of linear nonelastic material behavior. The crack propagates when a critical load is exceeded. In this case, successive unloading at different crack increments leads to plastic deformation as bridging and pull-out, fiber–matrix debonding, debris effects, and extensive matrix micro-cracking.

The total energy consumed now consists of the part necessary for the formation of new crack surface and the energy necessary for supplying all the irreversible damage mechanisms that appear.

The characteristic parameters now are the load P, the COD δ, the remaining displacement at each unloading step δ_{ir}, as well as the modified compliance

$$C^* = \frac{\delta - \delta_{ir}}{P} \qquad (4.35)$$

If the loops of the loading cycles are not linear, nonlinear, nonelastic behavior is assigned. Using the medians to the loops, one may apply the calculation procedure presented in the following.

The main reason for using the loading–unloading procedure is that the irreversible mechanisms are traced via the registration of the residual displacement δ_{ir}. The variation of compliance C^* between two successive cycles, $n - 1$ and n, permits an estimation of the crack increment using an experimentally well established formula,[17,37,38]

$$\alpha_n = \alpha_{n-1} + \frac{b_{n-1}}{2} \frac{C_n^* - C_{n-1}^*}{C_n^*} \qquad (4.36)$$

where $\alpha_n, \alpha_{n-1} =$ crack lengths at the two successive cycles n and $n-1$, respectively

$C_n^*, C_{n-1}^* =$ the corresponding modified compliances during the cycles n and $n-1$

$b_n = W - \alpha_n =$ uncracked ligament of the specimen after n-cycles

Thus, at each unloading stage one may assign an effective crack length α_{eff}. In general the total energy consumed for crack propagation from α to $\alpha + \Delta\alpha$ consists of two contributions:

- The energy Γ to create a new crack surface ΔA

- The energy U_{ir} dissipated to the irreversible mechanisms due to material structure

The energy contribution U_{ir} under some circumstances may exceed the energy necessary for the creation of the crack surface, Γ. Thus its magnitude is among the important parameters that must be determined for the characterization of the fracture behavior of a macroscopically nonelastic material. In Fig. 4.10, which is a $P-\delta$ curve of such a material, the definition of all the energy parameters involved in the procedure of fracture characterization are presented. The assumption made is that the crack propagates in a slow and stable way due to the external energy supply and the kinetic energy of the system is negligible.

The total mechanical energy, W, given to the system for crack growth has the form

$$W = U_e + U_{ir} + \Gamma \tag{4.37}$$

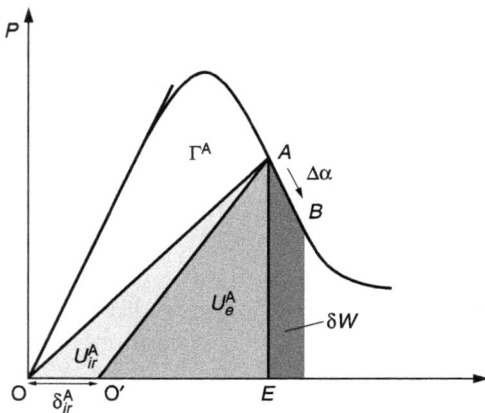

Figure 4.10 Definition of the energy parts involved in the fracture process of a linear non-elastic material using the Load-Displacement curve.

where W = the total work given to the system externally
U_e = the elastic energy
U_{ir} = the energy loss due to irreversible phenomena
Γ = the energy for the formation of a new crack surface

For quasi-static increment of crack surface $\Delta A = B\,\Delta\alpha$, G_R^*, is the elastic energy release rate, given by the rate of change of the area OAO'. As a result G_R^* is the rate of change of crack advance energy, Γ. For the elastic deformation, the strain energy component is given by

$$U_e = \frac{1}{2}P(\delta - \delta_{ir}) = \frac{1}{2}P^2\left(\frac{\delta - \delta_{ir}}{P}\right) = \frac{1}{2}P^2 C^* \tag{4.38}$$

The elastic energy release rate is therefore

$$\frac{dU_e}{da} = PC^* \frac{dP}{da} + \frac{P^2}{2}\frac{dC^*}{da} \tag{4.39}$$

According to[39] and Fig. 4.10, a fraction λ (%) of the total mechanical energy input W corresponds to the irreversible mechanisms energy U_{ir} at each crack growth stage, and is given as the ratio of the areas, OAO' (U_{ir}) and OAEO (W):

$$\lambda_i = \left.\frac{U_{ir}}{W}\right|_{\alpha_{eff}} \tag{4.40}$$

Differentiating Eq. (4.35) and taking into account Eq. (4.38), the modified potential energy release rate G_R^* is given by

$$G_R^* = \frac{(1-\lambda)}{B}\frac{P\,d\delta}{da} - \frac{PC^*}{B}\frac{dP}{da} - \frac{P^2}{2B}\frac{dC^*}{da} \tag{4.41}$$

It should be noted that the component $dP/d\alpha$ cannot be neglected since irreversible mechanisms are present, thus this quantity should be calculated for each loading–unloading cycle. Furthermore, neglecting the quantity $dP/d\alpha$ and assuming that ratio λ tends to zero, the LEFM approach is obtained as a special case of Eq. (4.41).

The energy rate Φ_{ir} coming from the nonelastic energy part is associated with the irreversible mechanisms due to development of the damage zone in the vicinity of the crack tip:

$$\Phi_{ir} = \lambda\frac{P\,d\delta}{B\,d\alpha} \tag{4.42}$$

According to the previous analysis, the experimental evaluation of the energy release rate values G_R^* and Φ_{ir} is based on the following steps:

1. Monitoring of the P–δ curve during loading-unloading cycles.
2. Calculation of δ, δ_{ir}, P, and C^* for each cycle.
3. Evaluation of effective crack increment using Eq. (4.36) for n, $n-1$ loading–unloading cycles.
4. Determination of the functions $P(\alpha_{\text{eff}})$ and $\delta\,(\alpha_{\text{eff}})$.
5. Numerical integration for the evaluation of the external work done on the system, W, and the nonelastic U_{ir} for the calculation of the energy ratio λ.

A typical loading–unloading curve from the tests performed so far for CF/PA 12 is presented in Fig. 4.11. Then a detailed analysis is conducted to calculate the load P, the crack opening displacement δ, the irreversible remaining displacement δ_{ir} at each unloading stage, and the modified compliance C^*. Based on this, the effective crack length, a_{eff}, at the nth loading–unloading cycle is evaluated using the recurrence relation of Eq. (4.36). From the data displayed in Fig. 4.12, it is very important to point out the very good agreement between the experimental measured crack length during the tests and the effective crack length obtained from the analysis described above.

Furthermore, based on the P–δ curve, the plastic dissipated energy ratio λ can be defined according to Eq. (4.40). The results obtained from the present analysis are given in Fig. 4.13.

It is evident that at the early stages of crack propagation the ratio of irreversible plastic energy dissipated to the total mechanical energy input is low, namely 10–15%. This becomes clear from the form of the P–δ curve together with the small value of the irreversible displacement δ_{ir} for each loading–unloading loop. When stable crack propaga-

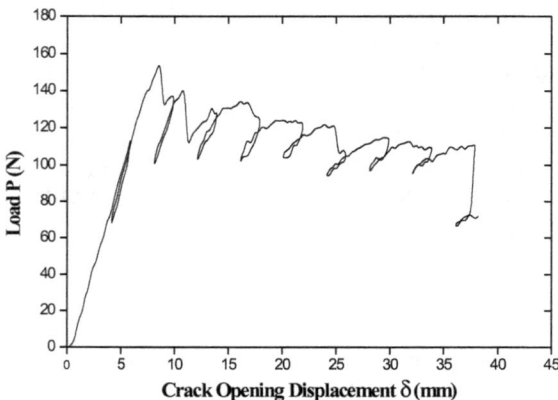

Figure 4.11 Typical Load versus crack opening displacement curve in the case of CF-PA12 DCB specimen.

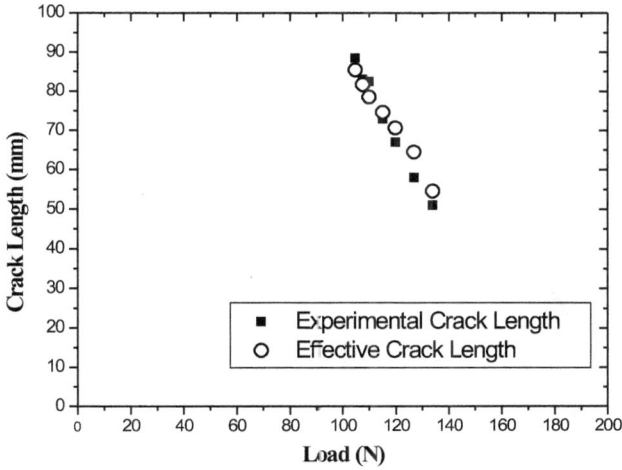

Figure 4.12 Experimentally measured and calculated effective crack length versus applied load.

tion initiates, the ratio λ increases and reaches a plateau value. This is due to the complete development of the fracture process zone, which mainly consists of cracked matrix, bridged by intact and/or failed fibers, which debond, slip, and pull out. Crack fiber bridging enhances material crack resistance by partially shielding the crack tip from the applied load. Figure 4.14 presents the elastic energy release rate G_R^* versus the effective crack extension $\Delta\alpha_{\text{eff}}$. Initially the energy

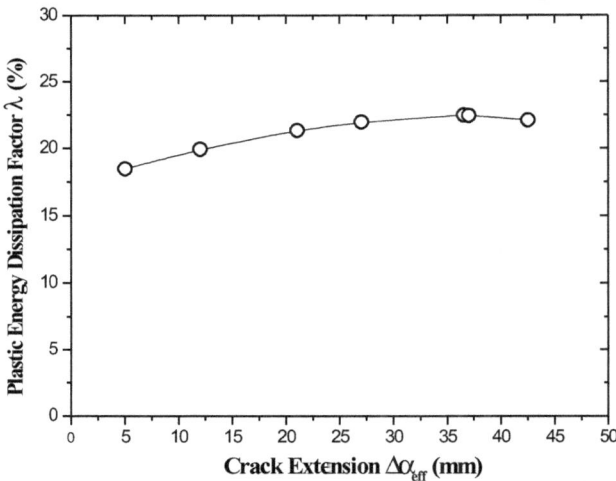

Figure 4.13 Energy dissipation rate λ versus effective crack length.

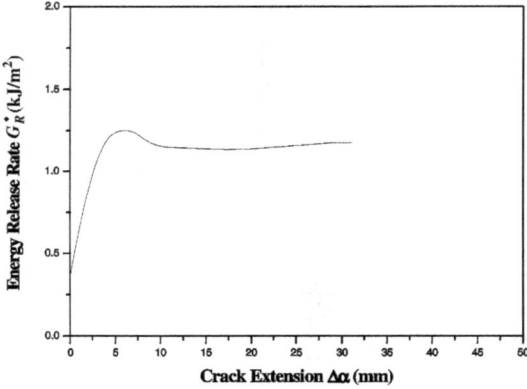

Figure 4.14 Energy release rate G_R^* versus effective crack length.

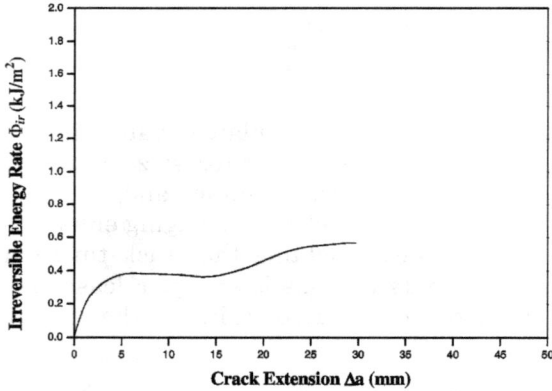

Figure 4.15 Plastic energy dissipation rate Φ_{IR} versus effective crack length.

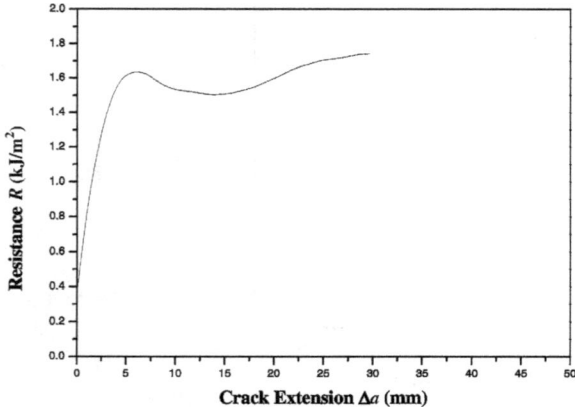

Figure 4.16 Resistance $R(= G_R^* + \Phi_{IR})$ versus effective crack length.

increases as the crack propagates for $\Delta\alpha_{\text{eff}} \cong 5\,\text{mm}$; then it remains almost constant at a plateau value of $G_R^* = 1.2\,\text{kJ/m}^2$.

In Fig. 4.15 a plot of the irreversible energy rate Φ_{ir} is given versus the effective crack extension $\Delta\alpha_{\text{eff}}$. It is shown that again there is a crack propagation period when the energy increases. Figure 4.16 shows the crack resistance R $(=G_R^* + \Phi_{ir})$ versus the effective crack extension, $\Delta\alpha_{\text{eff}}$.

4.3.3 Mode I fatigue loading

Only limited literature and test results are available for mode I tests under fatigue loading conditions.[9,40–43] Among other reasons for the limited amount of work in the field is that mode I delamination tests are mainly designed to serve quasi-static needs. In general, fatigue tests under mode I loading provide data for the fatigue crack propagation rate $(d\alpha/dN)$ as function of the variation of the energy release rate ΔG available for growth (or the applied variation stress concentration factor ΔK) at the crack tip.

Delamination growth depends upon a number of microstructural material parameters. In particular, the following general results can be derived[44]:

- For the same fiber–matrix combination, increase of the toughness of matrix material results in an improved fatigue crack propagation (FCP) resistance. However, tough composites show a relative susceptibility under fatigue loading conditions compared to their much higher quasi-static crack growth resistance.

- For a given fiber–matrix system, with increasing fiber volume fraction the FCP resistance also increases as a general trend.

- A poor fiber–matrix interface results in accelerated fatigue crack growth.

- The presence of fiber bridging enhances the FCP resistance.

- The crack orientation also plays a very pronounce role in the crack growth rate monitored during fatigue. As expected, cracks normal to the fiber direction exhibit much lower crack growth rate $(d\alpha/dN)$ compared to delamination type cracks parallel to the fiber direction for a given variation of the energy release rate available at the crack tip.

Figure 4.17 presents results for a glass fiber reinforced composite made of epoxy matrix for two different volume fractions (20% and 30%) for crack orientations parallel to the fiber direction. The remarks made above are now obvious. It is important to note here the relative decrease of FCP resistance monitored in the case of tough composites (thermoplastics) compared to their quasi-static crack growth resistance. Under fatigue conditions the energy release rate G available for the crack propagation is continuously changing. Under given fatigue conditions, G obtains its maximum and minimum values G_{max} and G_{min}, respectively, during a fatigue cycle. Thus we define the difference of $G_{max} - G_{min}$ as ΔG, or the variation of the energy release rate available for crack propagation during one loading cycle.

Using this definition of ΔG, one may see that in the case of tough composites significant propagation under fatigue may occur for $\Delta G/G_{Ic} = 0.1$, while in the case of thermosets (less tough composites) the ratio of $\Delta G/G_{Ic}$ for which fatigue crack propagation appears is about 0.5.[37]

This behavior is due to the following effects:

- Tough composites under mode I loading form a damage process zone ahead of the crack tip. The development of this area absorbs part of the energy necessary for the formation of new crack surface. In case of fatigue loading, the size of the damage zone is smaller than the size developed under quasi-static loading conditions. This makes the crack more susceptible in minor changes of the mechanical properties of local microstructure that control the crack growth process. Fatigue significantly affects the material properties in the vicinity of the crack tip, and thus matrix ductility and fiber–matrix interface become of crucial importance for the FCP resistance.

- Far-field effects, such as initiation of secondary cracks at fiber ends away from the crack front (distance up to 3 mm),[44] which may

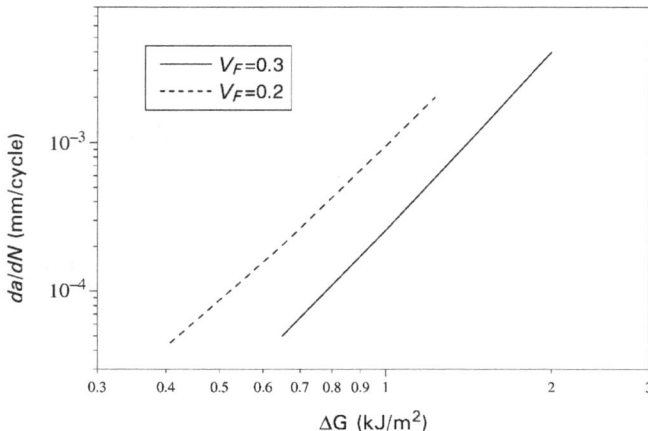

Figure 4.17 Fatigue crack propagation rate versus the applied difference of energy release rate as a function of the fiber volume fraction in a Glass fiber reinforced Epoxy.

promote the growth of secondary cracks, are dominant mechanisms of energy absorption. These mechanisms are activated under quasi-static loading and contribute to the value of G_{Ic}. However, the far-field effects are less critical for tough matrix systems since their propagation is more difficult. Thus, under fatigue loading conditions these energy absorption far-field mechanisms play an almost negligible role in the crack growth resistance.

In the following, four different models accounting for the crack growth rate under fatigue conditions are presented. All of them are of phenomenological nature and in a sense are generalizations of the Paris law in the case of anisotropic materials. They will be presented in a chronological order.

In 1985 Ramkumar and Whitcomb[40] presented, for mode I conditions, the model

$$\frac{d\alpha}{dN} = A_1 \left(\frac{Y_t^2}{E_y G_{\mathrm{Ic}}}\right)^{-1} \left(\frac{G_{\mathrm{Imax}}}{G_{\mathrm{Ic}}}\right)^{b_1} \qquad (4.43)$$

where Y_t is the tensile failure strength in the y direction of the composite laminate, E_y is the transverse modulus of elasticity of the laminate, G_{Imax} and G_{Ic} are the maximum applied and the critical energy release rates respectively, and A_1 and b_1 are the material constants that are determined from fatigue tests under mode I loading conditions at different G_{Imax} levels.

This model does not include the effect of the applied minimum energy release rate G_{min} and therefore is applicable only when $G_{\mathrm{min}} \approx 0$.

In 1987 Gustafson and Hojo[41] proposed a different model in which the effect of G_{min} is included. According to these authors:

$$\frac{d\alpha}{dN} = \frac{A_1}{(G_{\mathrm{Ic}})^{b_1}} \left(\frac{Y_t^2}{E_y G_{\mathrm{Ic}}}\right)^{-1} (\Delta G_{\mathrm{I}})^{b_1} \qquad (4.44)$$

where $\Delta G = G_{\mathrm{max}} - G_{\mathrm{min}}$, Y_t, E_y, and G_{Ic} are as defined above, and A_1 and b_1 are material constants.

In 1989 Russel and Street[42] introduced a modified approach to the problem. They proposed a simpler model of the following form:

$$\frac{d\alpha}{dN} = A_1 \left(\frac{\Delta G_{\mathrm{I}}}{G_{\mathrm{Ic}}}\right)^{b_1} \qquad (4.45)$$

where again A_1 and b_1 are material constants calculated from the fitting of the experimental data.

Finally in 1994 Dahlen and Springer[43] proposed an alternative for the determination of the crack propagation rate that combines the earlier existed models. Although the method seems more complicated, it provides reasonable agreement with experimantal data collected by different composites. Its form is

$$\frac{d\alpha}{dN} = A_1 \left(\frac{Y_t^2}{E_y G_{Ic}} \right)^{-1} \left(U \frac{G_{Imax}}{G_{Ic}} \right)^{b_1}$$ (4.46)

where $$U = \left(1 - \frac{G_{Imin}}{G_{Imax}} \right) \left[1 + \frac{G_{Imin}}{G_{Imax}} \left(1 - \frac{G_{Imax}}{G_{Ic}} \right) \right]^{u_1}$$ (4.47)

and A_1, b_1 and u_1 are material constants determined by experiment. The other parameters have been described earlier.

It is interesting to note here that in the case where $G_{Imin}/G_{Imax} \approx 0.01$ then $0.99 \leq U \leq 1$ and Eq. (4.47) reduces to the same form as that proposed by Ramkumar and Whitcomb.

All the above-mentioned works refer to mixed mode fatigue loading of delaminated specimens and in the present section we present the degenerated form, which is valid whenever pure mode I loading conditions are applied on the test specimen.

4.3.4 Miscellaneous

Tapered DCB test specimens have also been used for the measurement of the mode I delamination critical energy release rate.[45–47] The typical configuration of the tapered DCB specimen is shown in Fig. 4.18. The critical energy release rate is calculated using

$$G_{Ic} = \frac{12 P_c^2}{E_x t^3} \left(\frac{\alpha}{B} \right)^2$$ (4.48)

When the taper ratio (α/B) is constant, G_{Ic} is calculated directly using the critical load P_c and the flexural modulus of elasticity E_x.

Experiments may be performed again under monotonic or cyclic loading using displacement control. An representative crosshead velocity is 1 mm/min.

In the Table 4.2 some indicative values of G_{Ic} for DCB specimens are given. The last column of the table shows the source of data. In all of the cases the delamination is in the middle surface of the specimen and is found in between the 0° adjacent plies.

4.4 Mode II Interlaminar Fracture

Interlaminar fracture toughness under mode II loading is usually much higher than the fracture toughness under mode I loading for

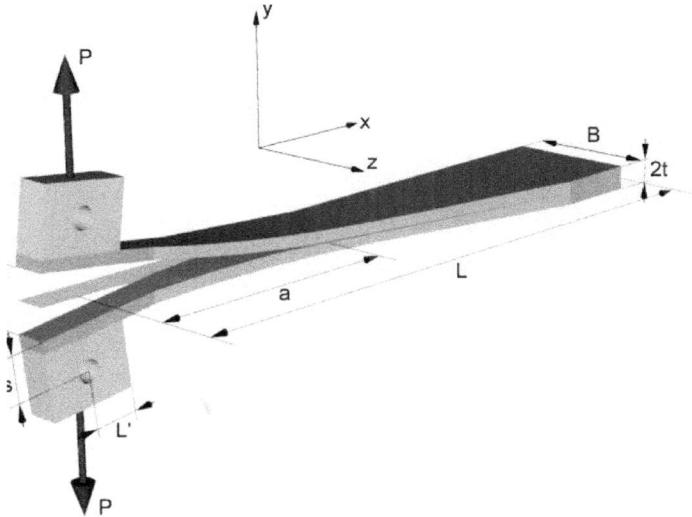

Figure 4.18 typical DCB width tapered test specimen.

TABLE 4.2   Literature Values of G_{Ic} from DCB tests

Material	Critical Energy Release Rate G_{Ic} (kJ/m^2)	Reference
CF-PEK-C	0.87	21
AS4-PA12	1.6	48
T300-PEI	1.02	49
T300-Polysulfone	1.13	49
AS4-PA (K-III)	1.8	50
IM6-PA (K-III)	1.3–1.7	50
AS4-PA (9)	2.72	51
AS4-PPS	0.82	52
AS4-PEEK	2.41–2.89	53
IM6-PEEK	2.5–3.3	54
AS4-LEXAN	1.6	55
Aramide fiber-EP	2.26	Unpublished data
AS4-3501-6	0.175	56
AS4-3502	0.19	55
AS4-2220-3	0.238	57
AS4-DOW P6	0.16	55
AS4/1908	0.321	60
T300-S208	0.1	56
T300-914	0.185	59
T300-3100	0.17	58
T300-Hx205	0.38	56
T300-BP907	0.38	55
Aramide fiber/PEI	1.53	Unpublished data

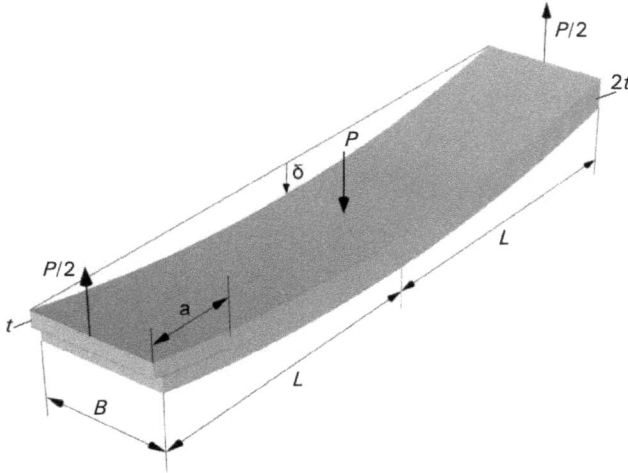

Figure 4.19 Schematical representation of the End Notched Flexure (ENF) laminate test specimen.

the same composite system. Since damage tolerance of composites is mainly a fracture toughness-driven behavior, it has been shown that mode II delamination resistance is the dominant property, for larger delaminations, as far as damage tolerance is concerned. Damage tolerance in composites is basically characterized using compression after impact (CAI) tests applied on laterally impacted panels. During these tests a very good correlation between the residual compressive strength after impact and the mode II interlaminar fracture toughness has been demonstrated.[61] Due to its critical importance for the damage tolerance, mode II interlaminar toughness has received considerable attention.

Several test configurations have been proposed for the measurement of the mode II interlaminar toughness including the end-notched flexure (ENF) test[62] and the end-loaded split (ELS) laminate test,[63] which are the most popular.

The ENF test configuration is shown in Fig. 4.19. It consists of a typical three-point bending test for a precracked specimen. The specimen is subjected to transverse shear and flexural loading that generates a shear crack driving force at the crack tip region.

At the interface, between the upper and the lower point of the delaminated arms, a tensile stress field exists at the lower surface of the upper arm and a compressive stress field exists at the upper surface of the lower arm due to bending. This stress discontinuity produces an interlaminar shear driving force for delamination propa-

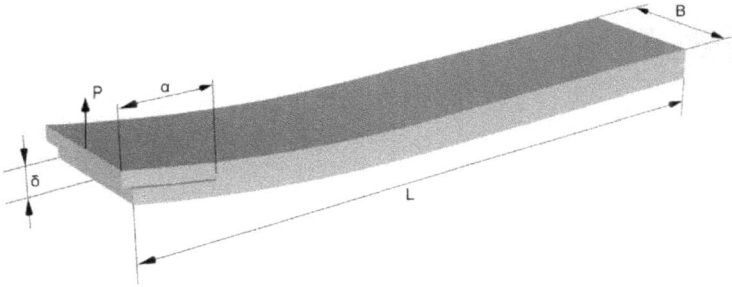

Figure 4.20 Schematical representation of the End-Loaded Split (ELS) laminate test specimen.

gation under mode II conditions. The ELS test is presented in Fig. 4.20. It also uses a cantilever delaminated composite beam loaded at the cracked end. Again the stress discontinuity produced due to bending at the cracked interface produces an interlaminar shear driving force.

In the present work both test configurations will be presented analytically. However, it is important to notice that ENF test has recently gained an advantage in its application since it has been proposed as the standard test specimen for mode II interlaminar fracture toughness characterization, although the ELS specimen offers the advantage of similar experimental set up for mode I and mode II fracture toughness measurements.

Other test configurations have been also proposed. Among them is the Arcan test method shown in Fig. 4.21 as modified,[64] which permits the pure mode I, mode II, and mixed mode interlaminar fracture studies using the same test jig and the same specimen geometry. However, because of intrinsic test problems (unstable crack propagation), the method is not widely used.

4.4.1 Edge-notched flexure (ENF) test

The application of the classical beam theory analysis as proposed[65] is valid only in the case of materials with large shear rigidity and relatively small thickness to crack length (t/α) ratio. In this case the middle point deflection δ of the beam is calculated as

$$\delta = \tfrac{1}{2}(\delta_1 + \delta_2 + \delta_3) \tag{4.49}$$

where δ_1, δ_2, and δ_3 are shown in Fig. 4.22. Due to the presence of the crack, the larger deflection is no longer at the middle point of the beam (where the load is applied) and the slope at the middle point is not zero. Thus, the beam is turned around (A) in such a way that the slope

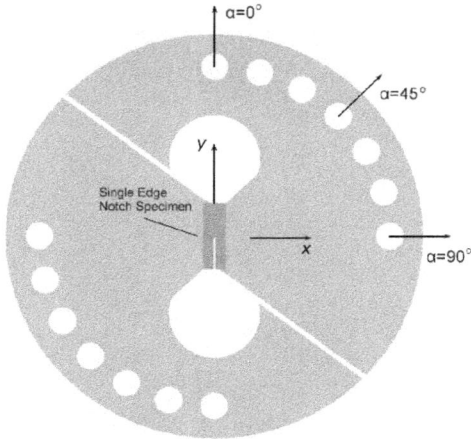

Figure 4.21 Schematical representation of the modified Arcam test specimen.

at the middle point becomes zero. Then we define δ_1, δ_2, and δ_3, which are calculated using classical beam theory analysis and assuming that beams $C'A$ and $C'D$ are cantilever beams at C'. The compliance in this case was found to be

$$C = \frac{2L^3 + 3\alpha^3}{8E_x Bt^3} \tag{4.50}$$

Taking into consideration the shear compliance of the beam and applying Timoshenko's beam theory, one obtains for the compliance:

$$C = \frac{2L^3 + 3\alpha^3}{8E_x Bt^3}\left[1 + \frac{2(1.2L + 0.9\alpha)t^2 E_x}{(2L^3 + 3\alpha^3)G_{xz}}\right] \tag{4.51}$$

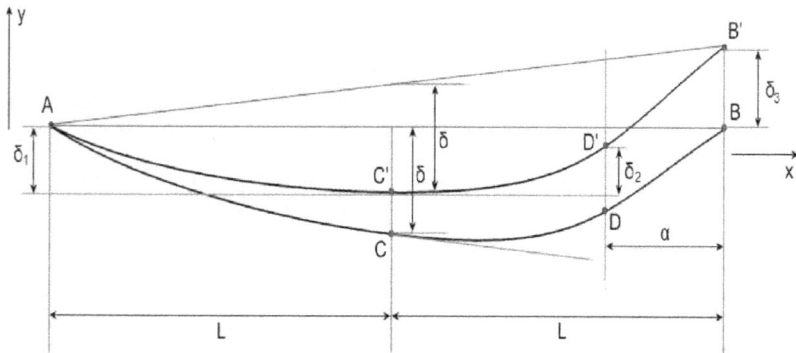

Figure 4.22 Definition of the displacement parameters in the use of ENF test specimen (classical beam theory analysis).

Using the Eq. (4.9) for the definition of the energy release rate, the mode II delamination critical energy release rate G_{IIc} is given by

$$G_{\mathrm{IIc}} = \frac{9\alpha^2 P_c^2}{16E_x B^2 t^3} \tag{4.52}$$

in the case of classical beam theory. P_c is the critical load for the crack propagation under mode II conditions. Applying the expression of compliance calculated using Timoshenko's beam theory, the critical mode II delamination energy release rate is given by

$$G_{\mathrm{IIc}} = \frac{9\alpha^2 P_c^2}{16E_x B^2 t^3}\left[1 + 0.2\frac{E_x}{G_{xz}}\left(\frac{t}{\alpha}\right)^2\right] \tag{4.53}$$

Recently, a further detailed analysis of the ENF test specimen has been proposed.[66] This analysis combines the solution of a beam problem on a generalized elastic foundation in order to incorporate the effect of crack tip compliance and a Timoshenko's beam solution of the problem to accommodate the transverse shear.

In the analytical approach to the problem, the compliance of the beam is given by[66]

$$C = \frac{2L^3 + 3\alpha^3}{8E_x Bt^3}\left\{1 + \left[2 + 3\left(\frac{\alpha}{L}\right)^3\right]^{-1}\left[\frac{2}{\kappa}\frac{E_x}{G_{xz}}\left(\frac{t}{L}\right)\right.\right.$$
$$\left.\left. + 4\left(\frac{\alpha}{L}\right)^3\frac{3}{2}\Psi\left(\frac{t}{a}\right) + 3\Omega\left(\frac{t}{\alpha}\right)^2 - \Gamma\left(\frac{t}{\alpha}\right)^3\right]\right\} \tag{4.54}$$

where κ is the Timoshenko shear coefficient and accounts for the nonuniform shear distribution over the beam cross section. For isotropic beams with rectangular cross section subjected to static load, $\kappa = 5/6$. However, for an orthotropic beam this value must be increased to $\kappa = 12$, although the compliance is not very sensitive to κ.

The parameters Ψ, Ω, and Γ are given by the expressions

$$\Psi = \frac{2}{\xi}\left[\left(2\frac{e^2}{\xi} - 1\right)\frac{\beta}{s} + e\right] \tag{4.55}$$

$$\Omega = \left[2\frac{e}{\xi^2} + \frac{1}{6}\left(\frac{1}{2} - \frac{e^2}{\xi}\right)\frac{1}{\kappa(e + \mu_0)}\frac{E_x}{G_{xz}}\right]\frac{\beta}{s} + \frac{1}{\xi} \tag{4.56}$$

$$\Gamma = \frac{1}{2\kappa}\frac{e}{\xi(e + \beta)}\frac{\beta}{s}\frac{E_x}{G_{xz}} \tag{4.57}$$

and

$$s = \left[\frac{(\beta - \nu_0)}{(e + \mu_0)} - 2e\frac{\beta}{\xi}\right] \tag{4.58}$$

Additionally, the quantities λ, γ and η exhibit the forms

$$\lambda_0 = \frac{n}{\kappa}\frac{E_y}{G_{xz}} \tag{4.59}$$

$$\gamma_0 = \frac{1}{12\kappa}\frac{E_y}{G_{xz}}\left(\frac{n}{n+1}\right) \tag{4.60}$$

$$\eta = \left[\frac{n}{12\kappa^2}\frac{E_xE_y}{G_{xy}^2} - \frac{2}{3}\right]\left(\frac{n}{n+1}\right) \tag{4.61}$$

where n is defined as a foundation parameter. Since here the assumption of elastic foundation has been adopted, $n = 2$. Now whenever

$$\lambda + \frac{1}{3\gamma_0} < 2\sqrt{\frac{\lambda_0}{\gamma_0}}$$

the parameters e, β, μ, v, and ξ are defined as

$$e = \frac{1}{t}\sqrt{\frac{1}{2}\sqrt{\frac{\lambda_0}{\gamma_0}} + \frac{1}{4}\left(\lambda_0 + \frac{1}{3\gamma_0}\right)} \tag{4.62}$$

$$\beta = \frac{1}{t}\sqrt{\frac{1}{2}\sqrt{\frac{\lambda_0}{\gamma_0}} - \left(\frac{1}{4} + \frac{1}{3\gamma_0}\right)} \tag{4.63}$$

$$\mu_0 = (\eta e + 3\gamma_0 e\beta^2 - \gamma_0\alpha^3) \tag{4.64}$$

$$v_0 = (3\gamma_0 e^2\beta - \gamma_0\beta^3 - \eta\beta) \tag{4.65}$$

$$\xi = (e^2 - \beta^2 - \lambda_0) \tag{4.66}$$

In the case where

$$\lambda_0 + \frac{1}{3\gamma_0} > 2\sqrt{\frac{\gamma_0}{\lambda_0}}$$

then e keeps the same form given in (4.62) while

$$\beta = \frac{1}{t}\sqrt{\frac{1}{4}\left(\lambda_0 + \frac{1}{3\gamma_0}\right) - \frac{1}{2}\left(\frac{\lambda_0}{\gamma_0}\right)} \tag{4.67}$$

and whenever

$$\lambda_0 + \frac{1}{3\gamma_0} = 2\sqrt{\frac{\lambda_0}{\gamma_0}}$$

the parameters e and b obtain the form

and
$$e = \frac{1}{t}\sqrt{\frac{1}{2}\left(\lambda_0 + \frac{1}{3\gamma_0}\right)} \qquad \beta = 0 \qquad (4.68)$$

and the expressions for μ, v, and ξ are not necessary for the determination of Ψ, Ω, and Γ.

In this case the critical mode II delamination energy release rate is obtained by

$$G_{IIc} = \frac{9}{16}\frac{P_c^2\alpha^2}{E_{xx}B^2t^3}\left\{1 + \frac{4}{3}\left[\Psi\left(\frac{t}{\alpha}\right) + \Omega\left(\frac{t}{\alpha}\right)^2\right]\right\} \qquad (4.69)$$

Another very interesting approach for the analysis of the ENF test instead of beam theory is the shear deformation plate theory. According to this approach, which has been proposed by Carlsson and Gillespie,[67] the beam is divided into four different regions, namely, the crack region, the singularity region, the left-beam theory region and the right-beam theory region. Applying appropriate boundary conditions at the interfaces and assuming that the length of the singularity region is λ_0 given as a function of the geometrical and material property by

$$\lambda_0 = wt\sqrt{\frac{E_x}{G_{xz}}} \qquad (4.70)$$

where ω is a geometrical parameter calculated using FE analysis then the middle-point compliance is given by the expression

$$C = \frac{1}{8Bt^3E_x}\left\{2L^3\left[1 + \frac{6}{5}\frac{E_x}{G_{xz}}\left(\frac{t}{L}\right)^2\right] + 3\alpha^3\left[1 + \frac{\lambda_0}{2\alpha}\left(1 + \frac{\lambda_0}{5\alpha}\right)\right]\right\} \qquad (4.71)$$

As one may see in Eq. (4.71), the middle-point compliance is in accordance with the results provided by the simple beam theory in Eq. (4.50) but it contains different correction terms for the length L and for the crack length α, which is not the case either in Eq. (4.51) or in Eq. (4.54), where the correction term is the same.

In this case the critical mode II delamination energy release rate is given by

$$C_{IIc} = \frac{9\alpha^2P_c}{16E_xB^2t^3}\left[(1 + 1.15)\left(\frac{\alpha}{L}\right)^3\right]^{-1}$$

$$\times \left\{1 + 1.2\left(\frac{E_x}{G_{xz}}\right)\left(\frac{t}{L}\right)^2 + 1.5\left(\frac{\alpha}{L}\right)^3\left[1 + \frac{\lambda_0}{2\alpha}\left(1 + \frac{\lambda_0}{5\alpha}\right)\right]\right\} \qquad (4.72)$$

Several proposals have been made for the analysis of ENF test using higher-order beam theory. Introducing higher-order theories there is no longer need for shear correction factors.

According to this approach the beam is divided in four regions and appropriate boundary conditions are assumed. However, in all such formulations no singularity region is considered.[67–69] A typical result for the middle-point compliance calculated based on the higher-order beam theory is given in[69] the form

$$C = \frac{1}{8E_x B}\left[2\bar{L}^3 + 3\bar{\alpha} + \frac{704\bar{L} + 51\bar{\alpha}}{340}\frac{E_x}{G_{xz}} + \frac{9\bar{\alpha}}{\beta_0^2}(1 + \beta_0\bar{\alpha})\right] \qquad (4.73)$$

where $\qquad \bar{L} = \dfrac{L}{t}, \qquad \bar{\alpha} = \dfrac{\alpha}{t} \qquad$ and $\qquad \beta_0 = 4\sqrt{\dfrac{14}{5}\dfrac{G_{xz}}{E_x}}$

and the G_{IIc} is given by the expression

$$G_{IIc} = \frac{9P_c\bar{\alpha}^2}{16E_x B^2 t}\left[1 + \frac{2}{\beta_0\bar{\alpha}} + \frac{1}{60\beta_0^2\bar{\alpha}^2}\left(\beta_0^2\frac{E_x}{G_{xz}} + 60\right)\right] \qquad (4.74)$$

The effect of friction.   In all the alternatives, for the calculation of the compliance presented above, the effect of friction was ignored. However, as is evident when the load is applied to the ENF test specimen, the crack surfaces are in contact. The friction between the crack flanks dissipates energy and increases the critical mode II delamination energy release rate. As the delamination propagates, the energy dissipated due to friction may increase. Thus the monitored mode II delamination critical energy release rate is an apparent quantity that contains two parts, namely, the energy release rate due to the material and the contribution of the friction[26,65]:

$$G_{IIc}^{ap} = G_{IIc} + \frac{3P_c^2\mu\alpha}{4EB^2 t^2} \qquad (4.75)$$

where μ is the friction coefficient. In Eq. (4.75), the expression for G_{IIc} resulting from the classical beam theory and given in Eq. (4.52) was used. Assuming $\alpha/t = 25$ and $\mu < 0.3$, the estimated error due to friction is less than 1.5%. It has been shown[70] that the contribution of frictional effects is negligible in the case of long delamination cracks ($\alpha/L \geq 0.5$) when $\mu < 0.3$.

Data reduction schemes.   It is necessary to calibrate the compliance experimentally in the case of ENF test for each test specimen. For this reason, the specimen is loaded and unloaded in the elastic range by sliding the specimen between the loading points in order to vary the crack length. Crack lengths of 0, 15, 20, 25, 30, 35, and 40 mm have to be used. During this procedure the load–displacement curves are obtained.

The shear precrack of the specimens is the next step. Using a crack length $\alpha = 40$ mm by appropriate mounting of the test coupon into the test rig, the specimen is loaded until the load starts to drop. At this stage the specimen is unloaded, and when the crack is arrested the crack tip is marked.

Finally, by relocating the specimen into the test ring, a ratio of $\alpha/L = 0.5$ is maintained. Then the specimen is loaded again and when the crack propagation is reinitiated and the load starts to drop, the specimen is unloaded.

Beam method. Accoding to this approach, the energy release rate given in Eq. (4.52) may be written as

$$G_{\text{IIc}} = \frac{9P_c^2\alpha^2}{2B(2L^3 + 3a^3)}C(\alpha) \tag{4.76}$$

Then knowing the expression for compliance $C(\alpha)$ and calculating it for a given crack length α, the mode II delamination critical energy release rate is calculated.

However, using the classical beam theory models it is evident that G_{IIc} is underestimated since transverse shear effects are not included. A correction procedure has been proposed[72] using finite element calibration and determining the necessary correction factors.

Compliance calibration. The expression for compliance given in Eq. (4.50) for the ENF test may be rewritten in the form

$$C(\alpha) = \frac{2L^3}{8E_xBt^3} + \frac{3\alpha^3}{8E_xBt^3} \tag{4.77}$$

or
$$C(\alpha) = d_0 + d_1\alpha^3 \tag{4.78}$$

where d_1 is the slope of the compliance curve determined from the curve fitting of the compliance versus α^3 plot.

In this case the mode II delamination critical energy release rate is given by

$$G_{\text{IIc}} = \frac{3P_c^2\alpha^2}{2B}d_1 \tag{4.79}$$

where again P_c is the critical load that causes the propagation of delamination. Taking into consideration the fact that the crack length is a numerator term of the form α^2, the errors in the measurement of the delamination length produce more extensive data scattering in the values of G_{IIc} compared to the relative values of G_{Ic} calculated using an analogue procedure.

Area method. In all the data reduction methods previously described, the critical energy release rate was calculated using the measured critical load P_c at the onset of crack growth and the crack length a prior to crack extension. Therefore, G_{IIc} is a measure of the energy required to initiate crack propagation. However, as the crack propagates, G_{IIc} does not necessarily remain constant; in this case a propagation and an arrest value of G_{IIc} can arise. In the area method, an alternate approach for evaluating fracture toughness, G_{II} is interpreted as the energy required to create a new cracked surface area. G_{II} can then be given by

$$G_{II} = \frac{U}{b\,\Delta\alpha} \tag{4.80}$$

where U is the area between the load–deflection curve for loading and unloading of a very small change in crack length, $\Delta\alpha$, as shown in Fig. 4.23(a). An important advantage of this method is that only elastic material behavior is required to predict G_{II}. Therefore, for geometrically nonlinear and/or nonlinear elastic material responses, this method gives an average energy release rate for the observed crack extension $\Delta\alpha$. For unstable crack growth, it gives an average value for G_{IIc} that typically falls between the initiation and arrest values measured for linear behavior. The load–displacement record should return to the origin to guarantee that no significant far-field deformation of damage is included in U in Eq. (4.80). The inclusion of energy dissipation in far-field damage in the fracture energy term U in Eq. (4.80) would give an erroneously high estimate of G_{IIc}. This is the case presented in Figure 4.23(b) where nonelastic deformation is shown. For materials in which G_{II} is independent of crack growth rate and crack growth distance (that is, systems with minimal fiber bridging and/or plastic wake), the average and initiation G_{II} should be identical.

4.4.2   End-loaded split (ELS) test

The second frequently used test sample for mode II delamination growth experiments is the end-loaded split test. As was mentioned earlier, the main advantage of this test is that one may use similar test rig as in the case of the DCB test; however, different boundary conditions must be applied at the supported edge. Applying classical beam theory and assuming a cantilever beam configuration with load

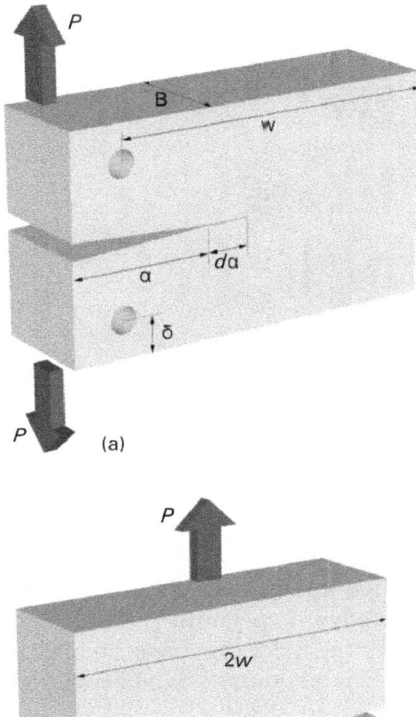

Figure 4.23 Load-Displacement curves for ENF specimens a) typical curve for an elastic material, b) non-elastic permanent deformation due to irreversible for field damage.

application at the free edge, as is shown in Fig. 4.20, the beam compliance is obtained by[72]

$$C(\alpha) = \frac{L^3 + 3\alpha^3}{2E_x Bt^3} \tag{4.81}$$

and the mode II delamination critical energy release rate is given by

$$G_{IIc} = \frac{9P_c \alpha^2}{4E_x B^2 t^3} \tag{4.82}$$

where again P_c is the critical load for crack propagation.

Equations (4.81) and (4.82) are the ideal linear elastic approach to the problem where the P–δ curve is assumed to be linear up to fracture. In practice this assumption is likely not to be satisfied either due to the material ductility or due to large deflection required to initiate the fracture.

Corrections both for the test specimen length and the crack length have been proposed[73,74] in an analog way to the corrections proposed in the case of DCB specimen.

According to[74] the expression for the compliance exhibits the form

$$C = \frac{3(\alpha + b_1 t)^3 + (L + 2b_2 t)^3}{2E_x B t^3}$$ (4.83)

where b_1 and b_2 are given by

$$b_i = \left\{ \left(\frac{1}{18 K_i} \frac{E_x}{G_{xz}} \right) \left[3 - 2 \left(\frac{\Gamma}{1 + \Gamma} \right)^2 \right] \right\}^{1/2}, \qquad i = 1, 2$$ (4.84)

where

$$\Gamma = \frac{1}{K} \frac{G_{xz}}{(E_x E_y)^{1/2}}$$ (4.85)

and for the calculation of b_1 b_2 the quantities of $18K_1 = 63$ and $18K_2 = 11$ are given, respectively.

The main problem in the ELS specimen configuration is the application of the load, which is along the lines of the DCB test and is applied to an end-block that is fixed at the beam surface. Thus the fixed edge of the specimen may be designed in such a way that although it applies $\delta = 0$ and $d\delta/dx = 0$ at the support edge, it permits small horizontal movement, since under typical fixed support of the edge the horizontal distance between the load line and the clamped end shortens as the specimen deflects.

Another interesting modification has been proposed[75,76] keeping the distance between the load line and the clamped edge fixed. However, for large deflections the additional constraint may affect significantly the P–δ curve.

The interested reader may find a nonlinear approach to the problem.[77]

Data reduction schemes. The expressions of the energy release rate and the compliance for the ELS test are analogous to those used in the case of ENF test and thus the data reduction schemes are along the same lines as the ones presented earlier.

Beam method. According to this approach the energy release rate given in Eq. (4.82) can be written in the form

$$G_{IIc} = \frac{9 P_c^2 \alpha^2}{2B(L^3 + 3\alpha^3)} C(\alpha)$$ (4.86)

Once the compliance $C(\alpha)$ is measured, the critical energy release rate for mode II delamination growth can be determined directly using Eq. (4.86) at the critical load P_c that causes delamination growth.

Compliance calibration.   The expression compliance in Eq. (4.81) for the ELS test may be rearranged in the following form:

$$C(\alpha) = \frac{L^3}{2E_x B t^3} + \frac{3\alpha^3}{2E_x B t^3} \tag{4.87}$$

$$C(\alpha) = d_0 + d_1 \alpha^3 \tag{4.88}$$

Plotting compliance data versus α^3 and fitting the data in a linear manner, both constants d_0 and d_1 are determined.

Then G_{IIc} is given by

$$G_{\mathrm{IIc}} = \frac{3 P_c^2 \alpha^2}{2B} d_1 \tag{4.89}$$

where again P_c is the critical load that causes the propagation of a delamination of length α.

At this point and before the presentation of the mode II fatigue delamination growth of this chapter, a very brief presentation of mixed mode delamination growth will be given.

Very frequently in composites, delamination grows under mixed mode I + II loading conditions and this represents a more realistic loading configuration. Several test specimens for the mixed mode have been proposed in the literature, but practical limitations appear in their use. In fact, the range of mixed mode ratio, G_{II}/G_T, (where the subscript T denotes the total energy release rate) is of critical importance. Some of the proposed test specimens such as the asymmetrical load double cantilever beam specimen,[78] the mixed mode flexure specimen[75,79] and cantilever beam opening notch specimen[80,81] do not permit large mixed mode ratios, while others such as cracked lap shear[82] and edge delamination tension[83] permit easy calculation of the total energy release rate and require detailed finite element analysis in order to determine the relative amounts that G_{I} and G_{II} contribute to the total energy release rate. Among the promising test specimens is the mixed mode bending (MMB) specimen, which allows characterization of the delamination initiation and growth for any value of G_{II}/G_T modal ratio.[84]

A typical configuration of the MMB test specimen and the related test ring are given in Fig. 4.24. The load is applied on a delaminated beam specimen using a lever–fulcrum system. As the distance e between the point of load application and the fulcrum varies, the ratio G_{II}/G_T is controlled. When $e = L$ pure mode I loading is applied, while $e = 0$ results pure mode II loading of the delaminated beam.

Equation (4.9) will be used for the calculation of the critical mixed mode energy release rate. Then using the compliance calibration data

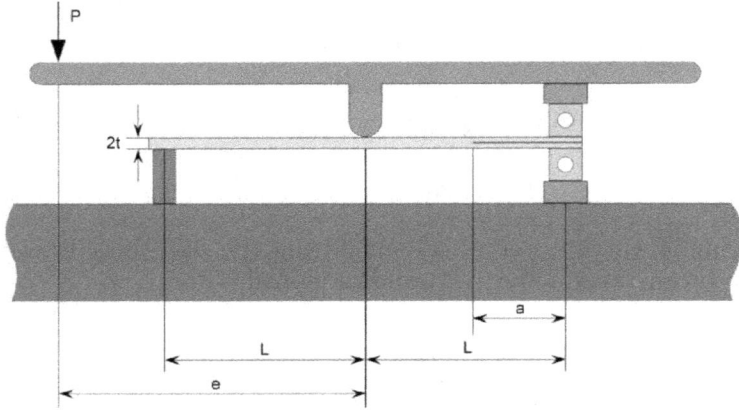

Figure 4.24 Schematical representation of MMB apparatus.

reduction scheme presented earlier, one may determine the coefficients d_0 and d_1 of the compliance expression

$$C(\alpha) = d_0 + d_1\alpha^3 \tag{4.90}$$

from the interpolation of the experimental data for compliance presented in a plot $C(\alpha)$ versus α^3. The expression for the mixed mode critical energy release rate is

$$G_{Tc} = \frac{3P_c^2\alpha^2}{2B}d_1 \tag{4.91}$$

As a general remark one must note that when increasing the contribution of mode II to the total energy release rate, the mixed mode critical energy release increases as well.

According to the recent literature, based on a semi-empirical expression, one may calculate the total critical energy release rate based on the results of mode I and mode II critical energy release rates. This works in the direction of a "mixed mode failure criterion," which communicates the criticality of a mixed mode loading level for a given material. The following expression has been proposed:

$$G_{Tc} = G_{Ic} + (G_{IIc} - G_{Ic})\left(\frac{G_{II}}{G_T}\right)^m \tag{4.92}$$

where the exponent m is a material parameter, which in the case of CFRP was found to be $m = 1.557$ and in the case of GFRP was found to be $m = 2.6$.[84]

Fatigue delamination growth. The propagation of delamination damage under fatigue loading is of critical importance for the structural integrity of a composite structure. Thus the prediction of delamination growth is very important

Delamination growth may be predicted either on the basis of experiments or by introducing models. From a designer's point of view, modeling is much more useful because it provides information on the delamination growth over a wide range of loading conditions.

In this section semi-empirical models developed for estimating the delamination growth in organic matrix composites under fatigue loading will be presented Although the subject of this section is the crack growth resistance under mode II loading conditions, the models that will be introduced, will focus on mixed mode fatigue loading. This is mainly for the completeness of the presentation.

In general, the expressions that describe the crack growth rate under fatigue loading provide the crack growth rate ($d\alpha/dN$) versus the variation of the energy release rate $\Delta G = G_{max} - G_{min}$ monitored at the extremities of a loading cycle. As mentioned, the equations that will be given in the following refer to mixed mode fatigue loading conditions. Nevertheless, one may obtain the corresponding expressions for pure mode II fatigue loading as degenerate cases. The general remarks given for fatigue delamination growth in Section 4.3 are still valid.

The four models introduced first in Section 4.3 under mode I fatigue loading will be presented in their complete form for mixed mode fatigue loading and the crack growth rate will be given over a wide variety of loading conditions.

In the following the delamination growth model proposed by Ramkumar and Whitcomb[40] is presented.

$$\frac{d\alpha}{dN} = A_1 \left(\frac{Y_t^2}{E_y G_{Ic}} \right)^{-1} \left(\frac{G_{Imax}}{G_{Ic}} \right)^{b_1} + A_2 \left(\frac{S^2}{E_y G_{IIc}} \right)^{-1} \left(\frac{G_{IImax}}{G_{IIc}} \right)^{b_2} \qquad (4.93)$$

where Y_t, E_y, G_{Imax}, G_{IIc}, A_1, and b_1 have been defined in Section 4.3, S is the longitudinal shear stress, G_{IImax} and G_{IIc} are the maximum applied and the critical energy release rate, respectively, under mode II loading conditions, and A_2, b_2 are material constants that are determined applying curve fitting to experimental fatigue crack growth data produced from fatigue tests performed under pure mode II loading conditions at different G_{IImax}max levels.

The main drawback of the above model is that it does not include the effect of the applied G_{Imin} and G_{IImin} energy release rates and thus is applicable whenever G_{Imin} and G_{IImin} are negligible.

A model that includes the effect of G_{min} in the expression of the delamination growth during mixed mode fatigue loading has been proposed by Gustafson and Hojo[41]:

$$\frac{da}{dN} = \frac{A_1}{G_{Ic}^{b_1}}\left(\frac{Y_t}{E_y G_{Ic}}\right)^{-1}(\Delta G_I)^{b_1} + \frac{A_2}{G_{IIc}^2}\left(\frac{S^2}{E_y G_{IIc}}\right)^{-1}(\Delta G_{II})^{b_2} \qquad (4.94)$$

where A_1, A_2, b_1, and b_2 are material constants calculated by curve fitting using experimental fatigue data and

$$\Delta G_I = G_{Imax} - G_{Imin}$$
$$\Delta G_{II} = G_{IImax} - G_{IImin}$$

Another model, proposed by Russel and Street,[42] provides an alternative expression for the fatigue crack growth under mixed mode loading:

$$\frac{da}{dN} = (A_1 f_1 + A_2 f_2)\left(\frac{\Delta G}{\Delta G_c}\right)^{f_1 b_1 + f_2 b_2} \qquad (4.95)$$

where ΔG is the applied mixed mode energy release rate and ΔG_c is the critical mixed mode energy release rate. A_1 and A_2 are material constants, which are determined using experimental fatigue data and

$$f_1 = \frac{G_I}{G_I + G_{II}} \qquad (4.96)$$
$$f_2 = 1 - f_1$$

Finally, Dalhen and Springer[43] proposed a more complete semi-empirical model, which distinguishes two different situations under mode II conditions, namely,

- there is no shear reversal or
- there is shear reversal.

Shear reversal is defined as the reverse displacement of the upper and the lower arms of the delaminated region, which produces reverse shear loading at the crack tip. The presence of shear reversal at the crack tip activates different failure mechanisms and for this reason the model takes this difference into account.

It is important to note here that the way of measuring the loading cycles in each case is different. When no shear reversal exists, one cycle is defined as the change from a total minimum energy release rate to a maximum value and back to a minimum. $G_{max,ns}$ and $G_{min,ns}$ are the maximum and the minimum values occurring during this cycle. On the other hand, when shear reversal exists, then a cycle is defined as the period where the energy release rate starts from a minimum value passes through a first maximum to a minimum and finally goes through a second maximum to a final minimum. The energy release rates that corresponds to the higher and the lower values during a cycle are defined as $G_{\max,s}$ and $G_{\min,s}$.[43]

The proposed expression for the evaluation of the delamination growth rate is

$$\frac{da}{dN} = \left[g_1 \left(\frac{D_y G_{Ic}}{Y_t^2} \right) + g_2 \left(\frac{E_y G_{IIc}}{S^2} \right) \right] A \left[U \left(\frac{G_{Imax}}{G_{Ic}} + \frac{G_{IImax}}{G_{IIc}} \right) \right]^b \quad (4.97)$$

where

$$g_1 = \frac{G_{Imax}}{G_{Ic}} \left[\left(\frac{G_{Imax}}{G_{Ic}} \right) + \left(\frac{G_{IImax}}{G_{IIc}} \right) \right]^{-1} \quad (4.98)$$

$$g_2 = \frac{G_{IImax}}{G_{IIc}} \left[\left(\frac{G_{Imax}}{G_{Ic}} \right) + \left(\frac{G_{IImax}}{G_{IIc}} \right) \right]^{-1} \quad (4.99)$$

$$A = (A_1)^{g_1} + (A_2)^{g_2} \quad (4.100)$$

$$b = g_1 b_1 + g_2 b_2 \quad (4.101)$$

U in the presence of shear reversal is given by

$$U_s = \frac{G_{min,s}}{G_{max,s}} + \frac{G_{max,s}}{G_c} \left(1 - \frac{G_{min,s}}{G_{max,s}} \right)^{g_2} \quad (4.102)$$

while, when no shear reversal exists,

$$U_{ns} = \left(1 - \frac{G_{min,ns}}{G_{max,ns}} \right) \left[1 + \frac{G_{min,ns}}{G_{max,ns}} \left(1 - \frac{G_{max,ns}}{G_c} \right) \right]^u \quad (4.103)$$

and

$$u = u_1 g_1 + u_2 g_2 \quad (4.104)$$

In the above expressions for U the terms $G_{\max}/G_c$ and $G_{\min}/G_{\max}$ are defined as

$$\frac{G_{max}}{G_c} = \frac{G_{Imax}}{G_{Ic}} + \frac{G_{IImax}}{G_{IIc}} \quad (4.105)$$

and

$$\frac{G_{min}}{G_{max}} = \frac{G_{Imin}}{G_{Imax}} + \frac{G_{IImin}}{G_{IImax}} \quad (4.106)$$

The coefficients A_1, A_2, b_1, b_2, u_1, and u_2 are material parameters, which are determined from cyclic tests performed under mode I and mode II delamination growth.

Miscellaneous. The specimen width for tests of this kind (both ENF and ELS) is 20 mm. Additionally, the thickness of the specimen and the support span is chosen so as to avoid large displacement and transverse shear effects, although correction coefficients may be used.

TABLE 4.3   Literature Values of G_{Ic} from ENF and ELS tests

Material	Test specimen	Critical Energy Release Rate G_{IIc} (kJ/m^2)	Reference
CF-PEEK	ENF	2.32	77
AS4-PEEK	ENF	1.11	59
AS4-PEEK	ELS	1.78	59
CF-PE1	ENF	1.88	Unpublished data
CF-PE1	ELS	1.59	Unpublished data
AF-PE1	ENF	1.313	Unpublished data
AF-PE1	ELS	1.62	Unpublished data
AS4 Dow P6	ENF	1.75	55
AS4 Dow P4	ENF	0.8	55
AS4-3501-6	ENF	1.15	55
AS4-Lexan	ENF	1.7	55
T300-985	ELS	0.697	57
T300-5208	ENF	0.865	85
T300-5208	ELS	0.716	57
T300-BO907	ENF	2.627	85
T300-BP907	ELS	1.423	57
Glass-DGEBA	ENF	1.715	59
Glass-DGEBA	ELS	2.11	59
AF-EP	ENF	1.437	Unpublished data
AF-EP	ELS	2.11	Unpublished data

Typical specimen thickness is of about 3 mm (24 plies), while the composites have unidirectional lay-up with the fiber reinforcement at 0°. The length is nominally 165 mm. A typical support span of 100 mm ($2L = 100$ mm) is used, while the crack length to half-span ratio is $\alpha/L = 0.5$ for all the mode II interlaminar fracture tests.

In this case, again, a precrack is necessary, which is accomplished by the insertion of a starter film of nominal thickness 7–13 μm during the manufacturing of the plate at a given position. The films should be sprayed with a mold release agent before lay-up. The cross-head velocity of the testing machine during the test must be 0.5 mm/min. A minimum of five specimens must be tested to obtain the value of G_{IIc}. In Table 4.3 some indicative results of G_{IIc} measured either using ELS or ENF specimens are given. The last column of the table shows the source of data.

4.5   Mode I Loading of Notched Composites

For the characterization of the mode I fracture toughness of notched composite laminates normal to the direction of reinforcement, two different test specimens are mainly used: the compact tension (CT) test specimen and the double edge notched (DEN) one. Both are presented in Fig. 4.25.

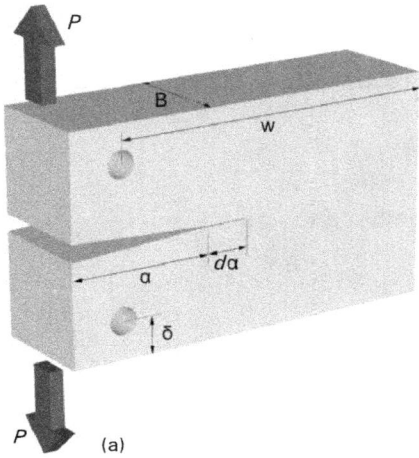

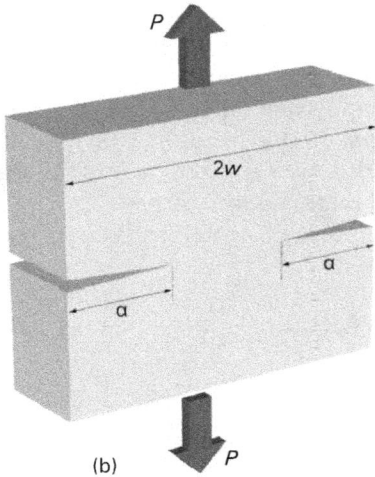

Figure 4.25 Sehcmatical representation of the test specimen used for fracture toughness measurements of notch composites normal to the plane of reinforcement. (a) Compact tension specimen; (b) double edge notch specimen.

The stress intensity factor K_I for the case of CT specimen is given by

$$K_I = \frac{P}{B\sqrt{w}} F_I\left(\frac{\alpha}{w}\right) \tag{4.107}$$

where $F_I(\alpha/w)$ is a geometrical shape function given by

$$F_I\left(\frac{\alpha}{w}\right) = \left(1 - \left(\frac{\alpha}{w}\right)\right)^{-3/2}\left[\left(2 + \left(\frac{\alpha}{w}\right)\right)\left(0.886 + 4.64\left(\frac{\alpha}{w}\right) - 13.32\left(\frac{\alpha}{w}\right)^2\right.\right.$$
$$\left.\left. + 14.72\left(\frac{\alpha}{w}\right)^3 - 5.6\left(\frac{\alpha}{w}\right)^4\right)\right] \tag{4.108}$$

for the case of an isotropic material.

The expressions (4.107) and (4.108) provide an accuracy of 0.5% for $\alpha/w > 0.2$.

The stress intensity factor for the case of DEN tension specimen is given by

$$K_{\mathrm{I}} = \sigma\sqrt{\pi\alpha}F_{\mathrm{I}}\left(\frac{\alpha}{w}\right) \qquad (4.109)$$

where $F_{\mathrm{I}}(\alpha/w)$ is again a geometrical shape function given by

$$F_{\mathrm{I}}\left(\frac{\alpha}{w}\right) = \left(1 - \left(\frac{\alpha}{w}\right)^{-1/2}\right)\left[1.122 - 0.56\left(\frac{\alpha}{w}\right) - 0.205\left(\frac{\alpha}{w}\right)^2\right.$$
$$\left. + 0.471\left(\frac{\alpha}{w}\right)^3 - 0.19\left(\frac{\alpha}{w}\right)^4\right] \qquad (4.110)$$

for the case of an isotropic material.

The expressions (4.109) and (4.110) provide an accuracy of 0.5% for any value of α/w.

However, in the case of composite materials the above expressions are not valid, since now the mechanical behavior of the material in the two-dimensional subspace is described by the constants E_x, E_y, G_{xy}, and v_{xy}. Therefore, a phenomenological approach for incorporating the material anisotropy in the expressions (4.107) and (4.108) is the modification of the geometrical shape function in order to introduce in the above equations the effect of the material anisotropy.

The definition of the anisotropy parameters c_1 and c_2 is given by the formulas

$$c_1 = \frac{E_y}{E_x} \quad\text{and}\quad c_2 = \frac{E_y}{G_{xy}} - 2v_{xy} \qquad (4.111)$$

The geometrical shape functions F_{I} are no longer functions of (α/w) but assume the following form:

$$F_{\mathrm{I}} = F_{\mathrm{iso}}\left(\frac{\alpha}{w}\right)F_{\mathrm{orth}}\left(\frac{\alpha}{w}, c_1, c_2\right) \qquad (4.112)$$

where $F_{\mathrm{iso}}(\alpha/w)$ is the same as in the case of isotropic material and is given by Eqs. (4.108) and (4.110) for the CT and the DEN specimens, respectively, and $F_{\mathrm{orth}}(\alpha/w, c_1, c_2)$ represents the so-called normalized factor of orthotropy.[86]

Furthermore, in the case of notched composites, where the notch is normal to the fiber reinforcement, the crack advance produces a fracture process zone, which induces stress redistribution and partial shielding of the crack tip from the applied load. The shape and size of the damage process zone depend upon the constituent materials of the composite, the form of the reinforcement (UD lamina, woven fabric, etc.), as well as the geometrical factors related to the test coupon.

The phenomenon is much more pronounced in the case of ceramic matrix composites where the size of the damage process zone is greater compared to that developed in the case of organic matrix composites. In all these cases the material behavior is much better given in the form of an R-curve instead of a fracture toughness (K_{Ic}) parameter.

Next the fracture behavior of a notched 2-D carbon/carbon composite will be investigated using CT specimens ($W = 40$ mm). The R-curve behavior was calculated using both monotonic tensile loading and cyclic loading conditions. Two different material thicknesses, 2.7 mm and 10 mm, were used to examine the thickness effect on the results.

The mechanical loading was applied through a pin-and-shackle system to prevent the specimen from bending. The crosshead velocity of the testing machine was 0.1 mm/min. The crack geometry was of V-shape. The notches were prepared using a diamond saw for the initial cut, while a razor blade was used for the final cut at the crack tip.

Monotonic tensile loading of the CT specimen. Thin and thick specimens were used for the determination of the R-curve behavior of 2-D C/C composites using linear elastic fracture mechanics (LEFM) analysis, adopting the compliance calibration technique. This approach is valid whenever small-scale energy-absorbing mechanisms are present. The following procedure was adopted for building up the R-curve.

Generation of the compliance calibration curve. It is well known that a macrocrack can be defined, in the case of ceramic composites, as a material region for which the level of damage is so extensive that its rigidity is negligible. Therefore, the development of a damage zone affects the specimen compliance in the same way as crack propagation. This approach justifies the use of a compliance calibration curve $C(\alpha)$, which correlates the propagation of the damage zone to the notch lengths of specimens, cut with a thin diamond saw (0.3 mm thickness) in order to retain the same compliance.

The calibration curve allows the determination of an effective macrocrack length from the compliance measurements. In the present work, compliance calibration curves were determined using crack mouth opening measured by a crack opening displacement (COD) transducer.

Application of the LEFM approach. In both cases, of thin and thick 2-D C/C composites, the LEFM was applied for the calculation of the strain energy release rate according to the well-known relation:

$$G_R = \frac{P_c^2}{2B}\frac{dC}{da} \qquad (4.113)$$

where P_c is the fracture load at each crack increment, B is the specimen thickness, and $dC/d\alpha$ is the slope of the compliance calibration curve at that point, which is described by the calculated crack length. In addition, the material fracture toughness K_{Ic} is calculated using the expressions (4.107) and (4.108).

Finally, as it is well known, within the bounds of LEFM the existing relation between G_R and K_{Ic} is given by

$$G_R = \frac{K_{Ic}^2}{E'} \qquad (4.114)$$

where $E' = E$ under plane stress and $E' = E/(1 - v^2)$ under plane strain conditions.

Plot of the energy release rate against the effective crack length. The change of crack length during the monotonic loading was computed from the change in compliance. This method provides a safe way to obtain the elastic strain energy release rate assuming the material is damageable elastic. The R-curve is obtained by plotting the strain energy release rate versus the apparent crack increment $\Delta\alpha$.

As already stated, this kind of approach to fracture behavior of 2-D C/C materials involves remarkable uncertainties, mainly in the case of thick specimens, where the presence of "irreversible damage mechanisms" affect the resulting R-curve, thus overestimating the fracture resistance of the material. Application of the above methodology leads to the compliance calibration curves for thin and thick specimens, which are presented in Figure 4.26a and b. Figure 4.26c illustrates the normalized compliance, with respect to the specimen thickness, for both thin and thick specimens. As shown, the two normalized compliance curves are very close, as was expected since the normalized compliance is referred to the same specimen geometry and could be considered as a kind of material constant.

The application of the compliance calibration procedure to calculation of the strain energy release rate G_{Ic} leads to the results illustrated in Fig. 4.27 for thick and thin specimens. In both cases, an initial notch size of $\alpha_0 = 17$ mm was used. The plateau value of the strain energy release rate for the thick specimen is 8 kJ/m^2, which is slightly increased compared to the relative value given by the thin specimen, which is close to 7 kJ/m^2. This difference is explained based on the following arguments. Since the thick specimen is more compliant for loads close to those for failure, this leads to higher effective crack length values. Conversely, taking into account that the slope of the compliance is an increasing function of the effective crack length, as shown in Fig. 4.26, higher effective crack values lead to higher values of the derivative $\partial C/\partial\alpha|_{\alpha_{\text{eff}}}$ and as a result to higher plateau values of the strain energy release rate.

The monitored plateau region in both cases of thin and thick specimen is of the same size, about 11 mm long, and after that a high increase rate of the energy release rate appears due to the compressive stress field that is developed due to the bending of the rear side of the

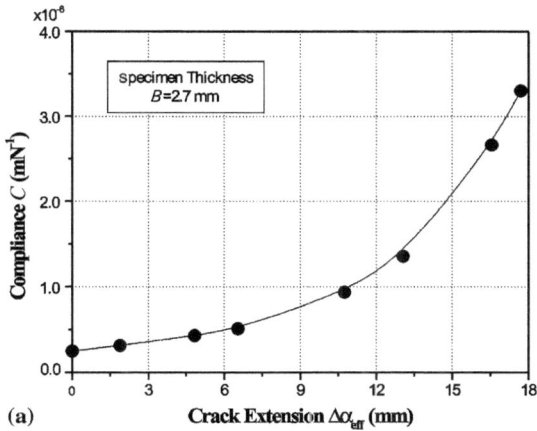

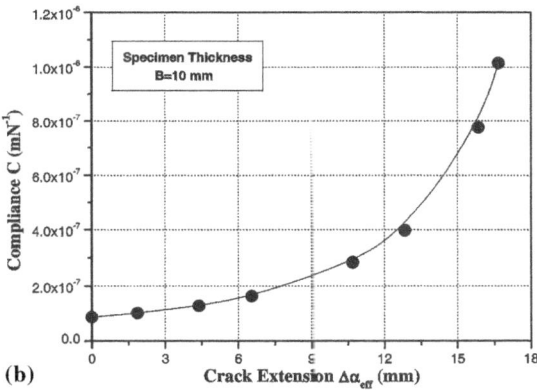

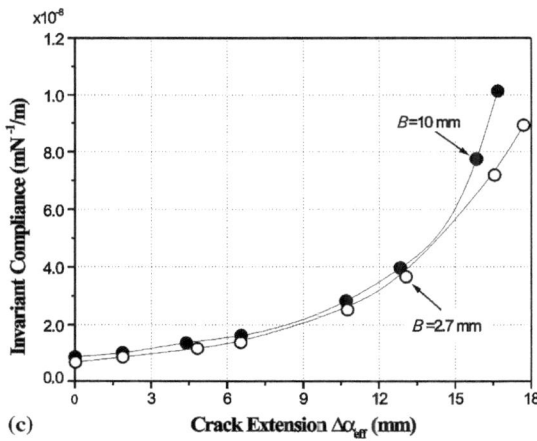

Figure 4.26 Compliance calibration curve versus crack extension. (a) $B=2$, 7 mm; (b) $B=10$ mm; (c) invariant compliances.

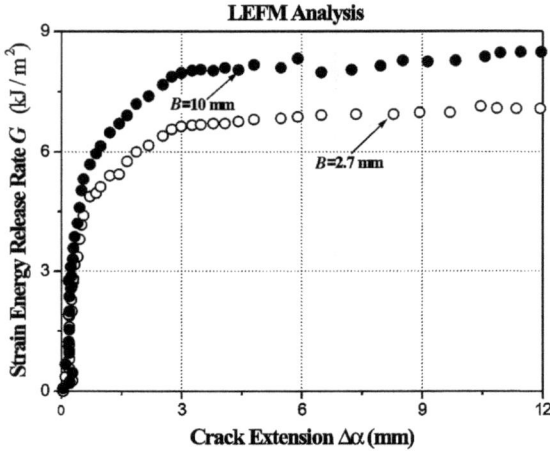

Figure 4.27 Strain energy release rate G_{IC}, using linear elastic fracture mechanics.

CT specimen. The size of the plateau is in agreement with calculations that have been made[87] and show that the compressive field begins after a normalized length of

$$\bar{x} = \left[\frac{x}{(W-\alpha)}\right] = 0.6$$

Assuming linear elastic behavior for the 2-D C/C material and using Eq. (4.107), the fracture toughness K_{Ic} is about $15\,\text{MPA}\,\sqrt{\text{m}}$ for both thin and thick specimens, which is very close to the value of the fracture toughness reported[88] ($K_{Ic} = 16\,\text{MPa}\,\sqrt{\text{m}}$). However the assumption of linear material behavior gives rise to Eq. (4.107), the application of which gives $K_{Ic} = 25\,\text{MPa}\,\sqrt{\text{m}}$ for the thick specimen and $K_{Ic} = 20.5\,\text{MPa}\,\sqrt{\text{m}}$ for the thin one. Although the procedure described for the characterization of the fracture behavior of 2-D C/C material is remarkably easy to apply, the results obtained demonstrate all the uncertainties involved in the method.

In the following, an alternative approach for the characterization of the fracture behavior of composites in general is proposed. It is again applied in the case of the same two 2-D C/C materials. Comparison of the R-curves obtained based on the application of the different methods of analysis is also presented.

The basic idea is along the lines of Section 4.3 in order to take into consideration the bridging effect during the interlaminar mode I fracture toughness measurements. To reveal the irreversible damage mechanisms occurring in the vicinity of the crack tip during the monotonic loading, a loading/unloading procedure has been applied under displacement control of the testing frame. Again the crack growth resistance is assumed to be the sum of two energy rate contributions:

- The nonlinear strain energy release rate G_R^*, which can be directly correlated to G_{Ic} fracture parameter, calculated by LEFM.

- The plastic energy dissipation rate Φ_{ir}, which is the energy consumed due to the process of the formation of the damage zone which surrounds the crack tip region.

Although the geometry of the test specimen is completely different, the procedure is exactly the same as that described in Section 4.3.2 (Fiber Bridging).

The application of this methodology in the case of the 2-D C/C composite and the results are presented next for both composite thicknesses.

A typical loading–unloading P–δ curve is presented in Fig. 4.28 for thick specimens. A detailed analysis is conducted in order to calculate the load P, the crack opening displacement δ, the irreversible remaining displacement δ_{ir} at each unloading stage, and the modified compliance C^*. Based on this, the effective crack length, α_{eff}, at the nth loading–unloading cycle is evaluated using the recurrence relation of Eq. (4.36). Furthermore, based on P–δ curves, the plastic dissipated energy ratio λ can easily be defined according to Eq. (4.40). The results

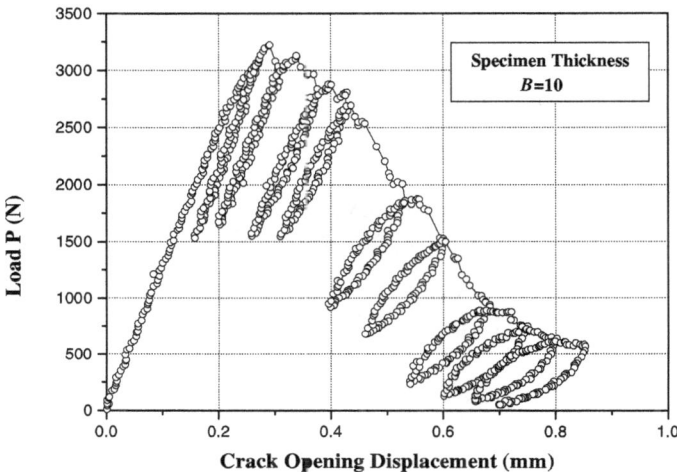

Figure 4.28 Typical loading-unloading-reloading curve for a thick specimen ($B = 10$ mm).

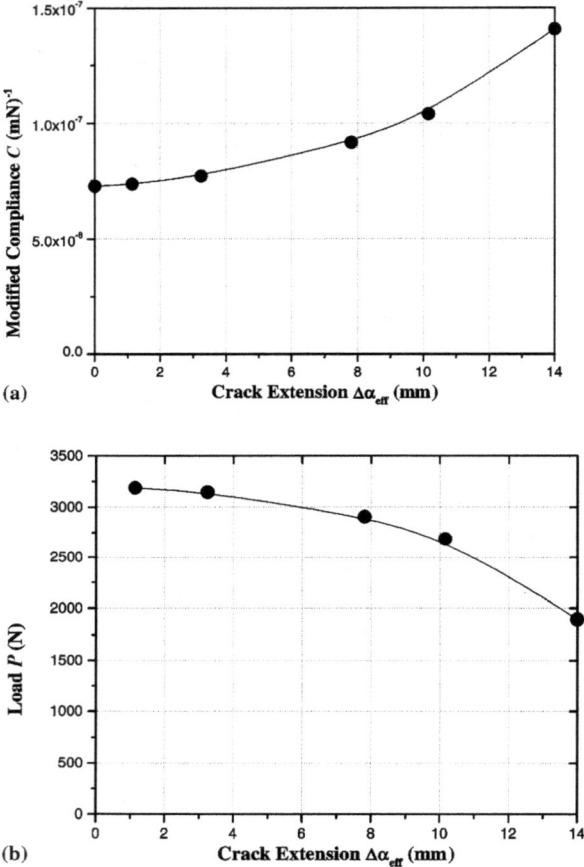

Figure 4.29 Parameters of the proposed analysis. (a) Modified compliance versus effective crack extension; (b) load versus effective crack extension

obtained from the above analysis are shown in Fig. 4.29a–d. In all the cases, data have been kept only up to a normalized distance

$$\bar{x} = \left[\frac{x}{(W - \alpha)} \right] = 0.6$$

After that the compressive field starts to act and the data are misleading.

It is quite interesting to focus for a while on the $\lambda(\alpha_{\text{eff}})$ curve illustrated in Fig. 4.29c. It is evident that in the early stages of crack propagation the ratio of irreversible plastic energy dissipated to the total mechanical energy input is very small, namely 10–15%,

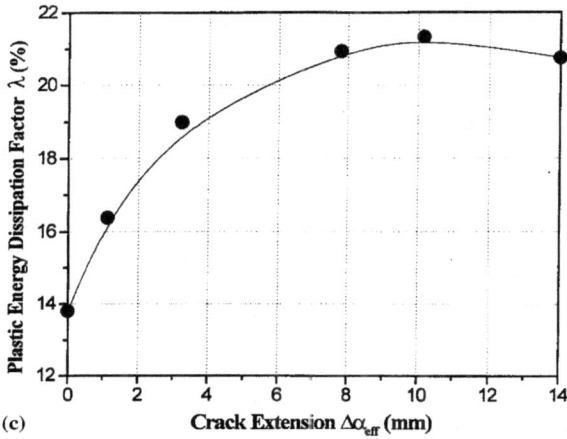

(c)

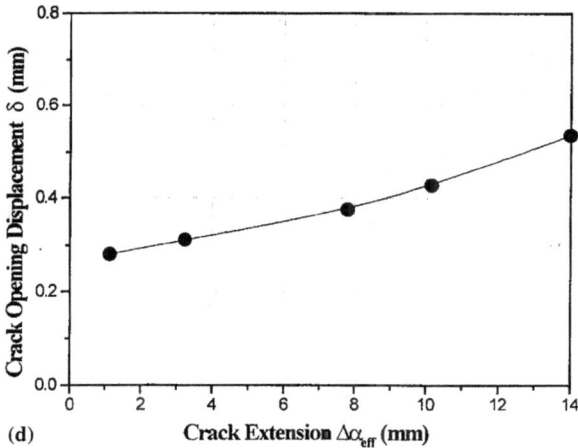

(d)

Figure 4.29 Parameters of the proposed analysis. (c) Energy dissipation rate λ versus effective crack extension; (d) crack opening displacement versus effective crack extensions.

and this becomes clear from the form of the P–δ curve together with the small value of the irreversible displacements δ_{ir} for each loading-unloading loop. For thin specimens the ratio of dissipated energy λ is lower, of the order of 6–10%. When stable crack propagation initiates, the ratio λ rapidly increases and reaches a plateau value. This is due to the complete development of the fracture process zone, which mainly consists of a cracked matrix, bridged by intact and/or failed fibers, which debond, slip, and pull out. Crack bridging enhances material crack resistance by partially shielding the crack tip from the applied load. Further crack advance over a critical length leads to the interaction of the fracture process zone with the compressive field generated from the bending of the rear surface of the CT sample. This interaction becomes dominant and a fast increase of the λ ratio appears. Figure 4.30a shows the elastic energy release rate G_R^* for both thin and thick specimens versus the effective crack extension $\Delta\alpha_{\text{eff}}$.

Initially the energy increases as the crack propagates, for $\Delta\alpha_{\text{eff}} \cong 2.5$ mm. This value is nearly the same for both thick and thin specimens. Then the energy remains almost constant, appearing slightly different between the plateau values that correspond to the thin and thick specimens. This plateau value is, for the thin specimen $G_R^{*\text{thin}} = 3\text{--}3.5$ kJ/m^2, and for the thick one $G_R^{*\,\text{thick}} = 3.5\text{--}4,5$ kJ/m^2. The length of the plateau region is 10 mm and appears to be common for both cases.

The associated stress intensity factor (SIF) (if one permits this kind of interpretation of the definition of SIF) is $K_R \cong 16$ MPa $\sqrt{\text{m}}$. This is in agreement with the value of SIF calculated earlier during the analysis of the monotonic tensile loading experiments when the LEFM was applied.

In Fig. 4.30b, plots of the irreversible energy rate Φ_{ir} for both thin and thick specimens are given versus the effective crack extension $\Delta\alpha_{\text{eff}}$. It is shown that again there is a crack propagation region where the energy increases and its length has nearly the same value as in Fig. 4.30a. However, in that case the plateau values for the thin specimen is $\Phi_{ir}^{\text{thin}} = 0.5\text{--}1$ kJ/m^2 and for the thick specimen configuration is $\Phi_{ir}^{\text{thick}} = 1.5\text{--}2$ kJ/m^2. Figure 4.31 shows the crack resistance $R(= G_R^* + \Phi_{ir})$ versus the effective crack extension, $\Delta\alpha_{\text{eff}}$. The total plateau value for the thin specimen is about $R^{\text{thin}} = 4$ kJ/m and for the thick one $R^{\text{thick}} = 6$ kJ/m^2.

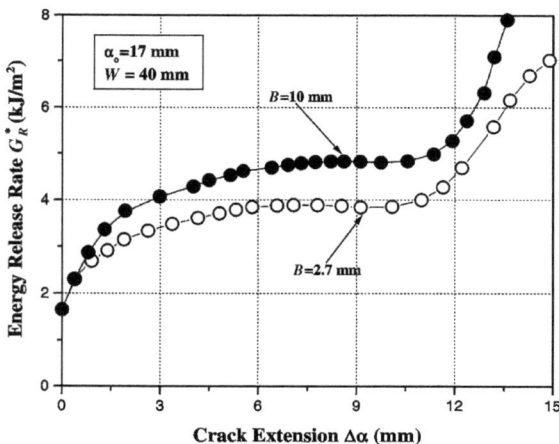

Figure 4.30a Elastic energy release rate G_R^*, versus effective crack extension, $\Delta\alpha_{\text{eff}}$, for thin ($B = 2, 7$ mm) and thick ($B = 10$ mm) specimens.

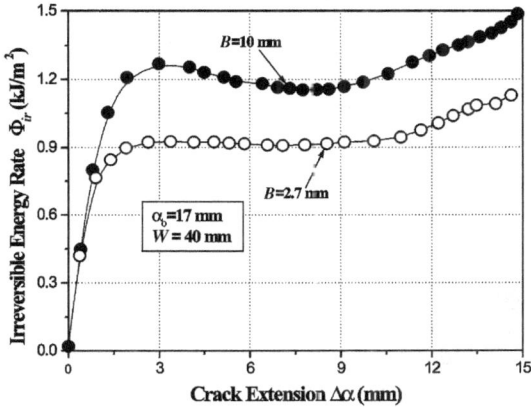

Figure 4.30b Plastic energy dissipated rate, Φ_{IR}, versus effective crack extension, $\Delta\alpha_{eff}$, for thin ($B=2$, 7 mm) and thick ($B=10$ mm) specimens.

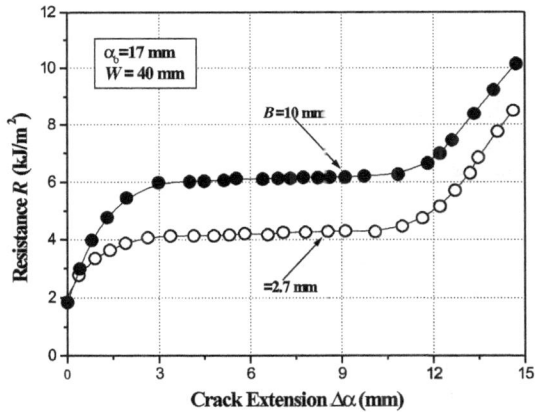

Figure 4.31 Energy resistance R versus effective crack extension $\Delta\alpha_{eff}$ for thin ($B=2$, 7 mm) and thick ($B=10$ mm) specimens.

These values are lower than those calculated by the application of the monotonic tensile loading on CT samples and the compliance calibration technique, and for the thick specimen there is a difference of 25%. This difference increases to 40% in the case of thin specimens.

4.6   Acoustic Emission for Characerization of Fracture in Composites

Acoustic emissions (AE) is defined as transient elastic waves generated by the rapid release of energy from localized sources within a material.[89] These localized sources of energy release may be the formation and the propagation of microcracks within the material structure or other defect-related deformation such as crack growth and plastic deformation. AE is a widely accepted method for the characterization of fracture mechanisms in composites. AE wave analysis has developed rapidly and special attention has been paid to the correlation of AE to the fracture mechanics analysis.

From the fracture mechanics point of view, it is of great importance to establish a nondestructive methodology that may provide in-situ information about the critical size of defects generated within the material structure. It is well known that the stress intensity factor and the energy release rate are dependent upon the applied load and the size of the defect. Analysis of AE-monitored signals during the loading of a structure in general may establish a relation between AE parameters and the stress intensity factor and/or the energy release rate available for crack advance. But this is the constant demand from any nondestructive evaluation of the residual strength of a specimen or structure.

General remarks

Acoustic emission waves are elastic waves emitted due to the propagation of microstructural defects. The defect may be considered as a point source that emits the released energy when the crack advances. The emission is not continuous but has the form of an impulse (with wide frequency band).

Based on the type of microfailure, the emitted elastic wave may be of longitudinal/pressure (P) and/or transverse/shear (S) form and, in the case of real structures where free traction surfaces always exist, surface Rayleigh waves (R) are present. No matter what the type of initial emission, as it propagates in the elastic host medium it is scattered by the material inhomogenities and/or the preexisting crack network, and due to the mode conversion effect all the three types of elastic waves are propagated in the medium.

The wave velocities of P, S, and R waves are defined as C_P, C_S and C_R and they are given for a homogeneous and isotropic medium by

$$C_P = \sqrt{\frac{\lambda + 2\mu}{\rho}}$$
$$C_S = \sqrt{\frac{\mu}{\rho}}$$
$$C_R \approx \frac{0.862 + 1.14v}{1 + v}\sqrt{\frac{\mu}{\rho}}$$

(4.115)

where λ and μ are the Lamé constants of the medium and v is its Poisson's ratio. In the set of Eqs. (4.115) the velocity of R-waves is given by Viktorov[90] and represents a very good approximation.

One must keep in mind that the amplitude of the R-wave (displacement field) decreases exponentially with depth below the free surface. The amplitude may be considered negligible at a depth of about two wavelengths below the surface.

Under specific conditions such as the presence of a fluid–solid interface or an interface between two flat solid materials, some other types of waves may be generated, namely, Love (Stoneley) waves and Lamb waves, respectively.

Returning to Eq. (4.115), it is clear that

$$C_P > C_S > C_R \qquad (4.116)$$

A schematic representation of the propagation modes of AE waves originated by a structural defect is given in Fig. 4.32. The AE waves propagated in a structure are detected using piezoelectric displacement transducers. Thus the surface displacement profile at the point of transducer placement is monitored. The monitored signals, as expected, are quite different from the waves emitted by the source due to multi-scattering process that they are involved in. The transducers used may be of narrow frequency band, which offers better signal/noise ratio for the selective frequency band, but their sensitivity out of the band is very small or of wide frequency band. The transducer characteristics also affect the monitored signal.

In composite materials, any single defect generated within the material structure during loading acts as a source of AE activity.

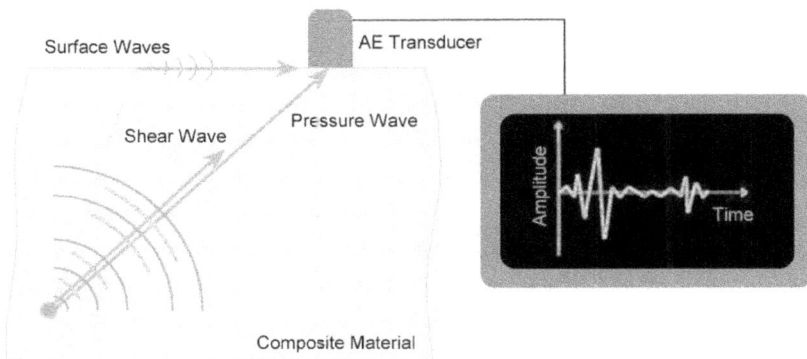

Figure 4.32 Schematic representation of AE waves generated by a structural defect.

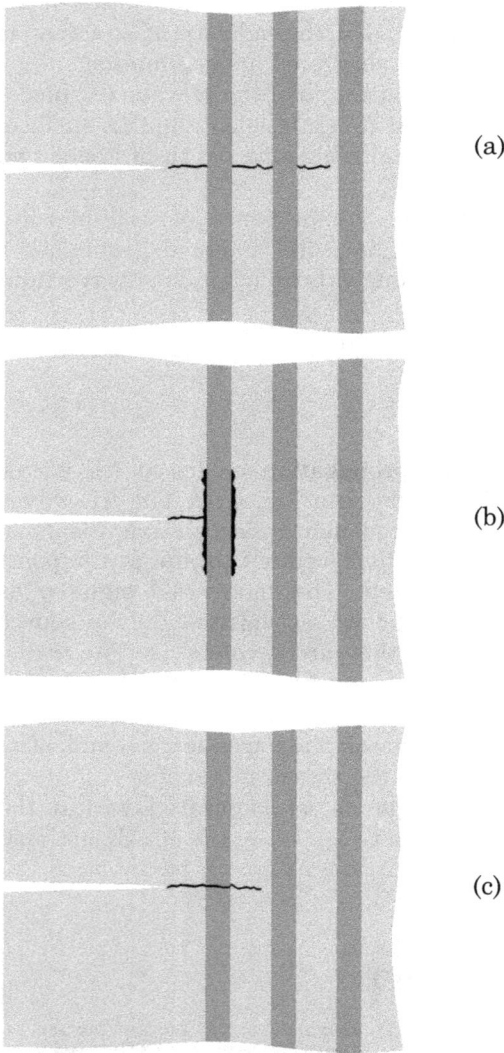

Figure 4.33 Failure modes in composites that produce AE activity: (a) matrix cracking; (b) fiber–matrix debonding; (c) fiber breaking.

Matrix cracking, fiber–matrix debonding, friction between cracked surfaces, fiber pull-out, and fiber breaking are only some of the damage modes that produce AE signals (see Fig. 4.33). All these microstructure damage modes emit almost simultaneously in the case of loading of a notched composite laminate and their sources are mainly located in the vicinity of the crack tip. In addition, as the load increases, the material is damaged progressively and the characteristics of the different signals coming form the same damage mode that are recorded by the AE system are different because the propagation medium alters.

A typical AE signal emitted by a point source is presented in Fig. 4.34. In this diagram the key characteristics of an AE signal are also defined.

The application of AE in the field of composite materials has two main objectives:

- The prediction of the fracture behavior of the material by identifying its residual strength.

- The identification of the failure modes developed within the structure of the material during loading.

Both topics are of specific interest, although they are very difficult to accomplish.

In the field of nondestructive evaluation of the residual strength there is only limited work in the literature, mainly due to the complexity of the problem.[91-95] On the other hand, failure mode identification in composites using AE has received considerable attention during the last 10 years. Various efforts have been made either using theoretical tools based on wave propagation and wave scattering in both time and frequency domains or by using parameters based on analysis applied to the signals recorded using AE equipment.

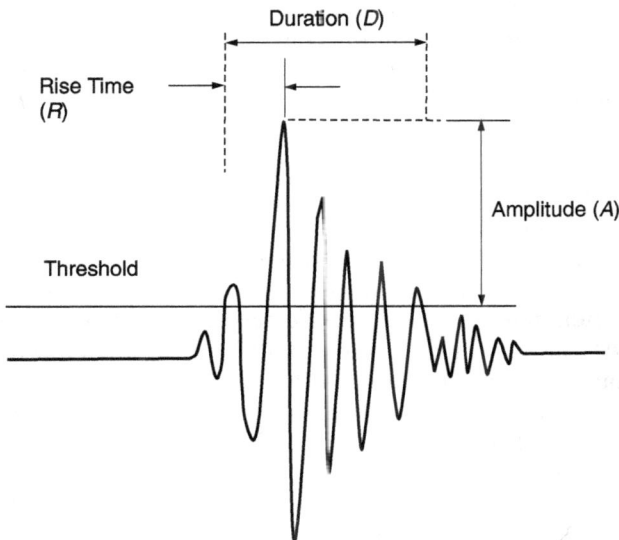

Figure 4.34 Typical form of AE signal and its main characteristics.

An example of damage mode identification will be presented next which is based on the analysis of the parameters of the signals monitored using AE equipment. The proposed damage mode identification (DMI) scheme was applied on a 2-D carbon/carbon notched composite. Among the aims of this work are:

- To establish a relation between the crack resistance curve of C/C composites and the AE data monitored during the loading–unloading–reloading procedure followed for the calculation of R-curves.
- To identify the activated damage mechanisms and their evolution with time (and load) during the unloading–reloading process by applying pattern recognition classification algorithms.

To increase the reliability of AE data analysis for the identification of damage mechanisms in composites, a multiparameter approach is necessary, and a specific algorithm has been proposed[94] that serves that end. This algorithm contains unsupervised pattern recognition classification of AE data, descriptor selection procedures, validation techniques, filtering and statistical analysis of AE data, taking into account the stochastic character of AE data and the knowledge of the damage mechanisms presented in CMCs, which have been determined using microscopical fractographic examination. Thus, preliminary correlation between the calculated clusters and specific damage mechanisms was achieved for the unloading–reloading cycles.

The AE activity was also correlated systematically to the fracture parameters and it was shown that the emission activity during the unloading–reloading cycles follows a decreasing pattern for the felicity ratio (see below), which, however, remains constant during the stable crack propagation stage. Because AE is produced by stress-induced deformation, it is highly dependent upon the stress history and the stress level.

In many cases like the present one, AE was carried out under rising load conditions. Normally, the first application of load results in a number of AE events, much more than produced in subsequent loadings. There is a group of materials where no emission occurs during subsequent loadings unless the previous applied maximum load is exceeded, a phenomenon called the Keizer effect. However, in most materials the AE activity during subsequent loadings initiates at load levels lower that the maximum applied load.

The ratio of the load at which AE begins again to the previous maximum load is called the felicity ratio (FR), and it has been determined as a pronounced structural integrity parameter.[96]

The DMI analysis was applied on a commercially available 2-D C/C material of 10 mm thickness made by Schunk.

A modified compact tension test specimen was used to permit a longer stable crack propagation zone. Cyclic mechanical loading was applied through a pin-and-shackle system to prevent the specimen from bending. The crosshead velocity of the testing frame was 0.2 mm/min. The crack geometry was of a V-shape. The notches were prepared using a diamond saw for the initial cut and a razor blade for the final cut at the crack tip. The edge radius of the razor blade was 3 μm. During the experiment, the crack mouth opening was measured using a crack opening displacement (COD) transducer. At the same time, AE activity was monitored continuously during the tests using two 150 kHz resonant transducers, which were placed laterally and at the rear side of the test coupon. The AE transducers were coupled using a medium-viscosity vacuum grease and held in place with adhesive tape. AE events were traced by a physical acoustic corporation SPARTAN 2000 system and the AE parameters amplitude (A), rise time (RT), energy (E), duration (D) and counts (C) of each event were stored. The test coupon geometry and the AE transducers are shown in Fig. 4.35.

To reveal the irreversible damage mechanisms occurring in the vicinity of the crack tip, a loading–unloading–reloading procedure was applied under displacement control.

Detail description of the methodology applied has been given in Section 4.5. Following that procedure and again assuming that the crack growth resistance of the material is the sum of two different energy rate contributions (a nonlinear elastic and a plastic one), the R-curve is calculated. It was found that the plateau values of G_R^*, Φ_{ir}, and $R\,(= G_R^* + \Phi_{ir})$ are 4.0–4.5 kJ/m^2, 1.0–1.5 kJ/m^2, and 6 kJ/m^2, respectively, while the length of the plateau region extends

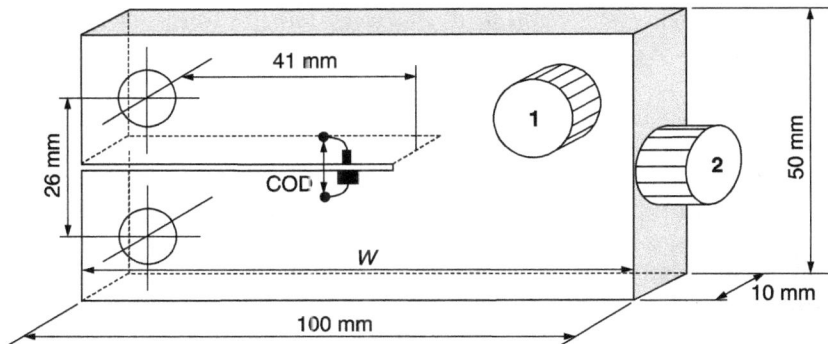

Figure 4.35 Schematic representation of the modified CT specimen geometry and the positions of the AE transducers.

between 0.48 and 0.6 of α_{eff}/W. This plateau region is slightly larger than that achieved by the application of the standard CT geometry, although the length of the used testing sample is double.

To correlate the fracture behavior of a 2-D C/C with its AE response during the cyclic loading, AE events and AE counts in addition to the felicity ratio (FR) for each unloading–reloading cycle will be used. According to recent literature,[96] FR is defined as the ratio of the load at which a loading structure starts to emit after reloading, over the maximum load that the structure had experienced during a previous loading phase. When FR = 1 the structure follows the known Kaiser effect.

The identification of the activated damage modes during the unloading–reloading cycles using AE is another target of the present work. Unsupervised pattern recognition analysis will be applied for the separation of different damage mechanisms using a set of AE parameters–descriptors wherever possible. The algorithm that will be used[94] of three different parts. For the application of the proposed algorithm, the following assumptions have to be made:

- All AE descriptors are random variables.[97,98]

- Each AE descriptor exhibits Gaussian distribution during the clustering.[99,100]

- Each material damage mechanism can be characterized by a single cluster in a unique way.[99,101]

- All the damage mechanisms are continuously active during loading with variable intensity.[94]

It was proven that the degree of fulfillment of the above assumptions affects the effectiveness of the algorithm.[93] However, a feedback control loop that verifies the validity of the assumptions has been built into the proposed algorithm, providing the analysis with a dynamic character and securing its efficiency. Special care has been taken for noise reduction of the AE data using special filtering techniques.

After the optimum classification of AE data (second part of the algorithm), the data set for every cluster is known. The next important step is then determination of the cluster activation in time. Activation of a cluster in time is defined as the time derivative of the normalized counts of a cluster. Combining the variation of the activation in time of each cluster and its characteristic AE parameters with the micromechanical observations of the material damage mechanisms, one may correlate in a very precise way clusters and damage mechanisms (third part of the algorithm). Nevertheless, the introduced variable of cluster activation in time (CAT) represents the progress of each material damage mechanism at each loading state, which is very important in practice. This clustering procedure will be followed for every unloading–reloading cycle in order to find relations between the R-curve analysis and the clusters represent damage mechanisms.

A typical loading–unloading–reloading (P–δ) curve is given in Fig. 4.36 (solid line), together with a typical quasi-static loading curve (dotted line). It is obvious that the unloading–reloading cycles do not significantly affect the general response of the notched specimens, from a fracture mechanics point of view.

A very informative diagram is presented in Fig. 4.37, where the cumulative AE counts for the both AE channels are plotted versus the stroke displacement. The dashed line shows the modified compliance C^* versus stroke displacement. A detailed examination of Fig. 4.37 shows that during the zeroth cycle, where a maximum load of 2600 N has been applied, the FR is almost unity and thereafter starts to decrease. It is important to note here that AE activity also exists during the initial stage of the unloading phase. Increasing the rate of growth of modified compliance, it is evident that the FR decreases significantly, and this is confirmed from both AE channels used.

In Fig. 4.38, the calculated FR for both AE channels versus stroke displacement is plotted. In the same figure, the normalized effective crack length is also given versus stroke displacement. It is shown that for stroke values of 1.3–1.9 mm there is a plateau value for the FR, which was calculated based on the AE data monitored from both AE channels. Within the same stroke displacement region, the variation of the α_{eff}/W parameter is almost linear. This linearity of α_{eff}/W reveals a

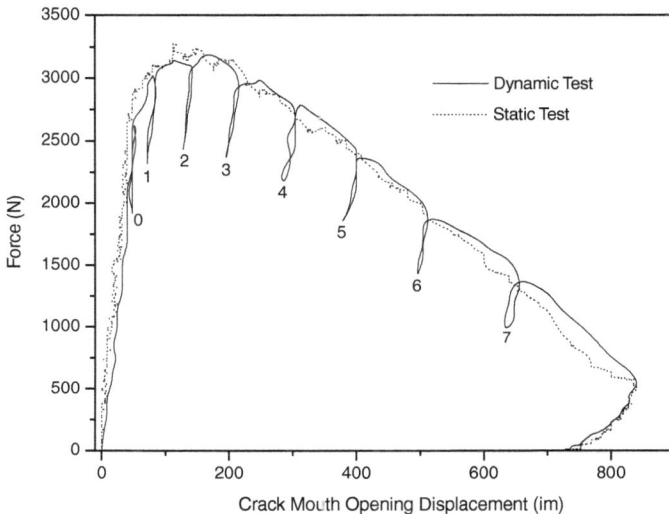

Figure 4.36 Typical loading–unloading–reloading diagram, presenting applied force versus crack mouth opening displacement.

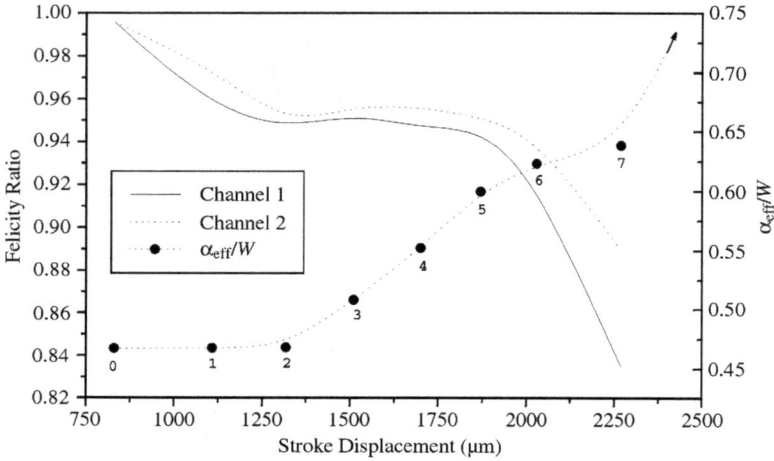

Figure 4.37 Cumulative AE counts for both AE channels versus stroke displacement. The dashed line shows modified compliance C^*. The numbers indicate unloading–reloading cycles.

stable crack propagation zone, and for this zone an almost constant value of the FR parameter was detected. For stroke displacement values smaller than 1.3 mm the formation and the development of the damage zone is completed before the crack advances, while for stroke displacement values higher than 1.9 mm the compression zone developed at the rear front of the test coupon seriously affects the stable

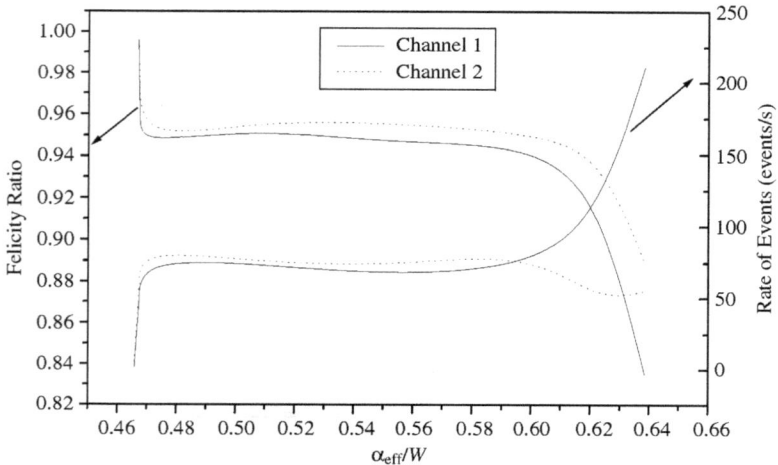

Figure 4.38 Felicity ratio monitored by both AE channels versus stroke displacement. The dashed line shows the normalized effective crack (α_{eff}/W).

crack propagation and the value of the FR parameter drops significantly. These conclusions could be also supported by C-Scan examination of the specimens after the cyclic loading. Thus, constant value of the FR, independently of the positioning of the AE transducer, indicates stable crack propagation.

Figure 4.39 correlates the rate of generation of AE events and the estimated FR using AE data monitored in both AE channels to the normalized effective crack length. Both parameters show a considerable constant value (plateau region) for effective crack length α_{eff}/W in the vicinity of 0.48–0.6. This region of α_{eff}/W coincides with the area where stable crack propagation occurs and is in complete agreement with the remarks made regarding Fig. 4.37. Thus, within the region of stable crack propagation, the FR exhibits an almost constant value of 0.95. The same FR value is deduced using data monitored by both AE channels.

As a general remark, based on the results given in Figs 4.37 and 4.38, one may say that after the initial phase of the development of the damage process zone ahead the crack tip, the crack propagates in a stable form and in a self-similar way activating the same set of damage mechanisms in each step.

A more detailed analysis of AE data monitored during the unloading–reloading cycles will reveal the damage mechanisms and their activation in time, confirming or not the conclusions drawn earlier, and this is the target of the next step. The importance of determining the contribution of each specific damage mode to the crack resistance

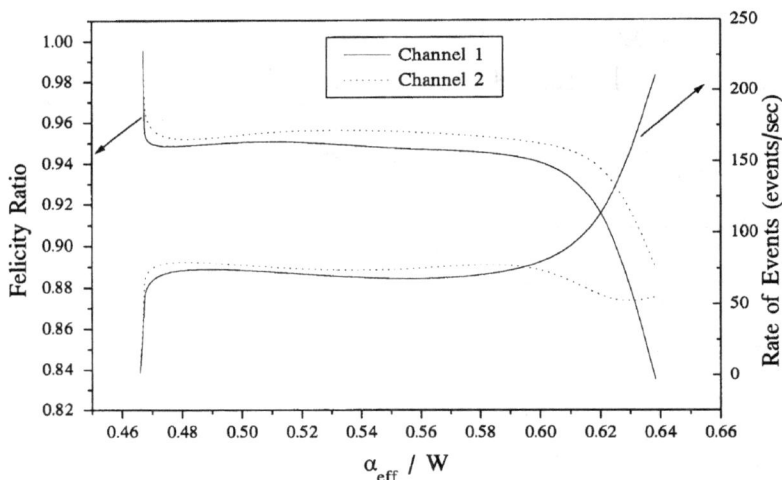

Figure 4.39 Felicity ratio and rate of generation of AE events for both AE channels versus normalized effective crack length (α_{eff}/W).

of the C/C material, or equivalently to the total energy release rate, is indisputable, even at the stage of optimazation of material properties. To that end, the analysis next performed intends to characterize the damage mechanisms involved in each loading cycle using the signature they provide through their AE activity. It is expected that the AE signature characteristics for the damage mechanisms will not be constant values independent of the loading cycle they monitor, since the crack tip distance from the AE sensors diminishes as the crack advances. Although in the stable crack propagation region macro-fracture characteristics appear that have constant values, the micromechanics of crack propagation is also affected by the local stress state.

Therefore, the first step of the analysis is the presentation of the AE emissions monitored at each unloading–reloading cycle, for both channels. Figures 4.39 and 4.40 present the rate of count generation during each cycle, of the cyclic loading process, for channel 1 and channel 2 respectively. The count generation rate has been shown to be a direct estimator of the damage activation within the material structure.[94]

According to Fig. 4.40, in the case of the first two cycles, AE activity is monitored during the first period of the unloading phase and the last period of the reloading phase. This can be explained taking into account that during these cycles the overall displacement is limited and no crack propagation has occurred. During the next loading cycles, the AE activity pattern varies significantly for a wider area of the unloading and the reloading cycles, which indicates that new damage mechanisms have been activated.

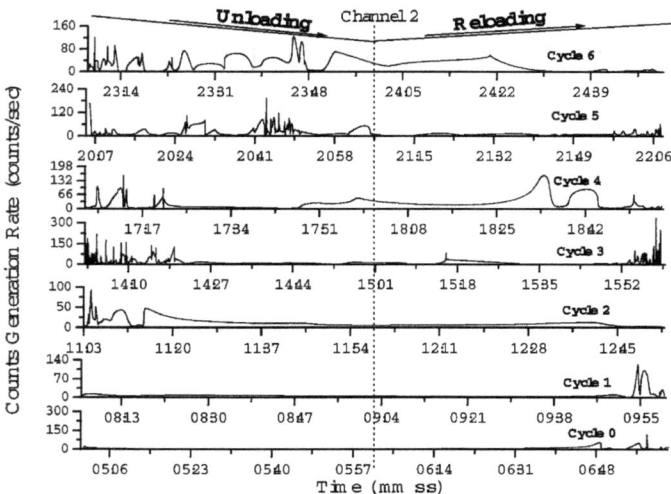

Figure 4.40 Count generation rate versus time, for the AE data monitored by channel 1, for each unloading–reloading cycle.

According to the fractographic examination, after the third cycle the damage process zone has been completely developed and the crack starts to advance in a stable, self-similar way. Now the bridging of the crack by intact or fractured fibres out of the crack region contributes to the stress redistribution and the AE activity due to the fibre pull-out and breaking is important. Finally, the AE pattern changes again after the 6th loading cycle, indicating that the stable crack propagation phase is over, due to the development of compressive stresses in the vicinity of the crack tip. Macroscopically, it appears in the form of irreversible deformation of the material during the unloading stage, where the AE activity is much higher than the relative AE activity during the reloading stage. Figure 4.41 shows the count generation rate monitored by channel 2 during each unloading–reloading cycle. The behavior follows the general pattern described earlier, with only some minor differences.

Further analysis of AE activity monitored during each unloading–reloading cycle was performed using unsupervised pattern recognition classification of AE data as has been described in detail elsewhere.[98] Actually, the k-means pattern recognition (PR) method was used as the standard algorithm, for each unloading–reloading cycle. It was found that for 2-D C/C material and for all seven cycles case, the optimum number of AE descriptors was five, namely amplitude, energy, duration, counts, and rise time. Using the above descriptors, the clustering determination algorithm reveals the presence of six different clusters for each unloading–reloading cycle. The activation in time of each individual cluster, based on the data monitored in channel

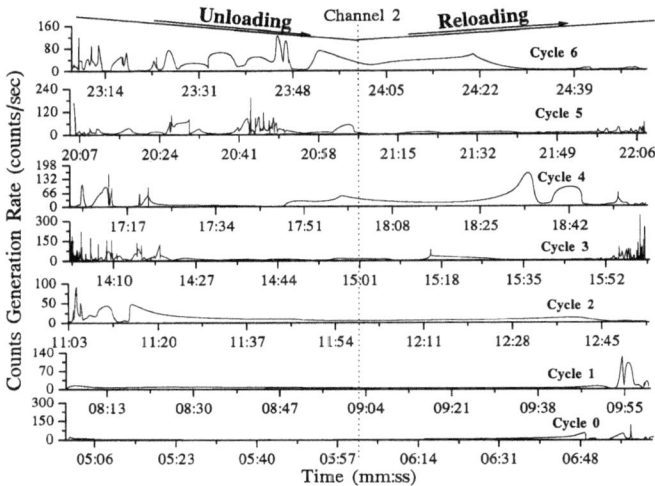

Figure 4.41 Count generation rate versus time, for the AE data monitored by channel 2, for each unloading–reloading cycle.

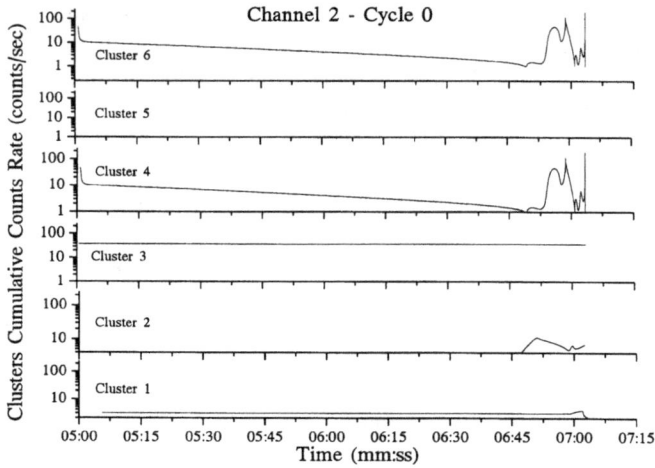

Figure 4.42 Activation in time of all the six clusters for cycle no. 0 (data monitored in channel 2).

2 is given in Figs. 4.42–4.44, for the unloading–reloading cycles 0, 3, and 5, respectively.

According to these plots, each cluster shows significant character-istic differences at different cycles. Comparing the activation of clusters in time, there is a correspondence between the results given for cycle 3 and cycle 5 with the exception of cluster number 5. An explanation could be that during both cycles 3 and 5, stable crack propagation has been assumed. This is not the case for the form of clusters in cycle 0, where a completely different behavior has been

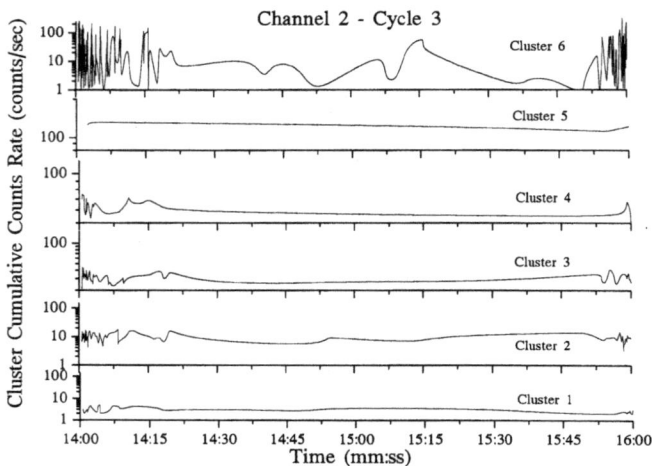

Figure 4.43 Activation in time of all the six clusters for cycle no. 3 (data monitored in channel 2).

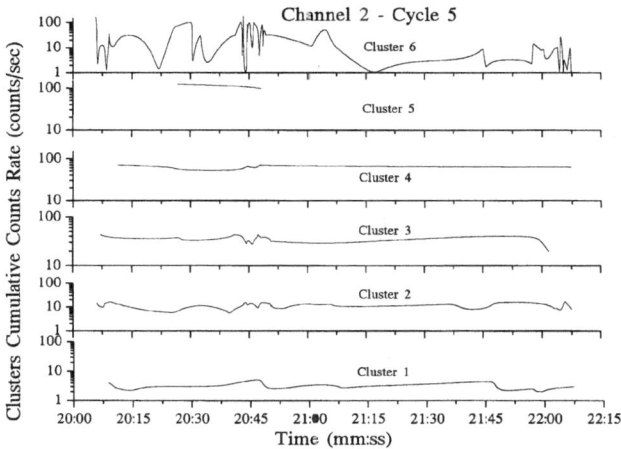

Figure 4.44   Activation in time of all the six clusters for cycle no. 5 (data monitored in channel 2)

calculated. The higher activation monitored for the same cluster as the cycle number increases is possibly due to the approach of the crack tip to the AE sensors.

Figure 4.45 represents the complete results of pattern recognition analysis of AE data monitored by channel 2 as far as hits per cluster for each cycle is concerned. The numbers in the columns of that graph indicate the percentage of total AE hits per cycle, for each cluster. During the third unloading–reloading cycle, where the crack starts to propagate in a stable way, all the clusters are active continuously. Furthermore, cluster 6 has steadily the highest amount of hits, almost half of the total hits per each cycle. Since the dominant stress redistribution mechanism in C/C composites is shear band formation (class III material), which is formed by matrix cracking normal to the crack direction, cluster 6 clearly represents the matrix cracking damage mechanism.

Additional structural characteristics of C/C material are the weak fiber–matrix interface, and the lower shear strength compared to tensile. Thus, matrix cracks, extending within the shear band zone, are combined with fiber–matrix debonding and single fibre failures, and lead to fibre pull-out with frictional sliding.

Using arguments that have been clarified in earlier works,[94] some other clusters may be correlated with confidence to some of the basic damage mechanisms described above. This identification of damage mechanisms was supported by microscopical examination of the tested coupons after each loading cycle and ultrasonic monitoring of the samples after the test.

Thus, in addition to cluster 6, which represents matrix cracking and fibre–matrix debonding for the formation of the shear band in the C/C

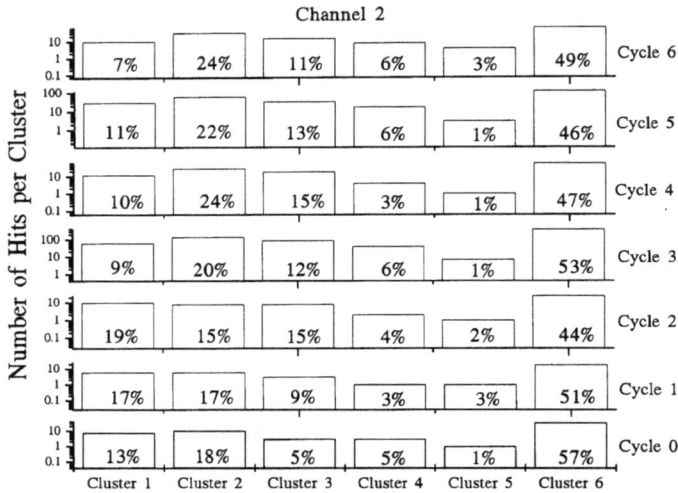

Figure 4.45 Number of hits per cluster for each unloading–reloading cycle.

composites, cluster 2 is related to long fibre fracture and cluster 4 represents frictional sliding and fibre pull-out/push-in phenomena.

As a general conclusion, one may state that damage mechanism identification may be performed in very complicated material structures, although it is not a trivial procedure and it demands an a priori knowledge at the microscale level of the damage modes activated within the material.

Acknowledgments

The author gratefully acknowledges the contribution of his colleagues G. C. Christopoulos, Y. Z. Markopoulos, and Y. P. Pappas to the material presented here. Special thanks to Dip. Engs. K. Lantavou, S. Tsantzalis, and D. Vlachos for the processing of the material and the preparation of the drawings. Without their generous help, this chapter would never have been completed.

References

1. G. C. Shi and E. P. Chen (1981) Mechanics of Fracture, in *Cracks in composite materials, Vol. 6*, Martinus Nijhoff, The Hague.
2. D. Broek (1986) *Elementary engineering fracture mechanics*, Martinus Nijhoff, Dordrecht.
3. M. F. Kanninen and C. H. Popelar (1985) *Advanced fracture mechanics*, Oxford University Press, New York.
4. H. M. Westergaard (1939) Bearing pressures and cracks, *Transactions of the American society of mechanical engineers*, **61**, 49-53.

5. G. R. Irwin (1957) Analysis of stresses and strains near the end of a crack traversing a plate, *Journal of Applied Mechanics*, 24, 361-364.
6. M. L. Williams (1957) On the stress distribution at the base of stationery crack, *Journal of Applied Mechanics*, 24, 109-114.
7. T. L. Anderson (1995) Fracture Mechanics, in *Fundamentals and Applications*, CRC Press, Florida.
8. E. J. Ripling, S. Mostovoy, and R. L. Patrickm (1964) *Measuring Fracture Toughness of Adhesive Joints*, ASTM STP-360, 129-134.
9. D. J. Wilkins, J. R. Eisenmann, R. A. Camin, W. S. Margolis, and R. A. Benson (1982) *Characterizing delamination growth in graphite-epoxy*, Damage in composite materials, ASTM STP-775, 168-183.
10. W. D. Bascom, J. L. Bitner, R. J. Moulton, and A. R. Siebert (1980) The interlaminar fracture of organic-matrix woven reinforced composites, *Composites*, 11, 9-14.
11. J. M. Whitney, C. E. Browning, and A. Mair (1974) *Analysis of the flexure test for laminated composite materials*, ASTM STP-546.
12. J. P. Berry (1963) Determination of fracture energies by the cleavage technique, *J. Appl. Phys.*, 34, 62-68.
13. J. M. Scott, D. C. Philips (1975) Carbon fibre composites with rubber roughened matrices, *J. Compos. Mater.*, 10, 551-562.
14. J. G. Williams (1989) Fracture mechanics of anisotropic material in application of fracture mechanics to composite materials, in *Composite Materials Series* Vol. 6, ed. K. Friedrich, Elsevier Science Amsterdam, 3-38.
15. Hult J. A. H., and McClintock F. A. (1957) Elastic-plastic stress and strain distributions around sharp notches under repeated shear, *Proceedings of the Ninth International Congress for Applied Mechanics*, University of Brussels, 51-58.
16. L. J. Hart-Smith (1973) *Adhesive-bonded scarf and stepped-lap joints*, National Aeronautics and Space Administration, Report CR 112237.
17. K. Kageyama, T. Kobayashi, and T. W. Chou (1987) Analytical compliance method for mode-I interlaminar fracture-toughness of composites, *Composites*, 18, 393-399.
18. X. N. Huang and D. Hull (1989) Effects of fiber bridging on G_{IC} of a unidirectional glass epoxy composite, *Compo. Sci. Technol.*, 35, 283-299.
19. M. N. Charalambides and J. G. Williams (1994) Mode I delamination of angle-ply epoxy/glass fibre laminates exhibiting permanent deformation during fracture, *Comp. Sci. Technol.*, 50, 187-196.
20. S. Hashemi, A. J. Kinloch, and J. G. Williams (1990) Mechanics and mechanisms of delamination in a poly(ethersulfone) fiber composites, *Comp. Sci. Technol.*, 37, 429-462.
21. J. Zhou, T. He, B. Li, W. Liu, and T. Chen (1992) A study of mode-I delamination resistance of a thermoplastic composite, *Comp. Sci Technol.*, 45, 173-179.
22. P. J. Hine, B. Brew. R. A. Duckett, and I. M. Ward (1998) The fracture behaviour of carbon fibre reinforced poly (Ether Ether cetone), *Comp. Sci. Technol.*, 33, 35-71.
23. Gehlen, P. C. Popelar, C. H., and M. F. Kanninen (1979) Modeling of dynamic crack propagation: I. Validation of one-dimensional analysis, *Int. J. Fract.*, 15, 281-294.
24. M. Ashizawa (1981) *Fast interlaminar fracture of a compressively loaded composite containing a defect*, 5th DoD/NASA Conference on Fibrous Composites in Structural Design, Naval Air Development Center, Report 81-096-60, New Orleans.
25. S. Hashemi, A. J. Kinloch, and J. G. Williams (1990) The Analysis of Interlaminar Fracture in Uniaxial Fiber-polymer Composites, *Proc. Royal Soc. London*, Vol. A 427, No. 1827, 173-199.
26. J. E. Grady (1992) Fracture toughness testing of polymer matrix composites, in *Handbook of Ceramics and Composites*, Vol. 2 Mechanical Properties and Specialty Applications, (N. Cheremisinoff ed.), Marcel Dekker, New York.
27. J. M. Whitney, I. M. Daniel, and R. B. Pipes (1984) in *Experimental Mechanics of Fiber Reinforced Composites*, Revised edition, Publ. By the Society of Experimental Mechanics, Prentice-Hall Inc., Englewood Cliffs, NJ.
28. J. R. Rice (1968) A path independent Integral and the approximate analysis of strain concentrations by notches and cracks, *J. Appl. Mech.*, 35, 379-386.
29. P. E. Keary, L. B. Ilcewicz, C. Shaar, and J. Trostle (1985) Mode I interlaminar toughness of composite using slender double cantilevered beam specimens, *J. Compos. Mater.*, 19, 154-162.

30. J. D. Landes and J. A. Begley (1972) *The J-integral as a fracture criterion*, Fracture toughness, Part II ASTM STP 514, 1-23.
31. J. D. Landes and J. A. Begley (1977) Recent development in J-ic testing, Development in fracture mechanics test methods standarization, ASTM STP 632 65-120.
32. M. R. Mili, D. Rouby, and G. Fantozzi, (1990) Energy toughness parameters for a 2D carbon fibre reinforced carbon composite, *Composite Science and Technology*, **3**, 207-221.
33. A. J. Russel (1987) Micromechanics of Interlaminar Fracture and Fatigue, *Polym. Compos.*, 342-350.
34. W. S. Johnson and P. D. Mangalgiri (1986) *Investigation of fiber bridging in DCB specimens*, NASA, Tech. Memo 87716.
35. P. Davis and M. L. Benzeggagh (1989) Interlaminar mode I fracture testing, in Applications of Fracture Mechanics to Composites, *Composite Materials Series*, (K. Friedrich ed.), Elsevier Science Vol. 6, Amsterdam, 81-112.
36. B. F. Sorensen and T. K. Jacobsen (1998) Large-scale bridging in composites R-curves and bridging laws, *Composites*, Part A, 29A, 1443-1451.
37. V. Kostopoulos, Y. P. Markopoulos, Y. Z. Pappas, and S. D. Peteves (1998) Fracture energy measurements of 2-D carbon/carbon composites, *Journal of Ceramic Society*, **18**, 69-79.
38. J. W. Cao and M. Sakai (1995) The crack face bridging of brittle matrix composites, *Proceedings of the 6th International Conference on Fracture Mechanics of Ceramics*, Stuttgart, Germany, 163-176.
39. Y. W. Mai and M. I. Hakeem (1984) Slow crack growth in cellulose fiber cements, *Mater. Sci.*, **19**, 501-511.
40. R. K. Ramkumar and J. D. Whitcomb (1985), *Characterization of mode-I and mixed mode delamination growth in T300/5208 graphite/epoxy*, Delamination and debonding of materials, ASTM-STP-876, 315-335.
41. C. G. Gustafson, and M. Hojo (1987) Delamination fatigue crack growth in unidirectional graphite/epoxy laminates, *J. reinforced Plastics Compos.*, **6**, 36-52.
42. A. J. Russell and K. N. Street (1989) *Predicting interlaminar fatigue crack growth rates in compressively loaded laminates*, Composite Materials: Fatigue and Fracture, ASTM STP 1002, 162-178.
43. C. Dahlen and G. S. Springer (1994) Delamination growth in composites under cyclic loads, *J. Compos. Mater.*, **28**, 732-781.
44. K. Friedrich (1983) Fracture of Polymer Composites, Lecture Notes, Center for Composite Materials, University of Delaware, CCM-83-18, Newark, Del..
45. W. D. Bascom, J. L. Bitner, R. J. Moulton, and A. R. Siebert (1980) The interlaminar fracture of organic matrix woven reinforced composites, *Composites*, **11**, 9-21.
46. G. Smith, A. K. Green, and W. H. Bowyer (1977), *Proceedings Conference Fracture Mechanics in Engineering Practice*, Sheffield, Elsevier Applied Science, Barking, UK, 271.
47. W. D. Bascom, G. W. Bullman, D. L. Hunston and R. M. Jensen (1984), *Proceedings 29th Nat. SAMPE Symp.*, 970.
48. G. C. Christopoulos, Y. P. Markopoulos, and V. Kostopoulos (1998) Interlaminar non-linear fracture energy measurements of thermoplastic composites using DCB specimen configuration, *Proceedings of ECCM/9*, Naples, Italy, 501-508.
49. N. J. Johnston, T. K. O'Brien, D. H. Morris, and R. A. Simonds (1983), *Proceedings 28th SAMPE Symp.*, 502.
50. R. J. Boyce, T. P. Gannet, H. H. Gibbs, and A. R. Wedgewood (1987) *Proceedings 32nd SAMPE Symp.*, 169.
51. K. B. Su (1985), *Proceedings 5th International Conf. On Composite Materials,* San Diego, 995.
52. J. E. O'Connor, A.Y. Lou, and W. H. Beever (1985), *Proceedings ICCM5*, San Diego, 963.
53. D. C. Leach, D. C. Curtis, and D. R. Tamblin (1987) ASTM-STP-937, 358.
54. R. M. Turner and F. N. Cogswell (1986), *Proceedings 18th SAMPE Tech. Conf.*, 32.
55. W. L. Bradley (1989) Understanding the translation of resin toughness into delamination toughness of composites, *Key Eng. Mater.*, **37**, 161-198.
56. D. L. Hunston, R. J. Moulton, N. J. Johnston, and W. D. Bascom (1987), ASTM STP-937, 74.

57. L. B. Ilcewicz, P. E. Keary, and J. Trostle (1988) Interlaminar fracture toughness testing of composite mode I and mode II DCB specimens, *Polym. Eng. Sci.*, **28**, 592-601.

58. S. Mall, W. S. Johnson and R. A. Everett (1984) *Adhesive Joints*, Plenum Press, New York.

59. Y. J. Prel, P. Davis, W. L. Benzeggagh, and F. X. de Charentenay (1989) *Mode I and mode II delamination of thermosetting and thermolastic composites*, composites materials: fatigue and fracture, ASTM STP-1012, 251-261.

60. M. Iwamoto, S. Araki, K. Kurashiki, and K. Saito (1993) Comparison between mode-I interlamina and intralamina fracture toughness of thin unidirectional graphite (AS4)/Epoxy(1908) laminaces, *Proceedings of the ninth International Conference on Composite Materials (ICCM/9)*, Madrid, Spain, 795-804.

61. D. Guedra-Degeorges, A. Dartus, A. Rodriguez, A. Horoschenkoff, P. Vautey, B. Degriny, and V. Kostopoulos, (1994) Development of improved damage tolerant carbon fibers organic composites, publishable synthesis report, BRITE EURAM BREU-0089-C(CD).

62. A. J. Russell and J. M. Street (1982) Factors affecting the interlaminar fracture energy of graphite/epoxy laminates, Progress in Science and Engineering of Composites, (T. Hayashi, K. Kawata, and S. Umekawa eds.), *Proceedings ICCM-IV*, Tokyo, 279-291.

63. W. L. Bradley and R. N. Cohen (1985) *Matrix deformationand fracture in graphite-reinforced epoxies*, delamination and debonding of materials, ASTM STP-876, 389-410.

64. R. A. Jurf and R. B. Pipes (1982) Interlaminar fracture of composite materials, *J. Compos. Mater.*, **16**, 386-396.

65. L. A. Carlsson, J. W. Gillespie Jr, and R. B. Pipes (1986) On the analysis and design of the end notched flexure (ENF) specimen for mode-II testing, *J. Compos. Mater.*, **20**, 594-604.

66. C. R. Corleto and H. A. Hogan (1995) Energy release rates for the ENF specimen using a beam on an elastic foundation, *J. Compos. Mater.*, **29**, 1420-1436.

67. L. A. Carlsson and J. W. Gillespie Jr. (1989) Mode-II interlaminar fracture of composites, in *Application of Fracture Mechanics to Composite Material*, composite materials series vol. 6, (K. Friedrich ed.), Elsevier science, Amsterdam, 113-157.

68. J. M. Withney (1988) Analysis of the end notch flexure specimen using a higher order beam theory based of Reissner's principle, *Proceedings American Society For Composites*, 3rd Technical Conference, Technomic Publishing, Lancaster, PA.

69. J. M. Withney (1989) Interlaminar Fracture Characterization of Composite Materials, in *Encyclopedia of Composites*, VCH.

70. S. Mall and N. K. Kochhar (1986) Finite elements analysis of the end-notch flexure specimens, *J. Compos. Technol. Res.*, **8**, 54-57.

71. J. W. Gillespie Jr, L. A. Carlsson, and R. B. Pipes (1986) Finite elements analysis of the end-notch flexure (ENF) specimen for measuring mode II fracture toughness, Compos. Sci. Technol., **27**, 177-186.

72. T. Vu-Khanh (1987) Crack-arrest study in mode I delamination in composites, *Polym. Compos.*, **8**, 331-341

73. J. G. Williams (1987) Large displacement and end block effects in the DCB inerlaminar test in modes I and II, *J. Compos. Mater.*, **21**, 330-348.

74. Y. Wang and J. G. Williams (1992) Corrections for mode II fracture toughness specimens of composites materials, *Compos. Sci. Technol.*, **43**, 251-256.

75. S. Hashemi, A. J. Kinloch, and J. G. Williams (1990) The effects of geometry rate and temperature on the mode I, mode II and mixed-mode interlaminar fracture of carbon-fibre/poly (ether-ether ketone) composites, *J. Compos. Mater.*, **24**, 918-955.

76. S. Hashemi, A. J. Kinloch, and J. G. Williams, (1989) Corrections needed in double cantilever beam tests for assessing the interlaminar failure of fibre composites, *J. of Mater. Science Let.*, **8**, 125-129.

77. H. Wang, T. Vu-Khanh, and V. N. Le (1995) Effects of Large Deflection on Mode II Fracture Test of Composite Materials, *J. Compos. Mater.*, **29**, 833-849.

78. W. L. Bradley and R. N. Cohen (1985) Matrix Deformation and fracture in graphite reinforced epoxies, Delamination and Debonding of Materials, ASTM STP 876, 389-410.

79. M. L. Benzeggagh, X. J. Gong, and J. M. Roedlanlt (1989) A mixed mode delamination specimen and its finite element analysis, *Proceedings of the 7th Conference on Composite Materials*, China, 210-219.
80. M. L. Benzeggagh, X. J. Gong, P. Davies, J. M. Roedlanlt, Y. Mourin, and Y. Prel (1989) A mixed mode specimen for interlaminar fracture testing, *Comp. Sci. Technol.*, **34**,129-143.
81. R. A. Jerf, R. B. Pipes (1982) Interlaminar fracture of composite materials , *J. Compos. Mater.*, **16**, 386-94.
82. D. J. Wilkins, J. R. Eisenmann, R. A. Camin, W. S. Margolis, and R. A. Benson (1980) *Characterising delamination growth in graphite-epoxy*, Damage in Composite materials, ASTM STP 77, 168-180.
83. T. K. O'Brien (1982) *Characterization of delamination growth in a composite laminate*, Damage in composite materials, ASTM STP 775, 140-167.
84. M. L. Benzeggagh and M. Kenane (1996) Measurement of mixed-mode delamination fracture toughness of unidirectional glass/epoxy composites with mixed-mode bending apparatus, *Compos. Scien. Technol.*, **56**, 439-449.
85. T. K. O'Brien, G. B. Murri, and S. A. Salpekar (1989) *Interlaminar shear fracture toughness and fatigue thresholds for composite materials*, composite materials: fatigue and fracture, Vol. 2, ASTM STP-1012, 222.
86. K. Kageyama (1989) Fracture mechanics of notched carbon/epoxy laminates in *Application of Fracture Mechanics to Composite Materials*, composite materials series vol. 6, (K. Friedrich ed.), Elsevier science, Amsterdam, 327-396.
87. V. Kostopoulos and Y. P. Markopoulos (1998) On the fracture toughness of ceramic matrix composites, *Materi. Sci. Eng.*, A 250, 313-319.
88. F. E. Heredia, S. M. Spearing, T. J. Mackin, M. Y. He, A. G. Evans, P. Mosher, and P. Brondsted (1994) Notch effect in carbon matrix composites, *J. Am. Ceram. Soc.*, **77**, 2817-2827.
89. ASTM E 610-82, Standard Definition of Terms Relating to Acoustic Emission.
90. I. A. Victorov (1967) Rayleigh and Lamb Waves, Penum Press, New York.
91. S. J. Vahaviolos (1974) Real time detection of microcracks in brittle materials using stress wave emission (SWT), *IEEE Trans.*, PHP-10, No. 3, 152-159.
92. V. Kostopoulos, Y. P. Markopoulos, Y. Z. Pappas and S.A. Paipetis (1998) Bridgig stresses in 2-D carbon/carbon composites using acoustic emission monitoring, *Proceedings of 8th European Conference of Composite Materials (ECCM8)*, Naples, Italy, 293-300.
93. Y. Z. Pappas, V. Kostopoulos, Y. P. Markopoulos, and S. A. Paipetis (1997) Failure mechanism analysis of ceramic matrix composites based on the pattern recognition approach of acoustic emision, *Proceedings of the 1st Hellenic Conference on Composite Materials and Structures*, Vol. II, Xanthi, Greece, 192-206.
94. Y. Z. Pappas, V. Kostopoulos, and Y. P. Markopoulos (1998) Failure mechanism analysis of 2D carbon/carbon using acoustic emission monitoring, *NDT. And E.International.*, **3**, 157-163.
95. T. P. Phillipidis, V. N. Nikolaidis, and A. A. Anastassopoulos (1998) Damage characterization of carbon/carbon laminates using neural network techniques on AE signals, *NDT. and E.International*, **5**, 329-340.
96. A. A. Pollock, (1995) Technical report, Physical Acoustics Corporation, Princeton.
97. T. W. Anderson (1984) An Introduction to Multivariate Statistical Analysis, Wiley, New York.
98. S. Botten (1993) *Use of pattern recognition as method to evaluate acoustic emission signals,* ASME. 257, 256-262.
99. K. Fukunaga (1990) Introduction to Statistical Pattern Recognition, Academic Press, New York.
100. B. D. Ripley (1996) Pattern recognition and neural networks, Cambridge University Press, London.
101. T. Y. Young (1974) Classification estimation and pattern recognition, Elsevier, Science, New York.

5

Notch Size Effects

C. Bathias

CNAM-ITMA, Paris, France

5.1 Introduction

It is well known that the residual strength of laminate is drastically reduced by notch effects. Since the local and average stress criteria were established,[1] several models—for examples, the progressive degradation model,[2] the point-strength model, and the minimal-strength model[3]—have been proposed to predict notched laminate strengths and explain the damage mechanism under monotonic uniaxial loading.

In the two models proposed by Tan,[3] the stress field around a hole in a laminate was described by an approximate and an elasticity solution.[4] In order to consider the influence of finite width on the stress distributions, first, isotropic finite-width correction factors were approximately applied to orthotropic laminates[3]; then, the infinite-width correction factors applicable to anisotropic or orthotropic laminates were derived[3]; more recently, the isotropic finite-width correction factors have been modified so that they can be more easily applied to anisotropic or orthotropic laminates.[5] So far, a numerical stress field around the hole in a laminate has not easily been adopted in the models. In fact, if the characteristic dimensions are not expressed as a function of the opening size, the results of the prediction are not very good, especially for orthotropic materials.

As the ply rigidity degraded, the local stress distributions (ply by ply) were recalculated and the characteristic dimensions of every critical ply were determined so that the models could be more reasonably applied to the lamination analysis and used to study the damage mechanism, including damage sequence, damage location, and when the damage appears. However, because the global stress distribution determined in the light of the laminate rigidity was not recalculated with ply rigidity degradation, and the isotropic finite-width correction factors were approximately applied to orthotropic laminates, the results of the prediction for orthotropic laminates are not very good.[6]

The most remarkable characteristic continues to be the fatigue strength of notched composite materials, whose ratio K_f between the unnotched and notched endurance limits is equal to or less than unity.[7] This means in general that these composite materials are not sensitive to the effect of the notch under cyclic loading. It is then reasonable to make two observations. First, composite materials, very vulnerable to the monotonic stress concentration, are no longer vulnerable to the cyclic stress concentration in fatigue. Second, compared to metals, these materials are of considerable interest in fatigue in the presence of a notch, since their endurance limit will be 2–3 times greater for a given ultimate tensile strength, and still more if specific strength is considered.

5.2 Premature Fracture in Tension Caused by a Notch

5.2.1 Notch effects

In the absence of brittleness, it is observed that the presence of a notch or a hole appreciably increases the apparent yield strength and even the ultimate tensile strength of metals. This phenomenon is due to the effect of triaxial stresses at the notch tip that result in a confinement of plastic deformation and slow the spread of plasticity. For example, the yield strength of a 2024T351 aluminum alloy, which is about 300 MPa, increases to 550 MPa in the presence of a notch with a theoretical stress concentration factor $K_t = 3.3$.

In carbon fibre composites the effect of a notch is totally different.[8] For quasi-isotropic stacking orientations, a hole or a notch cuts the ultimate tensile strength in half. In the case of isotropic stacking with an arrangement $(0/+45/90/-45)_s$, the residual strength falls to 260 MPa for an original tensile strength of 533 MPa. An increase in the number of layers at $0°$ significantly reduces this effect.

In a general way, the notch effect depends on the geometry. Figure 5.1 gives the fracture results for quasi-isotropic carbon fibre composites $(0,90,0 \pm 45,0)_s$, which show that the residual strength falls from 80% to 20% of the nominal strength when the size of the notch increases.[8] A notch having the same diameter as a hole is even more serious.

Fraction of unnotched
  tensile strength

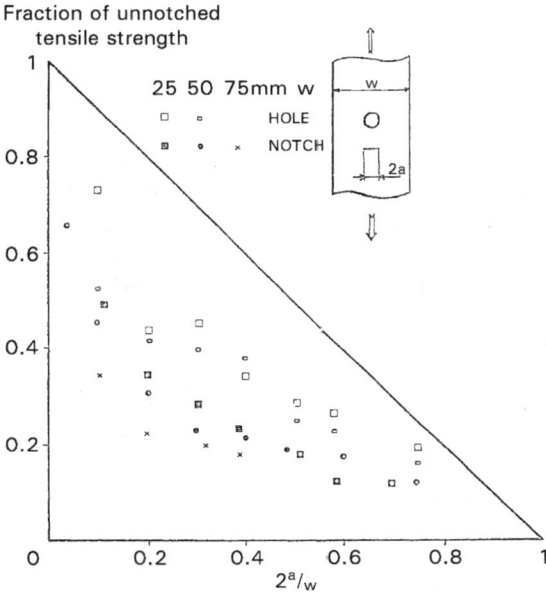

Figure 5.1 Notch effect in a CFRP $(0/90/0\pm45/0)_s$, in
tension[8]

Technologically, the substitution of a metallic alloy by a composite must consider that the admissible stress around a hole is about 2–3 times lower in the composite. It is worth mentioning that the behavior of the composite is distinctly different under cyclic loading. The fatigue behavior of notched composites is unaffected by the notch with respect to the static residual strength.

These observations show, therefore, that the substitution of a metal by a composite reverses the problem, the design of a metallic part requiring a careful fatigue study while the same part made of carbon fibre composites will have to be studied instead for resistance to static forces.

5.2.2 Toughness of composites materials

The application of fracture mechanics to notched laminates plates is more difficult. In some cases with a quasi-isotropic lay-up it is always possible to use fracture mechanics to predict the failure of notched plates. However, where a strong orthotropic effect appears, fracture mechanics criteria do not work very well. Other criteria such as the Nuismer and Whitney approach are recommended.

A priori to escape from the difficulties presented by anisotropy in composites, one preferably uses the dissipated energy G, from which one will possibly be able to extract a stress intensity factor. In both cases, it will be convenient to use the fracture criterion defined by the critical value G_c.

Nevertheless, the "macroscopic" anisotropy of the material must be such that fracture takes place in a plane of symmetry identical, macroscopically speaking, to the plane of the notch from which the fracture has originated. Any and all branching-off and/or change in mode results in the problem becoming even more complex, whether in metals or composites.

On reflection, it is clear, at the macroscopic scale, that metals are highly anisotropic because of the disorientation of the grains, the distribution of the symmetrical elements naturally being more random than in composites. Finally, it is anisotropy at the macroscopic scale that counts in the fracture analysis of the energy criterion G.

The "plasticity" effect for high-performance composites is small and more easily disregarded than in the case of metals. The anisotropic effect is not problematic for many quasi-isotropic composites and for the treatment of delamination. On the other hand, branching-off of fracturing in composites represents a big problem.

For carbon fibre composites having quasi-isotropic macroscopic properties, linear mechanics is quite applicable, especially in relation to the size and geometrical effect. Nevertheless, in the general case an orthotropic approach is necessary.

In carbon fibre composites with an isotropic stacking orientation, the critical stress intensity factor K_c (which is an indication of a materials fracture toughness) is between 40 and 50 MPa $\sqrt{m}$ for stacked woven plates 2 mm thick, with layers oriented at 45° intervals.

It should be noted that a high-strength aluminum alloy with the same 2 mm thickness attains a K_c of 80 MPa $\sqrt{m}$. A priori, the composite's toughness in the limit of small thicknesses is much less than that of rival aluminum alloys; the situation is different, however, for large thicknesses.

In effect, up to a thickness of 10 mm the value of K_c for the composite remains constant, while that of the alloy falls by about 50%. It follows that the toughness of composites compared with metals is small, and all the more so when the thickness of the plate is small. From the parts design point of view, a new difference appears between metal and composite: while fracture mechanics recommends small thickness to guarantee better toughness for metals, the same criteria applied to composites tend to recommend the opposite.

The materials that will best resist tearing for a given load are thick, stacked woven composites.

5.2.3 Criteria and models

Another approach to predicting the fracture of a notched plate has been proposed by Whitney and Nuismer using the stress at the tip of the notch. Since the point stress criterion and the average stress criterion were established by Whitney and Nuismer, a number of articles have been published concerning the strength prediction of notched composite laminates under uniaxial load. All the criteria employ a characteristic dimension that expresses the characteristic damage zone size. The characteristic dimension was considered as a universal constant; it was then treated as a function of the hole diameters. Although these criteria containing the characteristic dimension can provide good agreement with experimental data for the strength prediction, they leave much to be desired for the explication and study of the damage mechanism in notched laminates.

5.2.3.1 The point and average-stress criteria: PSC and ASC.

The two criteria are based on the stress distribution along the ligaments of notched laminates. According to PSC, the stress reduction factor (SRF) is given by

$$\text{SRF} = \frac{\sigma_N^x}{\sigma_0} \frac{2}{2 + \xi_1^2 + 3\xi_1^4 - (K_T^x - 3)(5\xi_1^6 - 7\xi_1^8)} \tag{5.1a}$$

$$\xi_1 = \frac{R}{R + a_0} \tag{5.1b}$$

According to the ASC:

$$\text{SRF} = \frac{\sigma_N^x}{\sigma_0} \frac{2(1 - \xi_2)}{2 - \xi_2^2 - \xi_2^4 - (K_T^x - 3)(\xi_2^6 - \xi_2^8)} \tag{5.2a}$$

$$\xi_2 = \frac{R}{R + a_1} \tag{5.2b}$$

where

$$K_T^x = 1 + \sqrt{\left[\frac{2}{A_{22}} \left(\sqrt{A_{11}A_{22}} - A_{12} + \frac{A_{11}A_{22} - A_{12}^2}{2A_{66}} \right) \right]} \tag{5.3}$$

and σ_N^x is the notched strength for an infinite-width laminate; σ_0 is the unnotched strength of the laminate; R is the hole radius; a_0 and a_1 are the characteristic dimensions that must be experimentally determined; [A] is the laminate rigidity matrix. The two criteria are shown in Fig. 5.2a; σ_y^x is the stress applied to the notched laminate.

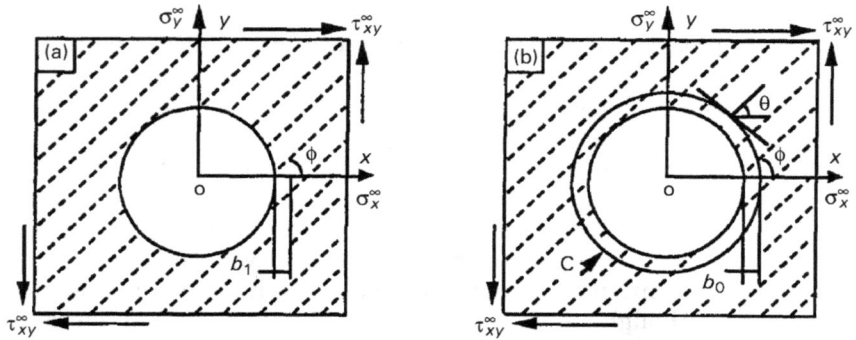

Figure 5.2 Damage models: (a) point-strength model; (b) minimum strength model.

5.2.3.2 The two-parameter criterion: TPC.

The characteristic dimension in the PSC was expressed as a function of the hole radius. Therefore, the PSC takes the following form:

$$\text{SRF} = \frac{\sigma_N^x}{\sigma^0} \tag{5.4a}$$

$$= 2/(2 + [f(R)]^{-2} + 3[f(R)]^{-4} + (K_T^x - 3)\{5[f(R)]^{-6} - 7[f(R)]^{-8}\}$$
$$\times f(R) = 1 + C^{-1}R^{m-1}) \tag{5.4b}$$

where C and m are the materials constants describing the notch sensitivity and which are determined by experiment.

5.2.3.3 The progressive-degradation model and minimum strength model: PSM and MSM.

The derivation of Eqs. (5.1) and (5.2) is based on an assumption about the stress distribution along the ligament adjacent to the hole. However, three stress components (two-dimensional) around the opening are considered in the PSM and MSM. A detailed description of the PSM and MSM has been given.[3–6] They will be described briefly as follows.

For an infinite symmetric anisotropic laminate with an opening, the stress distribution around the opening has been obtained by using a complex variable method.[4] The stresses have been derived by using a superposition principle with the assumption that the materials is homogeneous in each ply:

$$\sigma_i = \sigma_i^* + \sigma_i^0 \qquad (i = 1, 2, 6) \tag{5.5}$$

where 1 and 2 represent two principal axes of the opening; σ_i are the laminates stresses; σ_i^* indicate the stress components due to the opening; and σ_i^0 are the components due to the uniform stress field.

The coordinate systems is as shown in Fig. 5.2b, in which ϕ is the fiber orientation angle.

$$\sigma_1^* = \mathrm{Re}\left[\frac{\mu_1^2 f_1 g_2 - \mu_2^2 f_2 g_1}{\mu_1 - \mu_2}\right] \tag{5.6a}$$

$$\sigma_2^* = \mathrm{Re}\left[\frac{f_1 g_2 - f_2 g_1}{\mu_1 - \mu_2}\right] \tag{5.6b}$$

$$\sigma_6^* = \mathrm{Re}\left[\frac{\mu_1 f_1 g_2 - \mu_2 f_2 g_1}{\mu_1 - \mu_2}\right] \tag{5.6c}$$

where
$$f_j = \frac{(1 - i\mu_j\lambda)}{\beta\sqrt{\beta^2 - 1 - \mu_j^2\lambda^2} + \beta^2 - 1 - \mu_j^2\lambda^2} \tag{5.7a}$$

$$g_j = i\lambda\sigma_1^0 - u_j\sigma_2^0 - (1 - i\mu_j\lambda)\sigma_6^0 \tag{5.7b}$$

$$\beta = (1 + \alpha)\cos\theta + \mu_j(\lambda + \alpha)\sin\theta \qquad (j = 1, 2) \tag{5.7c}$$

$$\lambda = b/a \tag{5.7d}$$

$$\alpha = b_0/a \tag{5.7e}$$

and a and b are the semi-major and semi-minor axes of the elliptical opening, respectively; b_0 is the characteristic dimension at a distance b_0 from the opening contour; and θ is the angle between the normal vector of the characteristic curve and the 1 axis (Fig. 5.2b). The components of the uniform stress field are as follows:

$$\sigma_1^0 = \sigma_x^x \sin^2\psi + \sigma_y^x \cos^2\psi + \tau_{xy}^x \sin 2\psi \tag{5.8a}$$

$$\sigma_2^0 = \sigma_x^x \cos^2\psi + \sigma_y^x \sin^2\psi + \tau_{xy}^x \sin 2\psi \tag{5.8b}$$

$$\sigma_6^0 = (\sigma_y^x - \sigma_x^x)\sin\psi \cos\psi - \tau_{xy}^x \cos 2\psi \tag{5.8c}$$

where σ_x^x, σ_y^x and τ_{xy}^x are the stresses applied to the laminate, and ψ is the slant angle between the semi-major axis of the opening and the y-axis. The complex roots, μ_j, are the solutions of the following equation:

$$s_{11}\mu^4 - 2s_{16}\mu^3 + (2s_{12} + s_{66})\mu^2 - 2s_{26}\mu + s_{22} = 0 \tag{5.9}$$

where S_{ij} are the components of the compliance matrix of the laminate.

The two models are based on the evaluation of the notched first-ply failure (FPF) stress. In the first model (PSM), the ratio of notched to

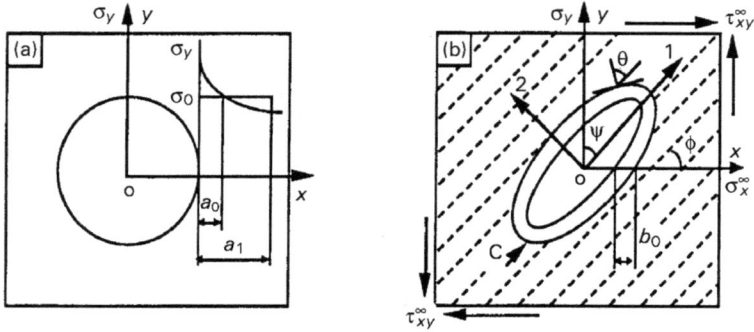

Figure 5.3 Coordinate system of an infinite laminate with an opening: (a) for the PSC and the ASC; (b) for the PSM.

unnotched FPF strengths is proposed to be the same as the ultimate SRF, i.e.,

$$SRF = \frac{\sigma_N^x}{\sigma^0}$$

$$= \frac{\text{Notched FPF strength at point } (d1, 0)}{\text{Unnotched FPF strength}} \qquad (5.10)$$

This model is shown in Fig. 5.3a.

In the second model (MSM), the stress distribution along curve C, which is parallel to the hole contour with a characteristic distance d_0 apart, is considered. The curve is called the characteristic curve. Along it, unnotched strength is proposed to be the same as the ultimate strength reduction factor, i.e.,

$$SRF = \frac{\sigma_N^x}{\sigma^0}$$

$$= \frac{\text{Notched minimum FPF stress at curve } C}{\text{Unnotched FPF strength}} \qquad (5.11a)$$

The characteristic curve C can be expressed

$$\frac{x_2}{(a+d_0)^2} + \frac{y^2}{(b+d_0)^2} = 1 \qquad (a = b) \qquad (5.11b)$$

This model is shown in Fig. 5.2b.

In our case, λ in Eqn (5.7d) is equal to 1 $(a = b)$; the slant angle, ψ, is equal $90°$; $\sigma_x^x = \tau_{xy}^x = 0$. The FPF strength is determined from the Tsai–Wu quadratic failure criterion containing the stress interaction term $F_{12}^* = -0.5$.

Experimental validation.   Laminate T300/epoxy is used, of which the stacking sequences are

$$X: (0/0/45/45/-45/-45/90/90)_s$$
$$J: (0/90/45/-45/-45/45/90/0)_s$$

All specimens are cut from a 300 × 600 plate. They have fixed ratio of width to diameter, equal to 5. They are respectively 15, 25, 40, 50 mm wide by 250 mm long including 75 mm of each end tap. Each ply has the same thickness, 0.125 mm. The machining of specimens is carried out with diamond tools. The elastic properties of each ply in the notched laminates are given.[9]

Comparisons between the ultimate strength predictions and the experimental data are illustrated in Figs. 5.4 and 5.5. All the predictions agree reasonably well with the experimental data. In particular, the prediction of the minimum strength model is more accurate than that of the others due to the consideration of stress distribution around the hole other than along the ligament.

It can be seen from these figures that the characteristic dimension is function of stacking sequences.

After the first laminate degradation (90° ply fails), the following +45° and −45° plies can be retraited as the so-called first failure plies and a new SFR–radius relation that can also give a good prediction result can be calculated by using the Tan models and classical lamination theory. Based on the same treatment for the 0° ply, after the second laminate degradation (+45° and −45° plies fail), another new SFR-radius relation can also give a good prediction.

Damage mechanism in tension.   The PSM and MSM can provide some damage information in notched laminates: (1) calculation of the safety coefficients in each ply permits us to know which ply will be damaged first; (2) determination of the damage angle in each ply, θ, permits us to know where the damage occurs the hole; (3) evaluation of the first-failure strength permits us to know when the first damage occurs.

The damage mechanism in the two nonwoven laminates (T300/914-1 and 2) has been elucidated. The damage processes are studied by acoustic emission (AE) and radiography. The AE rate as a function of the load applied, together with the strain given by the fifth gage (near the straight free-edge) at the ligament, are recorded at the same time. The notched FPF stresses can be obtained, but no marked knees, due to damage, in the load–strain curves are observed. However, as shown,[6] evident nonlinearity due to damage can be observed from the curve of the applied load versus the strain given by the first gage

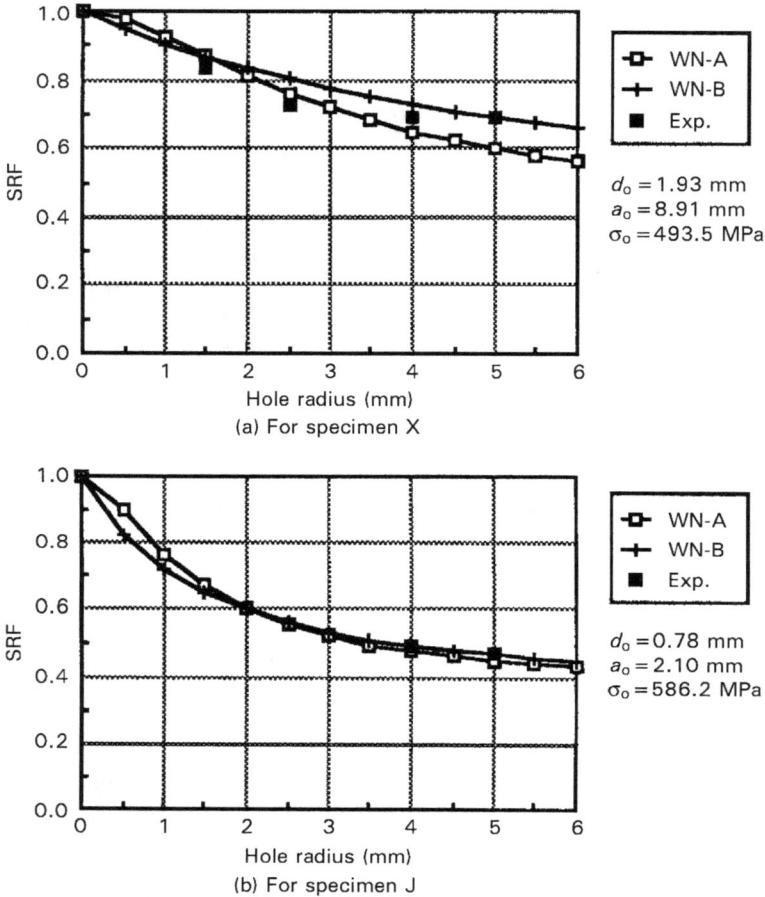

Figure 5.4 Comparison between the experimental results and the criteria WN.

(near the hole border) at the ligament, because the damage is very localized in the nonwoven notched laminate.

Radiography permits verification of the AE measurement results and helps to understand the various types of damage inside laminates in detail (Fig. 5.6).

The damage mechanism for two nonwoven laminates can be partially predicted by the PSM and MSM, and verified by experiment. Viewed experimentally, the T300/914-1 undergoes damage stages follows: (1) Cracks in the 90° plies, as the first intrinsic material damage is initialized at the hole border. The corresponding load is almost 40–55% of the ultimate values. (2) The cracks in 90° plies are

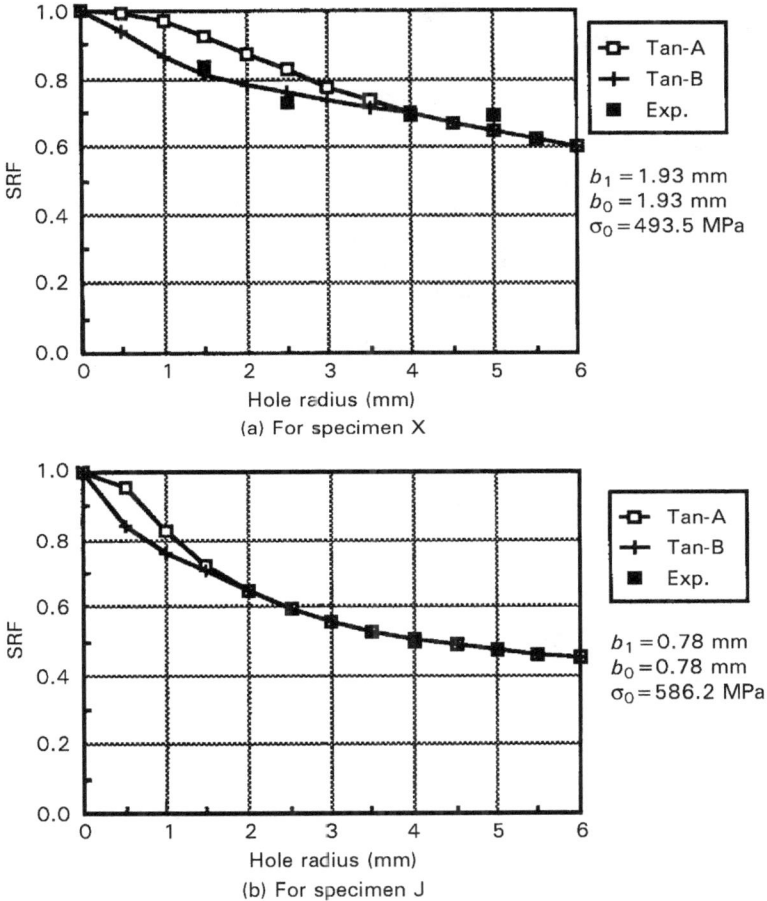

Figure 5.5 Comparison between the experimental results and the Tan criteria.

multiplied to the specimen length direction and evolve toward the straight free edges. Affected by the damage in the 90° plies, the ±45° plies are cracked both from the hole border and from the cracks in the 90° plies. Cracks appear also in the 0° plies near the hole border; in particular, delamination occurs around the hole and adjacent to the straight free edges. (3) The final fracture in each ply occurs at different angles around the hole. The fracture in the 0° plies is influenced by that in the 90° plies. From the point of view of prediction, the predicted ply damage order is the same as the experimental observation results; the predicted first failure strength is higher than that obtained from the experiment, because the former is for a point far from the hole

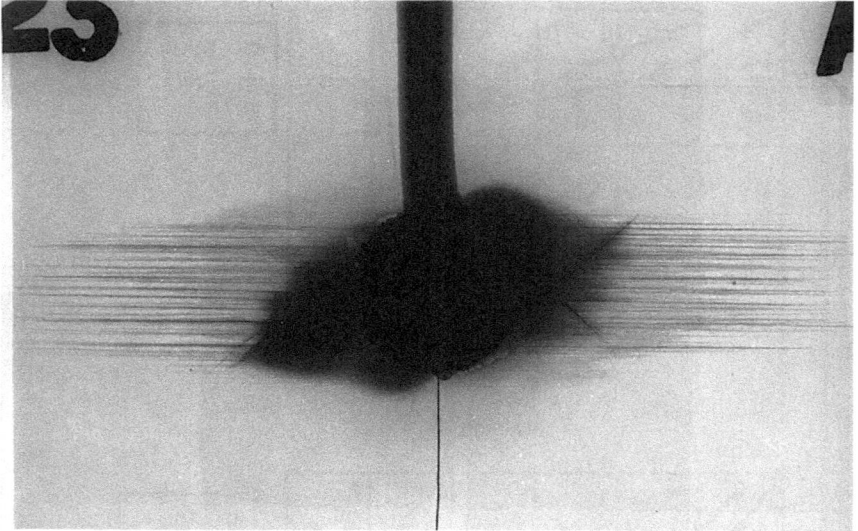

Figure 5.6 Radiography of damage zone around a hole in CFRP.

border but the latter is just for the border; in particular, the fracture angles in the 0° plies are influenced by those in the 90° plies.

For T300/914-2, the first damage may also be presented in the form of cracks in the 90° plies, initialized from the hole border. But they are neither evident nor numerous. Before the final fracture, damage in other plies and remote from the ligament is not observed, neither is any delamination near to the free edges. The fracture surface is smoother. This means that the stacking sequences influence not only the delamination adjacent to the free edges but also the damage development in the plies, as pointed out in.[10]

However, the damage mechanism of the three woven glass laminates is very different from that of the two non-woven laminates, and this is not yet understood. As indicated in,[6] although the specimens are loaded almost to 90% of the ultimate tensile strength, and a small impervious reservoir permits them to be penetrated by an opaque penetrant in a tensile state, neither cracks nor delamination can be observed by the usual experimental techniques such as microscopy and X-ray radiography. It is certain that damage occurs, because many acoustic emission signals appear very early, despite a lack of reproducibility. But the form in which damage occurs and the ways in which it can be observed and verified remain problems to be solved. At the laminate fracture surfaces, no visible cracks can be observed, even in the 90° plies, but there are evident delamination phenomena in the glass laminates.

It is the fabric structures that inhibit the damage from being observed by microscopy and X-ray radiography, and that delay the onset delamination in laminates favoring delamination so that the delamination appears at the same time as the final fracture. An explanation of the damage mechanism with reference to the models for the woven cases analogous to that for the nonwoven cases remains to be found.

5.3   Effect of a Notch on the Fatigue Strength

5.3.1   Fatigue strength in laminates

It is now well known that all metals are very notch sensitive and that their endurance limit then falls in significant proportions. In composite materials this phenomenon is practically unknown, which endows them with a definite advantage with respect to metals.

The most spectacular illustration is undoubtedly the Williams experiment,[7] in which the author shows that a quasi-isotropic, notched composite plate with carbon fibers exhibits an endurance limit greater than the residual strength of the material, greater even than the endurance limit of the unnotched material, when the load is applied incrementally (Fig. 5.7).

Endurance curves determined for the quasi-isotropic composites T300/5208 and 914 show that the fatigue limit in repeated tension of the notched material does not differ by more than 10% from that of the unnotched material. As we will see later, these results are explained by the stress relaxation caused by damage.

5.3.2   Effect of a notch in woven composites

Under cyclic loading, the behaviors for woven glass/epoxy laminates are not very well understood. So far, due to the complexity of composite materials, a fundamental understanding of the relationship between damage mechanisms and failure modes that dominate the composite fatigue behavior has not yet been established. Most studies on composite material fatigue behaviors have been carried out only with the help of phenomenological methods. Several statistical models[11] have been proposed to predict fatigue life and residual strength. The parameters in these models, obtained with various statistical methods, for example, the maximum likelihood estimation, are not easy to determine because it is necessary to have a number of specimens. In particular, the fatigue residual strength of a notched laminate does not always vary monotonically with the number of cycles. As pointed out,[7]

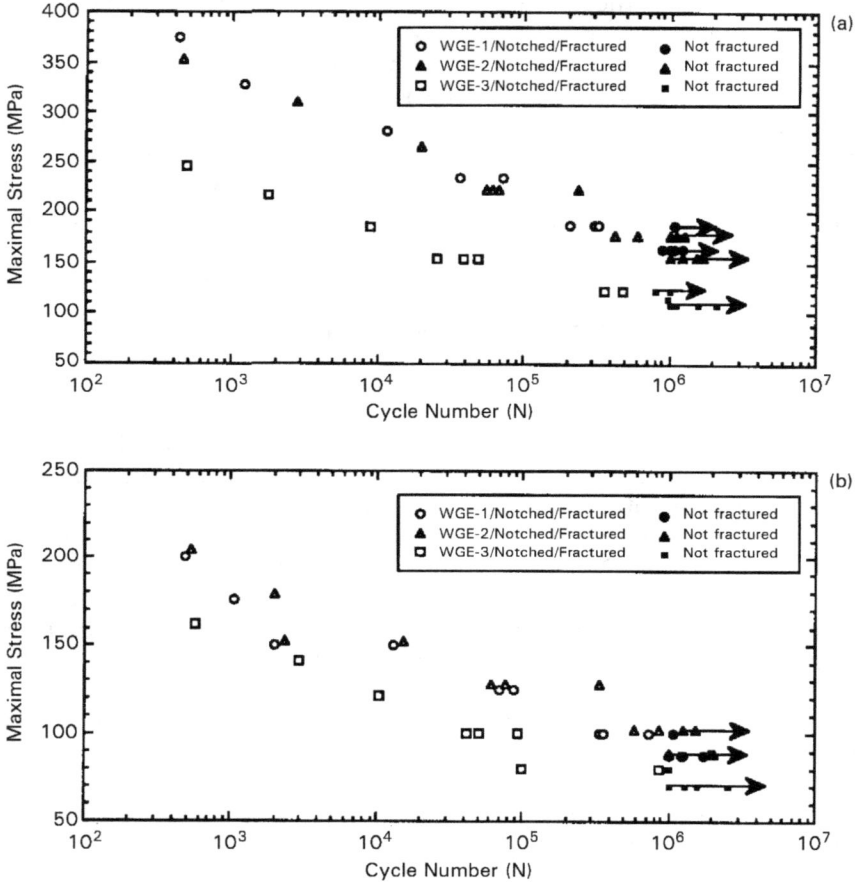

Figure 5.7 Influence of stacking sequence on fatigue strengths of (a) unnotched specimens and (b) notched specimens.

it cannot be preferentially chosen as a parameter characterizing the damage mechanisms. As a result, some researchers considered the apparent secant stiffness changes as parameters characterizing damage mechanisms and measured these changes under different test conditions.

Figures 5.7a and 5.7b show respectively unnotched and notched fatigue strength comparison among the three laminates of different stacking sequence. It can be seen that, as for static tensile strength, one cannot observe a significant fatigue strength difference between the two orthotropic laminates (Glass-1 and 2) although Glass-1 is favorable for delamination, and that Glass-3 is of a much lower fatigue strength than Glass-2 and Glass-3 because of its lower fraction of $0°$ plies (25%) than in the two others (50%).

Figures 5.8a and 5.8b show respectively and unnotched and notched fatigue strength comparison among the three laminates of different stacking sequence at the same ratios of the maximal fatigue stress to UTS. It seems that the Glass-2 has slightly longer fatigue life than the other two, especially at lower stress levels. Microscopic observation can reveal the damage evolution in the laminates. Damage propagation must overcome many more obstacles created by the interface between the plies at $90°$ and the other plies in Glass-2 than in the other laminates, so that the route of the damage propagation in this laminate is much more sinuous than that in the other laminates. This is why this laminate has both a greater residual stiffness during the fatigue tests and a longer fatigue life, as shown in.[10]

The notch effect on the behavior may be discussed. It seems that, at the same ratio of the maximal stress to UTS, for each stacking sequence, the notched specimen has a longer fatigue life than the unnotched specimen, as shown in Figs. 5.9a to 5.9c. The notched laminate specimens have the same ratios of the fatigue strength to UTS as the unnotched specimens, almost equal to 0.40. The ratios of the notched fatigue strength ($\sigma_{d,0}$) are almost equal to their respective statistical strength reduction factors. It can be considered that the fatigue strength for the Glass-3 is the least sensitive to the hole effect because of its higher value of $\sigma_{d,N}/\sigma_{d,0}$.

5.3.3 Fatigue in compression

(1) Fatigue of unnotched plates. Under simple loading, all composite materials present a compressive strength inferior to the ultimate tensile strength and decreasing as a function of the reinforcement: boron, carbon, glass, Kevlar, in that order.

Under cyclic loading, the behavior is the same. At the limit, when the fatigue cycle is entirely in compression, fracture can occur. It has long been shown[12] that the S_d/UTS ratio varies from 460/850 to 200/850 when the load ratio R goes from 0.1 to -1, in the case of quasi-isotropic carbon/epoxy composites. More recently, Williams[7] showed in similar composites (0 ± 30) that the S_d/UTS ratio tended toward 200/850 for a load ratio of $R = 10$, that is to say, with a cycle whose maximum value is zero and whose minimum value is negative (Fig. 5.10).

This last type of damage is unknown in metals and alloys, which do not break in fatigue in compression.

Knowing that composites are not sensitive to the effect of a notch in cyclic tension but that they damage more rapidly in alternating

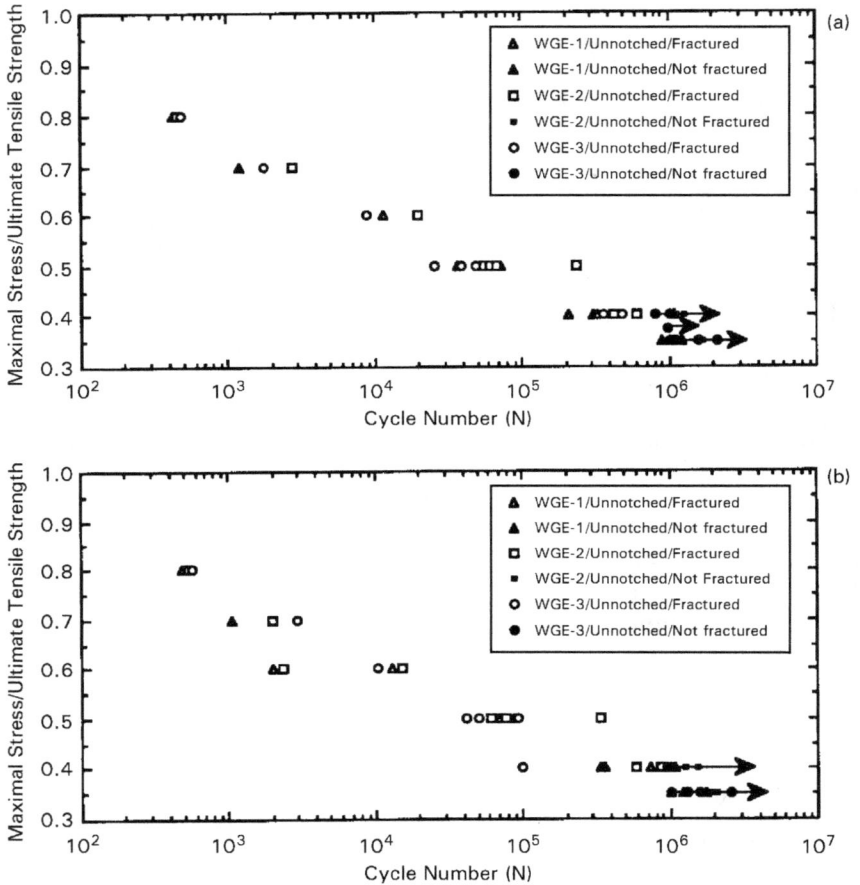

Figure 5.8 Influence of stacking sequence on the fatigue life of (a) unnotched specimens and (b) notched specimens.

tension, it is interesting to consider the fatigue in compression of notched composite plates. For example, in the presence of a 9 mm hole, Stinchcomb[13] finds in a T300/5208 quasi-isotropic composite an endurance limit of 250 MPa for a ratio $R = -1$; this means that in these loading conditions there is a marked notch effect or a marked compression.

To deepen the knowledge of the behavior of these composite materials in compression, we propose to use a modified compact specimen, which allows us to avoid the generalized buckling of the laminated plate when the compressive force is significant.

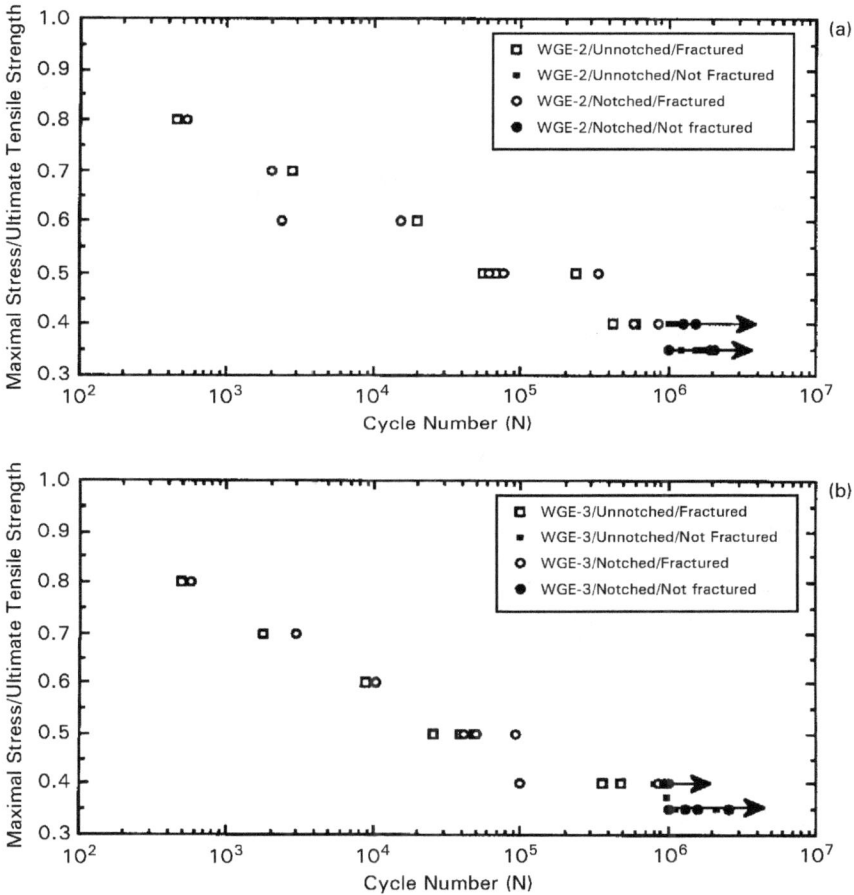

Figure 5.9 Influence of the notch on the fatigue life of the laminate.

(2) Fatigue in compression of notched plates.   The fatigue compression test with modified compact specimens is described elsewhere.[14] It is convenient to note first that the fracture load of the same type of T300/5208 quasi-isotropic composite specimen is greater in compression than in tension, which signifies to a first approximation that the toughness in tension is less than that in compression.

The results show that for a cyclic compression load, $R = 10$, the ratio S_d/UTS in compression–compression tends toward 0.4 while in tension–tension it may be close to 1

In the intermediate case where the loading is in alternating compression–tension, the S_d/UTS ratio is in the neighborhood of

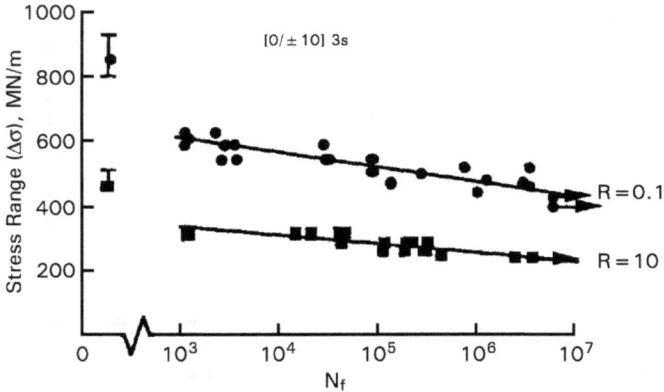

Figure 5.10 $S-N$ curves of $[0/\pm30]_{3s}$ graphite/epoxy. Static strengths are shown at 1/4 cycle.[7]

only 0.3 with respect to ultimate tensile strength but close to 0.5 with respect to compression strength. It emerges from these observations that there exist radically different behaviors in carbon epoxy composite materials according to whether or not a notch is present and whether the load is in tension or compression. The situation can be shown schematically as in Fig. 5.11.

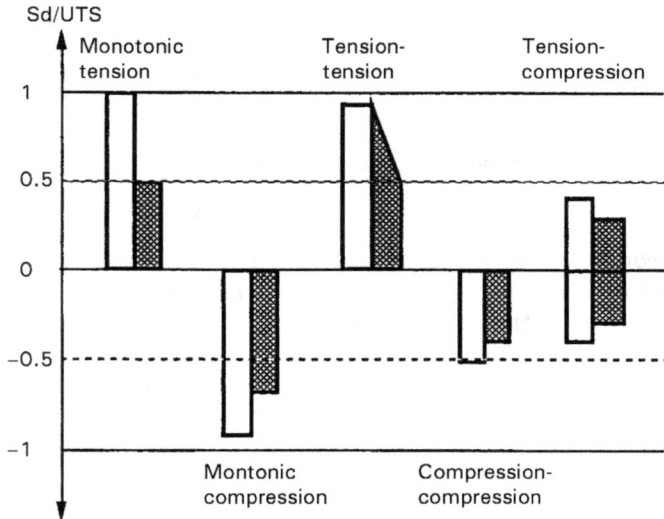

Figure 5.11 Schematic of the effect of loading and notch on the fatigue limit. The crosshatched rectangles represent the behavior of the notched plates.

For the unnotched material, the endurance limit in cyclic tension is between 0.5 UTS and UTS inclusive. It falls below 0.5 UTS in symmetric tension–compression.

For the notched material, the endurance limit is roughly equal to the static residual strength when subjected to cyclic tension–torsion. On the other hand, if the load is compressive, the endurance limit is clearly less than the residual compressive strength.

Finally, whether or not the composite material is notched, a uniaxial compression may result in fracture by fatigue. The process of damage by fatigue in compression results from the formation of delaminations in the external layers of the composite, which then propagate toward the interior until fracture. The stacking sequence, the nature of the resin, and the fibers thus have an important part to play.

References

1. J. M. Whitney and R. J. Nuismer (1974) Stress fracture criteria for laminated composite containing stress concentration, *J. Compos. Mater.*, **8** 253–256.
2. F. K. Chang and K. Y. Chang (1987) A progressive damage model for laminated composites containing stress concentrations, *J. Compos. Mater.*, **21**, 834–855.
3. S. C. Tan (1987) Laminated composites containing an elliptical opening. I. Approximate stress analysis and fracture models, *J. Compos. Mater.* **21**, 925–948.
4. S. G. Lekhnitskii (1986) *Anisotropic Plate*, Gordon and Breach.
5. N. K. Naik and P. S. Shembekar (1992) Notched strength of fabric laminates, I: Prediction, *Compos. Sci. Technol.*, **44**, 1–12.
6. J. Y. Xiao and C. Bathias (1993) Modified Tan's models for the strength prediction of woven laminates with circular holes, *Compos. Eng.*, **3**, 961–973.
7. S. V. Ramani and D. P. Willilams (1977) Notched and unnotched fatigue behavior of angle-ply graphite/epoxy composites, *STP* **638**, 27–46.
8. G. Dorey (1987) Impact damage in composites, *ICCM 6*, **3**, 1–3, 26.
9. J. Y. Xiao (1993) Etude de la prévision de la résistance à la rupture et des mécanismes d'endommagement en traction monotone et cyclique des stratifés composites non tissés et tissés contenant des trous circulaires, Doctoral thesis, CNAM, Paris, France.
10. J. Y. Xiao and C. Bathias (1994) Fatigue behavior of unnotched woven glass/epoxy laminates, *Compos. Sci. Technol.*, **50**, 141–148.
11. C. Bathias (1991) Fracture and fatigue of high performance composite materials, *Eng. Fracture Mech.*, **40**, Nos. 4–5, 757–783.
12. D. Schutz and J. J. Gerharz (Oct. 1987) Fatigue strength of a fibre reinforced materials, *Composites* 245–250.
13. C. E. Bakis and W. W. Stinchcomb, (1986) Response of thick, notched laminates subjected to tension compression cyclic loads, *STP* **906**, 314–334,
14. D. Lai and C. Bathias (1987) The compression fatigue resistance of a carbon–epoxy composite plate containing a hole, *ICM*, **5**, 1231–1238.

Constant and Variable Amplitude Fatigue Damage of Laminated Fibrous Composites

Sp. Pantelakis and G. Labeas

Laboratory of Technology and Strength of Materials,
Mechanical Engineering and Aeronautics Department,
University of Patras, Patras, Greece

6.1 Introduction

In recent decades, fiber reinforced composite laminates have been established as competitive candidates for use in a series of weight-sensitive applications such as aircraft primary structural components, lightweight fast ships, train wagons, and wind energy generators. Widespread use of composite materials is due to their excellent specific properties such as stiffness, strength, and fatigue and corrosion performance. In addition, composite laminates have been recognized as damage tolerant materials in the sense that fatigue damage that accumulates gradually during service does not seriously reduce the residual strength of the component. On the other hand, the design of primary composite structures by applying a fatigue damage tolerance philosophy is far from being formalized and introduced in practice.[1] A main reason for this serious drawback is the lack of fatigue damage functions suitable for use in damage tolerance concepts. It is no small task to formulate quantitatively "damage" that is accumulating gradually during fatigue. Fatigue damage mechanisms of laminated fibrous composites are extremely complex. Nonhomogeneity of the material microstructure causes local failures at certain favorable locations.

Defects are discrete at the microscopic scale level. Depending on the type of reinforcing fibers (glass, carbon, etc.) and laminate (unidirectional or multidimensional), as well as on the applied fatigue load (R-value, fatigue stress level), fatigue mechanisms are different and may be dominated by matrix microcracks arrested by the fibers, transverse matrix cracks, longitudinal matrix cracks, free edge and local delaminations, out of plane buckling of delaminated fibers or plies, fiber breakage, etc.[2]

One should note that even the definition of the word "damage" is not obvious and therefore not unique. Indeed, the fatigue-induced gradual degradation of the internal material integrity in form of matrix cracks, fiber debonds, delaminations, fiber breaks, etc., is a meaningful definition of fatigue damage. Yet consideration of damage as defined above in structural integrity analysis is impractical and this fact calls for more simplified models. In the present work, fatigue damage will be understood as the gradual loss of the material's ability to respond to mechanical loads. It is acknowledged that when adopting the above definition of damage, the rate of accumulation of damage may vary significantly from the degree of material structural changes that take place during fatigue; nevertheless, engineering-wise the question whether a component may retain its integrity is essential. This chapter is mainly concerned with the fatigue-induced degradation of mechanical properties and with the approaches for obtaining analytical formulation of this property loss. The suitability of the proposed analytical formulation for exploitation in fatigue damage tolerance design concepts of laminated composites will also be discussed.

The first section aims to give a short overview of the physical background of fatigue damage. This chapter will only be concerned with those fatigue mechanisms that have a direct effect on materials' mechanical properties or fatigue life. In Section 6.2 the different fatigue analysis concepts are briefly reviewed and are discussed with regard to their suitability for composite structures. In Section 6.3 the damage tolerance design of composite materials is discussed, with reference to current design practice. In Section 6.4 the different fatigue damage functions are categorized with regard to the damage parameter involved. In Sections 6.5 and 6.6 two different approaches to defining fatigue damage functions and performing the subsequent fatigue analyses are discussed extensively. Section 6.5 focuses on the presentation of characteristic fatigue damage functions and fatigue analysis concepts that are based on the exploitation of S–N curves. These concepts have attracted considerable interest due to the similarities to the fatigue analysis approaches applied on metallic structures, as well as due to their simplicity in application in the design; they are widely applied in today's design practice, particularly when the components are expected to perform under variable fatigue load service spectra. Discussed in Section 6.6 are concepts for assessing fatigue damage and material property degradation using results from nondestructive evaluation data. Emphasis is given to the presentation of the damage severity factor (DSF) approach. This approach provides a damage tolerance background for fatigue analysis of composite laminates. Finally, in Section 6.7 the issue of establishing fatigue damage functions under variable fatigue spectra is discussed. For the formulation of the required fatigue damage functions, a concept utilizing material property degradation D as damage parameter is proposed.

6.2 Fatigue Damage Mechanisms in Laminated Composite Materials

The fatigue damage mechanisms of engineering composite materials are very complex, and are therefore far from being fully understood. This complexity is due to the high inhomogeneity of composites, as well as to the existence of the "initial damage," already present before the exposure of the material to any loading. During fatigue all these defects interact, the interactions being dominated by different mechanisms. As indicated, defects do not coalesce into a dominant macrocrack, so that the commonly accepted fracture mechanics models associating the crack growth rate with the incremental change of the crack tip stress intensity factor range cannot be applied. Establishment of generic criteria for predicting local fatigue failure within a composite is very difficult; although such criteria have been advanced,[2] they remain very limited due to their excessive complexity for practical application. One should consider that even the calculation of local stresses within a composite material subjected to fatigue is not a small task, and in many practical cases is still not manageable at present. In addition, during fatigue the damage accumulates gradually, but for the purpose of fatigue analysis fatigue damage has to be quantified and accounted for.

To better understand the fatigue damage mechanisms, description of the complexity of fatigue damage should start from the simpler case of unidirectional (UD) composites loaded in tension parallel to the fibers. Even for this simplest case, the failure mechanisms can be divided into three basic modes as shown in Fig. 6.1.

Fiber breakage occurs when the local stress exceeds the strength of the weakest fiber, causing shear stress concentration at the fiber–matrix interface near the broken fiber tip. The interface area acts as a stress concentrator for the longitudinal tensile stress, which may exceed the fracture stress of the matrix, leading to transverse cracks

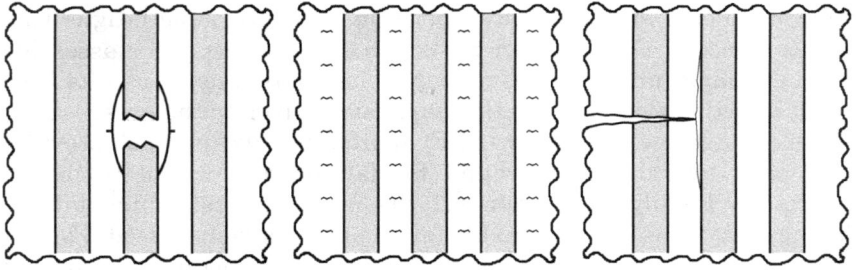

Figure 6.1 Basic fatigue damage mechanics in UD composites including fiber breakage/interfacial debonding, matrix cracking, and shear failure.

in the matrix. These cracks are randomly distributed and are initially restricted by the fibers. With the development of the fatigue process, the local strains exceed a certain threshold, resulting in fiber breakage and propagation of matrix cracks. During matrix crack propagation, the matrix–fiber interface will also fail due to severe shear stresses at the crack tip (Fig. 6.1). The final failure occurs when a sufficiently large crack (perhaps of the order of 1 mm) has developed. A typical fatigue life diagram is shown in Fig. 6.2.

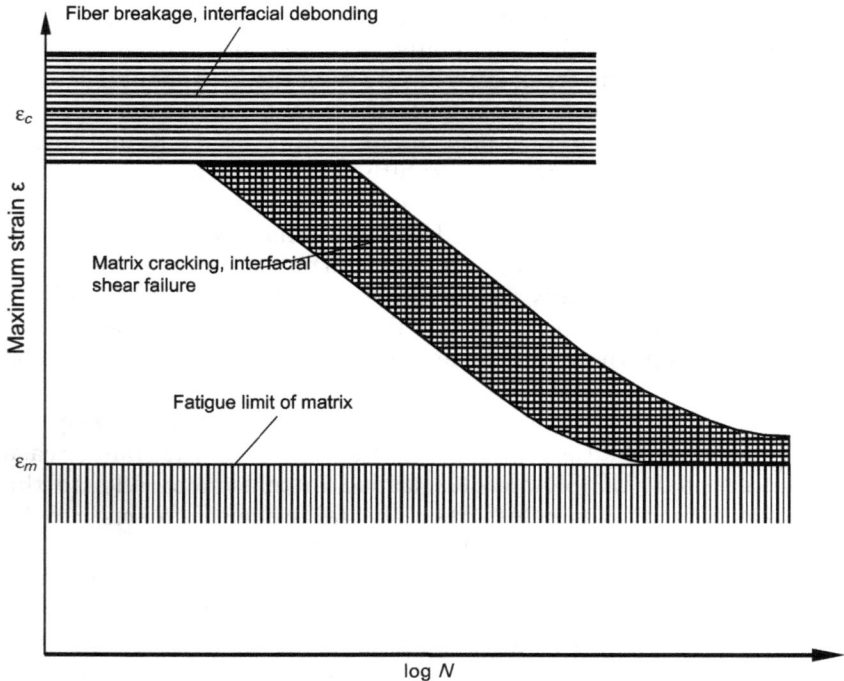

Figure 6.2 Typical fatigue life diagram for UD composites under load parallel to the fibers.

The lower strain limit for the matrix is characterized as ε_m, (see Fig. 6.2) and is the threshold strain below which the matrix cracks remain arrested by the fibers. This strain is observed to be approximately the fatigue strain limit of the unreinforced matrix material. The upper limit of the diagram is given by the strain to failure of the composite, ε_c, which is the strain to failure of the reinforcing fibers. The progressive damage mechanism is matrix cracking with associated interfacial shear failure, as described above, and this governs the fatigue life.

When the cyclic loading axis in inclined to the fiber axis at an angle of more than a few degrees, the predominant damage mechanism is matrix cracking along the fiber–matrix interface. For this case, the lowest fatigue limit is given by the strain for transverse fiber debonding (ε_{db}), which is strongly connected to the loading angle θ (see Fig. 6.3). This is reflected in the fatigue limit strain, which is plotted in Fig. 6.4 versus the off-axis angle.

For the case of bi-directional laminates the rate of damage progression of the off-axis plies is reduced due to the constraint provided by

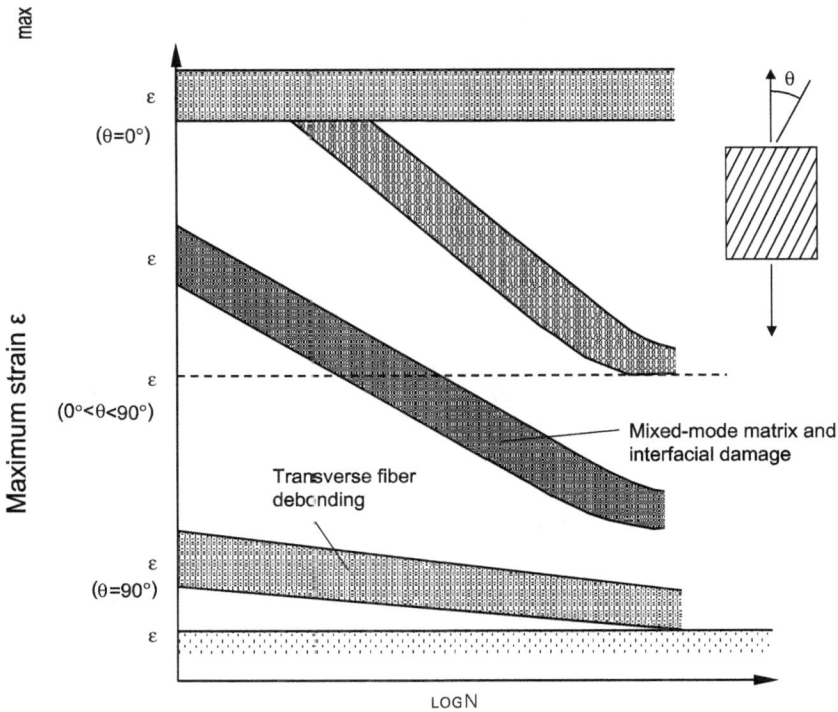

Figure 6.3 Fatigue life diagram of UD composite loaded at angle θ to the main material axis.

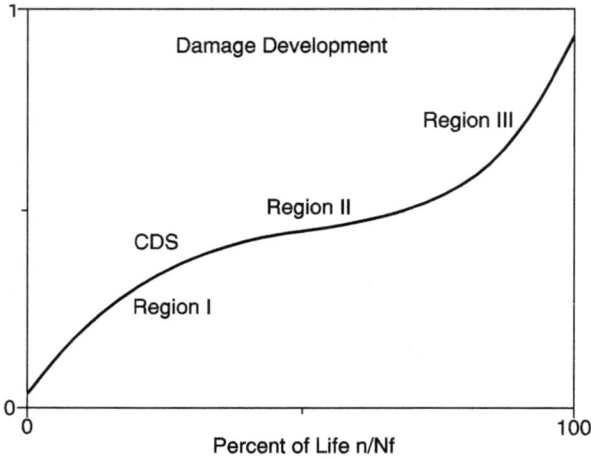

Figure 6.4 Fatigue damage development in multidirectional composites.

the on-axis plies. When two orthogonal directions are used in the lamination, the load on the laminate develops transverse cracks along the fibers. These cracks are constrained by fibers normal to the crack plane, which leads to a saturation of the crack density. Fatigue loading may direct these cracks to the ply-interfaces, introducing interlaminar cracks that will eventually cause delaminations.

In the case of general laminations, combinations of the above damage modes will occur. The early state of damage development is dominated by matrix cracking of the off-axis plies along the fibers, the number of which increases monotonically until a saturation density is reached (Fig. 6.4).

If crack saturation does not occur in all the off-axis plies, the resulting crack pattern is found to be characteristic of the laminate configuration and independent of the load amplitude. The damage state associated with the characteristic crack patterns is called the "characteristic damage state" (CDS) (Fig. 6.4). The next stage begins with initiation of cracks transverse to the primary cracks and lying along the plies adjacent to the plies with primary cracks. These are called secondary matrix cracks and are the initiators of interlaminar cracks. The resulting local delaminations lead to large crack delaminations. The final stage is dominated by fiber breakage and the final failure will occur when locally failed regions have sufficiently weakened the laminate, which can no longer withstand the maximum loads.

Failure mechanisms and modes of composite laminates also depend strongly on parameters such as stress amplitude, material system, loading frequency, and stress ratio. The sensitivity of the fatigue behavior and the understanding of failure mechanisms in relation on the fatigue stress level are fundamental. In Figs. 6.5 and 6.6 typical fracture surfaces from APC-2 and Fiberdux 6376-HTA structures, respectively, subjected to fatigue loading up to failure are presented. The magnification is 35 times. The APC-2 specimen of Fig. 6.5a was fatigued at 240 MPa stress reversals (low stress), while the specimen of Fig. 6.5b was subjected to 310 MPa stress reversals (high stress level). Similarly, the 6376-HTA specimen of Fig. 6.6a was subjected to stress reversals at 250 MPa (low stress level), while those of Figs. 6.6b–d were subjected to 280, 310, and 330 MPa (high stress), respectively. The fracture surfaces differ significantly. Specifically, those specimen subjected to low stresses fail due to extensive delaminations. This is not the case for specimens fatigued at high stress levels. C-scan examination of the failed specimens confirmed the optical observations. In Fig. 6.7 the C-scan readings of an APC-2 specimen before any fatigue damage as well as at the state of failure are shown. The blue color output range indicates 100% damage, while 100% signal reduction indicates the fully delaminated specimen; 0% signal reduction refers to the perfect specimen. Similar results are shown in Fig. 6.8, where C-scan outputs of 6376-HTA specimens that failed under fatigue loading at stress levels of 250, 280, 310, and 330 MPa are presented.

From Figs. 6.7 and 6.8, it is obvious that the specimen subjected to low mean stress are highly delaminated, while those subjected to high stresses are highly cracked.

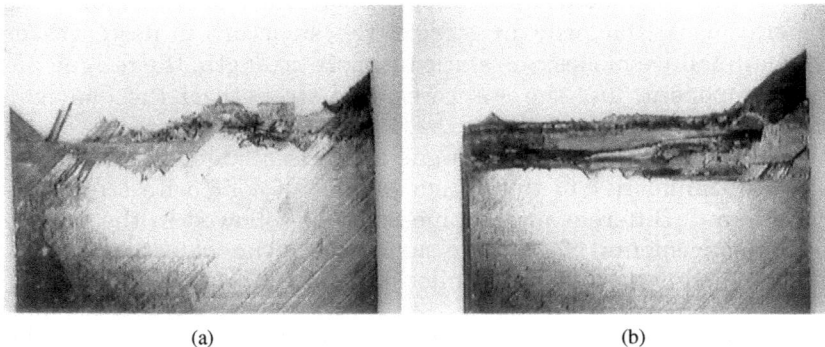

(a)   (b)

Figure 6.5 Fracture surfaces of APC-2 fatigue specimens under (a) 240 MPa and (b) 310 MPa.

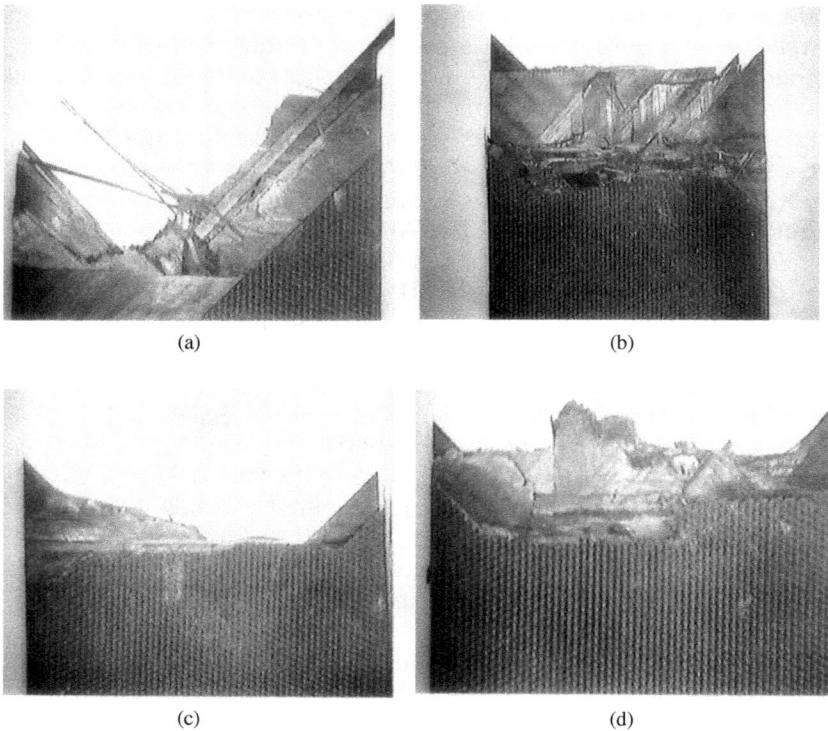

(a)   (b)

(c)   (d)

Figure 6.6 Fracture surfaces of 6376-HTA fatigue specimens under (a) 250 MPa, (b) 280 MPa, (c) 310 MPa, and (d) 330 MPa.

6.3 Fatigue Design Approaches to Composite Structural Components

To ensure that a structural component operating under fatigue loads will perform satisfactorily in service, the structure is designed for three main failure modes: the static ultimate strength, the fatigue life of the component, and the static residual strength of the damaged component.

Above requirements refer to fatigue analyses of metallic structures, but many similarities to the design process of composite structures may be found. Different philosophies may be followed in the fatigue design requirements of metallic structures. The oldest and most widely applied is the **fail-safe** philosophy, which requires that:

- The customer's service life requirements must be fulfilled for the operational loading conditions.

- During the service life, the structure must support 80–100% of the limit loads without any catastrophic failures.

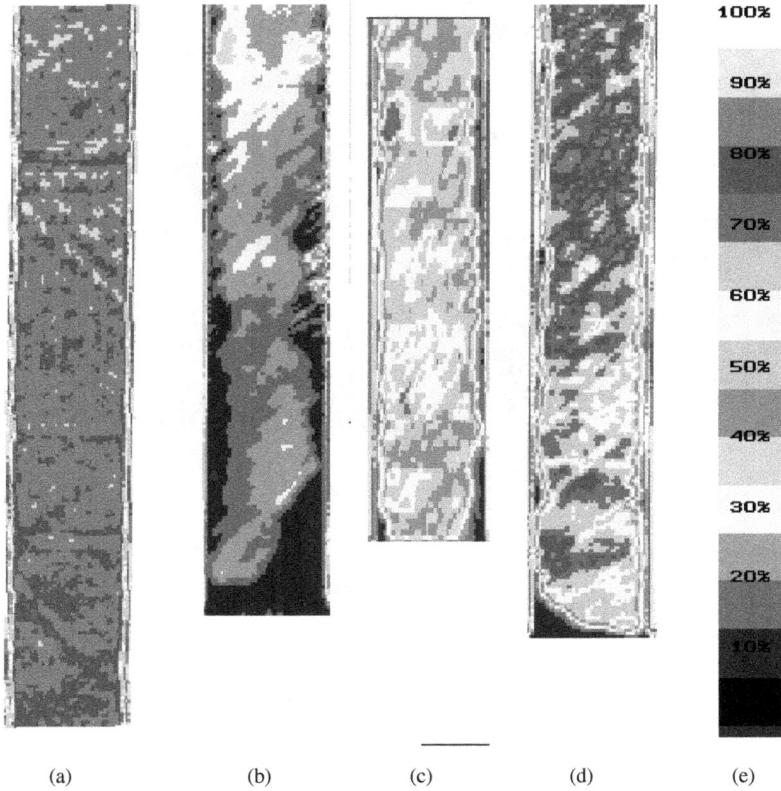

Figure 6.7 C-Scan graphs of APC-2 specimens: (a) reference, (b), (c), (d) fatigued under stress levels at 240 MPa, 280 MPa and 310 MPa; (e) the C-scan range.

- The inspection techniques and frequency must be specified so that the risk of catastrophic failure is minimized.

  From the practical point of view, a **fail-safe** component:

- May contain fatigue or other types of damage.
- Requires multiplicity of structural members, load transfer capability between members, and slow crack propagation properties.
- Needs inspections.
- Has maintenance problems arising from fatigue damage.

  In current practice, composite structures are designed following the above principles. However, this leads to "heavy" constructions that do

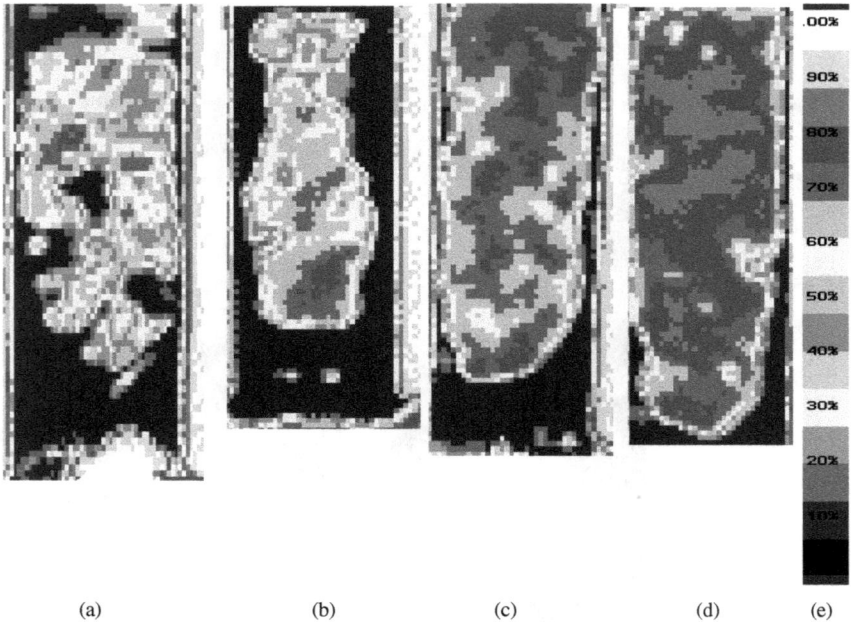

(a)   (b)   (c)   (d)   (e)

Figure 6.8 C-Scan graphs of Fiberdux 6376-HTA specimens; (a), (b), (c), (d) fatigued under stress levels of 250 MPa, 280 MPa, 310 MPa and 330 MPa; (e) the C-scan range.

not take advantage of the well-known damage tolerance of composites and that in practice diminish the advantage of the good specific properties of the materials.

Application of the **safe-life** philosophy for metallic structures requires that:

- The components remain crack free in service. Replacement time must be specified for components designed for limited life.

- The inspection techniques and frequencies as well as the replacement time must be specified in such a way that the probability of failure due to fatigue cracking is extremely remote.

The practical considerations for **safe-life** design of a component are:

- Resistance to damaging effects of variable load environment.

- Knowledge of environment, fatigue performance, and fatigue damage accumulation.

- Fatigue of the structure is a safety problem.

While the use of terms **fail-safe** and **safe-life** may indicate that there are two paths for building certifiable structures, there is really no such clear boundary between them. Both philosophies are equally necessary for creating a structurally safe and operationally satisfactory structure. Just as the ultimate strength design must consider and combine all sources of uniaxial and biaxial normal or shear stress for the assessment of strength, so must adequate fatigue evaluation include both the resistance of the structure to fatigue damage *initiation* and the resistance of the structure to fatigue damage *accumulation* to avoid the point of catastrophic failure. Thus, the designer has really but one overall objective in relation to fatigue design philosophy: To design a structure that has a high degree of structural reliability during the intended service life.

It is well known that fatigue failure is a progressive process and material degradation initiates practically with the first cycles. In metallic structures, the degradation/damage accumulation includes first local plasticity phenomena until one or more finite cracks have nucleated and then propagation of the crack(s) until the failure process culminates in a complete failure of the structure. The total life can be divided into three stages:

1. The **initial life interval**, during which a complete failure can occur only when the applied load exceeds the design ultimate strength. This is the period in which a crack will reduce the ultimate strength capability of the structure. This time interval is usually defined as the safe-life interval.

2. The **life interval**, which is the period during which a complete failure may occur even when the applied load is below the ultimate design load and the reduction of the load carrying capability of a structure due to a small crack, is a function of the material fracture toughness property and the stress intensity factor at the critical locations.

3. The **final life interval**, during which a complete failure will occur even when the applied load is below the ultimate design load and the reduction of the load carrying capacity of the structure, is a function of the material fracture toughness properties, the area reduction due to a growing crack, and the stress intensity factor at the critical locations.

A typical progressive failure diagram of metallic materials is shown in Fig. 6.9. The intervals life interval and the final life interval form a time interval that may be called the "fail-safe" life. The 'fail-safe" life corresponds to the time interval between inspections. This means that

Progressive failure process:	Failure damage accumulation	Visible crack appearance and initial crack propagation	Crack propagation
Complete-final failiure load:	Final failure load		Design ultimate Design limit 80% limit
Final failure due to:	Exceeding design ultimate load Fatigue life (safe life interval)	Failure load≤ult. Strength reduction primary due to material fracture toughness properties (Fail-safe life interval)	Failure load<ult. Strength reduction: Combination of material fracture toughness properties and area reduction

LifeTime

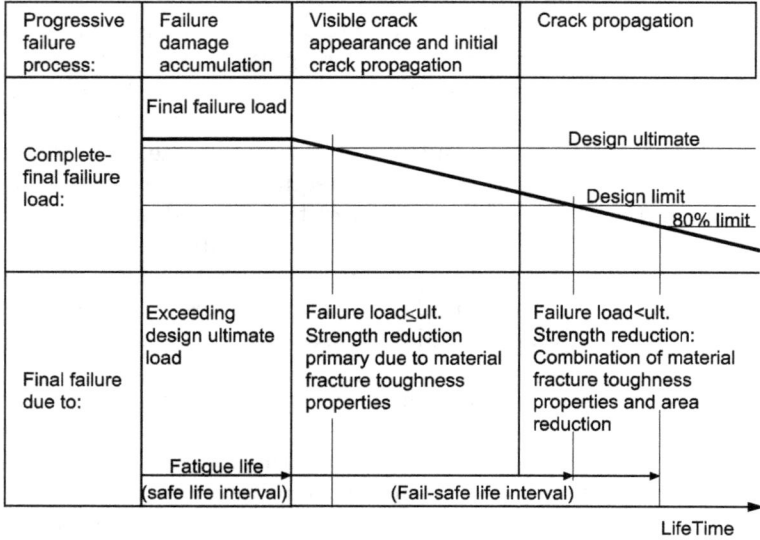

Figure 6.9 Progressive failure of a metallic structural element.

a crack that may initiate after an inspection should not propagate to the critical length until the next inspection; the residual strength should not decrease below the fail-safe design load before the next inspection during which the crack should be detectable. This approach indicates that the structure obtains a degree of **damage tolerance**. For aircraft structures the **damage tolerance** requirements are stated in FAR 25-571, which states that "it must be shown by analysis, or tests, or both that catastrophic failure or excessive structural deformation that could adversely affect the flight characteristics of an airplane are not probable after fatigue failure or obvious partial failure of a single principal structural element." To achieve this goal in the case of composite structures and apply a "damage tolerance design philosophy," it is necessary to quantify the interactive fatigue damage accumulation processes. This is a complex and difficult task, as stated before, because the defects do not always coalesce into a dominant crack. Therefore, classical fracture mechanics models cannot be applied.

For the application of the damage tolerance philosophy in the case of composite materials in a manner analogous to that of isotropic materials, the following are necessary:

■ The identification of one or more parameters that are suitable for use as a measure of the damage in the composite material.

- The further development of the nondestructive evaluation methods and systems, in order to enable the reliable measurement of the damage parameters.
- The quantification of functions for the damage development.
- The definition of critical values for the damage parameters so that the failure can be predicted.
- The correlation between damage parameters and residual mechanical properties of the material.

6.4   Fatigue Damage Functions

With regard to the definition of fatigue damage adopted here, the material mechanical properties that are used in structural integrity analyses seem to be suitable quantities for reflecting fatigue effects in composite materials. A complete consideration of fatigue effects in composites should account for the gradual reduction of remaining strength, stiffness, interlaminar shear strength etc., as well as for failure after a certain number of stress cycles. From this viewpoint, each of the above quantities is suitable for use as a damage parameter. By defining the cycle-dependent degradation of the selected quantity as well as the dependency of the derived relation on fatigue stress, fatigue damage functions may be formulated, but they reflect only a partial fatigue effects in composites. It is the reduction of the entirety of all these quantities that has to be accounted for in a generic fatigue damage characterization. The fatigue damage functions may be categorized according to the selection of the damage parameter and the procedure followed in the definition of the damage function. This categorization is more or less arbitrary. In this chapter, damage functions that have been proposed in the literature and that fit the view of present work will be briefly discussed and, for overview purposes, classified following the above arbitrary criteria.

6.4.1   Fatigue theories based on S–N curves

The most common approach for fatigue life prediction is based on the empirical S–N curves. In these theories the consumed fatigue life of the structure may be considered as the damage parameter. The main advantage of these theories is their simplicity in application. The theories may be adapted for the cases of multiaxial loading and spectrum loading, which are most common for real structures. These advantages have been responsible for their widespread use, although they cannot be directly related to damage tolerance concepts; for weight-sensitive structures the latter shortcoming is a serious disadvantage. In Section 6.5, characteristic fatigue damage accumulation concepts based on S–N curves, and their extensions to multiaxial loading are described.

6.4.2 Fatigue models based on residual stiffness and residual strength

Mechanisms of fatigue damage in composites result in cracks of various orientation in the volume of the material, as mentioned in Section 6.3. This leads to the degradation of the overall material properties, including the stiffness and strength in various directions. The crack initiation, multiplication, and propagation mechanisms that begin early in the fatigue process lead to reduction in modulus (Fig. 6.10) and variation in Poisson's ratio which are proportional to the crack density. Therefore, stiffness has been established as the most popular quantity, which is supposed sensitive enough to account for accumulating fatigue damage.

However, stiffness reduction of composites during fatigue varies with the type and stacking sequence of laminate. Unidirectional laminates do not show much change in stiffness until immediately before final failure. Cross-ply laminates show a gradual reduction in stiffness as the fatigue process continues. The difference in the variation of stiffness between these two stacking sequences lies in the mutual influence that one ply exerts in the neighboring plies. Similar remarks are valid for the material strength. A typical plot of the

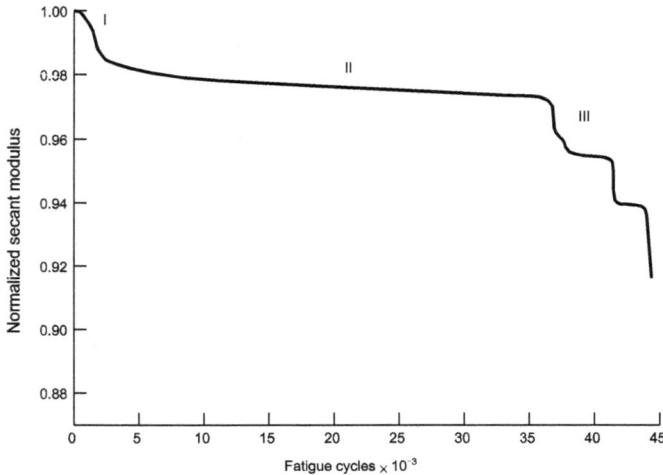

Figure 6.10 Degradation of the elastic modulus of cross-ply laminates of C/epoxy under fatigue.

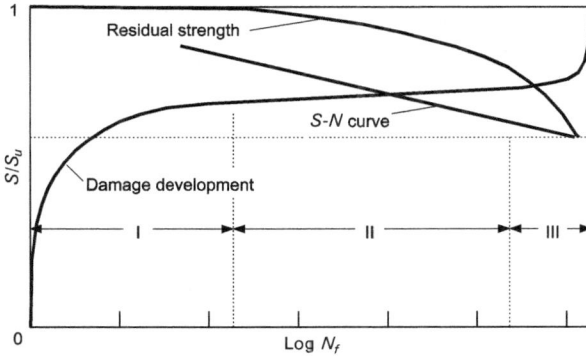

Figure 6.11   Typical reduction of the residual strength in multidirectional composites in different damage development stages.

expected strength reduction with advancing fatigue damage is shown in Fig. 6.11 for a multidirectional compilation. Analytical and even numerical prediction of the expected stiffness and strength reduction is difficult. Presently no generic method exists to solve the above problem. Proposed models are mostly numerical and refer to specific cases.[5] Thus extensive macroscopic characterization of fatigue damage using stiffness or strength as damage parameters has been made.[1,13,14,18–22] Experimental results, however, have shown that for some very important stacking sequences, such as the quasi-isotropic lay-up, neither stiffness not strength is sensitive enough to the accumulating fatigue damage for utilization as a fatigue parameter.[27,35] Obtained stiffness degradation for two quasi-isotropic laminates of the engineering composites, APC-2 thermoplastic and Fiberdux 6376-HTA (thermosetting), for different stress amplitudes is plotted in Fig. 6.12.

A clear trend of stiffness decrease may be seen, but the decrease obtained does not exceed 5–10%; this makes impossible the use of this quantity for formulating fatigue damage functions that also account for the stress amplitude dependency. Similar observations were made when using the residual tensile or compression strength as damage parameter.

Several other mechanical properties have been suggested that are supposed sensitive to damage increase. These include the specific damping capacity[21] and the energy density. Despite of the fact that all the latter approaches can be applied with relative ease, the models are case specific. In addition, application to structural components of parametric fatigue damage characterization by means of a given property degradation is oversimplified and may become critical, since no indication is given of the impact of accumulating damage on

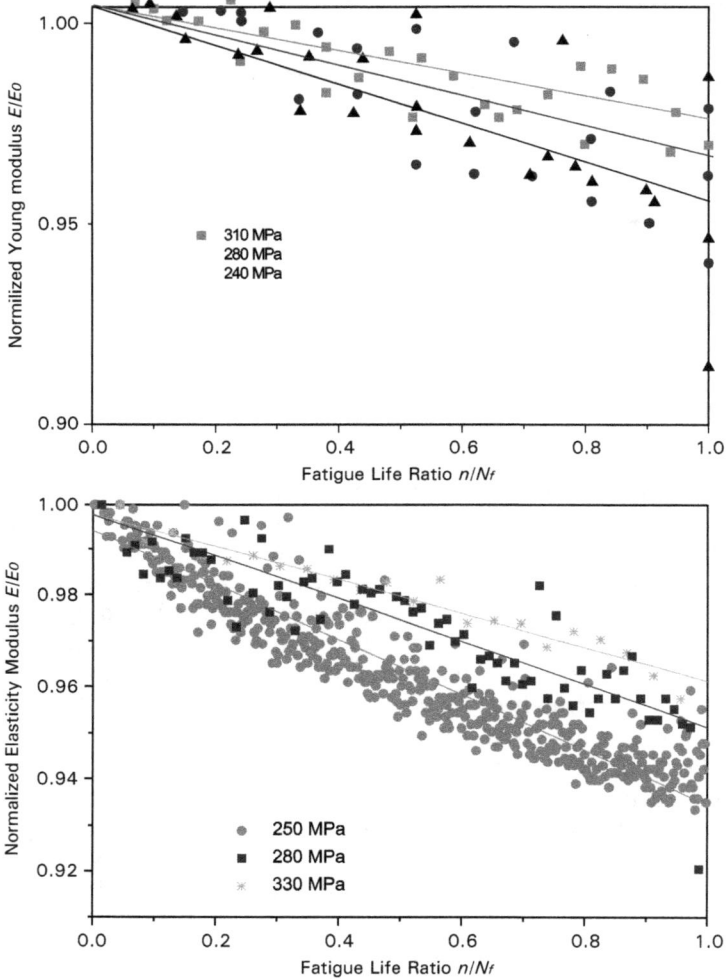

Figure 6.12 Stiffness degradation for APC-2 (thermoplastic) and Fiber-dux 6376-HTA (thermosetting) composite material for the different stress amplitudes.

other mechanical properties used for designing the component under consideration.

6.4.3 Tensorial damage functions

A common feature in these theories is the development of a damage tensor based on damage mechanics approaches. Micromechanical considerations focus on a representative volume element, in which the material configuration details (e.g., fiber diameter and spacing, fiber orientation, ply thickness), as well as the damage configuration details (e.g., size, shape, orientation of cracks, voids, debonds) are incorporated. Solution of the boundary value problem that is formulated for the representative volume element yields the response functions of interest. Macroscopically, damage may be characterized by a set of internal variables representing the damage state of the material. The objective of this approach is to develop a set of phenomenological relationships between the response function and the internal state variables.

Both approaches have advantages and disadvantages. The representative volume element approach characterizes damage directly; its accuracy is related to the accuracy in solving the associated boundary value problem. Still, the boundary value problem may become highly complex even for very simple damage cases, requiring large-scale computing facilities. Although these facilities are becoming increasingly available, the risk of losing insight into the basic physical phenomena must be considered. The internal variable approach depends on the choice and number of variables. The connection of damage to the overall response appears as a set of phenomenological constants that are to be determined experimentally. The physical insight into the basic phenomena is retained through interpretation of the phenomenological constants. The feasibility of the approach depends largely on the choice of the internal variables and their correlation to the actual damage details.

The most representative examples of the tensorial damage approach are those of references.[6–10] The damage is characterized by second-order tensor-valued internal state variables representing locally averaged measures of specific damage states such as matrix cracks, fiber–matrix debonding, etc. In Reference 6 relationships between the overall stiffness properties and the intensity of damage in the individual intralaminar (matrix cracking) and interlaminar (interior delamination) damage modes were determined, using a vectorial representation of damage as internal variables in a phenomenological theory. In Reference 7 a specialized constitutive model is developed for the case of matrix cracks only and its application is limited to the prediction of stiffness degradation. Even in that case use of the model is associated with essential experimental effort and several simplifications are needed to utilize linear elastic fracture mechanics for the stiffness calculations. In References 8 and 9 locally averaged history-dependent constitutive equations are constructed utilizing constraints imposed from thermodynamics with internal state variables. In that generic form the model is very complex and difficult for engineering application because of the difficulty of determining and measuring the state variables. Further to damage tensor-based fatigue theories, criteria have been advanced as well, to account for composite material fatigue damage accumulation that is associated with failure of critical material elements[11]; the application of these criteria for damage tolerance analysis is involves Some effort. The same applies to probabilistic approaches that have been also proposed.[12]

6.4.4 Fracture mechanics models

Composite material fatigue models that are based on fracture mechanics have their origin in the damage tolerance approach for metallic structures. As stated before, in metals the damage is represented by a crack that grows with cycling loading. By predicting the crack growth rate and taking into account the capabilities of nondestructive methods, inspection intervals are determined to ensure fail safety. In laminated composite structures, delamination represents the most commonly observed macroscopic damage mechanism that causes failure. Accordingly, efforts have been made[6,16–19] to develop for composite materials procedures that are similar to crack growth calculations. The rate of delamination growth with fatigue cycles, da/dN, has been expressed as a power law relationship in terms of the strain energy release rate, G, associated with delamination growth. It may be expressed as (c and n are material constants)

$$\frac{da}{dN} = cG^n \tag{6.1}$$

The analogy of Eq. (6.1) to the well known Paris equation for metals is obvious. The practical drawback of the fracture mechanics approach is that whereas in metals the rate of fatigue crack growth may be described over as much as two orders of magnitude in G, in composites the growth for delamination rates is characterized over barely one order of magnitude in G. Therefore, a small uncertainty in calculations of applied load may lead to an order of magnitude of mistake in the estimation of delamination growth. Furthermore, unlike crack tip plasticity, additional composite damage mechanisms, such as fiber-bridging and matrix cracking, do not always retard delamination growth to the same degree, which means that the generic value of such a characterization is questionable.[6]

6.4.5 Fatigue damage assessed from nondestructive evaluation

Nondestructive evaluation has been used extensively to characterize composite material damage. Techniques applied involve ultrasonic and acousto-ultrasonic acoustic emission, thermography, and radiography. In many cases damage evaluation refers to the microscopic material scale. The macroscopic scale has been exploited as well. In most case nondestructive evaluation data refer to damage caused by static loads. In addition, they have been rarely correlated to residual mechanical properties of the damaged material. With regard to fatigue, no model at present exists to directly relate nondestructive test data to material property degradation and final failure. Nevertheless, nondestructive evaluation may provide a means to obtain fatigue damage function. A concept for correlating material degradation due to fatigue with results from nondestructive monitoring has been proposed.[27,28] The methodology for achieving determination of such a damage severity factor (DSF) is described in Section 6.6. Up to now the formulation of the DSF has been based on ultrasonic C-scan readings of fatigued material. DSF values obtained refer to a material area that may be selected individually, which accounts for the varying severity of the damage at the different locations. A fatigue model based on DSF has been proposed.[27] The model fits well with the damage tolerance concepts. In addition, DSF seems to have the potential to be introduced in structural integrity analysis. The concept of the damage severity factor is far from being developed for application on real structures but, as it is an innovative and promising approach it will be discussed extensively in Section 6.6.

6.5 Fatigue Life Prediction Based on S–N Data

The prediction of fatigue life of a structural component is currently usually made by applying models based on the use of S–N curves.[29-35] In this case as the number of applied fatigue cycles is considered damage parameter. Knowing the S–N curve of the material, the statistically expected fatigue life at any stress level can be estimated. Engineering structures operate usually under multiaxial stress loading conditions, however, and in most cases fatigue loads vary with time as well. Fatigue life assessment under these complex loading conditions can be made manageable using S–N curves and applying the following approach:

1. Derivation of the S–N curves of the material for different stress ratios and various stacking sequences

2. Mathematical description of the S–N curve by means of a suitable equation

3. Application of a failure rule (e.g., Miner).

It is apparent that this procedure is not consistent with the damage tolerance philosophy, but it remains the most commonly applied approach, even for the case of lightweight aircraft structures. The main reason is the lack of suitable experimental data for application of a damage tolerance philosophy, as well as the lack of analytical tools for performing a damage tolerance design in composite materials. One of the advantages of the theories based on S–N curves for predicting fatigue life is that they may be adopted for assessing fatigue life under multiaxial loading conditions. On the other hand, use of these theories in multiaxial loading is associated with essential experimental effort, which limits the validity of the derived relations only to the cases investigated. The application of S–N curves in the prediction of multiaxial fatigue behaviour uses empirical failure theories. These theories are based on the principles of the well-known generalized static failure theories; they are adapted appropriately so that the stress ratio ($R = \sigma_{min}/\sigma_{max}$), the fatigue loading frequency, and the fatigue cycles become parameters of the model. Some characteristic models that could be applied for the design of composite materials will be discussed below. Although it is acknowledged that the following overview is far from being complete, the following theories have found widespread interest.

6.5.1   Hashin–Rottem fatigue failure theory

The Hashin–Rotem theory is one of the first theories describing the fatigue damage behavior of composite materials. The theory[38] is based on the separation of the different fatigue failure modes that the composite material can exhibit. For unidirectional materials the failure modes considered are either matrix failure or fiber failure. The theory predicts failure if one of the following relationships is valid:

$$\sigma_A \geq \sigma_A^u \tag{6.2a}$$

$$\left(\frac{\sigma_T}{\sigma_T^u}\right)^2 + \left(\frac{\tau}{\tau^u}\right)^2 \geq 1 \tag{6.2b}$$

where, σ_A^u, σ_T^u, τ^u are the three S–N curves of the material in uniaxial loading parallel to the fibers, perpendicular to the fibers, and in shear of the principal plane, respectively, while σ_A, σ_T, τ are the fatigue stresses in the above three directions.

The S–N curves of the material are of the form

$$\sigma^u = \sigma^s f(R, N, \theta) \tag{6.3}$$

where σ^s is usually the static strength, while the function $f(R, N, \theta)$ is a nondimensional function. In accordance with the strengths σ_A^u, σ_T^u, τ^u, the fatigue functions in the fiber direction, in the direction perpendicular to the fibers, and in shear are $f'(R, N, \theta)$, $f_T(R, N, \theta)$, and $f_\tau(R, N, \theta)$, respectively. Based on these definitions, the transition angle from the fiber failure mode to the matrix failure mode is

$$\tan\theta_c = \frac{\tau^u}{\sigma_A^u} = \frac{\tau^u f_\tau(R, N, \theta)}{\sigma_A^u f'(R, N, \theta)} \tag{6.4}$$

When the angle between the fibers and the loading is less than θ_c, a fiber failure will occur, otherwise the relationship (6.3) should be applied. In the latter case the fatigue function $f''(R, N, \theta)$ for any angle θ is

$$f''(R, N, \theta) = f_\tau \sqrt{\frac{1 + \left(\dfrac{\tau^u}{\sigma_\tau^s}\right)\tan^2\theta}{1 + \left(\dfrac{\tau^u f_\tau}{\sigma_\tau^u f_T}\right)^2 \tan^2\theta}} \tag{6.5}$$

The Hashin–Rotem theory as described above is valid for UD laminates. In Reference 36 the theory is extended for laminated structures of different specific laminations. In Reference 37 the theory is adapted in such a way that an additional failure mode, specifically the interlaminar fatigue failure mode, can be predicted as

$$\left(\frac{\sigma_d^c}{\sigma_d^u}\right)^2 + \left(\frac{\tau_d^c}{\tau_d^u}\right)^2 \geq 1 \tag{6.6}$$

where the exponent C stands for the applied fatigue stresses, the exponent U stands for the ultimate strengths, and d stands for delamination of two adjacent plies.

The experimental work supporting the Hasin–Rotem theory consists of fatigue tests in different off-axis angles of E-glass/epoxy unidirectional laminate, reaching up to 10^6 fatigue cycles. The tests have been performed at frequencies of 3, 19, and 1.8 Hz and stress ratio $R = 0.1$. The S–N curves for different off-axis angles are shown in Fig. 6.13.

6.5.2 Fawaz–Ellyin fatigue failure theory

The Fawaz–Ellyin theory can predict fatigue failure under multiaxial loading. The basis of the model is that a reference function exists in the form

$$S_r = m_r \log N + b_r \tag{6.7}$$

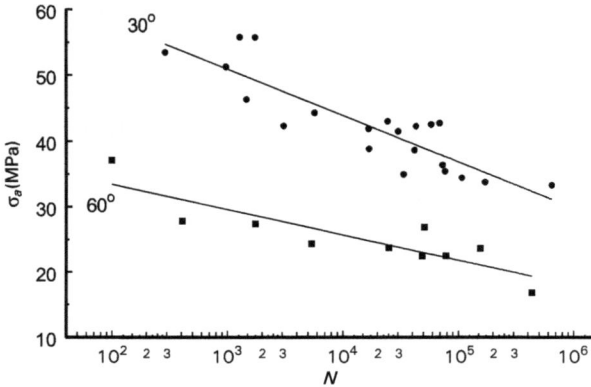

Figure 6.13 $S-N$ curves at 30° and 60° off-axis angles for E-glass/epoxy UD laminate.[42]

which can be the basis for the calculation of all the off-axis $S-N$ curves in the form

$$m = f(a_1, a_2, \theta)g(R)m_r \tag{6.8}$$

where
$$b = f(a_1, a_2, \theta)b_r$$
$$a_1 = \frac{\sigma_y}{\sigma_x}, \qquad a_2 = \frac{\sigma_z}{\sigma_x} \tag{6.9}$$

and R is the stress ratio.

The function $f(a_1, a_2, \theta)$ is defined as

$$f(a_1, a_2, \theta) = \frac{\sigma_x(a_1, a_2, \theta)}{X_r} \tag{6.10}$$

where $\sigma_x(a_1, x_1, \theta)$ is the static strength in the θ direction under multiaxial loading, and X_r is the ultimate strength in the reference direction. The function $g(R)$ is given as

$$g(R) = \sigma_{\max}(1 - R)/[\sigma_{(\max)r} - \sigma_{(\min)r}] \tag{6.11}$$

It is obvious that when $R = R_r$, $g(R) = 1$ and if $R = 1$ then $g(R) = 0$. A minimum experimental effort of one $S-N$ curve is required; however, the prediction of the fatigue behavior is very sensitive to the selection of the reference $S-N$ curve.

6.5.3 Owen and Griffiths fatigue failure theory

The Owen and Griffiths theory is based on the assumption that the experimental fatigue results can be approximated by static failure theories that are properly adapted for this purpose. The theory is not dependent on the fiber direction of the plies that compose the laminated structure. The experimental work supporting the theory consists of fatigue tests on cylindrical woven specimen of glass/polyester. The biaxial loading is imposed by applying both axial pressure loads on the cylindrical specimen at a frequency of a 1.8 Hz. For specific materials that have the same properties in the axial and the circumferencial directions, only two different $S-N$ curves are required for the formulation of the tensional failure function. Based on the two $S-N$ curves for $0°$ and $45°$ degrees, which are shown in Fig. 6.14, the failure curves for different ratios of the axial to the circumferencial stress are presented in Fig. 6.15.

6.5.4 Failure theory based on a tensorial polynomial theory (FTP)

The general theory of strength of anisotropic materials based on quadratic failure theories was initially proposed by Tsai.[39] The form of this failure theory is:

$$F_{11}\sigma_1^2 + F_{22}\sigma_2^2 + 2F_1F_2\sigma_1\sigma_2 + F_1\sigma_1 + F_2\sigma_2 + F_{66}\sigma_6^2 \leq 1 \qquad (6.12)$$

where
$$F_{11} = \frac{1}{XX'} \qquad F_{22} = \frac{1}{YY'} \qquad F_{66} = \frac{1}{S^2}$$
$$F_1 = \frac{1}{X} - \frac{1}{X'} \qquad F_2 = \frac{1}{Y} - \frac{1}{Y'} \qquad (6.13)$$

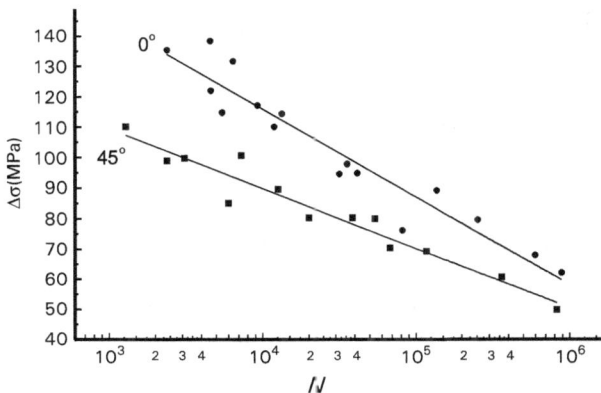

Figure 6.14 $S-N$ curves of woven glass/polyester cylindrical specimen.[43]

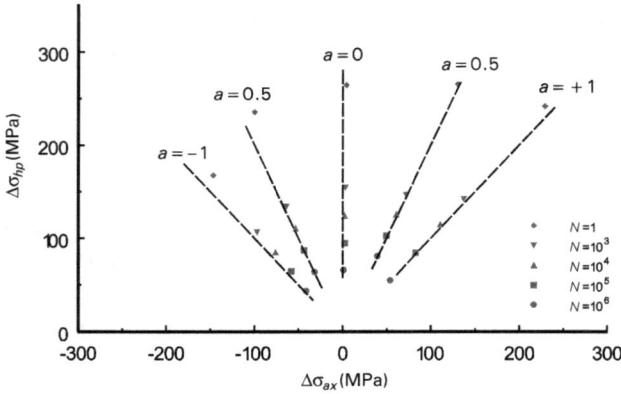

Figure 6.15 Failure curves of glass/polyester cylindrical specimen for different ratios a of circumferential/axial stress
$(\Delta\sigma_{hp}/\Delta_{xx})$.[44]

The selection of the value of the interaction term F_{12} leads to different
failure theories[40] of the form of Eqs. (6.6) to (6.12). In the present work
the interaction term is selected as

$$F_{12} = -\frac{1}{2}\sqrt{F_{11}F_{22}} \tag{6.14}$$

although it has been shown that the definition

$$F_{12} = \frac{1}{2}(F_{33} - F_{11} - F_{22}) \tag{6.15}$$

leads to theoretical values that are in better agreement with the
experimental results. The reason for the definition of F_{12} in the case
of fatigue is that the components of the failure tensor F_i and F_{ij} should
be functions of the number of cycles N, the stress ratio R, and the
loading frequency f. For the determination of the tensional components the static strengths X, X' Y, Y', and S of Eqs.. (6.13) and (6.14)
are replaced by the S–N curves of the material under the same
loading conditions.

Experimental tests on multidirectional E-glass/polyester specimens
under uniaxial loading are performed in Reference 36 and are
compared to the predictions of the tensorial polynomial theory
(FTP). The experimental work includes static and fatigue tests according to ASTM 3039-76. The specimen are cut from a $[0/(\pm45)_2 0]_s$
reference laminate. The off-axis loading directions are 0°, 30°, 45°,
60° and 90°. The experimental results of the static tests are presented
in Fig. 6.16, while the results of fatigue tests and predictions by FTP
are presented in Fig. 6.17.

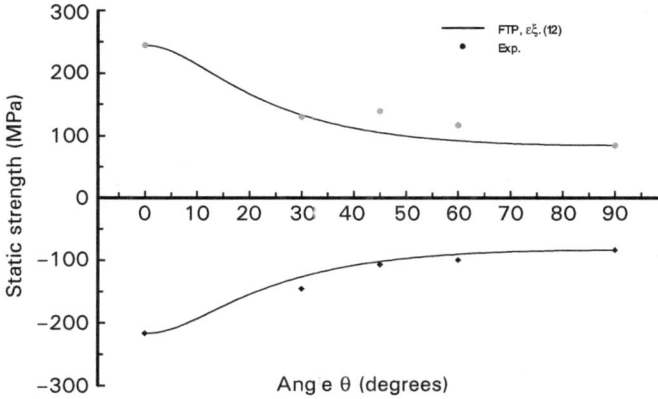

Figure 6.16 Off-axis static strength of E-glass/polyester MD flat coupons, from experiments and predictions by the FTP.[41]

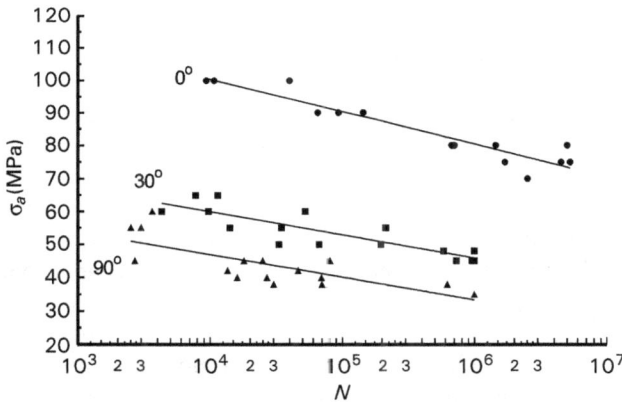

Figure 6.17 $S-N$ curves of E-glass/polyester MD flat coupons, from experiments and predictions by the FTP.[41]

6.6 Fatigue Damage Assessed from Material Degradation and Nondestructive Evaluation Data

The assessment fatigue damage of composites from material degradation and nondestructive evaluation data requires the combination of (a) nondestructive testing (NDT) techniques suitable for detecting the several types of local damage modes evaluating during fatigue and (b) models to relate accumulating fatigue damage to the resulting material degradation. Serious advances have been made in the first requirement. Knowledge has been accumulated concerning the type of damage modes that each NDT method is sensitive to. However, at present no model exists to directly relate nondestructive evaluation data to material property degradation and final failure.

The nondestructive test techniques applied to evaluate composite material fatigue damage include ultrasonics,[56] acoustic emission,[57] acousto-ultrasonics,[58] radiography,[59] eddy-currents and several thermal techniques.[60] Often material evaluation requires the use of more than one of these techniques. Some characteristic examples are overviewed in brief below. The selection of examples is arbitrary, but the purpose here is not to review the many important works presented to date on this subject but rather outline some of the methodologies applied to deal with nondestructive evaluation data. The most commonly applied technique in evaluating composites is the ultrasonic technique. Ultrasonics are sensitive to material porosity, matrix cracking and crazing, as well as delaminations. The method may be used to evaluate damage either at microscopic or at macroscopic level. For instance, scanning acoustic microscopes give high-resolution microscopic images of local failure surfaces.[62] For component structural analysis those methods are more relevant that provide parameters able to integrate the whole of the damage. Such techniques are standard C-scan, D-scan, and acousto-ultrasonics. Efforts have also been made, to derive correlations between ultrasonic parameters and mechanical properties such as stiffness and strength.

In Reference 56, all five elastic moduli of orthotropic composites were determined using ultrasonic signals, by applying longitudinal and transverse transducers at the specimens, so that the respective longitudinal and transverse waves were generated and their respective velocities were measured. Later, a series of works were presented to determine the macroscopic elastic moduli of composite laminates utilizing ultrasonics following different procedures.[64,70] Efforts have also been made to obtain correlations between ultrasonic signal parameters and the strength of composite laminates.[62,70,71] In Reference 62, correlation was obtained between the degree of attenuation of ultrasonic waves and material strength. To express this relation, a parameter called the stress wave factor (SWF) was introduced. SWF was found to be sensitive to the tensile strength variations and interlaminar shear strength. The observed sensitivities of SWF to the tensile strength variation represent a first step towards obtaining fatigue damage functions by utilizing ultrasonic signals to formulate a generic parameter and relating its mechanical property variation. It is assumed in this that strength degradation is a suitable quantity to express the accumulating fatigue damage and that SWF may be formulated to a global parameter referring to the entire specimen.

The work,[71] as well as some other work by same principal researches, has been the background for developing the well-established acousto-ultrasonic technique. The acousto-ultrasonic technique has advantage over ultrasonics because it requires accessibility to only one side of the examined specimen. In addition, the technique provides a quantitative parameter that can more easily be correlated to mechanical properties of the material. The acousto-ultrasonic method was further studied in a series of works.[58,73] In Reference 73 the results of an investigation focusing on the problem of the reproducibility of SFW, the correlation between microstructure, mechanical properties and SWF, and the comparison of SWF to other NDT data for subcritical damage states caused by mechanical loading were presented. In Reference 64, an alternate method for quantifying SFW is proposed in which the amplitude–time signal detected at the receiver transducer is captured by a digitizing transient recorder and is stored. This signal is transformed by fast Fourier technique to frequency–domain representation. Finally, comparison between the derived parameters and stiffness degradation occurring during fatigue were made, using SFW measurements at certain intermediate stages of the fatigue test.

In addition to ultrasonics and acousto-ultrasonics, acoustic emission is also used extensively at present to investigate the damage development of polymer matrix composites. It was early recognized that different failure mechanisms—i.e., fiber fracture, matrix fracture, or debonding—result in different characteristic acoustic emission signal outputs. The monitoring of structural integrity of composite materials for the case of pressure vessels[57] has been very successful. An interesting procedure for assessing the fatigue durability of laminated composites is proposed.[74] In recent years attempts have been made to introduce acoustic emission as a standard testing method for structural integrity analysis in applications such as energy generators and large structural components, e.g., aircraft fuselage.

X-ray radiography has been used entirely to detect microstructural damage. For instance, in reference 59, X-ray radiography technique was applied to obtain detailed information on the microscopic damage modes developed during loading. The thermal methods are based on thermography, a technique in which isothermal contours are mapped over a surface and the assumption is made that a local defect will disturb these isothermals. Several techniques were developed on this basis. An important advantage of the principle of this technique is the possibility of monitoring in situ damage evolution of a relatively large area. For instance,[74] the technique was successfully utilized to obtain delaminations on a large graphite/epoxy laminated filament-wound structure. Continiuing investigations aim to increase the application of the technique in aircraft composite damage monitoring. Some investigators have applied other NDT techniques to monitor damage. For instance, eddy currents were used for the investigation of the damage modes of composites. However, these techniques have not been used to the same extent as the techniques described above. It is important to realize that the NDT methods described above are limited to the correlation of obtained signals to static material properties, with very few exceptions relating NDT data to the fatigue performance. But, for a damage tolerance treatment of composite structures, the correlation of a suitable nondestructively measured damage parameter to fatigue life and residual mechanical properties of the material is necessary. The complexity of the fatigue damage problem, already discussed, can be significantly reduced when formulating a generic parametric characterization of fatigue damage that accounts also for the damage state inhomogeneity at the several locations within a component, but derivation of such a fatigue damage parameter is difficult. In addition to material inhomogeneity, the fatigue damage accumulation processes depend on both stress amplitude and ratio, as well as on laminate stacking sequence. As was discussed in Section 6.2, at relatively low and medium stress levels extensive delamination of the composite specimen is observed, while at higher stress levels fiber breakage rather than delamination produces the final failure. Furthermore, even for the same experimental conditions (loading type, stress level, stress ratio, frequency), the results demonstrate a high degree of scatter. In Fig. 6.18 the C-scan graphs refer to APC-2 specimens fatigued at 260 MPa for 180 000 fatigue cycles. Although the C-scan graphs of Fig. 6.18b, c, and d correspond to the same fatigue life ratio, the damage state within the specimens is quite different, as indicated by the differences in the respective C-scan graphs.

6.6.1   Definition of the Damage Severity Factor DSF

To assess composite material fatigue damage accumulation, signals of materials C-scan evaluation were used[27,67,68] to define a generic macroscopic damage parameter, DSF; it is understood as the damage severity factor. The aim of introducing the quantity DSF is to quantify the qualitative information obtained from the C-scan results. The potential of this damage severity factor to reflect the entire fatigue damage is limited by the capabilities of ultrasonic methods to monitor damage and depends additionally on the available ultrasonic test instrument. The sensitivity of C-scan evaluation data to accumulating damage during fatigue is demonstrated in Fig. 6.18 and 6.19. Recall that ultrasonics are sensitive to porosity, matrix crazing, matrix cracking, and delaminations, but it is not likely that they are sensitive to fiber breakage. As mentioned in the previous paragraph, the graphs in Fig. 6.18 indicate the expected deviations of accumulated damage, and consequently of fatigue life, when same material is subjected to same fatigue conditions; the evaluation demonstrates the sensitivity of the method to account for these material-related deviations. The graphs in Fig. 6.19 refer to specimens made from Fiberdux 6376-HTA material and fatigued at 280 MPa; they have been taken for different fatigue life ratios (n/N_f) of the specimen and demonstrate the sensitivity of C-scan to accumulating fatigue damage. N_f denotes the fatigue life of the specimen.

As it can be seen in Fig. 6.19, delaminations begin from certain locations of the specimen edges, propagate all over the edge area, and then expand on the other sites of the specimen. The areas where damage initiates are preferentially those of initial material imperfections. During the entire damage process, the C-scan signal gradually decreases; from the initial value of 85%, which is typical for undamaged specimens, to a value of 58%.

For simplicity it was assumed that ultrasonic signals and the severity of local damage reflected in each signal may be correlated into one-to-one relationship. Thus, a linear correlation is made between damage severity and the qualitative damage classification, which is characterized by different colors in the C-scan graphs. Based on this assumption, the damage severity is characterized by values

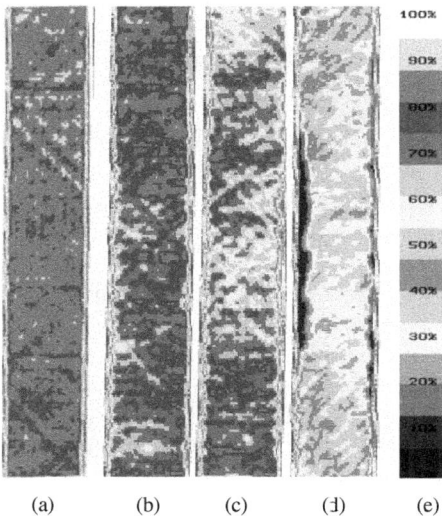

Figure 6.18 C-Scan graphs of APC-2 specimens: (a) reference; (b), (c), (d) specimens fatigued at $\sigma_x = 260$ MPa for 180 000 cycles; (e) the C-scan range.

(a)   (b)   (c)   (d)   (e)

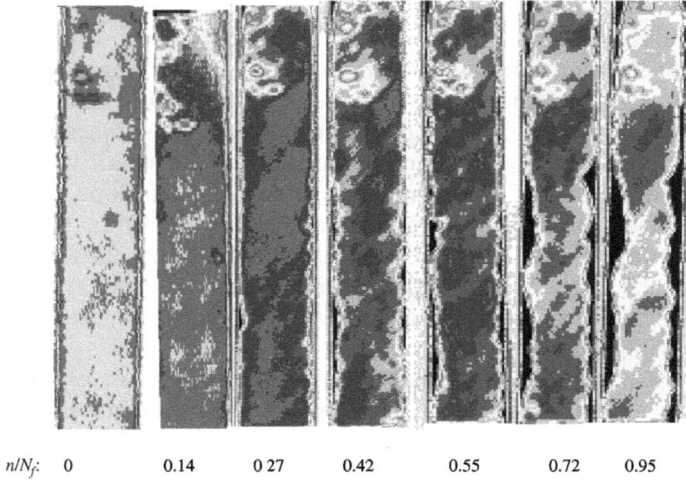

n/N_f: 0   0.14   0 27   0.42   0.55   0.72   0.95

Figure 6.19 C-Scan graphs of Fiberdux 6376-HTA specimens fatigued at 280 MPa stress reversals for different n/N_f ratios.

ranging between 0% and 100%. Similarly, the introduced damage parameter DSF ranges between 0 and 1. The value of 0 is assigned to the ideal material, which outputs a completely white C-scan graph. In fact, composites include a small amount of damage by definition. Therefore, it is expected that the initial DSF values will be slightly different from 0. Accordingly, the factor 1 is assigned to the completely damaged material. The damage severity factor D is given by

$$D = \sum \frac{A_i}{A}(1 - k_i) \qquad (6.16)$$

where A is the total specimen area, A_i is the area which, according to the C-scan reading, demonstrated same damage state, and k_i is the damage severity (state) of the area A_i. Using the above definition for the damage severity factor, the DSF values that refer to the respective C-scan graphs may be determined.

The graphs in Fig. 6.20 refer to three unfatigued carbon/epoxy specimens. Due to the defects that always exist in engineering composites, the derived DSF values are not zero but have a certain positive value. In addition, different damage factors correspond to each specimen. The graphs in Fig. 6.21 are derived from three different Fiberdux 6376-HTA specimens subjected to 230 000 stress reversals at 280 MPa. The damage severity factor and the consumed life percentage n/N_f are also indicated in the figure. N_f refers to the expected fatigue life at the

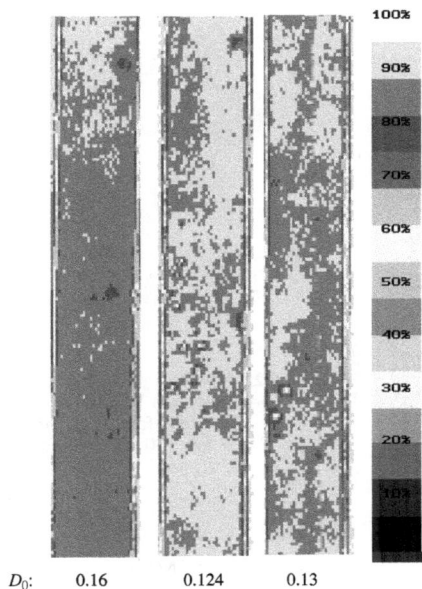

Figure 6.20 Initial damage of three Fiberdux 6376-HTA reference specimens.

D_0:	0.16	0.124	0.13

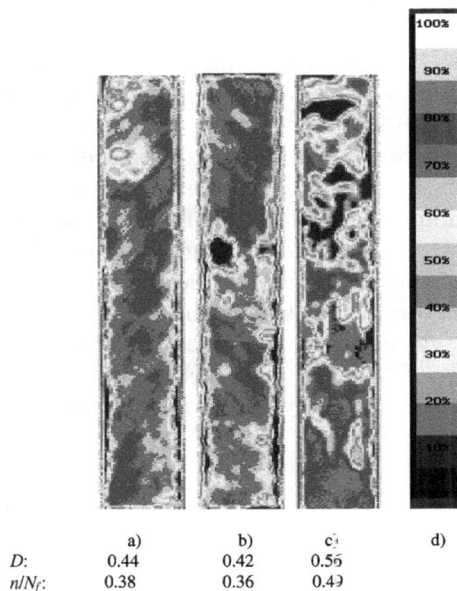

Figure 6.21 C-Scan graphs of Fierdux 6376-HTA specimens fatigued at $\sigma_\alpha = 280\,\mathrm{MPa}$ for $230\,000$ cycles (a,b,c) and the ultrasonic range (d).

	a)	b)	c)	d)
D:	0.44	0.42	0.56	
n/N_f:	0.38	0.36	0.40	

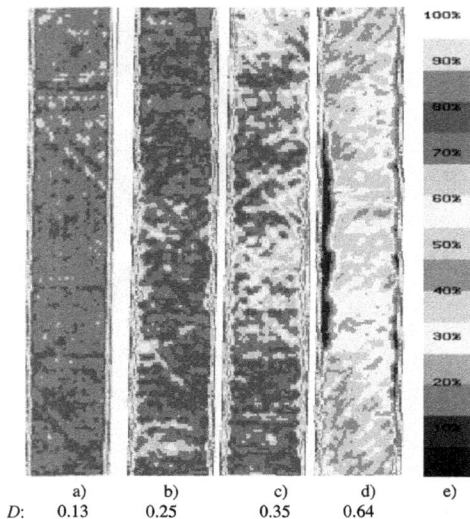

Figure 6.22 C-Scan graphs of APC-2 specimen: (a) before fatigue; (b), (c), (d) fatigued at $\sigma_x = 260\,\mathrm{MPa}$ for 160 000 cycles; (e) ultrasonic range.

	a)	b)	c)	d)	e)
D:	0.13	0.25	0.35	0.64	

stated fatigue conditions according to the respective $S-N$ curve of the material. It is observed that the number of applied stress reversals corresponds to higher consumed fatigue life for the specimen with the higher damage factor. The different values of DSF derived for same fatigue exposure explain the deviations obtained by many researchers in the residual mechanical properties after fatigue. The derived damage severity factors DSF for the four APC-2 specimens of Fig. 6.22, fatigued under the same conditions (260 MPa) and scanned after the same number of fatigue cycles (160 000), were 0.25, 0.35, and 0.64, respectively. The initial DSF value for the unfatigued specimen was 0.13. The same remarks as for Fig. 6.21 can be made for the case of Fig. 6.22. All DSF values measured from the above readings are mean values for the entire specimens. In fact, to each area of the specimen there corresponds a different damage severity factor. Practical application of this concept to characterize fatigue damage will require accurate definitions of the area of the component to be scanned. For a certain area, e.g. for the entire area of the specimen, damage is characterized by the mean DSF value and the respective standard deviation. The standard deviation of the derived DSF value characterizes the extent of fatigue damage severity deviation within the examined area.

6.6.2 Derivation of fatigue damage functions

To make use of the defined damage severity factor for defining fatigue damage functions, the fatigue tests were interrupted at specified intervals and C-scan graphs were taken. C-Scan readings of a Fiber-dux 6376-HTA specimen, fatigued up to failure under stress reversal at 250 MPa, are shown in Fig. 6.23. The graphs have been taken at intermediate stages of the specimen fatigue life. N is the number of reversals corresponding to each graph; N_f is the fatigue life of the specimen. Derived values for the DSF evaluation during fatigue may be plotted over the consumed fatigue life to obtain fatigue damage function. Fatigue damage functions using DSF as damage parameter are shown in Fig. 6.24, for the material Fiberdux 6376-HTA. Due to the various initial values of DSF in unfatigued conditions, the plots in Fig. 6.24 have been made using the quantity $DSF' = DSF - DSF_0$ where DSF is the current value of the damage severity factor and DSF_0 is the respective value of the unfatigued specimen. As shown in Fig. 6.24, the experimental results may be fitted well by an exponential function of the form.

$$D' = a[1 - \exp(-b(n/N_f))] \tag{6.17}$$

where the constants a and b depend on the material system and the stress levels. The damage curves of Fig. 6.24 cover the whole range of the S–N curve, as the stress reversals considered refer to fatigue lives from 3×10^4 to 2.3×10^6 cycles. It can be observed that the curves corresponding to lower stresses are located at the "upper" side of the diagram; this can be explained by considering that the higher fatigue

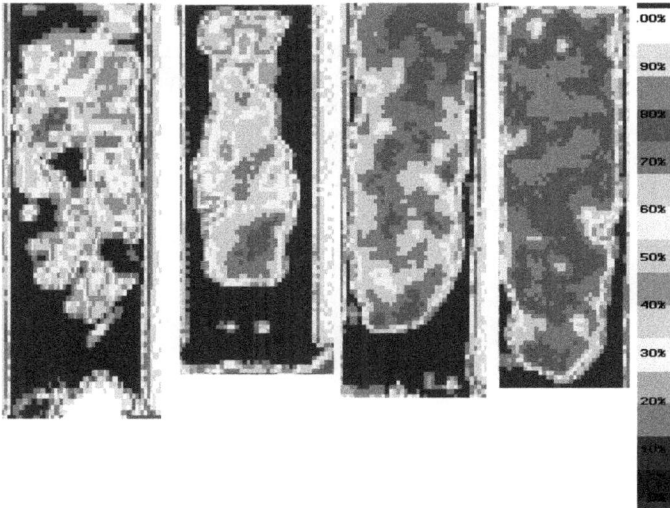

N/N_f: 0.32   0.56   0.78   0.86

Figure 6.23   C-Scan readings of a Fiberdux 6376-HTA specimen fatigued at 250 MPa.

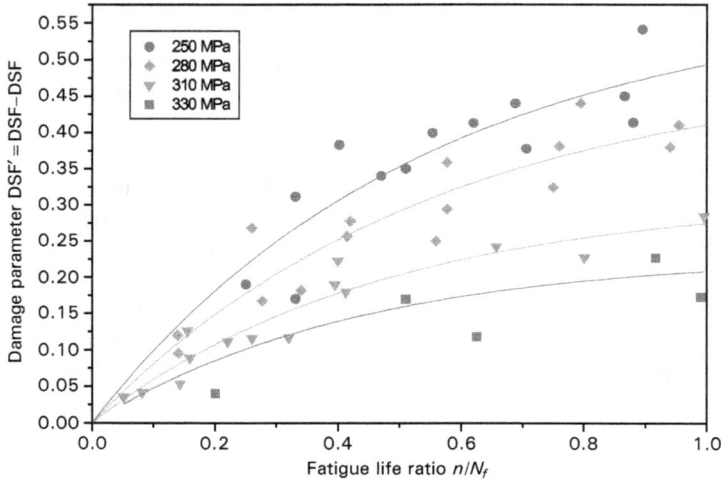

Figure 6.24 Damage parameter D' versus fatigue ratio n/N_f for stress amplitudes 250, 280, 310, and 330 MPa.

load results is less delamination and therefore lower values of DSF. A fatigue damage function involving stress amplitude dependence may be formulated when the relations between the constants a and b of Eq. (6.17) and the respective stress amplitude are known. The experimentally derived relations for the constants a and b are shown in Figs. 6.25 and 6.26, respectively.

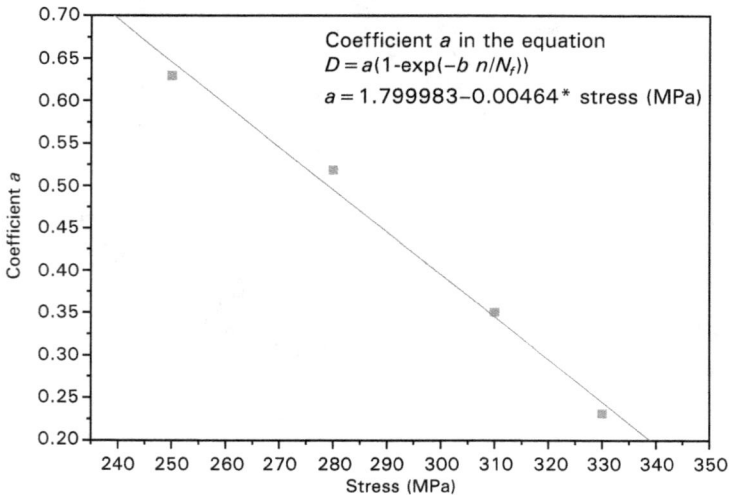

Coefficient a in the equation
$D = a(1-\exp(-b\, n/N_r))$
$a = 1.799983 - 0.00464 *$ stress (MPa)

Figure 6.25 Dependence of the constant a of Eq. (6.17) on stress level.

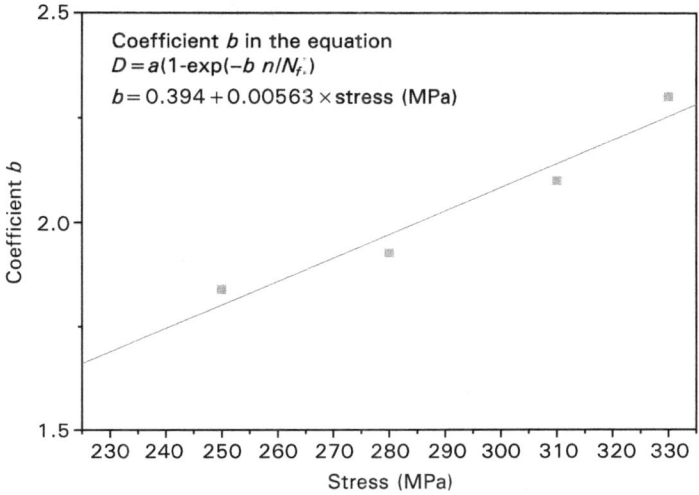

Figure 6.26 Dependence of the constant b of Eq. (6.17) on stress level.

By replacing the constants a and b in Eq.(6.17), by the linear expression shown in Figs. 6.25 and 6.26, the empirical fatigue damage function:

$$D'_f = 1.42498 - 00369\sigma_a \tag{6.18}$$

has been obtained, which may be used for the derivation of damage curves at any stress level lying within the ranges experimentally investigated.

6.6.3 Relation of the damage parameter DSF′ to the laminate residual properties

The determined fatigue damage severity factor DSF may be related to the residual properties of the composite laminate. These relationships would serve as a background for a damage tolerance treatment of composite structures. At present, establishment of such relations is at an early stage. Formulation of analytical expressions for the dependence of the mechanical properties axial stiffness, tensile strength, and interlaminar shear strength on the damage severity factor is described below. Figures 6.27 and 6.28 show the determined dependence of the axial stiffness on the damage severity factor for the materials APC-2 and Fiberdux 6376-HTA, respectively. In Fig. 6.27, E_0 is the stiffness of the unfatigued specimen and E is the stiffness of the fatigued specimen; the ratio E/E_0 is shown as a function of the damage severity factor DSF. By definition, as can be seen in Fig. 6.27 there is a trend of stiffness decrease with increasing damage severity factor value, but the measured normalized stiffness decrease is very small and does not exceed 5% for the entire fatigue life at any stress level investigated. This makes the derivation of the stiffness decrease dependence on the stress amplitude very difficult and, for the case investigated, insignificant in engineering terms. The results obtained may be fitted well using a master curve that is a straight line drawn from initial point (D_0, E_0^*), which refers to the unfatigued material, by means of regression analysis. The determined dependence is given by

$$E^* = 1 + b(D - D_0) \tag{6.19}$$

where b stands for the slope of the straight line; its value is determined to 0.16 for APC-2 and 0.19 for Fiberdux 6376-HTA.

The standard deviations were found to be 0.05 and 0.07, respectively. For $D = 1$ the value of the normalized modulus of elasticity E^*, for APC-2 material is 0.86. Recall that the value DSF $= 1$ refers to the fully delaminated material. The determined value $E^* = 0.86$ for DSF $= 1$ lies very close to the value $E^* = 0.82$, which refers to the stiffness of the zero plies divided by their number in the laminate; for the case investigated this was 4. Using the pair of points with $(D_0 = 0.14, E^* = 1)$ and $(D = 1, E^* = 0.82)$, respectively, the experimentally determined master curve is presented in Fig. 6.27 by the dotted line. The same procedure has been applied to calculate the

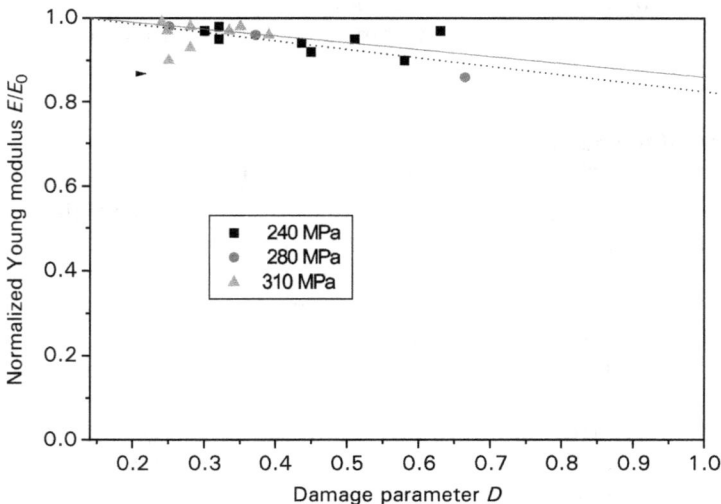

Figure 6.27 Dependence of the axial stiffness degradation on the damage severity factor DSF for an APC-2 specimen.

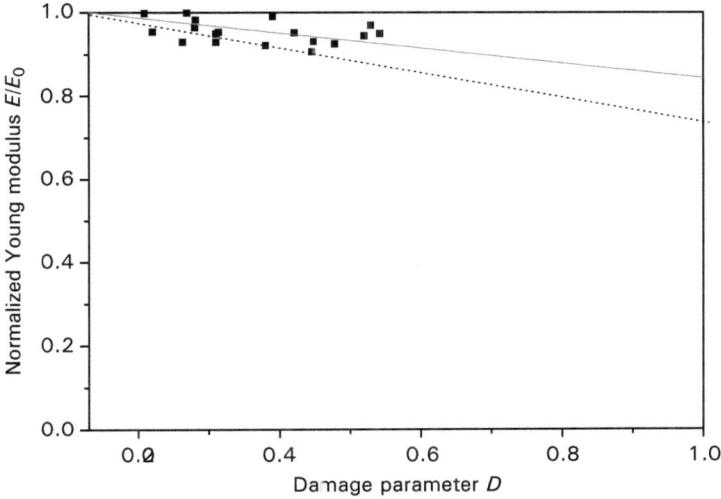

Figure 6.28 The determined dependence of the axial stiffness on the damage severity factor DSF for Fiberdux 6376-HTA.

dotted line in Fig. 6.28. These curves may be used for assessing the dependence of E^* on D without performing fatigue tests. This result refers solely to the examined case, however, and has to be further validated with other material systems and laminate stacking sequences.

Similar experimental results are derived for the dependence of residual stiffness degradation on the damage severity factor, as shown in Figs 6.29 and 6.30. R_m^* in these figures stands for the quantity R_m/R_{m0} with R_m and R_{m0} being the tensile strength of fatigued and reference specimens, respectively. A trend of residual strength decrease with increasing damage severity factor was determined. Measured residual strength decrease reached values up to 30%, but one should note that the residual strength values show appreciable deviations. With the limited number of experiments currently available, it is difficult to conclude the appropriate fitting functions for the dependence residual strength on the damage severity factor and even more so for the stress amplitude dependence. For simplicity, the experimental results have also been fitted using a straight line as a master curve; it has been drawn from the initial point (D_0, R_{m0}^*), which stands for the unfatigued material, using regressions analysis. The standard deviation was calculated as 0.59. The plot may be expressed analytically by

$$R_m^* = 1 + d(D - D_0) \qquad (6.20)$$

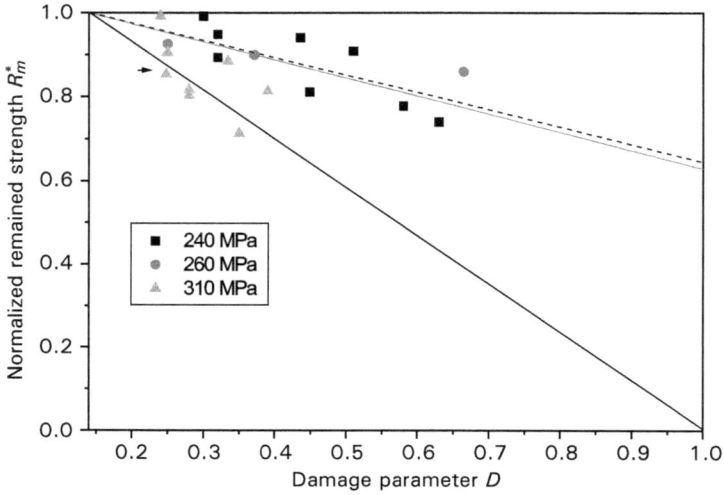

Figure 6.29 The dependence of the dimensionless residual strength degradation on the damage severity factor for APC-2.

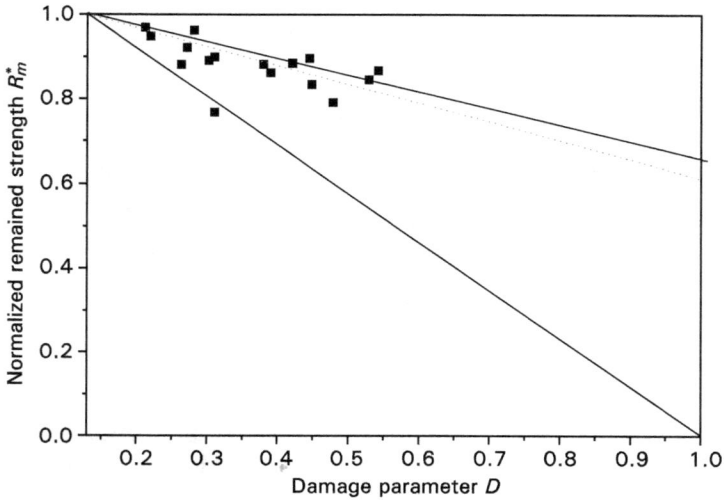

Figure 6.30 The dependence of the dimensionless residual strength degradation on the damage severity factor D, for Fiberdux 6376-HTA.

where d is the slope of the straight line. The value of d was determined to 0.43 for APC-2 and 0.54 for Fiberdux 6376-HTA materials, respectively.

Using this procedure to assess residual stiffness after fatigue based on tensile test data for the unfatigued material, the dotted lines in Figs. 6.29 and 6.30 were calculated. These give a good estimation of the expected residual strength decrease after fatigue exposure when the damage severity factor is known. Again, this simple relation has to be validated for other laminate types and material systems.

In Figs. 6.31 and 6.32, the interlaminar shear strength dependence on the damage severity factor for APC-2 and Fiberdux 6376-HTA, respectively, has been plotted. In the figures, ILSS* denotes the quantity $ILSS/ILSS_0$ where ILSS and $ILSS_0$ are the interlaminar shear strength values determined at several locations of the fatigued and reference specimen, respectively. The experimental results in Figs. 6.31 and 6.32 show a clear trend of ILSS strength degradation that becomes significant at values of DSF higher than 0.65. The measured experimental data are fitted well by

$$ILSS^* = 1 - D^m \tag{6.21}$$

with m being a material-dependent constant. Recall that D is defined on the basis of C-scan evaluation; it is sensitive to delamination but not to fiber breakage. To investigate the sensitivity of the formulated factor D to quantify the damage state at higher stress levels, ILSS

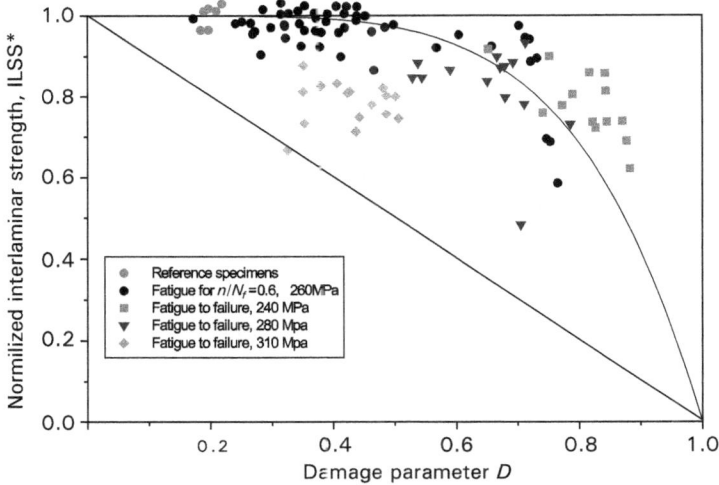

Figure 6.31 Interlaminar shear strength dependence on the damage severity factor DSF for an APC-2 specimen.

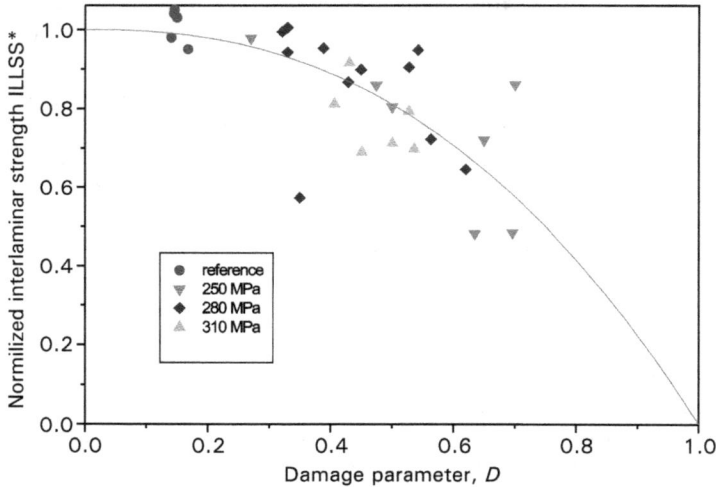

Figure 6.32 Interlaminar shear strength dependence on the damage severity factor DSF for a Fiberdux 6376-HTA specimen.

strength was determined on specimens cut from coupons fatigued at 310 MPa which is a high stress level for both materials. As can be seen in Figs. 6.31 and 6.32, for $\sigma_\alpha = 310$ MPa the determined damage severity factor underestimates the occurrence of fatigue damage appreciably. In that case, C-scan has to be replaced by more suitable nondestructive evaluation techniques (e.g., acoustic emission).

6.7 Damage Functions for Variable Amplitude Fatigue Loading

Structural components are usually expected to perform under variable fatigue load spectra, yet available experimental data are based on constant amplitude laboratory fatigue tests. Assessment of composite materials, fatigue behavior under variable amplitude loading has been made manageable by exploiting the S–N curves of the material and the Miner rule[47–49] or its modifications.[50] Application of these approaches to real components is not an obvious task; it requires experimental effort to derive all necessary S–N curves for the different loading conditions as well as empirical correction factors to fit the Miner rule to the experimental results. Calculation results and experimental data may differ significantly; correlation depends essentially on the statistical reliability of the available experimental data. In addition, these approaches are not compatible with damage tolerance design concepts. This drawback cannot easily be tolerated in weight-sensitive applications, as both reliability and cost efficiency of these structures are associated with the accuracy of assessing the expected fatigue life during service. Nonetheless due to the lack of tools for assessing fatigue life of composite materials subjected to service loading, S–N based fatigue life calculations are today practically the only means to support fatigue analysis, which remains experimental. Although it is acknowledged that design of composite structures by application of damage tolerance concepts is far from being formalized implemented, this section will be only concerned with the necessary background toward this end.

In principle, damage tolerance design can be made manageable by following procedures similar to those applied for the fatigue design of metallic structures, which are expected to perform under irregular load spectra. Such procedures comprise the following basic steps:

- Transformation of the irregular load spectrum into a regular spectrum of distinguished cycles.

- Derivation of the damage increment caused by each distinguished cycle.

- Accumulation of the damage according to a suitable damage accumulation rule.

As is obvious, the principal issue in the above procedure is the establishment of fatigue damage functions that may also be used for cases of material exposure to variable fatigue loading. One should recall that in the present work damage accumulation is defined as the gradual loss of the material's ability to respond to mechanical loads. This definition directly relates fatigue damage accumulation to the structural integrity of an engineering construction. With that in mind, this section focuses on the application of classical nonlinear damage accumulation approaches established for metals to the case of composite materials as well.

6.7.1 Transformation of an irregular fatigue spectrum into distinguished cycles

The simplest approach for transforming a random spectrum into a spectrum of full distinguished cycles is to consider each reversal causing tension as a full distinguished cycle. Unloading and compressive reversals may simply be neglected. This simplified approach is consistent with the postulates of linear-elastic fracture mechanics, but fatigue damage of composite materials is mainly associated with the compressive reversals. Application of this methodology to composite structures may result in significant overestimation of their fatigue life.

Therefore, for the transformation of the random stress spectrum into a sequence of distinguished stress cycles, one of the well-known counting methods such as the range pair method, the rainflow method,[48] etc., is recommended. For certain components operating under random loading conditions (e.g., for the turbine blades of a wind energy generator), the rainflow method has found widespread acceptance.[23] An advantage of this method when transforming a random spectrum is the consideration of all reversals namely, loading and unloading reversals, in both tension and compression. On the other hand, a serious drawback of the technique is that no attention is paid to the relative position of each reversal within the loading spectrum. This oversimplification implies the assumption of a linear damage accumulation during fatigue. This is justified when evaluating fatigue life on the basis of the Miner rule but not in models where damage accumulation is considered nonlinear. This shortcoming has been confronted by discretizing the original loading spectrum into n loading blocks consisting of a small number of reversals (Fig. 6.33). The rainflow method is then applied to each of these blocks separately.

Using this modification of the rainflow method the assumption of linear fatigue damage accumulation is limited within the range of each small loading block. Mathematically, it represents the approximation of a power function, which usually fits well a damage accumulation curve, by a sufficient number of successive, small straight lines of increasing slopes (Fig. 6.34). Proposed modification reduces computing costs by increased accuracy.

6.7.2 Fatigue damage accumulation under variable amplitude loading

To express fatigue damage as it is understood in this work, one should be concerned with the gradual reduction of all material mechanical

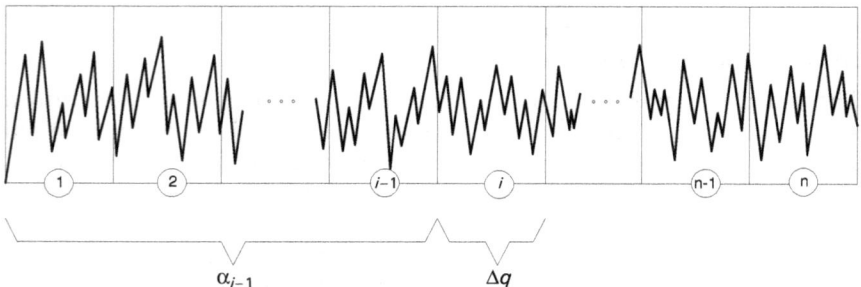

Figure 6.33 Discretization of the irregular loading spectrum.[23]

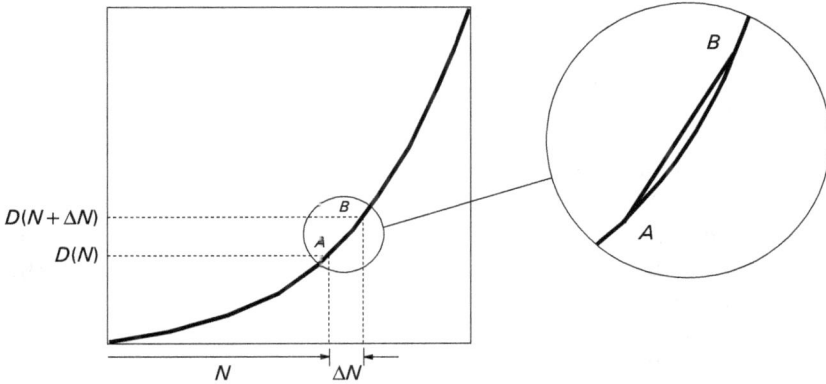

Figure 6.34 Approximation of the fatigue curve by linear damage increments.

properties that are involved in structural integrity analyses of engineering constructions. In most models that associate fatigue damage accumulation with the gradual decrease of mechanical properties, fatigue damage assessment is made by considering remaining strength or stiffness as damage parameter. Their evolution with the consumed fatigue life yields, also with regard to the fatigue stress dependence of the evolution, the fatigue damage function. Residual tensile or compressive strength based models have attracted considerable attention for assessment of fatigue damage under variable loading conditions.[51,52] For the relation of the residual strength to the remaining fatigue life, various phenomenological models have been proposed. The simplest is the linear rule of the form[52]:

$$S = S_{\text{ult}} - (S_{\text{ult}} - S_{\text{mas}}) \frac{n}{N_f} \qquad (6.22)$$

where S is the remaining static strength, S_{ult} is the strength of unfatigued specimens, $S_{\text{mɛs}}$ is the maximum fatigue stress, n is the number of fatigue cycles, and N_f is the fatigue life of the specimen when fatigued until final failure under the maximum stress S_{max}. Residual strength curves of specimens under different stress amplitudes are therefore required in order to use the residual strength as damage parameter in variable amplitude loading. A common characteristic of all the residual strength approaches is that specimens having the same residual strength are considered to suffer the same damage, irrespectively of the stress level and the fatigue cycles that caused the damage. The failure is predicted when the maximum fatigue stress equals the residual specimen strength. The residual strength procedure is based on a wide experimental database, which gives to the advantage of a high degree of reliability of the methodology and at the same time the disadvantage of high "cost." An additional drawback is the requirement of the destructive evaluation of the damage parameter. As will be shown below, in principle, all these damage functions may be used for calculating fatigue damage accumulation under variable amplitude fatigue loading. Another suitable parameter for calculating variable fatigue loading damage accumulation is the damage severity factor (DSF) discussed in the previous section. In the use of latter approach it is advantageous that the evolution of the damage severity factor, which may be derived nondestructively, can be directly related to the entirety of the remaining mechanical properties of the composite.

Let the apparent value of the fatigue damage of a composite that is subjected to fatigue be expressed by the simple relation

$$D = 1 - (\tilde{Q}/Q_0) \qquad (6.23)$$

In Eq. (6.23), D stands for damage and Q for a quantity suitable to reflect damage as has been discussed above (e.g., remaining strength or stiffness, the damage severity factor DSF, etc.). For the damage D, only the boundary values are known; they are zero for the unfatigued material and unity for the completely damaged material. The value 1 gives the failure condition with regard to the selected quantity Q. In expression (6.23), the magnitude $\tilde{Q}$ is the apparent value of the selected quantity; it corresponds to a certain damage state caused by a certain number of fatigue cycles N. Q_0 is the initial value of the property; it refers to the unfatigued material. The expression $\tilde{Q}/Q_0$ is considered to be a function of the load cycle N normalized with reference to N_f that corresponds to the fatigue life at the stress amplitude. The latter function is usually experimentally defined. A common form of this function is

$$\tilde{Q}/Q_0 = (1 - R)^m \qquad (6.24)$$

where R equals n/N_f. The exponent m is assumed to depend on the fatigue stress amplitude σ_0 and material constants. For simplicity, a linear relation for m may be taken,

$$m = A + B\sigma \qquad (6.25)$$

with A and B being material-dependent constants. Equations (6.24) and (6.25) may be put into equation (6.23) to obtain the damage accumulation function:

$$D = 1 - (1 - R)^{A+B\sigma} \qquad (6.26)$$

Equation (6.26) applies for the cases of constant and of variable amplitude loading. Now assume an irregular stress spectrum as shown in Fig. 6.33. Using the rainflow technique as proposed in previous paragraph, the load spectrum is transformed into blocks of distinguished cycles. Accumulation of damage can be done as is shown schematically in Fig. 6.35. For the mathematical formulation of the damage accumulation, let n denote the number of blocks and ΔD_i the damage increase corresponding to the ith loading block. The damage increase ΔD_i depends on the relative position of the ith loading block within the n-block spectrum as well as on the material damage state before its application. The damage of the material accumulated within the ith block can be expressed by means of a macroscopic parameter D, defined as:

$$D = D_i + \Delta D_i \tag{6.27}$$

where $D = 0$ and $D = 1$ correspond to the initial undamaged state and to the final fatigue failure respectively, and D_i accounts for the damage state before the ith load block application. For the first load block, D_i may be set equal to zero. When discretizing sufficiently small load blocks, the influence of the relative position of each load cycle within a certain block on the damage increment corresponding to that cycle is small and negligible. Therefore, the damage accumulation within each block can be considered as linear and is independent of the load sequence within the block.

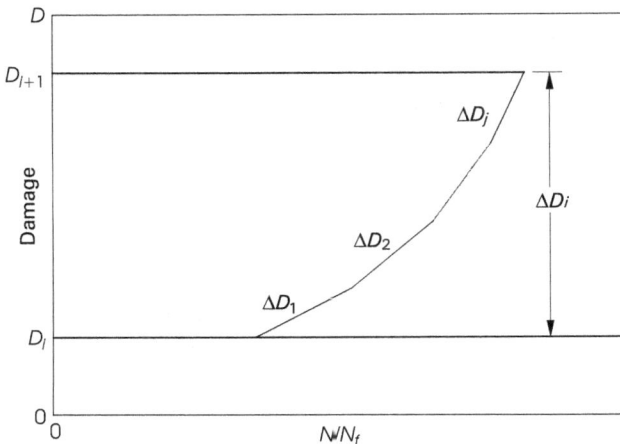

Figure 6.35 Derivation of damage accumulation under variable stress fatigue loading.

Thus, the damage increase ΔD_i is obtained by the summation

$$\Delta D_i = \sum_{j=1}^{\lambda} \Delta D_j \qquad (6.28)$$

with λ the number of cycles corresponding to the ith block and ΔD_j the damage increase due to the application of the jth load cycle. To further simplify the problem, the damage increase ΔD_j may be expressed by the difference

$$\Delta D_j = D_j - D_i \qquad (6.29)$$

where D_j corresponds to the material damage after application of the load cycle j on a material characterized by the damage state D_i. This simplification is equivalent to considering each cycle j as the first cycle of the ith block. In Fig. 6.35 the derivation of the damage increase ΔD_i caused by the ith block is shown explicitly.

Inserting into Eq. 6.26 the values $R = [(N_j/N_{fj}) + (1/N_{fj})$ and $m = m_j$ for the parameters R and m, the following expression for D_j is obtained:

$$D_j = 1 \left[1 - \left(\frac{N_j}{N_{fj}} + \frac{1}{N_{fj}} \right) \right]^{m_j} \qquad (6.30)$$

The quantity N_j/N_{fj} can be expressed as a function of the known quantity D_i after putting into Eq. (6.26) the values $R = N_j/N_{fj}$ and $m = m_j$ rewriting to obtain

$$\frac{N_j}{N_{fj}} = 1 - (1 - D_i)^{1/m_j} \qquad (6.31)$$

Using the above expression, the damage accumulated after the n load blocks of the spectrum follows by the linear summation of the quantities D calculated from Eq. (6.27) for each block. The procedure is repetitive and may easily be computed.

6.7.3 Application

In the present work, two alternative quantities are used as damage parameter D: the residual strength degradation and the damage severity factor DSF. The former choice of damage parameter was made for the case of carbon fibre reinforced thermoplastic PEEK (APC-2) of lamination $[-45,0,45,90]_{2s}$ and thickness 2 mm. Fatigue specimens were prepared according to specifications.[55] The specimens were subjected to irregular tension–compression fatigue loads under

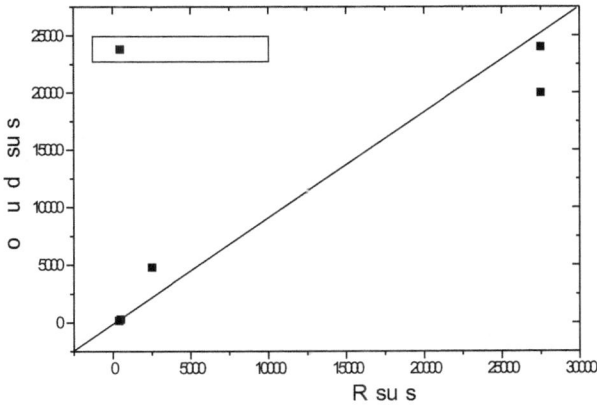

Figure 6.36 Correlation between computed and experimental results obtained on APC-2 specimens subjected to irregular fatigue loading.

normal laboratory conditions. Computed (solid line) and experimental fatigue lives of specimens subjected to irregular stress history are shown in Fig. 6.36 for APC-2. As can be seen, good correlation between estimation of fatigue life and experimental results is indicated.

The damage severity factor DSF was used as damage parameter for Fiberdux 6376-HTA material loaded by a random spectrum of zero mean stress value. Maximum fatigue stress values used for the loading spectra ranged between 270 and 360 MPa. The maximum stress values of the spectrum are plotted in Fig. 6.37 against the ratio of experimental to estimated failure cycles. Calculation of the failure cycles was made using both the damage severity factor concept and the Miner rule.

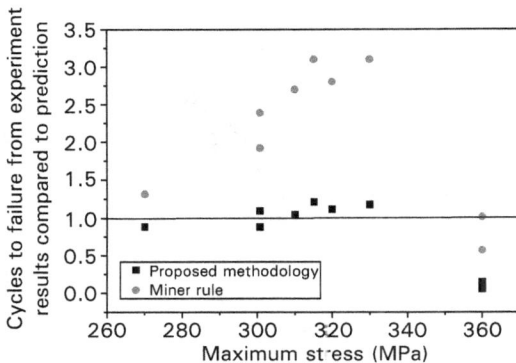

Figure 6.37 Failure ratio for different maximum stresses, calculated by DSF concept and Miner rule (Fiberdux 6376-HTA).

The straight line in the figure indicates complete agreement between experimental and calculated results. Deviations from the straight line characterize the accuracy of the calculation method. It can be concluded from Fig. 6.37 that the DSF concept gives good failure predictions up to maximum stresses of 330 MPa. At the stress level of 365 MPa the fatigue life is overestimated by a factor of 10. The basic reason for this disagreement is that the damage functions utilized above were developed for lower stress ranges.

6.8   Conclusions

The main conclusions arising from the work described concerning the constant and variable fatigue damage of laminated fibrous composites may be summarized as follows.

The capabilities existing today for fatigue analysis and design of engineering lightweight structures is quite limited and difficult to developed mainly for the following reasons: (a) The damage mechanisms of fiber reinforced composite materials are case-specific; it is therefore difficult to predict them, even for the simplest case of uniaxial constant amplitude loading. (b) The concept of fatigue damage for composite materials is difficult to quantify and to date there is no commonly accepted damage parameter analogous to the case of metals, where the crack length is considered as the damage measure. (c) Damage tolerance, which is the current design philosophy applied to metallic structures, cannot easily be extended to composites. (d) The available theoretical tools and the existing numerical codes for fatigue design of composites cover only simple and specific cases and are not generally applicable to the majority of composite fatigue design problems.

The investigation described here has demonstrated the need to consider the degradation of the entirety of mechanical properties that are involved in the analysis when evaluating structural integrity. The damage severity factor (DSF) discussed here is included among the concepts that aim to assess fatigue damage and material property degradation using results from nondestructive evaluation data. The DSF approach quantifies the severity of fatigue damage using C-scan graphs of fatigued specimens and provides a damage tolerance background for fatigue analysis of composite laminates. Further investigation is needed to determine whether correlations between the proposed damage index DSF and the residual mechanical properties might yield the derivation of generalized fatigue damage functions for composite materials.

References

1. T. K. O'Brien (1990) *Towards a Damage Tolerance Philosophy for Composite Materials and Structures*, ASTM STP 1059, American Society for Testing and Materials, Philadelphia, 7–33.
2. G. C. Sih (1988) Microstructure and damage dependence of advanced composite material behaviour, *Proceedings of Workshop on Composite Material Response: Constitutive Relations and Damage Mechanisms*, Glasgow, U.K., 1987, Elsevier Applied Science, 1–23.
3. K. L. Reifsnider (ed.) (1991) *Fatigue of Composite Materials*, Elsevier.
4. T. Adam, N. Gathercole, and B. Harris (1994) Life Prediction for fatigue of T800/5245 carbon-fibre composites: I. Constant-amplitude loading, Fatigue, **16**, 523–532.
5. T. Ireman (1999) Design of Composite Structures containing bolt holes and open holes, Doctoral thesis, Royal Institute of Technology, Sweden.
6. R. Talreja (1991) Damage characterization, in *Fatigue of Composite Materials*, (K. L. Reifsnider, ed.), Elsevier, 79–103.
7. D. H. Allen, C. E. Harris, and S. E. Groves (1987) A thermomechanical constitutive theory for elastic composites distributed damage—I. Theoretical development, *Int. J. Solids Structures*, **23**, 1301–1318.
8. D. H. Allen, C. E. Harris, and S. E. Groves (1987) A thermomechanical constitutive theory for elastic composites distributed damage-II. Application to matrix cracking in laminated composites, *Int. J. Solids Structures*, **23**, 1319–1338.
9. J. W. Lee, D. H. Allen, and C. E. Harris, (1989) Internal state variable approach for predicting stiffness reduction in fibrous laminated composites with matrix cracks, *J. Compos. Mater.*, **23**, 1273–1291.
10. T. W. Coats, and C. E. Harris (1995) Experimental verification of a progressive damage model for IM7/5260 laminates subjected to tension-tension fatigue, *J. Compos. Mater.*, **29**, 280–305.
11. K. L. Reifsnider (1986) The critical element model: a modeling philosophy, *Eng. Frac. Mech.*, **25**, 739–749.
12. H. Fukunaga, T. W. Chou, P. W. M. Peters, and K. Schulte (1984) Probabilistic failure strength analyses of graphite/epoxy cross-ply laminates, *J. Compos. Mater.*, **18**, 339–356.
13. K. L. Reifsnider, K. Schulte, and J. C. Duke (1983) Long-term fatigue behavior of composite materials, ASTM STP 813, 136–159.
14. A. L. Highsmith and K. L. Reifsnider (1982) Stiffness-reduction mechanics in composite laminates, in *Damage in Composite Materials*, ASTM STP 775, 103–117.
15. Wen-Fang-Wu, L. J. Lee, and S. T. Choi (1996) A study of fatigue damage and fatigue life of composite laminates, *J. Compos. Mater.*, **30**, 123–137.
16. A. S. D. Wang, P. C. Chou, and S. L. Lei (1984) A stochastic model for growth of matrix cracks in composite laminates, *J. Compos. Mater.*, **18**, 239–254.
17. S. C. Tan and R. J. Nuismer (1989) A theory for progressive matrix cracking in composite laminates, *J. Compos. Mater.*, **23**, 1029–1047.
18. Ogin, P. A. Smith, and P. W. R. Beaumont (1985) Matrix cracking and stiffness reduction during the fatigue of (0/90) GFRP laminates, *Compos. Sci. Technol.*, **22**, 23–31.
19. N. Laws and G. J. Dvorak (1988) Progressive transverse cracking in composite laminates, *J. Compos. Mater.*, **22**, 901–916.
20. A. Poursartip and P. W. R. Beaumont (1988) the fatigue damage mechanics of a carbon fibre composite laminate: II: Life prediction, *Compos. Sci. Technol.*, **25**, 283–299.
21. Sp. G. Pantelakis, Th. P. Philippidis, and Th. B. Kermanidis (1995) Damage accumulation in thermoplastic laminates subjected to reversed cyclic loading, *Proceedings of the 4th International Symposium on "High Technology Composites in Modern Applications,"* Corfu, University of Patras Public, 156–164.
22. A. Rotem (1986) Fatigue and residual strength of composite laminates, *Eng. Fract. Mech.*, **25**, 819–827.

23. Sp. G. Pantelakis, Th. B. Kermanidis, and D. G. Pavlou (1994) Fatigue life assessment of composite specimen subjected to random loading, *Proceedings of the International Conference on Design and Manufacturing using Composites (ATMAM 94)*, Montreal, Canada, 456–457.
24. DIN 65 586, Fatigue Strength Behaviour of Fiber Composite Under One-Stage Loading, Vorlage (April 1992).
25. ASTM D2344, Apparent Horizontal Shear Strength of Reinforced Plastics by Short Beam Method, ASTM Part 36 (1976), American Society for Testing and Materials, Philadelphia.
26. ASTM D3039, Tensile Properties of Fiber–Resin Composites, ASTM Part 36 (1976), American Society for Testing and Materials, Philadelphia.
27. Sp. Pantelakis, M. Kyriakakis, and P. Daglaras (1997) Towards the derivation of composite specimen fatigue functions, in *Proceedings of the 1st Hellenic Conference on Composite Materials and Structures*, Xanthi, Greece, 529–544.
28. T. Adam, N. Gathercole, and B. Harris (1994) Life prediction for fatigue of T800/5245 carbon-fibre composites: II. Variable-amplitude loading, *Fatigue*, **16**, 533–547.
29. Hwang, K. S. Han (1986) Cumulative damage models and multi-stress fatigue life predictioin, *J. Compos. Mater.*, **20**, 125–151.
30. S. S. Manson and G. R. Halford (1986) Re-examination of cumulative fatigue damage analysis—an engineering perspective, *Eng. Fract. Mech.*, **25**, 539–571.
31. T. Adam, N. Gathercole, and B. Harris (1994) Life prediction for fatigue of T800/5245 carbon-fibre composites: I. Constant-amplitude loading," *Fatigue*, **16**, 523–532.
32. T. Adam, N. Gathercole, and B. Harris (1994) Life prediction for fatigue of T800/5245 carbon-fibre composites: II. Variable-amplitude loading, *Fatigue*, **16**, 533–547.
33. M. S. Found and M. Quaresimin (19xx) "Fatigue damage of carbon fibre reinforced laminates under two-stage loading.
34. J. Lee, B. Harris, D. P. Almond, and F. Hammet (1997) Fibre composite fatigue-life determination, *Composites*, **28A**, 5–15.
35. M. Kawai, M. Morishita, K. Fuzi, T. Sakurai, and K. Kemochi (1996) Effects of matrix ductility and progressive damage on fatigue strengths of unnotched and notched carbon fibre plain roving fabric laminates, *Composites*, **27A**, 493–502.
36. A. Rotem and Z. Hashin (1976) Fatigue failure of angle ply laminates, *AIAA J.*, **14**, 868–872.
37. A. Rotem (1979) Fatigue failure of multidirectional laminates, *AIAA J.*, **17**, 271–277.
38. Z. Fawaz, and F. Ellyin (1994) Fatigue failure model for fibre-reinforced materials under general loading conditions, *J. Compos. Mater.*, **28**, 1432–1451.
39. S. Tsai and E. Wu (1971) A general theory of strength of anisotropic materials, *J. Compos. Mater.* **5**, 58–80.
40. P. Theocharis and T. Philippidis, (19xx) On the validity of the tensor polynomial failure theory with stress interaction terms omitted, *Compos. Sci. Technol.*, **40**, 181–191.
41. T. Philippidis and A. Vassilopoulos (1999) Fatigue of composite laminates under off-axis loading, *Int. J. Fatigue*, **21**, 253–262.
42. Z. Hashin and A. Rotem (1973) A fatigue criterion for fiber reinforced materials, *J. Comp. Mater.*, **7**, 448–464.
43. M. Owen and J. Griffiths, Evaluation of biaxial stress failure surfaces for a glass fabric reinforced polyester resin under static and fatigue loading, *J. Mater. Sci.*, **13**, 1521–1537.
44. Th. Phillipidis, A. Vasilopoulos, and I. Kolaxis (1997) *Macroscopic Criteria for the Fatigue Behavior Prediction of Fiber Reinforced Composite Materials*, CRES report, [in Greek].
45. T. O'Brien (1993) Local delamination in laminates with angle ply matrix cracks, Part II: Delamination fracture analysis and fatigue characterization, ASTM STP 1156, American Society for Testing and Materials, Philadelphia, 507–538.
46. K. L. Reifsnider, K. Schulte, and J. C. Duke (1983) "Long-term fatigue behavior of composite materials, ASTM STP 813, American Society for Testing and Materials, Philadelphia, 136–159.

47. R. S. Whitehead (1987) Certification of primary composite aircraft structures, in *Proceedings of the 14th ICAS Symposium*, (D. L. Simpson, ed.) EMAS Ltd., Warley, U.K., 585–617.
48. Hayes J. E. (1965) *Fatigue Analysis and Fail-Safe Design, Analysis and Design of Flight Vehicle Structures* (E. F. Bruhn, ed.), Tri-State Offset Co., Cincinnati, OH, C13–1.
49. R. Talreja (1991) *Fatigue of Composite Materials: Analysis Testing and Design*, Seminar, Basel, Switzerland.
50. G. C. Sih (1972) A special theory of crack propagation, in *Method of Analysis and Solutions to Crack Problems*, (G. C. Sih, ed.), Wolters-Noordhoff.
51. A. R. Bunsel, D. Laroche, and D. Valentin Damage and failure in carbon-fiber-reinforced epoxy resin, ASTM STP 813, American Society for Testing and Materials, Philadelphia. 38–54.
52. P. Ouellette and S. Hoa (1983) Fatigue of glass-reinforced polyester laminates, *Polym. Compos.*, **7**, 64–68.
53. R. Bheshlehurst, Analysis and modeling of damage and repair of composite materials in aerospace, in *Analysis and Modeling of Composite Materials*.
54. Sp. G. Pantelakis, Th. P. Philippidis, and Th. B. Kermanidis (1995) Damage accumulation in thermoplastic laminates subjected to reversed cyclic loading, *Proceedings of the 4th International symposium on "High Technology Composites in Modern Applications,"* Corfu, University of Patras Public, 156–164.
55. Wen-Fang-Wu, L. J., Lee, and S. T. Choi (1996) A study of fatigue damage and fatigue life of composite laminates *J. Compos. Mater.*, **30**, 123–137.
56. T. R. Tauchert and A. N. Guzelsu (1972) An experiment study of dispersion of stress waves in a fiber reinforced composite, *J. Appl. Mech.*, **39**, 98–102.
57. T. J. Fowler (1977) Acoustic emission of fiber reinforced plastics, ASCE Fall Convention and Exhibits, Oct. 1977.
58. R. Talreja, A. Covada, and E. Henneke (1984) Quantitative assessment of damage growth in graphite epoxy laminates by acousto-ultrasonic measurements, *Review of Progress in Quantitative Non-destructive Evaluation*, Plenum Press, New York, 1099–1106.
59. R. D. Jamison (1982) Advanced fatigue damage development in graphite epoxy laminaes, PhD dissertation College of Engineering, Virginia Polytechnic Institute and State University, Blackburg, VA.
60. D. Burleigh (1989) Thermographic NDT of graphite epoxy filament wound structures, *International Conference of Thermal Infrared Sensing for Diagnostics and Control*, Proc. SPIE 1094, 175–181.
61. A. Briggs (1985) *An Introduction to Scanning Acoustic Microscopy*, Microscopy Handbooks, vol. 12, Oxford University Press, Royal Microscopical Society, Oxford.
62. A. Vary and K. Bowles (1977) *Ultrasonic Evaluation of the Strength of Unidirectional Graphite Polyamide Composites*, NASA Technical Memorandum, NASA TM X-73646.
63. J. Reddy (1994) *Composite Structures, Testing, Analysis and Design*, Springer.
64. B. Tang and E. Henneke (1989) Lamb wave velocity measurement for evaluation of composite plates, *Non-destruct. Testing Commun.*, **4**, 109–120.
65. M. Sundaresan, M. Henneke, and A. Gavens (1989) NDE procedure for predicting the fatigue life of composite structural members, *Proceedings of the World Meeting on Acoustic Emission*, Charlotte, NC.
66. R. Talreja (1986) Stiffness properties of composite laminates materials with matrix cracking and interior delamination, *Eng. Fract. Mech.*, **25**, 751–762.
67. M. Kyriakakis (1999) Fatigue damage parameters for polymer composites, Ph.D. dissertation, Mech. Eng. Dept, Univ. Patras, Greece.
68. S. Pantelakis and M. Kyriakakis (19xx) Fatigue damage of APC-2 composite assessed from material degradation and nondestructive evaluation data, *Theor. Appl. Fract. Mech.*, to appear.
69. W. Rose and S. Roklin (1979) Elastic moduli of transversely isotropic graphite fibers and their composites, *Exp. Mech.*, **19**, 41–49.

70. D. Hayford, E. Henneke, and W. Stinchcomb, (1977) The correlation of ultrasonic attenuation and shear strength in graphite polyamide composites, *J. Compos. Mater.*, **11**, 429–444.
71. A. Vary and K. Bowles, (1979) An ultrasonic acoustic technique for non-destructive evaluation of fiber composite quality, *Polym. Eng. Sci.* **19**, 373–376.
72. R. Jones (1975) *Mechanics of Composite Materials*, McGraw-Hill, New York.
73. H. Henneke, J. Duke, W. Stinchcomb, W. Govada, and A. Lemascon, (1983) *A Study of the Stress Wave Factor Technique for the Characterization of Composite Materials*, NASA contractor report, 3670.
74. D. Russel and E. Henneke (Feb. 1984) Dynamic effects during vibrothermographic NDE of composites, *NDT Int.*, **17**, 19–25.

Fatigue Damage Tolerant Design in Unidirectional Metal Matrix Composites

Chris A Rodopoulos

SIRIUS-Department of Mechanical Engineering,
University of Sheffield, Sheffield, UK

7.1 Introduction

Although a substantial amount of research and development on metal matrix composites (MMCs) has been carried out since the beginning of space administration (the mid-1950s),[1] MMC materials are still considered to be in their infancy. This is basically because the industrial utilization of MMCs is characterized by extremely costly processes, as a result of difficulties in the production of raw materials (especially fibers), manufacturing, fabrication, and jointing. However, it is generally accepted that the utilization cost of MMCs would be offset if their response under loading were less complicated and unpredictable.

In general, design is always a problem with composite materials. In Chapter 1 it was mentioned that problems regarding the nonlinear behavior of composites can be compromised if we assume some type of pseudo-linearity of the constituent phases. This approach is proven to work successfully for the prediction of the mechanical properties and strength (e.g., Law of Mixtures). However, where damage mechanics issues are involved, for example, in the case of a fatigue crack, the hypothesis of pseudo-linearity was found to produce dangerous results.[2] This is because classical theories of fracture are unqualified to simultaneously treat complex and numerous damage mechanisms. As will be discussed later in this chapter, the fatigue cracking of MMCs demonstrates a variety of different damage modes that vary from an initial pseudo-linear behavior at short crack lengths to a highly nonlinear behavior at longer lengths (where the plastic zone becomes large enough to override crack tip singularities).

It is therefore important, before we begin with damage tolerance design principles, to examine in detail the *"material's response under fatigue,"* in order to determine the operational limits of the material. To enhance the usefulness of the chapter, each aspect is analyzed in conjunction with popular MMCs. Only unidirectional MMCs are discussed here, since particulate and whisker reinforced versions do not exhibit any particular superiority of fatigue resistance.

7.2 Fatigue Damage of MMCs

7.2.1 The subject

It is difficult to define the so-called *"material's response under fatigue"* for composite materials.[3] Work on monolithic materials has shown that the principles of mechanics, mathematics, and physics can be combined to simulate and predict experimental results. The development of linear elastic fracture mechanics (LEFM) is an example of such collaboration. However, in composite materials and especially in MMCs, the material's response under fatigue goes beyond previous understanding. That is because MMC composite materials are highly anisotropic and therefore the stress around a flaw could differ significantly from that in an isotropic medium. Consequently, while monolithic materials normally fail by the nucleation and propagation of a crack, composite materials exhibit a variety of different failure modes including, matrix grazing, fiber failure, etc. These failure modes basically result from randomly distributed flaws within the composite able to initiate a number of different failure events in their vicinity. For example, the growth of voids generally leads to matrix microcracking, while flaws in the vicinity of the fiber–matrix interface may initiate a debonding process (see Chapter 2). Additionally, since MMCs contain various microstructural constituents whose performance can differ widely under long-term application of external influences such as fatigue loads, properties like stiffness and strength could become time-dependent tensors, since they may significantly change with loading history.

Therefore, in contrast to monolithic materials in which fatigue failure is basically dictated by the length of the propagating crack, characterization of fatigue failure of unidirectional MMCs is based on a quantitative analysis of the damage generated by the four different damage modes that occur during the fatigue process, namely, matrix microcracking, interfacial debonding, fiber bridging, and fiber failure ahead of and behind the crack tip.[4–6]

Except for stress level, which will be discussed during the damage tolerant design section, factors like mechanical properties of the interface, the presence of flaws and notches, matrix ductility, and fiber volume fraction are observed to dictate the fatigue resistance of the material by controlling the severity of the above failure modes and therefore cannot be excluded.

7.2.2 Fiber–matrix interface

There are several reported works on the influence of the fiber–matrix interface on the fatigue performance and failure modes of growing matrix cracks.[7,8] However, the complex chemical, physical, and mechanical interfacial phenomena as indicated by the presence of reaction zones[9] (Fig. 7.1) have constrained the classification of the role of the interface on fatigue into two types, considering their resistance to debonding (interfacial shear strength), (Fig. 7.2). The debonding process is presented in detail in Chapter 2.

The first type considers that the interface is weak in comparison to the matrix yield strength[11] and so the stresses near the tip of the matrix crack as this approaches the fiber could cause the fiber to

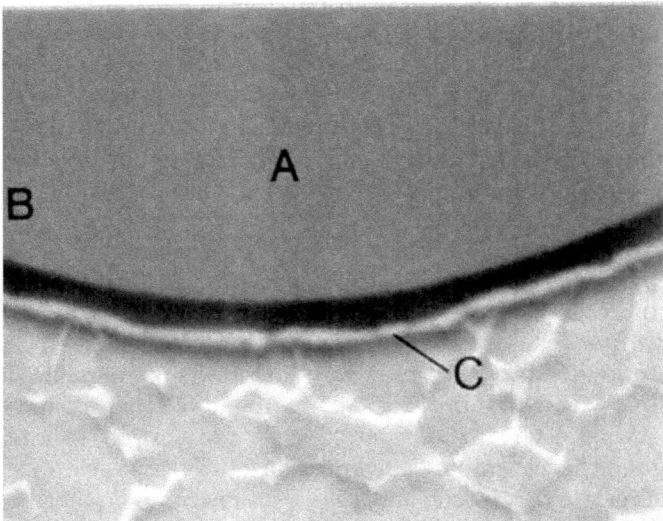

Figure 7.1 A typical reaction zone around a fiber. (*Photo taken from Reference 10, Reprinted with permission from Elsevier Science.*)

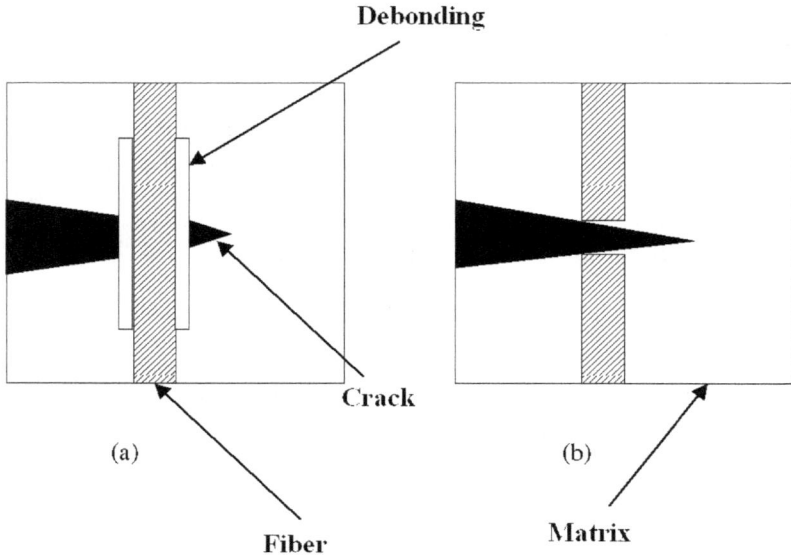

Figure 7.2 The effect of interface on crack propagation of unidirectional composites. (a) The crack propagates through weak interfaces by the debonding process; (b) propagation through strong interfaces is achieved by fiber failure.

debond from the matrix. This mechanism was first proposed by Cook–Gordon[12] (and is also known as Cook–Gordon effect), who suggested that the matrix crack deviates along the interface due to debonding propagation. During debonding, the stress and deformation fields developed at crack tips located at or near the interface crack (bimaterial since fiber and matrix posses different mechanical properties), induce a mixed-mode fracture character, $K = K_I + K_{II}$ (where K_I, K_{II} represent the strength of stress singularities in tensile and in-plane shear loading, respectively), which in turn reduces the crack tip driving force and consequently increases the fracture toughness of the material.[13–15] It was reported that the duration of the debonding propagation is increased in cases where high residual stresses prevail.[16] Clearly, high residual stresses reduce the effective stresses acting at the interface. Matrix cracks may revert back to mode I matrix cracking ($K_{II} = 0$) when the stress field ahead of the interface crack cannot overwhelm the bond strength.[17,18] Moreover, He and Hutchinson[19] argued that it is much easier for an interface crack to be initiated by a microstructural defect in the reaction zone than for a matrix crack to deflect and propagate by the mixed-mode fracture. Generally this type of fatigue damage was found especially relevant to long-fiber reinforced MMCs where notches or other discontinuities may developed high interfacial shear and peeling stresses.[20]

High values of fiber volume fraction were reported to enhance the possibility of interfacial debonding since the number and consequently the probability of weak interfaces ahead of the crack tip is increased.[21] On the other hand, fiber orientations between 45° and 90° were found to promote deflecting crack tip behavior.[22] Furthermore, in terms of fracture morphology, weak interfaces show extended fiber pull-out, giving an extremely irregular fracture surface[23] (Fig. 7.3). This is due to a physical correlation between the debonding and fiber bridging process, as detailed later. The degree of irregularity of the fracture surfaces may be used as a quality control factor.

When the fibers are well bonded to the matrix (strong interface), debonding may not occur as the crack approaches the interface, but instead fiber failure in a brittle manner is expected.[25] This type of damage behavior is explained by considering that for high values of the interfacial shear strength, τ, the stress required to debond the interfaces is higher than the ultimate tensile strength of the composite (which is usually defined by the failure of the critical fiber length).[26] Although this type of damage reduces the fracture toughness and the fatigue resistance of the material,[20] numerical work has shown that strong interfaces enhance the mechanical properties of the material in

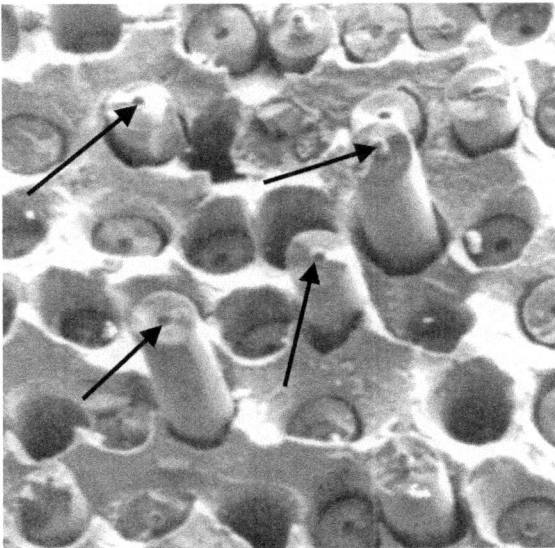

Figure 7.3 The number and density of fiber pull-out sites observed in fatigue fracture surfaces can indicate the strength performance of the interface.[24] Arrows indicate pulled-out fibers.

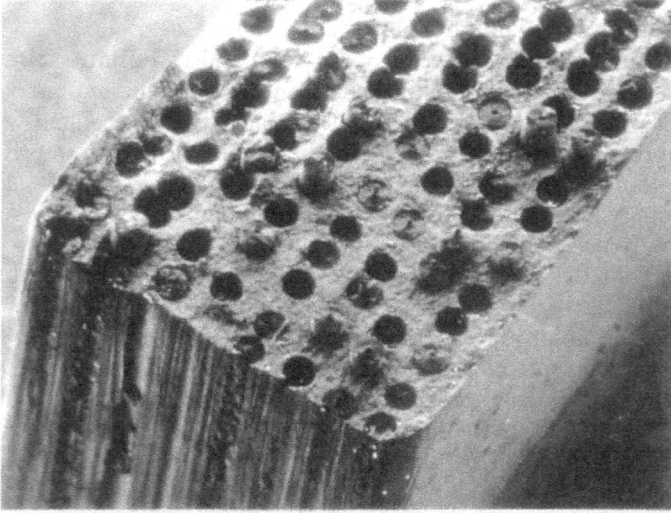

Figure 7.4 Typical fracture surface of a strong interface. Most of the fibers have failed close to the crack plane. Minimum fiber pull-out can be expected.

the transverse direction.[27,28] The multiple fiber fracture observed in this type of interface comes as a consequence of the loss of stiffness due to fiber failure. In detail, when the fibers ahead of the crack tip fail, the load supported by the broken fibers is transferred to the matrix, which in turn increases the stress field ahead of the crack tip. As a result, more fibers are expected to fail.[29] From the fracture surface morphology of this type of interface (Fig. 7.4), it is evident that most of the fibers have failed close to crack plane and therefore the surface exhibits a flat morphology.

However, in the case of titanium based MMCs, an intermediate interface type has been reported (Fig. 7.5). This type was found to be

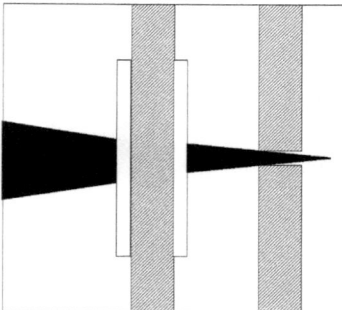

Figure 7.5 Crack propagation through an intermediate-strength interface.

sufficiently weak to allow some debonding, while the crack had failed to deflect to an extensive mixed-mode propagation as a result of large matrix plasticity in the mode I crack plane (the stress intensity in mode I was much higher than that at the interface, K_{II}, which has been developed from some interfacial defect).[21,30] This interface type exhibits small amount of fibre pull-out.[29]

7.2.3 Fatigue crack growth of plane specimens

Fatigue propagation modes in plane MMC specimens have been related to the fatigue properties of the constituent materials and interface.[31] Based on strain-failure criteria for the fibers, matrix, and interface, the fatigue behavior of MMCs has been categorized into four types: (1) *matrix dominated damage*—the cyclic strain required by the matrix to initiate damage is less than that of the fibers; (2) *fiber dominated damage*—fiber failure can take place at lower value of cyclic strain than that required by matrix; (3) *self-similar damage growth*—both fiber and matrix have similar cyclic strain to initiate damage; and (4) *interfacial failure dominated damage*—during off-axis loading, damage can initiate from interfacial failures. These damage initiation modes are shown schematically in Fig. 7.6.

In 1989 Johnson[32] suggested that in the case of SCS-6/Ti-15-3 laminates fatigue failure is fiber dominated. This argument emerged from experimental *S–N* data that revealed that during the first loading cycles the material experiences a stiffness loss due to fiber/matrix separations and fiber failures. However, as the loading history continues, the stiffness and therefore the fiber strain tended to a stabilized state (Fig. 7.7).

The stabilized fiber strain was then multiplied by the fiber elastic modulus (400 GPa for Textron SCS-6 fiber) and the resulting fiber stress was compared with that of the matrix material. From such interpretations (Fig. 7.8), it appeared that for the same cyclic strain the fibers would fail first, since the matrix posses a much lower elastic modulus (90–110 GPa).

Based on further observation, Johnson[34] has argued that, especially for the case of [0]$_8$ SCS-6/Ti-15-3, fatigue failure is characterized by self-similar damage, since both phases posses similar endurance limit strains. This was supported by examinations of fracture surfaces, which exhibited minimum fiber pull-out. Johnson concluded that fibers fail simultaneously with the matrix as a result of the high interfacial shear strength. Although experimental results had characterized the SCS-6/Ti-15-3 interface as weak,[32] Johnson believed

Matrix Dominated Damage

Fiber Dominated Damage

Self-Similar Damage Growth

Interfacial Dominated Damage

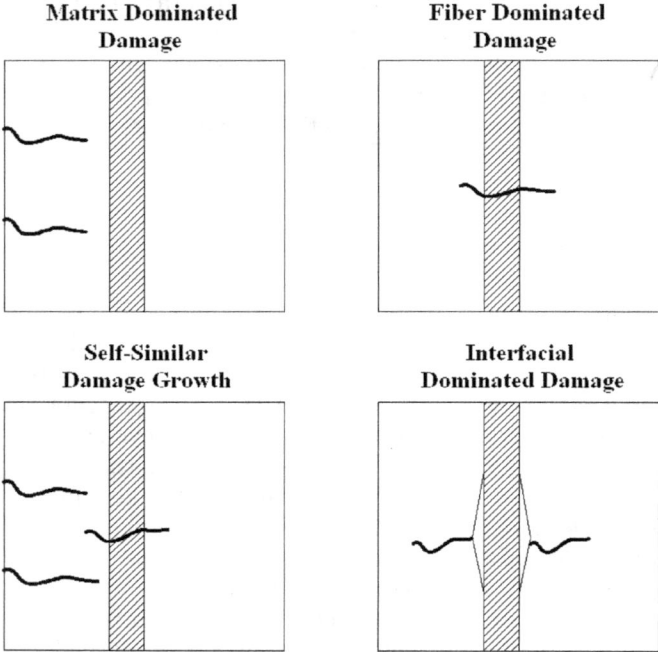

Figure 7.6 The four different types of fatigue damage initiation in MMCs.

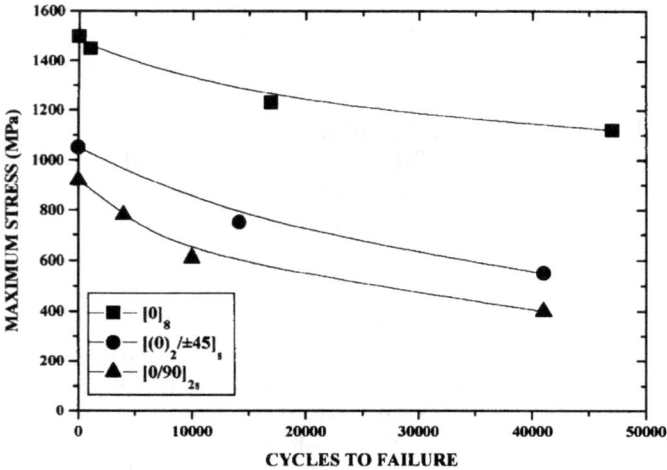

Figure 7.7 $S-N$ curves for three different configurations of reinforcement for a SCS-6/Ti-15-3 laminate in the as-fabricated state. (*Plot reproduced from Reference 33.*)

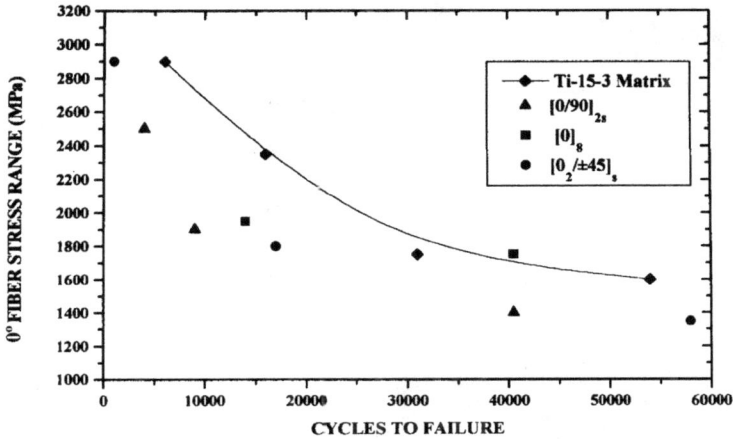

Figure 7.8 Comparisons between several composite configuration and the corresponding matrix material. The matrix stress has been calculated to reflect the equivalent stress at the 0° fibers. (*Plot reproduced from Reference 33.*)

that the resistance to interfacial displacement is increased by the clamping stress developed due to high radial residual stresses. The damage initiation classification mentioned above was basically derived from data taken from boron, SiC, and Al_2O_3 fiber reinforced aluminum alloy composites. These materials do not form a reaction zone of substantial size and therefore the initiation is limited to matrix cracking or fiber failure.[35]

A detailed study on plane SCS-6/Ti-6-4, SCS-6/Ti-15-3, and SCS-6/Ti-25-10 specimens[36] showed that both fatigue initiation and failure are controlled by the cyclic stress level. At high stress level (greater than 80% of the quoted tensile strength), the fracture surface of the SCS-6/Ti-15-3 specimen exhibited a flat pattern without significant fiber pull-out. Fatigue damage was reported to be constrained around the interface and randomly distributed throughout the specimen. The fracture surface was also found similar to that obtained by tensile loading. This suggested that failure was dictated by the accumulation of fiber breakage. Similar conclusions were drawn for SCS-6/Ti-64 and SCS-6/Ti-25-10. At medium stress levels (40–80% of the quoted tensile strength), SCS-6/Ti-15-3 specimens exhibited significant amount of pulled-out fibers. The pull-out lengths were reported to increase as the applied stress was decreased. The fracture surface was found to be composed of a number of fatigue-cracked regions that were not located close to the crack plane. Cracks were seen to initiate at broken fibers on machined edges and from breaks of the reaction layers. Cracking was also observed in the center of specimens. It is important to note

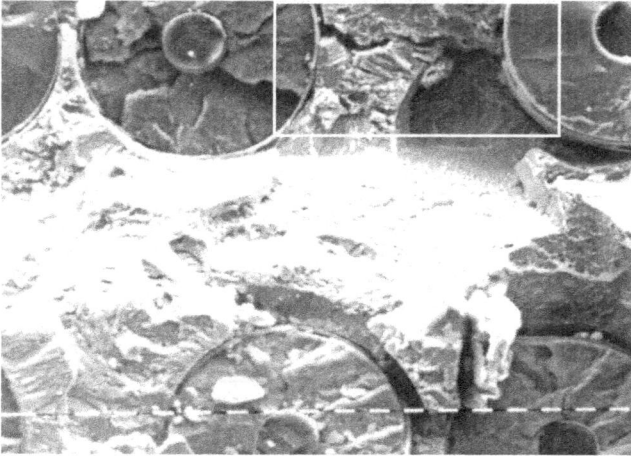

Figure 7.9 Matrix crack originated by broken interface in 32% SCS-6/Ti-15-3 tested at 800 MPa.[24]

that at these stress levels matrix cracks were bridged by intact fibers, evidence that interfacial debonding occurred before the matrix cracks reached the fiber. Analogous findings have been reported by Rodopoulos[24] (Fig. 7.9).

SCS-6/Ti-6-4 and SCS-6/Ti-25-10 also exhibit similar fatigue damage behavior. However, central cracking and secondary cracking in front of the main crack were not observed. Consequently, fatigue failure at this stress level is considered as interfacial damage dominated. At low stress levels (below 40% of the quoted tensile strength), where the applied strain was much less than the fracture strain of the interface, crack initiation in SCS-6/Ti-15-3 was found to be confined to the specimen edges only. In contrast to medium stresses, no cracks were observed to precedence the main crack. Multiple matrix cracking with almost identical cracking lengths was also observed. Similar damage characteristics have been reported for ceramic matrix composites under static loading.[36] For SCS-6/Ti-6-4 and SCS-6/Ti-25-10, behavior similar to that at medium stresses was observed.

Variations in the stress level response were attributed to differences in the reaction layer thickness. In detail, the thickness of the SCS-6/Ti-6-4 interface was quoted to be 1.7 μm with average strength of 768 MPa, while for the SCS-6/Ti-15-3 interface values of 2.43 μm and 551 MPa were measured for the interfacial thickness and the average strength, respectively. Therefore, the SCS-6/Ti-15-3 is characterized by lower resistance to interfacial layer breakage than the SCS-6/Ti-6-4, which mainly suppresses crack initiation from the interface. The

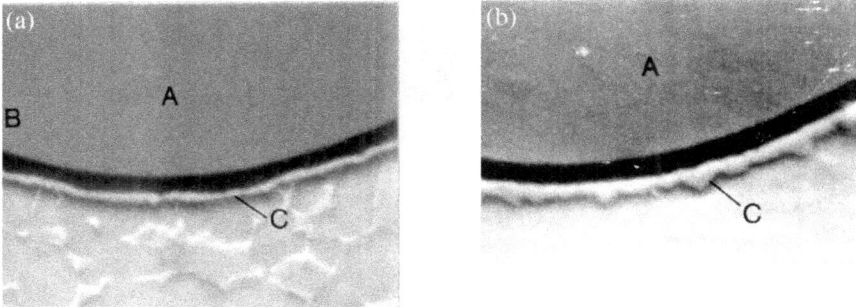

Figure 7.10 SEM micrographs of the reaction zone. (a) SCS-6/Ti-15-3; (b) SCS-6/Ti-6-4. *(Photos taken from Reference 10, Reprinted with permission from Elsevier Science.)*

difference of the reaction layer thickness between the SCS-6/Ti-15-3 and the SCS-6/Ti-6-4 can be seen in Fig. 7.10.

7.2.4 Fatigue crack growth of notched specimens

Fatigue behavior near stress concentrations is characterized by significant differences in the crack initiation and damage growth process. This is basically because: (a) the rate of stiffness degradation of the material increases; (b) depending on the corresponding stress intensity factor, fiber failure instead of debonding may take place, because the high strain zone is increased and therefore the area where fiber defects may exist; and (c) the crack arrest capacity of the material to the propagation of secondary cracks is reduced.[24]

Based on extensive research conducted by several workers in single-edge notched specimens (SCS-6/Ti-6-4),[37,38] mode I matrix cracks were found to propagate following three different modes (Fig. 7.11).

If the applied stress or the stress field in the vicinity of the crack is high, compared to the ultimate tensile strength of the notched material, the propagation of the crack follows a catastrophic mode I manner[39]—mode A. The crack propagation rate, da/dN, has been reported to follow a continuous steep acceleration, similar to that observed in monolithic material at high stress levels.[39] This occurs because some of the fibers fail ahead of the crack tip as a result of extensive localized damage,[30] while the rest fail immediately after entering the crack wake and therefore crack bridging is considered ineffective,[30,40] (Fig. 7.12).

With reduction of the nominal loading conditions, fibers do not fail immediately after entering the crack but produce a closure stress as a result of the relative sliding against the crack flanks. This causes the

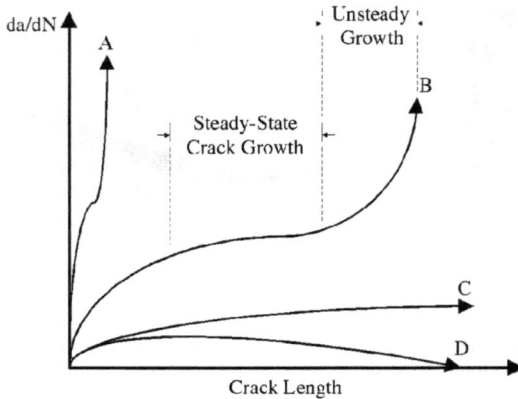

Figure 7.11 Different crack propagation modes with nominal loading conditions.

development of a closure stress that reduces the full effect of the far-field stress on the crack tip. It has been reported that fiber bridging promotes the development of a steady-state crack growth period at high stress levels[10]—mode B. Rodopoulos[24] suggested that the stress field ahead of the crack tip of the composite in the steady-state period is identical to that of the monolithic matrix at similar growth conditions. As the crack propagates further, the fibers closer to the crack mouth may fail (the causes are explained later). Subsequently, the load carried by the broken fibers is redistributed to other neighboring fibers, which might also fail. Extensive failure of bridging fibers would cause a rapid increase in da/dN[40] (Fig. 7.13). Walls et al.,[37] reported that the steady-state period of crack growth decreases for longer notch lengths. This is attributed to (a) the stress at the bridging fibers being higher (as a reaction to the sliding and the corresponding bridging

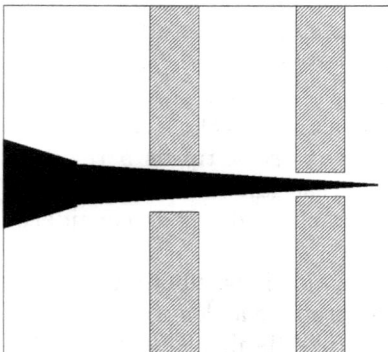

Figure 7.12 The schematic representation of mode A crack propagation.

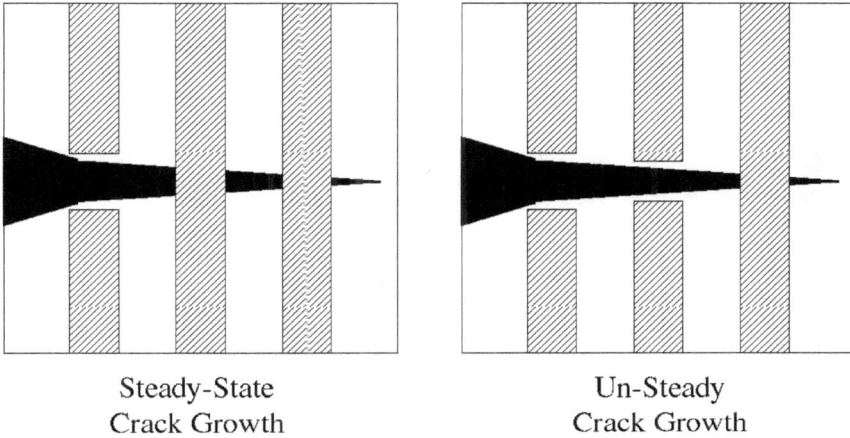

Steady-State   Un-Steady
Crack Growth   Crack Growth

Figure 7.13 During steady-state crack growth the number of intact fibers bridging the crack is larger than or equal to the number of broken fibers. As the crack advances, the number of broken fibers relative in those unbroken increases, signifying rapid crack propagation (unsteady crack growth).

stress, the fiber is subjected to tensile loading) and consequently the fibers failing at lower nominal stress levels[41] and (b) long notch lengths reducing the ability of the crack wake to accept more fibers before entering the unsteady crack growth period.[41] Ibbotson et al.[38] observed that pull-out stresses, produced by the sliding of broken fibers to the crack wake, are not significant (most of the bridging fibers have failed close to the crack plane).

At even lower stress, the crack growth rate experiences a continuous decrease, as the crack tip advances further from the notch.[7] This behavior is attributed to the large number of intact fibers bridging the crack wake (Fig. 7.14). The crack growth will either attain a long period of steady-state propagation (mode C) or it will become minimum, signifying conditions of crack arrest (mode D). Such behavior has been experimentally observed in the case of SCS-6/Ti-15-3.[42]

However, in contrast to SCS-6/Ti-6-4, where fatigue crack initiation is in the form of a single crack propagating perpendicular to the fibers, SCS-6/Ti-15-3 composites show a more complicated behavior. Apart from the main crack, several secondary cracks emanating from the notch were found to propagate along to the fiber–matrix interface. These cracks, after propagating a certain distance, deflected and continued to propagate perpendicular to the fibers[30] (Fig. 7.15).

For low nominal stress levels, the growth of these secondary cracks was observed to decline with crack extension and finally to halt (crack arrest) due to fiber bridging.[30] Similar secondary cracking was also

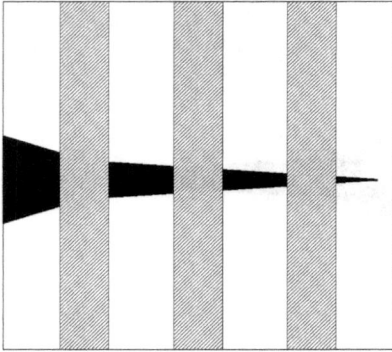

Figure 7.14 Mode C or D crack propagation. Maximum bridging stresses are expected due to the increasing number of intact fibers within the crack wake.

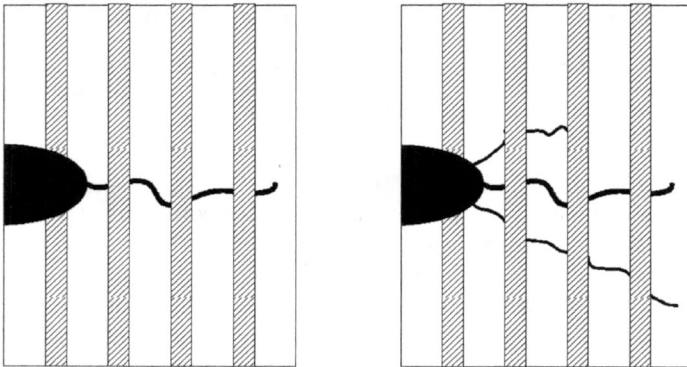

a)   b)

Figure 7.15 Fatigue crack initiation in (a) SCS-6/Ti-6-4 and (b) SCS-6/Ti-15-3.

observed in the case of circular holes.[43] In general, multiple cracking was observed to initiate at the points of maximum shear stress.[44]

7.3 Modeling Crack Bridging in MMCs

Conventional fracture mechanics characterization of mode I fatigue crack growth is generally achieved by empirical collaborating the stress intensity factor, ΔK, and the crack propagation rate, da/dN. The most known empirical collaboration is the Paris et al.[45] power law relationship,

$$\frac{da}{dN} = C\,\Delta K^n \tag{7.1}$$

where n and C are experimentally determined material-dependent constants and ΔK is the crack driving force in terms of linear elastic fracture mechanics (LEFM). A typical expression of ΔK is given by Eq. (7.2),

$$\Delta K = Y \Delta\sigma\sqrt{\pi a} \tag{7.2}$$

where $\Delta\sigma$ is the applied far-field stress range, a is the crack length, and Y is a function of the crack length to width ratio (a/W) and the geometrical configuration of the damage.

Although Eq. (7.1) is widely accepted as successfully simulating fatigue damage in monolithic materials with addressing appropriate values of the parameters C and n, experimental work conducted in unidirectional reinforced composites revealed that, as the crack propagates further, the crack growth rate could be much smaller than that predicted by Eq. (7.1) for the nonreinforced material[46–48] (Fig. 7.16). Hence Eq. (7.1) cannot successfully characterize these materials.

In 1971 Aveston et al.[6] suggested that even though crack growth takes place primarily within the matrix material, and consequently the parameters C and n of the neat matrix material (nonreinforced matrix) can be used, the applied stress intensity factor, Eq. (7.2), cannot solely represent the nominal damage conditions in the composite. In the same work,[6] Avenston and his co-workers suggested that when intact fibers are located within the crack wake (this phenomenon is known as bridging), the fibers can still carry load and thus

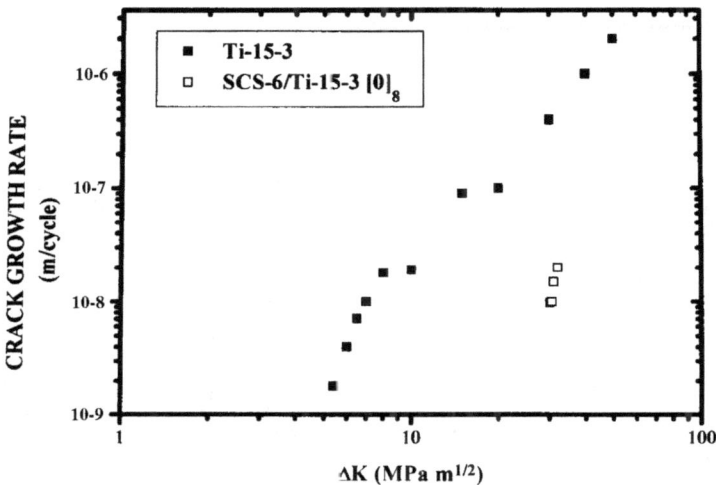

Figure 7.16   Crack growth rate as a function of ΔK. The crack growth rate of the neat matrix material is much higher than that of the reinforced composite. (*Data reproduced from Reference 26.*)

shield the crack tip from the full application of the far-field stress range. Consequently, the crack driving force, dictating crack growth, is reduced and should be defined within the limits of an effective driving force,[49] ΔK_{eff}, as in the case of similar closure mechanisms. This can be expressed by

$$\Delta K_{eff} = \Delta K_{app} - \Delta K_{bridg} \tag{7.3}$$

where ΔK_{appl} is the applied stress intensity factor (Eq. (7.2) and ΔK_{bridg} represents the effect of the closure stress due to bridging.

From Eq. (7.3) it is clear that substantial values of ΔK_{bridg} may significantly reduce the effective driving force to a value even lower than the threshold value (ΔK_{th}) required for crack propagation, signifying conditions for crack arrest.[50]

Several crack bridging models have been developed to described crack growth in MMCs,[6,51–53] especially during the early stages of research. In general, these models assume that when the fiber is positioned in the crack wake, a constant frictional shear stress acts along a distance where sliding of the matrix crack wake against the fiber is taking place. This distance, also known as slip length or sliding distance, l_s, has been shown experimentally[26] to increase as the crack tip propagates further from the bridging fiber (Fig. 7.17). In fact, when a debonded fiber enters the crack, the debonding process (bimaterial interface crack) reinitiates ($K_{II} \neq 0$). The gradual increases in slip length correspond to similar increases in the crack opening displace-

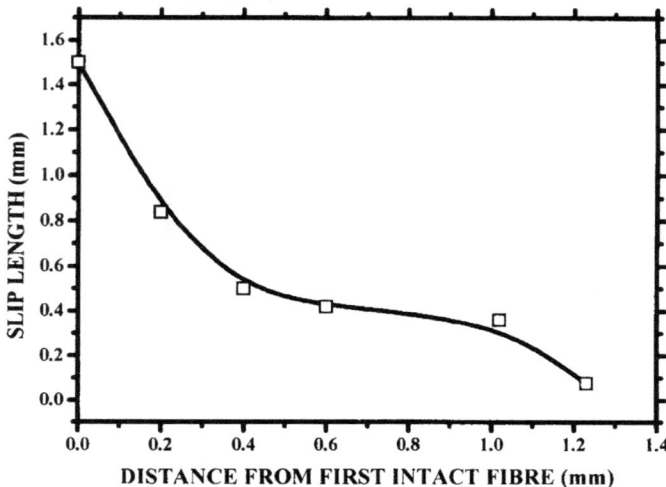

Figure 7.17 Slip length pattern for a 33% SCS-6/Ti-15-3 MMC. (*Data reproduced from Reference 26.*)

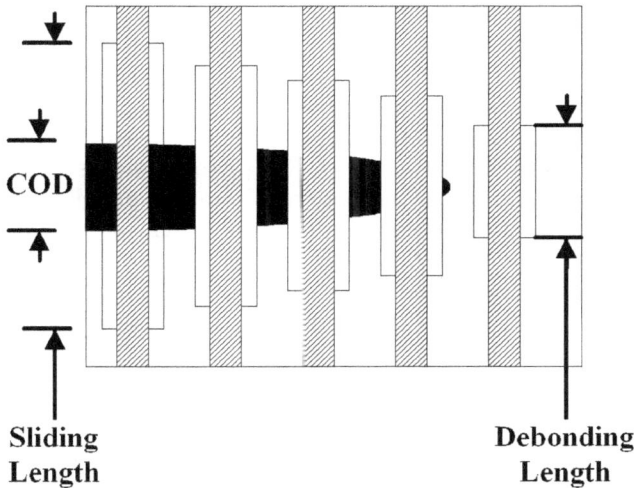

COD

Sliding
Length

Debonding
Length

Figure 7.18   Schematic representation of the debond and slip lengths.

ment (COD) at the position of the fiber[24] (Fig. 7.18). Therefore, the closure stress should be a function of COD, the interfacial shear stress acting along the sliding distance, and the mechanical properties of the constituents.

7.3.1   The steady-state fiber bridging models

These models are based on the assumption that the stress intensity factor ΔK_m experienced by the matrix material should be more appropriate to represent ΔK in Eq. (7.1). The first work to consider the steady-state crack growth of matrix cracks (see Fig. 7.13) in a ceramic matrix composite during monotonic loading, was published by Avenston et al. in 1971.[6] The model, also known as the ACK model from the names of the authors, was based on the determination of the ΔK_m through a strain energy balance before and after cracking. Using large-slip and no-slip shear-lag assumptions (both concepts can be found in every composite materials handbook), and uniform crack bridging, they determined that for conditions of steady-state cracking the stress intensity factor is independent of the crack length and is controlled only by the distance between the crack tip and the onset of steady-state cracking. In terms of the model, the matrix is considered stress free since the fibers support full load. Additionally, in cases where the crack is partially bridged, the contribution of the unbridged crack portion to the stress intensity factor is negligible.

Using a similar approach to ACK analysis, Budiansky et al.[52] suggested a new energy balance approach to describe crack growth. In terms of the model, the fibers are initially bonded to the matrix until the passage of the crack. These two models are termed steady-state fiber bridging (SSFB) models.[26] Due to the inherent deficiencies of both models in describing an advancing bridged crack, no further analysis is given.

7.3.2 Generalized fiber bridging or shear-lag models

Another class of models was introduced in 1985 and 1987 by Marshall et al.[51] and McCartney,[53] respectively. These models combine continuum fracture mechanics principles and micromechanics analysis to determine stress intensity factor solutions for a matrix crack of arbitrary size, subjected to monotonic loading. According to these models (also known as generalized fiber bridging models, GFB[26]), the friction stresses developed by the intact fibers within the crack are idealized by an unknown uniform closure pressure. The evaluation of the closure pressure in the GFB models is obtained by combining crack opening displacement solutions from continuum fracture mechanics and from micromechanics analysis. Although the models differ from each other in the methodology for relating these two issues, identical steady-state solutions (as derived from the SSFB models[6,52]) are used as boundaries to characterize K_m. The formulation of the GFB models as applied to fatigue loading was developed by McMeeking and Evans.[54]

According to the GFB models, the presence of the fiber within the crack wake causes a reduction in both the COD and the crack tip stresses. Based on the Marshal et al. analysis[51] (also known as MCE), the composite stress intensity factor is defined by the superposition of the normal stress intensity factor due to the remote stress on an unbridged crack and that due to the friction stresses due to fiber bridging. Using micromechanical analysis, the friction stresses are idealized as continuous, but with varying distributed crack flank pressure. The friction stress in relation to the fiber stress is given by

$$p(x) = \sigma_f(x)V_f \qquad (7.4)$$

where $p(x)$ is the crack flank pressure (Fig. 7.19), $\sigma_f(x)$ is the fiber stress at a given distance x from the crack mouth, and V_f is the fiber volume fraction. Equation (7.4) is considered valid only in cases where at least one fiber is positioned within the crack.

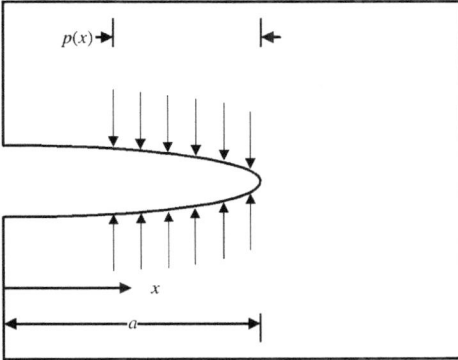

Figure 7.19 Modeling of crack bridging by idealizing the load carried by the fibers as crack flank pressure.

The friction stress is then related to the stress intensity factor, K_{tip}, through the modified Sneddon–Lowengrub equation,[55,56] which describes in a convenient form cracks in infinite bodies that are loaded by arbitrary crack flank pressure distributions. In the case of a straight crack embedded in an infinite medium, K_{tip} is written as[26]

$$K_{tip} = \sqrt{\frac{4a}{\pi}} \int_0^1 \left(\frac{\sigma_\infty - p(X)}{\sqrt{1 - X^2}} \right) dX \tag{7.5}$$

where a represents the crack length, $p(X)$ is the friction stress at X, X is the normalized distance along the crack length, defined as $X = x/a$, and σ_∞ is the far-field stress.

The value of K_{tip} in Eq. (7.5), is determined by the application of the shear-lag analysis on the basis that friction stresses are related to the crack opening displacement, COD, $p(x) \propto \sqrt{u(x)}$. It is assumed that there is a strain compatibility between the fiber and the matrix in the slip regime, while outside the slip regime the effect of the shear stress is negligible. According to MCE the closure pressure is given by

$$p(x) = 2 \left(\frac{u(x)\tau V_f^2 E_f E_c}{R(1 - V_f)E_m} \right)^{1/2} \tag{7.6}$$

where $u(x)$ is the COD at x, τ is the shear resistance of the interface, R is the fiber radius and E_f, E_c, E_m represent the elastic moduli of the fiber, composite, and matrix, respectively.

As an improved solution to the MCE closure pressure, the shear-lag model was further modified by McCartney[53] in order to make the model energetically consistent. The MCE closure pressure was calculated as

$$p(x) = 2\left(\frac{u(x)\tau V_f^2 E_f E_c^2}{R(1 - V_f)^2 E_m^2}\right)^{1/2} \qquad (7.7)$$

Close examination of Eqs. (7.6) and (7.7) reveals that, for the same τ, the crack opening displacement predicted by McCartney can be written as

$$u(x)_{\mathrm{McCartney}} = \sqrt{\frac{V_m E_m}{E_c}} u(x)_{\mathrm{MCE}} \qquad (7.8)$$

Based on Eq. (7.8), Cox and Lo[57] suggested that closure pressure can be written as

$$p(x) = 2n\sqrt{\frac{u(x)\tau V_f^2 (1 - V_f) E_f E_m}{Re_c}} \qquad (7.9)$$

where $n = 1$ for monotonic loading and $n = 1.41$ for fatigue loading. The distinction between the two loading conditions is attributed to the irregular nature of closure during reverse loading and the relaxation of axial residual stresses. The crack opening in the presence of a closure pressure, according Eqs. (7.6), (7.7), and (7.9), can be calculated by employing the Sneddon–Lowengrub equation, weight functions, Green's functions or finite element analysis (FEA).[58–60] According to Sneddon–Lowengrub, the crack opening in terms of the normalized position coordinate, X, is given as

$$u(X) = \frac{4(1 - v_c^2)a}{\pi E_c}\int_x^1 \frac{s}{\sqrt{s^2 - X^2}}\left\{\int_0^s \frac{(\sigma_\infty - p(t))}{\sqrt{s_2 - t_2}} \, dt\right\} ds \qquad (7.10)$$

where s, t are normalized position coordinates and v_c is the Poisson's ratio of the composite (details can be found elsewhere[61]).

If weight functions are to be used, the Bueckner[62] formulation is considered as the most appropriate. The crack opening profile in the

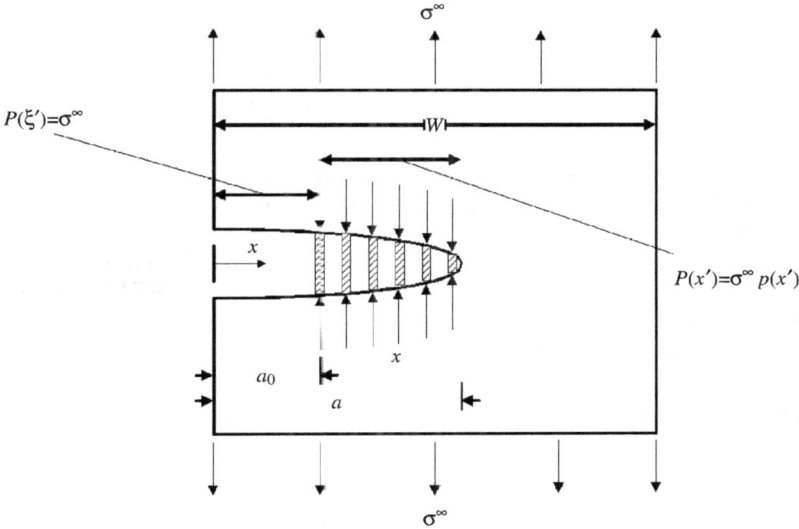

Figure 7.20 A single edge notch with a partially bridged crack.

presence of bridging for a single edge notch in a finite geometry specimen (Fig. 7.20) is given as

$$u(x) = \frac{2(1 - v_c^2)}{E_c \pi} \left\{ \int_x^a \frac{H(a, x)}{\sqrt{a - x}} \left[\int_0^a \frac{P(x')H(a, x')}{\sqrt{a - x'}} \, dx' \right] da \right\} \qquad (7.11)$$

$$P(x') = \begin{cases} \sigma^\infty & \text{for } 0 < x' < a_0 \\ \sigma^\infty - p(x') & \text{for } a_0 < x' < a \end{cases}$$

where
$$H(a, x') = 1 + m_1 \frac{a - x'}{a} + m_2 \frac{(a - x')^2}{a^2} \qquad (7.12)$$

and m_1 and m_2 are functions of the ratio of crack length over the width of the specimen and are written as

$$m_1 = 06147 + 17.1844 \frac{a^2}{w^2} + 8.7822 \frac{a^6}{w^6}$$

$$\qquad (7.13)$$

$$m_2 = 0.2502 + 3.2889 \frac{a^2}{w^2} + 70.0444 \frac{a^6}{w^6}$$

7.3.3 Fiber pressure model

In Reference 21 it is argued that residual stresses cause a permanent crack opening displacement, which results in a hysterisis of the bridging stresses, Fig. 7.21. Moreover, recent numerical comparisons between the MCE and McCartney models have revealed that the crack opening displacement profile obtained from Eq. (7.7) is identical to that determined by Eq. (7.6) if the shear resistance is reduced by a factor of 3.2.[63] Additionally, the lack of a standard method for obtaining the correct shear resistance induced Ghosn et al.[50] to suggest an alternative solution (known as the fiber pressure model, FPM) for the determination of the closure pressure. The closure pressure according to FPM is assumed to be equal to the stress carried by the fibers in the bridged region averaged over the total bridged area $(a - a_0)$. The problem of the shear stress parameter is then overcome by suggesting

$$p(x) = \sigma_\infty \left(\frac{w}{w - a_0} + \frac{6wa_0[0.5(w - a_0) - (x - a_0)]}{(w - a_0)^3} \right) \qquad (7.14)$$

where w is the specimen width, a_0 and a are the initial notch length and total crack length, and x is the distance to the bridged area measured from the free surface. Equation (7.14) is valid only for partially bridged cracks. It should be noted that Eq. (7.14) in contrast to Eqs. (7.6), (7.7), and (7.9), is not a function of the interfacial shear stress or crack opening displacement. In its simple form, Eq. (7.14) reflects a stress equilibrium condition between the bending and applied stresses acting along the bridged region.[64]

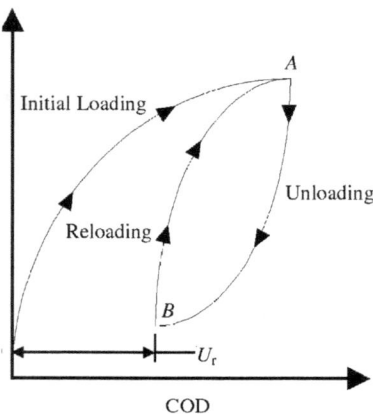

Figure 7.21 The effect of residual stress on crack opening displacement. U_r is the residual COD.

7.3.4 Evaluating the effective ΔK

Knowledge of COD or ΔK at the crack tip is important in determining the crack propagation rate using a Paris type relationship. However, exact analytical solutions for $u(x)$ can only be obtained if a numerical iterative method between the closure pressure and the crack opening displacement at each point along the crack length is utilized. Consequently, the process will require a tremendous amount of computing power. To overcome this, it has been suggested that it could be appropriate to assume that the near-tip opening displacement of the bridged crack is similar to that of the unbridged crack, especially when the crack length is less than that required to attain steady-state growth conditions (steady-state conditions as defined by Avenston et al. and McCartney). Clearly, such an assumption implies that the closure pressure at the opening of a short crack is so small as to be considered negligible, as in the case of a long unbridged crack. Therefore

$$\text{COD}_m = \text{COD}_{\text{tip}} \tag{7.15}$$

McCartney[53] reported that COD_{tip} actually represents the change in displacement within the slip length and thus should be related to COD_m as follows

$$\text{COD}_m = \text{COD}_{\text{tip}}^m \tag{7.16}$$

where the exponent is

$$m = \left(1 + \frac{E_f V_f}{E_m(1 - V_f)}\right)$$

Applied stress intensity factors are related to the effective stress intensity factors at the crack tip, K_{tip}, by[65]

$$K_{\text{eff}} = K_{\text{tip}} = K_{\text{appl}} - K_{\text{bridg}} \tag{7.17}$$

where K_{appl} represents the applied stress intensity factor and K_{bridg} is the stress intensity factor due to bridging. K_{bridg} can be written as

$$K_{\text{bridg}} = \sqrt{\frac{16A}{9\pi}} K_{\text{tip}} \alpha^{3/4} \left(2 - \frac{\alpha}{a}\right)^{3/4} \tag{7.18}$$

where α is the length of the bridging zone ($\alpha = a - a_0$, Fig. 7.20), and a is the full crack length including any unbridged length.

In the case of a fully bridged crack, Eq. (7.18) reduces to

$$K_{\text{bridg}} = \sqrt{\frac{16A}{9\pi}} K_{\text{tip}} a^{3/4} \qquad (7.19)$$

where

$$A = \frac{8(1 - v_k^2)\tau V_f^2 E_f(1 + U)}{[E_m(1 - V_f) + E_f V_f]R\sqrt{\pi}}, \qquad U = \frac{E_f V_f}{E_m(1 - V_f)}$$

and v_m is the matrix Poisson's ratio.

Marshall et al.,[51] using the ratio of the matrix elastic modulus to the composite elastic modulus, related the continuum effective stress intensity factor of the composite to that of the matrix

$$K_m = \frac{E_m}{E_c} K_{\text{tip}} \qquad (7.20)$$

Equation (7.20) suggests near-tip strain compatibility between the composite and the matrix. Such compatibility is valid only in cases of monotonic loading and steady-state growth (steady-state growth in this case is considered when there is no fiber failure ahead of the crack tip).

In a different approach, McCartney[53] used a fracture energy balance to relate K_{tip} and K_m. The fracture criterion proposed was that the fracture surface energy released by the matrix is equal to that released by the composite

$$\gamma_c = (1 - V_f)\gamma_m \qquad (7.21)$$

where γ_c and γ_m are the fracture surface energy of the composite and the matrix, respectively. Using the plastic strain work for crack extension as proposed by Irwin[66] ($\gamma = K_{\text{tip}}^2/E$), Eq. (7.21) can be rewritten in terms of the stress intensity factor as

$$K_m = \sqrt{\frac{E_m}{(1 - V_f)E_c}} K_{\text{tip}} \qquad (7.22)$$

Equation (7.22) has been utilized successfully in the case of fatigue crack growth in SCS-6/Ti-6-4 MMC.[67] Budiansky[68] proposed a similar approach based on the J-integral.

Another energy balance criterion was proposed by Gao et al.[69] In their work the strain energy of the crack tip released by crack extension in the composite, G_c, is assumed to be equal to that released by a similarly sized crack advancing in the monolithic matrix alone, G_m. For plane strain conditions they suggested

$$G_c = \frac{1 - v_m^2}{E_c} K_{\text{tip}}^2 = G_m \qquad (7.23)$$

If the Poisson's ratio effect of the matrix is neglected, Eq. (7.23) yields[60]

$$K_m = K_{\text{tip}} \sqrt{\frac{E_m}{E_c}} \qquad (7.24)$$

Results identical results to Eq. (7.24) were derived by assuming that the crack growth rates of the composite and the matrix are proportional to the crack opening displacement.[70]

A completely different approach has been proposed by McMeeking et al.[71] By defining the path of the J-integral at fracture where fibers and matrix fracture simultaneously by the passage of a crack (no bridging), they proposed an alternative solution to the stress intensity factor of the composite

$$K_{\text{appl}}^2 = \frac{(1 - V_f)E_c}{E_m} K_{\text{tip}}^2 + \frac{dV_f(1 - V_f)E_m S^3}{6(1 - v_m^2)E_f \tau} \qquad (7.25)$$

where d is the fiber diameter and S is the fiber strength. Equation (7.25) can be used to predict residual strength in terms of crack length.

7.3.5 Micromechanical modeling of fatigue crack growth

In 1995 de los Rios et al.[72] argued that the assumption defined by Eq. (7.15) can provide accurate prediction of the crack propagation rate if all the damage parameters affecting COD_{tip} are integrated into the modeling. In the same work they suggested that fiber bridging, interface debonding, fiber failure ahead of and behind the crack tip, and also the spread of crack tip plasticity should be considered. A direct connection between debonding and crack tip plasticity spreading has been detected experimentally by Schulte et al.[73] They reported that the propagation of the crack tip plasticity is not exclusively controlled by the resistance to plastic deformation (yield stress) of the matrix materials but also by the high stiffness fiber as the plasticity approaches an intact interface. As a result, a continuous tensile stress builds up at the fiber as the crack approaches the interface. When this tensile stress, also referred to as the constraint effect, achieves debonding of the interface by producing an interfacial shear stress equal to the interfacial shear strength, the restrained plasticity expands rapidly (jumps) to the next intact interface by propagating round the debonded fiber. This rapid expansion of the crack tip plasticity has been attributed to the fact that, after debonding, the resistance to plastic deformation drops to a value controlled by the matrix yield stress. These aspects are shown schematically in Fig. 7.22.

In 1995 de los Rios et al.[72] proposed an elasto-plastic model, known as the three-zone micromechanical model (TZMM). The TZMM takes into account, along with other damage events, the spread of crack tip plasticity. The TZMM, originally developed for monolithic materials by Navarro and de los Rios,[74] represents the crack and its plastic zone by means of dislocations subjected to an applied stress, σ, as first developed by Bilby et al.[75] in 1963. In 1992,[76] the model was extended further and a third zone was introduced to represent cases where the plastic zone is blocked by grain boundaries. Such an approach was argued to be more realistic in physical terms, since in the two-zone system an infinite stress level is sustained by the grain boundary. In terms of the three-zone system, the plastic zone size (slip band ahead of the crack tip) is blocked by the grain boundary and remains blocked until the stress in the third zone, i.e., the grain boundary, attains the required critical level for dislocations to cross this zone.

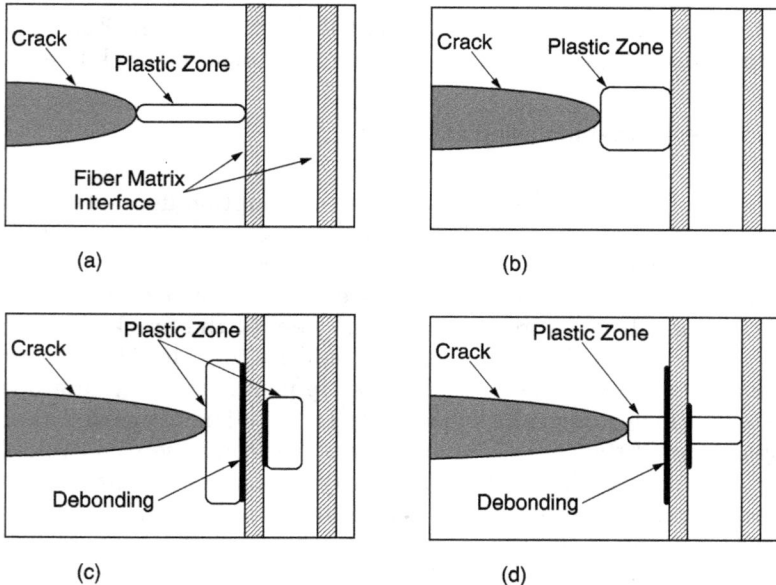

Figure 7.22 The four stages of crack propagation. When the crack plastic zone touches the intact interface (a), the "free" movement of crack tip plasticity is blocked by the high stiffness fiber. As a result, the plastic zone condenses and a tensile stress gradually builds up around the interface (b). Debonding (c) will result in a rapid release of the plastic zone to the next interface (d).

The TZMM was adopted to characterize crack propagation in the case of unidirectional MMCs[72,77] by considering the third zone to represent the constraint effect provide by an intact interface. Such a representation was shown to be sufficiently accurate in cases where only one or two matrix grains are situated between two successive fibers and therefore the effect of the grain boundary is negligible.[24] In terms of the model, the three zones of the crack system in a MMC are the crack zone, the plastic zone, and the plasticity constraint zone at the interface (Fig. 7.23).

In the case where the crack is subjected to an applied stress σ in mode I, the stresses in each zone are as follows: (a) σ_1 in the crack zone (friction stress due to fiber bridging); (b) σ_2 is the flow response in the plastic zone; and (c) σ_3 in the fiber zone ahead of the plastic zone (constraint effect provided by the intact interface to matrix crack tip plasticity). A typical stress distribution along the crack system is presented in Fig. 7.24. The stress σ_1 is considered zero for a non-bridged crack and nonzero for a bridged crack. The plastic zone is assumed to be blocked by the fibers, which impedes any matrix plastic displacement at the fibers. The effect of this constraint is the development of stress σ_3 in the matrix between the fibers of a row, which on

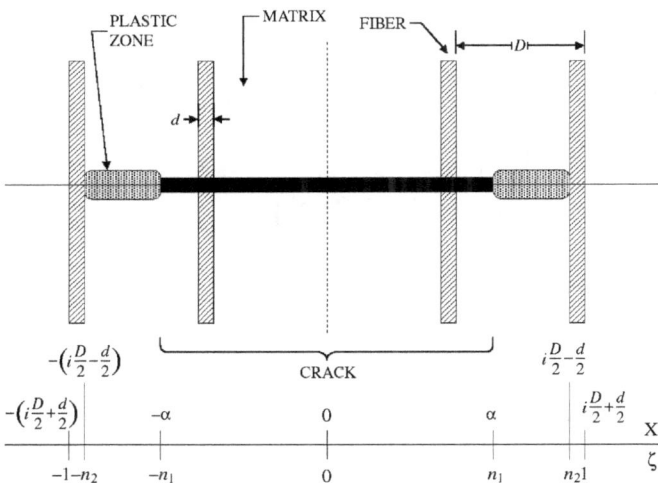

Figure 7.23 Schematic representation of the crack system. The crack length is denoted by $2a$, the fiber diameter by d, and the fiber spacing by D. Considering only the positive coordinates side, the crack tip is positioned at a, the plastic zone extends to the next fiber ahead of the crack tip at $iL/2 - d/2$, and the fiber plastic-constrained zone at $iD/2 + d/2 = e$, $i = 1, 3, 5 \ldots$. The dimensionless coordinate ζ describes position throughout, in particular, $\zeta = n_1$ at the crack tip, $\zeta = n_2$ at the plastic zone, and $\zeta = 1$ at the end of the fiber.

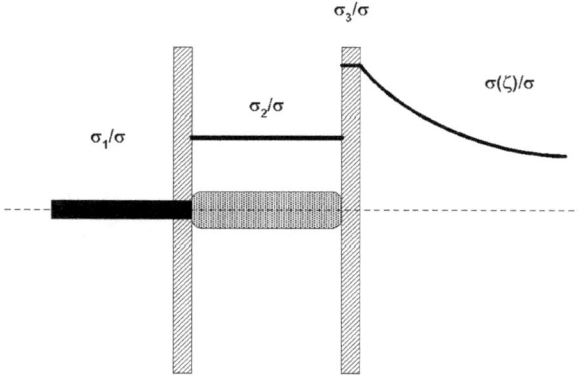

Figure 7.24 The stress distribution along the three-zone crack system. The stress at the crack due to fiber bridging is σ_1, the stress at the plastic zone is σ_2, and the stress at the fiber plastic-constrained zone is σ_3; $\sigma(\zeta)$ is the elastic stress distribution.

achieving a critical value, resolved along the fiber–matrix interface, will cause debonding.

Assuming an equilibrium of all the forces, internal and external, acting in the three-zone system, expressions for the COD $\equiv \phi$ over the entire crack system, and for the stress σ_3, can be written as

$$
\begin{aligned}
\phi = \text{COD} = \frac{bc}{\pi^2 A}\Bigg\{ &(\sigma_2 - \sigma_1)\bigg[(\zeta_b - n_1)\, \cosh^{-1}\bigg(\bigg|\frac{1-n_1\zeta_b}{n_1-\zeta_b}\bigg|\bigg)\\
&- (\zeta_b + n_1)\, \cosh^{-1}\bigg(\bigg|\frac{1+n_1\zeta_b}{n_1+\zeta_b}\bigg|\bigg)\bigg]\\
&- (\sigma_2 - \sigma_1)\bigg[(\zeta_a - n_1)\, \cosh^{-1}\bigg(\bigg|\frac{1-n_1\zeta_a}{n_1-\zeta_a}\bigg|\bigg)\\
&-(\zeta_b + n_1)\, \cosh^{-1}\bigg(\bigg|\frac{1+n_1\zeta_a}{n_1+\zeta_a}\bigg|\bigg)\bigg] + (\sigma_3 - \sigma_2)\\
&\times \bigg[(\zeta_b - n_2)\, \cosh^{-1}\bigg(\bigg|\frac{1-n_2\zeta_b}{n_2-\zeta_b}\bigg|\bigg)\\
&-(\zeta_b + n_2)\, \cosh^{-1}\bigg(\bigg|\frac{1+n_2\zeta_b}{n_2+\zeta_b}\bigg|\bigg)\bigg] - (\sigma_3 - \sigma_2)\\
&\times \bigg[(\zeta_a - n_2)\, \cosh^{-1}\bigg(\bigg|\frac{1-n_2\zeta_a}{n_2-\zeta_a}\bigg|\bigg)\\
&- (\zeta_a + n_2)\, \cosh^{-1}\bigg(\bigg|\frac{1+n_2\zeta_a}{n_2+\zeta_a}\bigg|\bigg)\bigg]\Bigg\}
\end{aligned}
\tag{7.26}
$$

$$
\sigma_3 = \frac{1}{\cos^{-1} n_2}\left((\sigma_2 - \sigma_1)\, \sin^{-1} n_1 - \sigma_2\, \sin^{-1} n_2 + \frac{\pi}{2}\sigma\right)
\tag{7.27}
$$

where b is the Burgers vector, $A = Gb/2\pi$ for screw dislocations or $A = Gb/2\pi(1 - v_m)$ for edge dislocations, G and v_m are the shear modulus and the Poisson's ratio of the matrix respectively, and σ is the applied stress.

If crack growth is considered to be a function of the crack tip opening displacement (CTOD), ϕ_i, through a Paris type relationship, then Eq. (7.26) determines da/dN when $\zeta_a = n_1$, $\zeta_b = 1$, where n_1 represents a dimensionless measurement of crack length, i.e., $n_1 = a/(iD/2 + d/2)$. In addition, Eq. (7.27) establishes the condition for crack propagation across the fiber row when the axial stress σ_3 acting at plasticity constraint zone is equal to the stress required for debonding, σ_{3d}. Clearly, $\sigma_3 = \sigma_{3d}$ acknowledges the condition when the clamping stress provided by the fibers to plastic displacements within the plastic zone is removed, since no or minimal interfacial shear stress is acting along the debond length.

Assuming that debonding is not a continuous process (propagation of a bimaterial interface crack) then the stress at the plasticity constraint zone required to debond a particular fiber length can be written as

$$\sigma_{3d} = \frac{\sigma_d E_c}{E_f} \tag{7.28}$$

where σ_d is the tensile stress at the fiber required to debond a particular embedded fiber length, and E_c and E_f are the elastic moduli of the composite and the fiber, respectively. Equation (7.28) is obtained by considering a simple force balance in a fiber push-out test and strain compatibility between the fiber and the composite in the plasticity constraint zone. If the interfacial shear strength is taken as constant along the fiber–matrix interface and the shear-lag analysis is utilized, the stress applied at the fiber to cause debonding is obtained as

$$\sigma_d = \frac{4\tau l}{d} \tag{7.29}$$

where l is the embedded fiber length, d is the fiber diameter and τ is the interfacial shear strength.

Published results in the literature indicate values for interfacial shear strength of SCS-6/Ti-15-3 and SCS-6/Ti-6-4 in the ranges 124–148 MPa and 138–156 MPa, respectively.[78,79] Furthermore, experiments have shown that the interfacial shear strength increases for longer embedded fiber length or thicker specimens and asymptotically approaches a constant value for fiber lengths approximately 4 to 5 times the fibre diameter ($d = 140$ μm for the Textron SCS-6[80,81]); see Fig. 7.25.

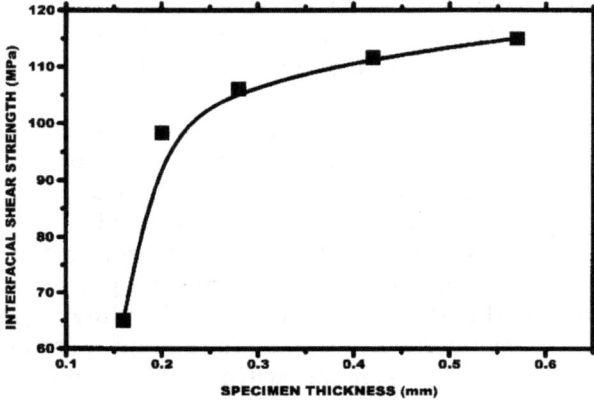

Figure 7.25 Interfacial shear strength vs. specimen thickness for an SCS-6/Ti-15-3 interface. (*Data reproduced from Reference 81.*)

The condition for crack propagation is achieved as follows: With the crack tip positioned between two fibers, the level of the stress σ_3 is given by Eq. (7.27). On further crack growth, the stress level of σ_3 increases due to an increase in n_1^i until σ_3 attains a value required for debonding, given by Eq. (7.28). The crack tip position at the critical point where debonding is achieved, n_1^{ic}, is obtained by substituting Eq. (7.28) into Eq. (7.27) and solving for n_1^i. At this point the crack tip plasticity constraining effect is overcome and the plasticity is allowed to pass round the debonded interface and to be constrained once again by the next intact interface.

The application of Eq. (7.27) requires the determination of the composite flow stress. In the case of unidirectional MMCs, the flow response ahead of the crack tip is controlled by the matrix yield stress and by the proportion of the higher-stiffness phase present in the plastic zone. Assuming an elastic-perfectly plastic matrix material and an isostrain condition between the fiber and the matrix within the plastic zone, the flow response of the composite can be written as,[40]

$$\sigma_2 = \frac{\sigma_{ym}}{E_m} E_f V_f + \sigma_{ym}(1 - V_f) \tag{7.30}$$

where σ_{ym} is the matrix yield stress. For a rigorous determination of σ_2, Eq. (7.30) is subjected to further modifications to include matrix residual stresses and fiber strength distribution. Analytically, the matrix material is subjected to residual thermal stresses developed from the mismatch of the thermal expansion coefficients (CTE) between the matrix and the fiber. In most MMCs, the matrix CTE,

(a)

(b)

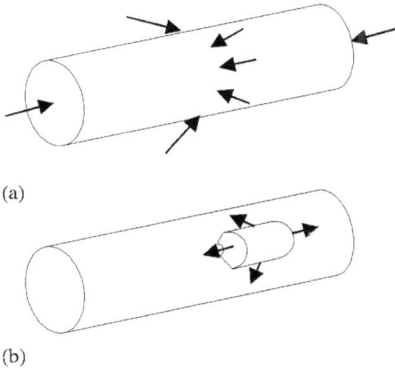

Figure 7.26 Residual thermal stresses after cooling: (a) stresses on the fiber and (b) stresses on the matrix.

α_m, exceeds that of the fiber, α_f. Thus, after cooling from the processing temperature, the matrix is subjected to a residual axial tensile stress σ_m^r, (Fig. 7.26). The residual thermal stresses can be evaluated as[40]

$$\sigma_m^r = \frac{V_f E_m E_f \, \Delta\alpha \, \Delta T}{E_c}, \qquad \Delta\alpha = \alpha_m - \alpha_f \tag{7.31}$$

where ΔT is the effective temperature change range.

A typical value of ΔT for the case of Ti-based MMCs is $50°C$.[82] Consequently, in Eq. (7.30) the matrix yield stress is considered to be

$$\sigma_{ym}^t = \sigma_{ym} - \sigma_m^r \tag{7.32}$$

Similarly to shear-lag models, the TZMM determines the closure stress due to fiber bridging by implementing a strain compatibility between the fiber and the matrix within the region where sliding of the matrix crack faces over the fiber is taking place (this area is known as slip length or sliding distance) (Fig.7.27). The closure stress produced by each bridged fiber row in the crack zone is calculated as follows[77]:

$$c_1 = \frac{\text{COD } G}{(l_s + \text{COD})NN} \tag{7.33}$$

where NN is the number of fibers per row, l_s is the sliding distance, and G represents the matrix shear modulus.

Equation (7.33) is obtained considering displacement (strain) compatibility between the fiber and the matrix at the interface. A schematic representations of the closure stress is shown in Fig. 7.28. Additionally, Eq. (7.33) acknowledges that all the fibers in the same

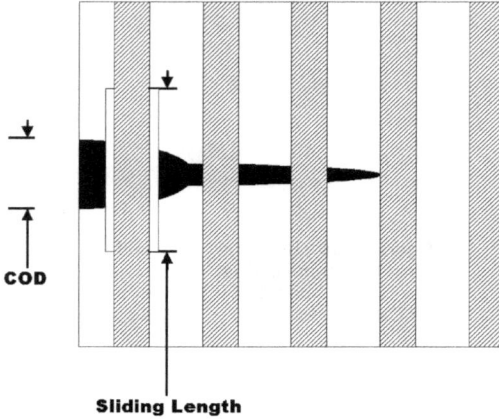

Figure 7.27 Strain compatibility between the fiber and matrix within the sliding distance could provide the necessary environment for determination of closure stress.

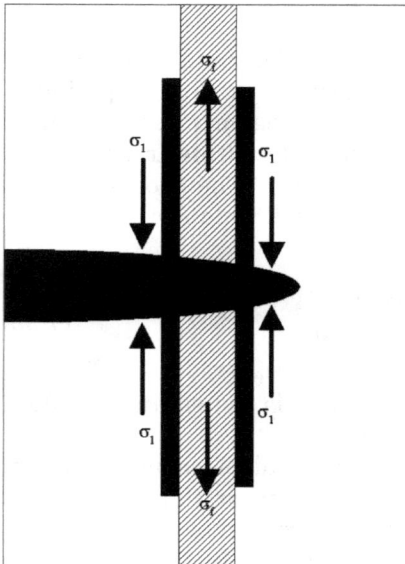

Figure 7.28 During bridging, closure stress σ_1 is generated by the relative sliding of the crack faces along the fiber within the sliding distance. A reaction to σ_1 is the tensile stress σ_f acting at the fiber.

row are subjected to an equal strain. An expression for the sliding distance, l_s, is given in[83]:

$$l_s = 2\sqrt{\frac{\text{COD } V_m E_m E_f d}{4\tau E_c}} \tag{7.34}$$

where E_c is the Young's modulus of the composite (calculated by the rule of mixtures), V_m is the volume fraction of the matrix, and τ is the interfacial shear stress.

However, Marshall et al.[51] indicated that, when the interface is damaged by the passage of the crack, the interfacial shear stress is reduced, (a typical reason for degradation is the failure of fiber coating), and values close to the sliding resistance are more realistic. By utilizing Eqs. (7.26), (7.33), and (7.34), the closure stress can then be determined. Since the COD depends on the value of σ_1, a numerical iterative method is required for the calculation of σ_1. Once the closure stresses have been determined, the stress at each fiber row, σ_f, can be evaluated[84]:

$$\sigma_f = \frac{4\sigma_1 l_s E_c}{d V_m E_m} \tag{7.35}$$

Equation (7.35), clearly shows that the closure stress has a maximum limit that is controlled by the strength integrity of the fiber. Recalling Chapter 1, the average strength of fibers of a given strength distribution and a particular gage length and fiber diameter is calculated by the Weibull function for average strength and is written as[85]

$$\sigma_{fr} = \sigma_0 \left(\frac{L}{d}\right)^{-1/m} \Gamma\left(\frac{m+1}{m}\right) \tag{7.36}$$

where m is the Weibull modulus, L is the gage length, σ_{fr} is the average fiber strength, σ_0 is the normalizing factor and Γ is a tabulated Gamma function. Masson and Bourgain[86] proposed the following estimation of the Weibull failure probability: If σ_{1L_1}, σ_{2L_2} are the average strengths at a given failure probability for gage lengths L_1, L_2, then the definition of m may be written as

$$\sigma_{2L_2} = \sigma_{1L_1} \left(\frac{L_1}{L_2}\right)^{1/m} \tag{7.37}$$

However, in some cases the *in situ* average strength of the fibers differ from that calculated by single fiber measurements. It has been understood that this difference results from the sliding resistance of the fiber–matrix interface. Thus, realistic measurements can be achieved by extracting fibers from the composite and measuring their tensile strengths.[27] The frictional coupling effect between the fiber–matrix interface can be characterized by a constant shear stress τ (equal to the sliding resistance) acting along the debonded interface. The length over which this effect takes place is the sliding distance calculated from Eq. (7.34). In this region, shear tractions are developed at the fiber ends, allowing stress to be transferred from the matrix to the fibers. Such behavior increases the probability of failure within the sliding distance, while outside of this region fiber strength can be considered independent of the frictional coupling. Additionally to frictional coupling, the crack plastic displacement, COD, at the fiber should also be considered in the sliding distance calculations (Fig. 7.28). Although this region defines a different statistical behavior, due to the lack of the interfacial shear stress τ for reasons of simplicity it is assumed that the interfacial shear stress τ also acts along the COD (COD very small compared to l_s). Thus, the average strength of the debonded fibers in the sliding region can be evaluated by considering a gage length equal to a sliding distance, $L = l_s + \mathrm{COD}_{cr}$ (COD_{cr} defines the critical displacement at the time of fiber failure). Assuming that $\sigma_f = \sigma_{fr}$, where σ_{fr} is the fiber fracture strength, and applying an iterative method between Eqs. (7.33) to (7.37), the critical COD necessary to fracture the fibers is derived as a function of the Weibull modulus m. In Fig. 7.29 the COD_{cr} is presented as a function of the Weibull modulus.

Metallographic observations of fracture surfaces on several unidirectional MMCs[10,81] have indicated that fatigue damage is limited to crack propagation within the matrix. Accordingly, the crack propagation rate of the composite can be considered as an "effective" crack propagation rate of the matrix alone, which includes fiber constraints and fiber bridging. Considering that the matrix crack propagation rate can be written as

$$\frac{da}{dN} = C\,\mathrm{CTOD}^f \tag{7.38}$$

Eq. (7.38) has been used successfully to predict the crack propagation rate in a MMC by substituting the CTOD of the monolithic material with the CTOD calculated from Eq. (7.26). In Eq. (7.38) the parameters C and f are considered material constants. Typical values of C and f for the matrix material may be derived by transforming da/dN–ΔK data into CTOD–a data. In Fig. 7.30, comparison between experimental and predicted crack propagation rate is presented.

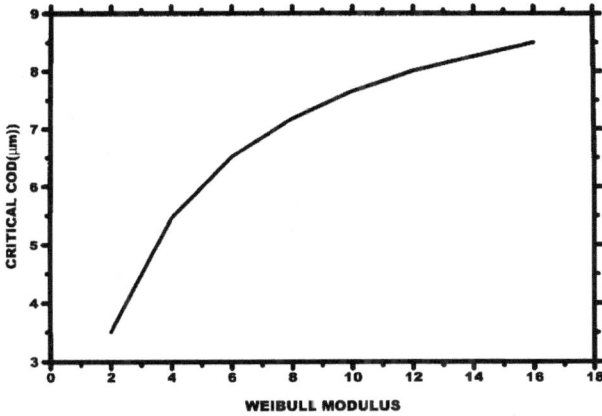

Figure 7.29   The calculated COD_{cr} (without frictional coupling) as a function of Weibull modulus for the case of 32% SCS-6/Ti-15-3 $[0]_8$ composite. An interfacial shear stress of 81 MPa was used. The difference due to frictional coupling is generally negligible.

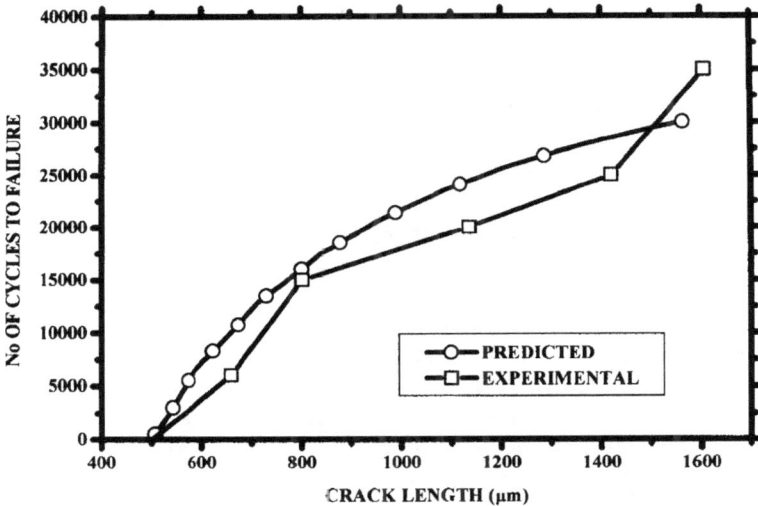

Figure 7.30   Predicted (TZMM) and experimental $a-N$ curves for a 32% SCS-6/Ti-15-3$_{[8]}$ composite tested at 800 MPa. In both cases an initial single edge notch of 0.5 mm was considered. (*Experimental data reproduced from Reference 24.*)

7.3.6 The evaluation of fiber bridging models

It is probably unfair to compare models developed at different times during the research history of MMCs, especially when advanced damage monitoring devices such as scanning electron microscopy and acoustic emissions were not always available. However, it is our scientific duty to evaluate and to determine their advantages and limitations. Before we proceed further, it must be acknowledged that most of the models presented in the previous section do capture the basic mechanisms operating during fatigue damage progression.

The SSFB models definitely reflect their age, their minimum knowledge regarding damage mechanics in composites, and, above all, that they have been created to represent stationary cracks in ceramic composites. However, they do recognize the effect of matrix cracking and interfacial debonding on stress intensity factor range. Undoubtedly, their limitations are apparent for the shear-lag analysis and in their deficiencies in application to cases of partially bridged cracks.

On the other hand, GFB models have been shown useful in simulating crack progression and for interpreting experimental results.[26] Additionally, the proclaimed effect of the interfacial shear stress on stress intensity factor, in the sense that strong interfaces promote fiber failure while weak interfaces promote crack propagation and bridging, has been verified experimentally.[87] There are definite limitations arising from the construction and application of these models. Those worth noting are: (a) the closure stress is idealized as uniform pressure acting along the crack flanks; (b) the closure pressure is related to the COD through a simplified one-dimensional analysis that does not consider crack tip mechanisms[26]; (c) the toughening mechanism due to crack tip plasticity constraint is not considered[73]; and (d) the utilization of ΔK in cases of immense crack tip plasticity is unrealistic.

Since the FPM follows a similar approach to the GFB models, especially in the way of introducing the closure stress into ΔK, similar advantages and limitations apply for it. However, it should be noted that, in terms of rigidity, the determination of the closure stress by neglecting the effect of shear stress is debatable. In contrast to all other models, the TZMM is based on an integrated elastoplastic approach capable of treating simultaneously different damage events. The basic advantages of the MTZZ are that: (a) the effect of all damage events on crack driving force is considered; (b) the model includes statistical implementations of the fiber bridging process; and (c) the model is able to operate under conditions of large-scale yielding. On the other hand, the limitations of the model that are: (a) the model considers interfacial debonding as an instantaneous event and therefore the toughening capacity of the material is underestimated; and (b) the model supposes a constant debonding length (constraint effect) for the freely propagation of crack tip plasticity for the whole range of crack lengths; and (c) it requires immense computational power.

7.4   Damage Tolerant Design of MMCs

7.4.1   The subject

There are three widely used philosophies for fatigue design according to the Royal Society,[88] namely, "safe-life," "damage tolerant," and "fail-safe." The "safe-life" method uses $S-N$ curves to determine the statistical mean life of specific load spectra and therefore to establish the limits at which structure can be considered free of fatigue damage. In general, the method suggests that when fatigue damage (a crack) is evident, the component should be replaced or retired. To accommodate load interaction effects (overloads), corrosion, etc., usually a safety factor of 3 to 5 is used. Consequently, the service life of a component is underestimated and the method itself is not considered "cost-effective."

The solution to the cost-effectiveness problem came with the introduction of nondestructive inspection (NDI) and the development of LEFM. Design engineers, especially in the aerospace industry, understood that by coupling proper inspection procedures and new design methodologies, the economic service life of a component could be extended beyond the "safe-life" limits. The "damage tolerant" method takes into account the unavoidable presence of imperfections created either by manufacturing or machining, the crack growth rates, and the conditions of final failure. Using the above data, the engineer is able to design components and inspection schedules to allow crack growth within specified limits. As a consequence, the service life of a component is predetermined (design life) and therefore should not be considered a material property.

To provide some tolerance to accidental damage and inspection failures, the "fail-safe" method was introduced. Fundamentally, the method implies that failure of one component does not mean failure of the whole structure.

Obviously in the case of MMCs only the "damage tolerant" design methodology can apply, since fatigue damage is always present (see Section 7.1) and because our knowledge in jointing components made by MMCs is limited. In this section we try to establish the service life limits of MMC components. The section is divided into three parts:

initial cracklike defects; conditions for crack growth; and conditions of failure.

7.4.2 Initial cracklike defects

To quantify crack nucleation in MMCs, it is necessary first to identify (a) the average size of inherent cracklike defects and (b) the role of the residual stresses in promoting fatigue damage.

7.4.2.1 Inherent cracklike defects. Many workers[10,89] have observed that cracklike defects in MMCs are characterized by varying sizes and types. On the basis of metallographic observations, fatigue cracks may initiate due to: (a) fiber breaks at the edges (damaged during machining); (b) voids in the interface; (c) broken reaction layers; (d) warts on the fibers; and (e) fiber touching.

Undoubtedly, broken fibers at the edges offer the most effortless type of defect examination. This is because broken fibers are easily observed due to their relatively large size, while most of the breaks are visible at the surface and therefore minimum surface preparation is required. The number of broken fibers per surface unit (the surface unit consisting of a rectangle with fiber centers at the corners) can provide useful information regarding the uniformity and concentration of such defects. In addition, the stress concentration effect of broken fibers can easily be evaluated by representing breaks as notches having circular, semicircular, or quarter-elliptical shape.[10,24,90,91]

Voids of different sizes at the interface have been reported, especially in the case of SCS-6/Ti-6-4 material.[10] To reveal voids, surface etching to the foil-to-foil layer is required. Although the origin of these defects is not yet known, the presence of voids has been attributed to the manufacturing process. Recent work[10] has suggested that their shape may reflect the presence of gas bubbles due to vaporization of contaminants. The average size of voids can be determined by calculating the average void size for each foil-to-foil layer. It should be noted that in the case of SCS-6/Ti-15-3 material the presence of voids is rare.[10]

Broken reaction layers are rarely pre-exist in a virgin material. However, it has been reported that, during the application of the first loading cycle,[64] breaks at the reaction layers may occur, especially at changes of the reaction layer thickness (stress concentration).[10] The measurement of such breaks is time consuming and requires specific examination techniques. Broken reaction layers by themselves are unlikely to significantly reduce the fatigue endurance of the material, unless they result in fiber breaks.

Fiber warts were also reported to play the role of crack initiators, independently of whether fibers have failed at the warts or not. Growth of cracks from warts without fiber failure occurs in a similar manner to that of cracks growing from voids or broken reaction layers. The number of warts can be estimated by examining electrochemically extracted fibers.

Touching fibers, along with the broken fibers at the edges, represent the most severe case. Touching fibers have been shown to exist in a wide range of MMCs verifying their manufacturing birthplace. Many workers[89,92,93] have reported that in cases where the center-to-center distance, of fibers within the same row, varies by a factor of two, cracking of the fiber coating or debonding within the area of contact is common. Generally, large number of touching fibers may lead to significant stiffness reduction. The distance of fibers centers perpendicular to the row is unlikely to vary by more than 10%. The density of the touching fibers can be easily examined by Scanning Electron Microscope (SEM). The severity of the initial crack like defects is shown in Fig. 7.31.

The evaluation of the maximum initial crack like defect size, is a puzzling and difficult task. This is attributed to the following reasons: (a) co-existence of all types of defects is possible; (b) the true severity of some defect types is revealed after the application of the first loading cycle and (c) depending on the density and the severity of the defects, the mechanical properties of the material, like stiffness, could change. The above, quite reasonable, constituent the idea that MMCs can not safely perform under "safe-life" requirements. Unfortunately, specific methodology leading to the evaluation of the maximum initial crack

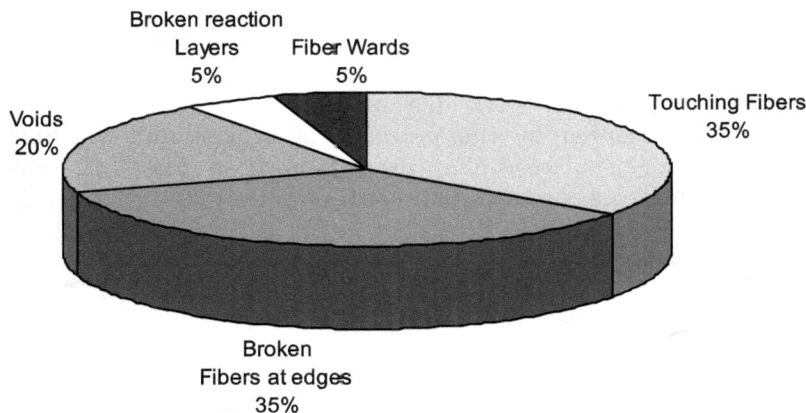

Figure 7.31   The severity of the initial crack like defects in a pie-chart.

like defect size in MMCs does not exist. Consequently, any attempt at evaluation should inevitably be based on empirical observations and theoretical assumptions.

On the empirical observation basis, for example, metallographic and fractographic examinations revealed, that depending on the MMC system, different matrix crackings exist. A typical example is the case of secondary cracking observed in the SCS-6/Ti-15-3 and not in the SCS-6/Ti-6-4. Considering that the presence of voids is rare and secondary cracking in the central region is unconnected to fiber breaks for the SCS-6/Ti-15-3,[10] it is rational to accept that broken reaction layers play a significant role as initial cracklike defects. However, if we attach theoretical assumptions, in this case short crack growth theory,[94] it is easy to derive that broken reaction layers may not have such a significant effect on the final fatigue endurance of the material. The worst-case scenario is that a broken reaction layer results in fiber failure immediately after the application of the first loading cycle. Therefore, an initial cracklike defect exists with size equal to that of the fiber diameter (140 μm for the case of SCS-6). From Reference 95 it is known that the growth of short cracks is dictated by microstructrural barriers (grain boundaries), especially when the cracks are approximately one to four grain diameters in length. In the case of the SCS-6/Ti-6-4 MMC, an initial cracklike defect of 140 μm is approximately equal to 16 alpha grain diameters. Therefore, the growth rate of cracks emanating from such defects is no longer subject to such obstacles and therefore should be included in our evaluation. In the case of SCS-6/Ti-15-3, similar size defects are represented by two or three grain diameters. Since slow propagation rates or even crack arrest characterizes the growth of such defects, their presence can be ignored. It should be noted that since the fiber volume fraction dictates the number of grains within two successive fibers, a metallurgical examination to reveal the exact number of grains is recommended.

However, experimental work has suggested that the worst-case scenario is represented by the presence of broken and touching fibers at the edge.[30] The configuration, shown in Fig. 7.32 may result in a high stress concentration area.

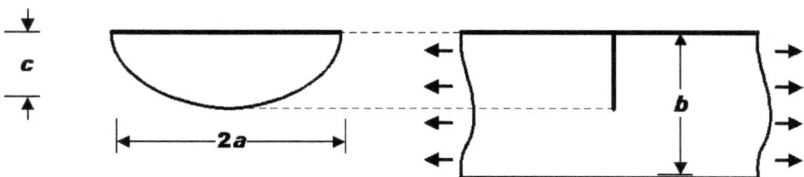

Figure 7.32 The simultaneous presence of broken and touching fibers can form a semielliptical edge crack. A stress concentration factor up to 1.6, depending on the ratios c/a and c/b, is possible.[96]

7.4.2.2 Residual stresses Composite materials are composed of constituents that usually possess different thermal properties. In the case of
titanium–SiC composites, the coefficient of thermal expansion (CTE)
of the matrix is substantially higher than that of the fiber. As a result,
on cooling from the consolidation temperature, tensile residual stresses will develop in the matrix and compressive ones in the fibers[97–99]
(see Fig. 7.26).

A number of workers[98,100] have analyzed the development of the
residual stresses in unidirectional MMCs by assuming an elastic
response of the matrix material throughout the temperature range.
Undoubtedly, such an approach yields a significant overestimation of
the residual stresses acting, since phenomena such as stress relaxation due to viscoplastic flow or creep are usually neglected. Johnson[31]
suggested that the effect of creep can be simply included into the
analysis by assuming that all residual stress developed at temperatures higher than half the melting point of the matrix will be relieved.
Similar assumptions have been reported by Mall et al.[101] If strain
compatibility between the fiber and the matrix is assumed within the
concentric cylinder model,[52] the thermal strain due to CTE mismatch
is given as

$$\Omega = (\alpha_m - \alpha_f)\, \Delta T \tag{7.39}$$

where α_m, α_f are the CTE for the matrix and fiber, respectively, and ΔT
represents the temperature range.

For the prescribed strain (Eq. 7.39), the axial residual stresses in the
fiber and the matrix are given by

$$\sigma_f^r = -(1 - V_f)\frac{E_f E_m (E_f + E_c)}{E_c (E_f + E_c(1 - 2v))}\Omega \tag{7.40}$$

$$\sigma_m^r = \frac{V_f E_f E_m (E_f + E_c)}{E_c (E_f E_c(1 - 2v))}\Omega \tag{7.41}$$

where v is Poisson's ratio (assumed to be the same for fiber and
matrix). The radial and hoop stresses acting on the fiber are

$$\sigma_f^R = \sigma_f^\vartheta = \frac{(1 - V_f)E_m E_f}{E_f + E_c(1 - 2v)}\Omega \tag{7.42}$$

All of the above equations are based on the assumption that the
fibers are distributed uniformly throughout the composite. A more
complete residual stress analysis in which parameters like the cooling
rate and creep rates are included can be found in References 102–104
It should be noted that residual stresses at the surface in the
transverse direction are negligible.[105]

In terms of structural integrity, residual stresses are responsible for two basic damage events: the creation of radial cracks, and a residual COD. Let us now consider that a broken single long fiber is embedded in an infinite matrix. If the matrix material contracts more than the fiber ($\alpha_m - \alpha_f > 0$), the matrix experiences an axial tension that can be simulated by a Lame distribution

$$\sigma_m^9 = 0.5\sigma_m^r \left(\frac{R}{x}\right)^2, \qquad x \geq R \tag{7.43}$$

where R is the fiber radius.

Equation (7.34) proclaims that the maximum stress due to the relaxation of the residual stresses is situated at the broken fiber radius, $x = R$. Furthermore, Eq. (7.43) can be expressed in terms of stress intensity factor by including the geometric features of the broken fiber at the edge (semielliptical edge crack) and the projected length of an initial crack

$$K = Y\sigma_m^r \sqrt{\pi a_{in}} \tag{7.44}$$

If the right part of Eq. (7.44) is set equal to the matrix fracture toughness, then, a_{in} will represent a statistical average of the emanating crack length due to residual stress relaxation

$$K_m^c = Y\sigma_m^r \sqrt{\pi a_{in}} \tag{7.45}$$

Figure 7.33 shows schematically the nucleation of radial cracks originating from broken fibres due to matrix residual stresses.

In 1992, Davidson,[83] after conducting extensive research into the field of fatigue failure of Ti-based MMCs, concluded that the residual stresses are responsible for the development of a permanent COD. In detail, Davidson supported the idea that due to the passage of the crack, the axial residual stresses are relieved and consequently the

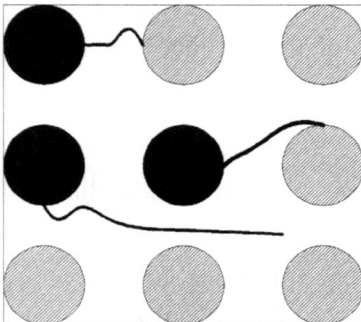

Figure 7.33 The nucleation of radial cracks from fiber breaks (especially at the edges) due to the relaxation of the matrix residual stresses.

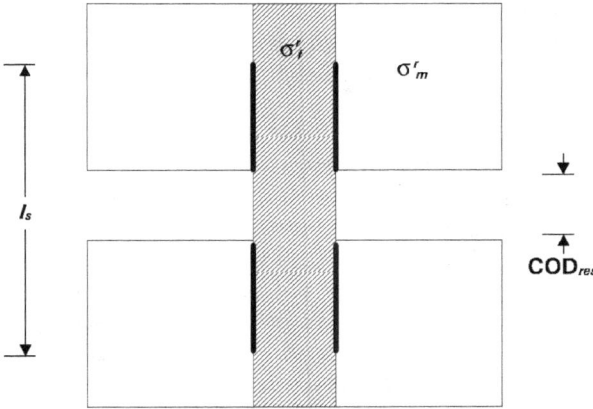

Figure 7.34 The residual COD due to residual stress relaxation.

fibers are allowed to elongate while the matrix contracts. These opposing displacements cause a permanent (residual) COD. The true effect of such finding can be appreciated in the determination of the COD_{cr} (Section 7.1). The COD_{res} can be calculated by relating the elongation of the fiber and the contraction of the matrix. In the case of a bridging crack (Fig. 7.34), the COD_{res} is calculated as

$$COD_{res} = l_f + l_m \qquad (7.46)$$

where
$$l_f = l_s \left(\frac{\sigma_f^r}{E_f} \right), \qquad l_m = l_s \left(\frac{\sigma_m^r}{E_m} \right) \qquad (7.47)$$

A solution for the slip distance l_s is given in Eq. (7.43).

7.4.3 Conditions for crack growth

In monolithic materials the propagation of a fatigue crack to catastrophic failure is achieved when the value of the stress concentration ahead of a slip band is sufficiently high to overcome the constraint effect provided by the grain boundary.[106] In most textbooks the above boundary condition is included in the term for the fatigue limit[107] or the threshold stress intensity factor, K_{th}.[108]

In unidirectional MMCs, the evaluation of a similar boundary condition is a puzzling task due to the number of parameters involved in the fatigue damage process. During the early stages of research, many workers supported the idea that crack arrest in MMCs can be defined in a way similar to crack arrest in monolithic materials.[109,110] In detail, they assumed that when the crack growth rate is approximately 10^{-8} mm/cycle and no crack progression is detected for at least 10^7 cycles, then conditions of crack arrest prevail. Undoubtedly, such an empirical approach is not able to provide numerical solutions and consequently information for design. In 1996, de los Rios et al.[111] published a study in which crack arrest in MMCs is defined in such a way that the governing damage mechanisms are included. According to their work, crack arrest is achieved when the stress concentration ahead of the slip band is unable to overcome the constraint effect provided by the fiber. According to TZMM (see the previous section), crack propagation is achieved only when the crack can developed shear stress at the interface equal to the interfacial shear strength (debonding). On such a basis, if the crack stress field is unable to produce debonding, then conditions of crack arrest should be assumed. The above is shown schematically in Fig. 7.35.

In (a) in Fig. 7.35, the plasticity constraint effect starts at the instant when the plastic zone contacts the fiber. In (b), (c) further propagation of the crack against the fiber results in plastic zone condensation and higher shear stress at the interface. The condition of crack arrest, (d), is defined when the crack contacts the fiber (negligible plasticity) and the shear stress at the interface is still lower than the interfacial shear strength. The mathematical modeling of crack arrest is achieved using Eq. (7.27) and the following two boundary conditions

(a)
$$n_1 = n_2 \approx 1$$

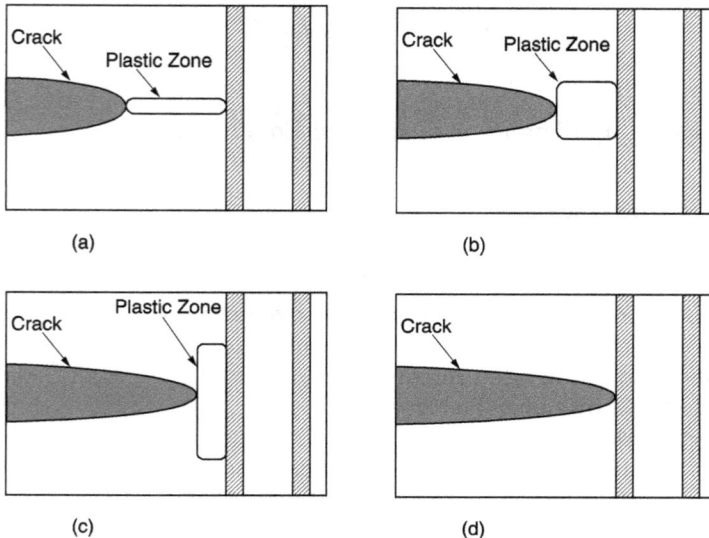

Figure 7.35 The hypothetical conditions of crack arrest.

which states that the crack tip plasticity is minimum, and

(b) $\sigma_3 \leq \sigma_{3d}$

which states that the developed shear stress at the interface is lower than the interfacial shear strength. Since $n_1 = n_2 \approx 1$, then $\sin^{-1} n_1 \approx \sin^{-1} n_2 \approx 1$ and $\cos^{-1} n_2 \approx \sqrt{2}\sqrt{d/(a + d + a_{in})}$. Using these approximations in Eq. (7.27), the maximum allowed applied stress that would still lead to crack arrest of a particular crack length, a, yields

$$\sigma_{\text{arr}} = \frac{2\sqrt{2}}{\pi} \sigma_{3d} \sqrt{\frac{d}{a + d + a_{in}}} + \sigma_1 \tag{7.48}$$

Equation (7.48), shown in Fig. 7.36, represents a theoretical Kitagawa–Takahashi (K-T) curve for unidirectional MMCs. However, in contrast to monolithic materials, where the true fatigue limit is the highest stress level that is unable to transform a fatigue flaw into a fatigue crack, in MMCs the true fatigue limit corresponds to the inability of an already established fatigue crack to propagate beyond one or more fiber rows.

From Eq. (7.48) it is clear that the first part (right-hand side) of the equation decreases with crack length while the second part (friction stress) increases with crack length since more fibers are now within the crack wake. It is therefore rational to accept that crack arrest in MMCs is governed by two factors: (a) the constraint effect provided by

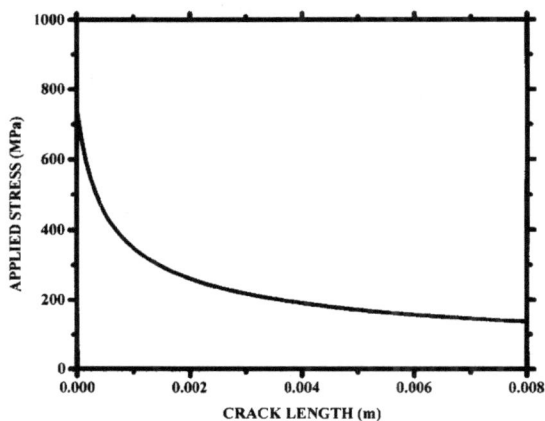

Figure 7.36 Typical maximum allowed stress for a 32% SCS-6/Ti-15-3. The friction stress was taken as negligible.

the intact interface (assumed constant along the crack plane), and (b) the friction stress provided by the bridging fibers. Fiber bridging has been of primary concern to many researchers who developed models to calculate it (see bridging models). In Reference 40 it was shown that fiber bridging cannot provide sufficient closure at high and medium stress. This has been attributed to the relatively large open displacement experienced by the crack wake and to the structural integrity of the fiber. More precisely, the maximum bridging stress that a single fiber row can produce is defined by the maximum open displacement the fibers can sustain without failure, COD_{cr}. In Reference 24 it was argued that, since failure of the bridging fibers (especially those closer to the crack mouth) is possible, the value of σ_1 fluctuates. Consequently, the determination of the effective driving force should include an iterative relationship between the applied stress and closure. As a result, a rigorous representation of σ_1, by means of a simple mathematical expression, is almost impossible. The problem, however, can be simplified if an upper bound to define the region where maximum closure can be expected is determined. If the value of COD_{cr} is known, it is rational to assume that the maximum σ_1 is achieved when the fiber row closer to the crack mouth is subjected to the COD_{cr}. This is shown schematically in Fig. 7.37.

The configuration in Fig. 7.37 suggests that the maximum σ_1 is obtained when $COD = COD_{cr}$ at $x = d/2$. The $\sigma-a$ area where maximum σ_1 is expected can be evaluated by employing the crack-opening solution given by Tada et al.[112]

$$COD(x) = \frac{4(\sigma - \sigma_1)}{E_c}(a^2 - x^2)^{1/2}F\left(\frac{x}{a}\right)$$ (7.49)

where $$F\left(\frac{x}{a}\right) = 1.454 - 0.727\left(\frac{x}{a}\right) + 0.618\left(\frac{x}{a}\right)^2 - 0.224\left(\frac{x}{a}\right)^3$$

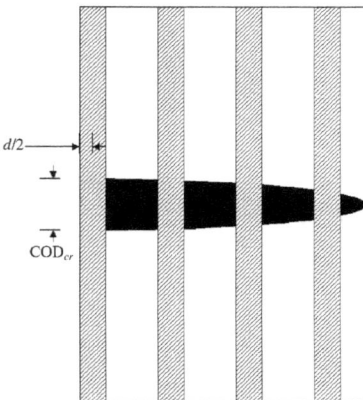

Figure 7.37 The theoretical maximum closure stress that can be achieved during bridging.

Solving Eq. (7.49) for crack length and considering that the closure stress is negligible, the $\sigma-a$ coordinates for maximum closure stress can be evaluated (Fig. 7.38). It should be noted that such an approach underestimates the true crack length for maximum bridging and thus is considered as conservative.

A value of $COD_{cr} = 7.45$ μm was used for the Textron SCS-6. Fibre The shaded boxes represent positions of the fibers. At high and medium stress levels, the maximum closure stress is produced by one or two fiber rows. At lower stress levels, the maximum closure stress takes considerable values as the number of bridging fiber rows increases.

Eq. (7.48) in terms of stress intensity factor can be written as

$$K_{th} = \frac{2\sqrt{2}}{\pi} \sigma_{3d} \sqrt{\frac{\pi da}{a + d + a_{in}}} \qquad (7.50)$$

where K_{th} is the threshold stress intensity factor for MMCs. In Eq. (7.50) the effect of the closure stress σ_1 is disregarded. The effect of different values of fiber volume fraction on K_{th} is presented in Fig. 7.39.

In Fig. 7.39 the area below the curves defines the limits for probable crack arrest. Although it has been constructed considering $\sigma_1 = 0$, it is expected than in real situations the bridging fibers will be able to develop friction stresses. Friction stresses along with the inherent constraint provided by the nondebonded fibers, will further minimize the spread of plasticity ahead of the growing crack, leading to earlier crack arrest. Substantial friction stress is only likely to develop in cases where the applied stress is relatively low and in cases where the crack length allows fiber bridging (Fig. 7.38). When the arrest of short

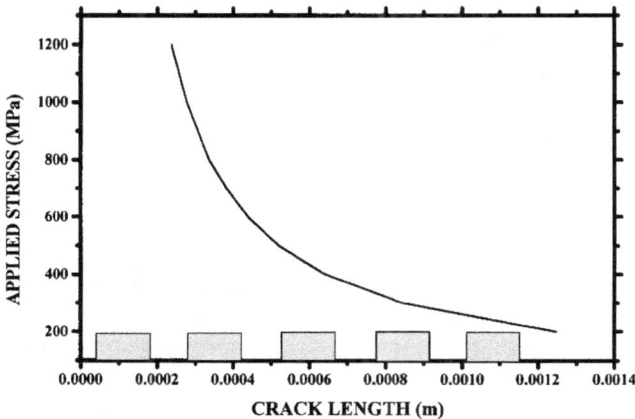

Figure 7.38 Crack length for maximum bridging as a function of applied stress for a 32% SCS-6/Ti-15-3.

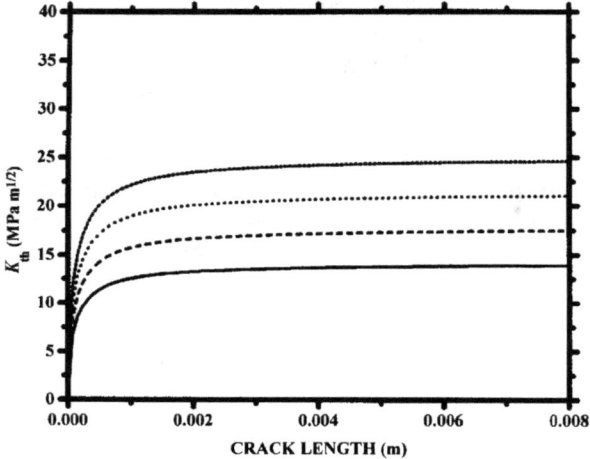

Figure 7.39 The effect of V_f on K_{th} for a SCS-6/Ti-15-3 MMC.
—, $V_f = 10\%$; – – –. $V_f = 20\%$,; · · · · · ·, $V_f = 30\%$; ———,
$V_f = 40\%$.

cracks is in question, the designer should only consider the application of Eq. (7.50). Although the distinction between short and long cracks in MMCs is not clear, it would be safer to consider as short cracks those with lengths that could not allow fiber bridging. In the case of long cracks, both Eqs. (7.49) and (7.50) should be employed. The validity of Eq. (7.48) is shown in Fig. 7.40.

7.4.4 Crack growth and conditions of failure

To perform the simplest evaluation of the fatigue endurance of a component, the design engineer should have accurate knowledge of four key factors: (a) initial cracklike defect; (b) crack propagation rate (da/dN); (c) the bound of the operational life of the material; and (d) final crack length or crack length at failure. If these are known, then the fatigue life is determined by integrating the equation for crack propagation within the limits defined by the initial and the final crack lengths. Although this process provides realistic results in the case of monolithic and homogeneous materials, its application in unidirectional reinforced MMCs may lead to dangerous results. This is considered to be due to the complexity characterizing each key factor. For example, in the previous section it was shown that MMCs exhibit microstructural defects that, except for the large variety of sizes and shapes, are able to form high stress concentrations. Stress concentra-

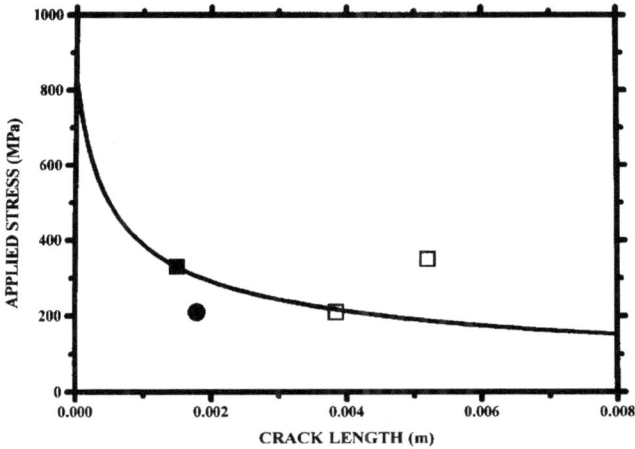

Figure 7.40 Experimental and predicted crack arrest for a SCS-6/Ti-15-3. The solid line has been constructed according to Eq. (7.48) with $V_f = 39\%$ and $\sigma_1 = 0$. ■ $V_f = 35\%$[109]; ● $V_f = 36\%$[170]; □ $V_f = 39\%$.[30]

tions in monolithic components can only be found in areas defined by design. However, the major drawback in a reliable endurance evaluation is the time-dependent character of some factors that control the damage mechanisms taking place during fatigue. Typical time dependent factors (TDF) are as follows.

(i) *The flow resistance of the material.* In References 24, 111 the flow resistance (yield stress) of the MMC was shown to follow a four-step pattern (Fig. 7.41). When a crack is short and no fibers are within the plastic zone, then it is reasonable to assume that the flow resistance of the MMC is dictated by the flow resistance of the matrix material alone, step $A-B$. With the introduction of fibers within the plastic zone, the flow resistance shows a rapid increase due to the high stiffness phase of the fibers (Eq. 7.30), step $C-D$. The magnitude $B-C$ is controlled by the elastic modulus of the fiber and the fiber volume fraction. With further crack propagation, the stress field ahead of the crack tip increases, while the material starts to exhibit degradation phenomena due to cycling, i.e., wear of the interfacial asperities (degradation of the shear stress), supplementary fiber failure, etc. All of these factors are responsible for the failure of fibers within the crack plastic zone. Once the first fiber within the plastic zone has failed, the flow resistance of the material begins a steep descent. As a result, the stress field ahead of the crack tip increases and failure of more fibers is imminent, step $D-E$.

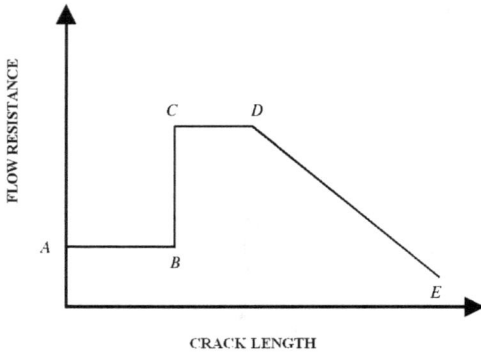

Figure 7.41 The fluctuation of the flow resistance of a MMC with crack length (loading cycles).

If all the fibers within the plastic zone have failed, then the flow resistance of the materials is given as

$$\sigma_2 = \sigma_{ym}(1 - V_f) \tag{7.51}$$

Equation (7.42) defines the minimum flow resistance (point E) that the MMC can develop. Point E is considered below that defining by the flow resistance of the matrix material alone (point A). This is because the plastic zone is now considered to include punctures and stress raisers in the place of the broken fibers. The evaluation of the number of cycles required from the point of the first fiber failure until the point where no intact fibers are located within the plastic zone has no engineering interest. In cases like this, the component is assumed to be close to catastrophic failure, since high crack propagation rates and close to general yielding conditions prevail. Yet, for a sufficient design, the evaluation of the onset of fiber failure ahead of the crack tip and the corresponding degradation of the flow resistance of the material is of great importance. Accurate knowledge of σ_2 could be elementary in terms of revealing the composite potentials. Although a rigorous evaluation should be based on σ_2 versus dN data (such data can be derived from stress–strain tensile tests in fatigued specimens), Rodopoulos[41] suggested a simple computational method based on elastoplastic fracture mechanics (EPFM). To predict the failure of the first fiber within the crack plastic zone, the method suggests a simplification of the COD. According to EPFM, COD extends from the crack mouth until the end of the plastic zone (Fig. 7.42). A straight line (dashed) is assumed to replace the original (solid) COD from the crack tip (CTOD) until the end of the plastic zone. With respect to this and by employing some simple trigonometry, the CTOD that corresponds

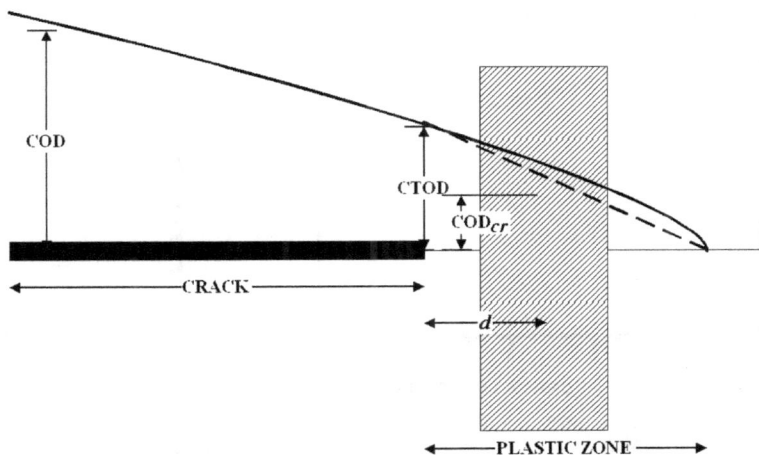

Figure 7.42 The simplified distribution of COD used to predict the onset of flow resistance degradation.

to a COD_{cr} (see Section 7.3.5) at a distance d (fiber diameter), is given as

$$CTOD = \frac{PLASTIC\ ZONE \times COD_{cr}}{PLASTIC\ ZONE\ -d} \qquad (7.52)$$

To calculate the plastic zone, the method makes use of a computer program written to predict crack growth according to TZMM. A distance of d to represent the position of the fiber middle ahead of the crack tip has been quoted as realistic.[41] A typical $\sigma - a$ curve for the onset of flow resistance degradation is presented in Fig. 7.43.

(*ii*) *The fiber bridging process.* Fatigue experiments conducted in plane and notched SCS-6/Ti-15-3 MMC,[24,38,50,64,111] showed a distinct change in the slope of the fatigue crack growth rate (Fig. 7.44). The fact that this change takes place well before the final failure of the specimen prompted the researchers to assume that such behavior is dominated by fiber failure in the crack wake and not ahead of the crack tip. Ibbotson et al.,[38] concluded that this change may mark the initiation of an unstable crack growth. A similar hypothesis proposed by de los Rios et al. in.[111]

In Reference [111] it was suggested that negligible closure stress and substantial crack length could signify a hypothetical lower bound for fatigue failure (or a typical bound of the operational life of the material). In Reference 41 it was reported that the onset of closure

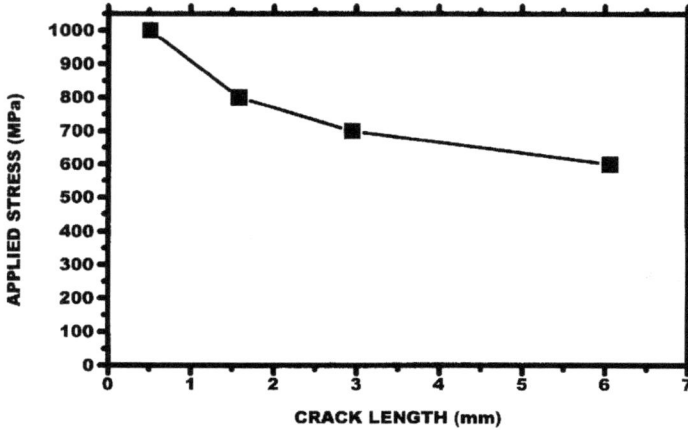

Figure 7.43 Predicted onset of flow resistance degradation curve for a 32% SCS-6/Ti-15-3 MMC with initial cracklike defect of 240 μm. At stress levels lower than 600 MPa, the curve tends asymptotically to final failure.

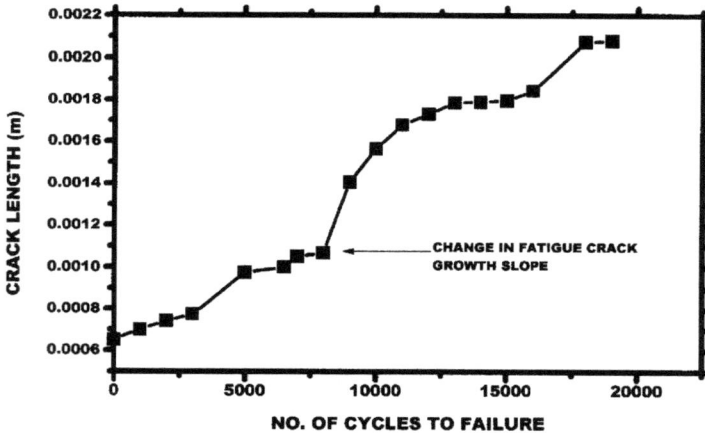

Figure 7.44 Experimental curve of crack length versus loading cycles for a 32% SCS-6/Ti-15-3. The change in slope represents minimum fiber bridging. (*Data taken from Reference 24*)

stress degradation can be predicted by assuming that the COD close to the crack tip is equal to the COD_{cr} (Fig. 7.45) and employing a modified version of the crack-opening solution given by Tada et al.[112]:

$$COD = \frac{4\sigma}{E_c}(a^2 - x^2)^{1/2}F\left(\frac{x}{a}\right) \qquad (7.53)$$

where $F\left(\dfrac{x}{a}\right) = 1.454 - 0.727\left(\dfrac{x}{a}\right) + 0.618\left(\dfrac{x}{a}\right)^2 - 0.224\left(\dfrac{x}{a}\right)^3$

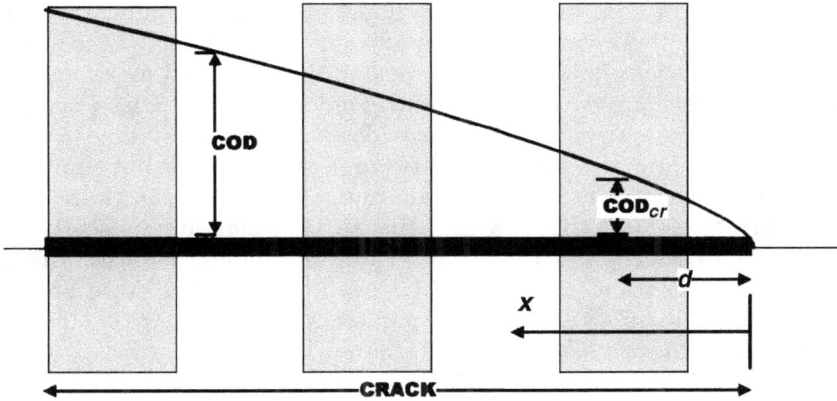

Figure 7.45 If the COD at a distance $x = d$ is equal to the COD_{cr}, failure of all bridging fibers could be assumed.

If the COD close to the crack tip is equal to the COD_{cr}, then it is rational to assume that all the fibers within the crack wake have failed. A predicted lower bound for fatigue failure is presented in Fig. 7.46.

(*iii*) *Constraint effect degradation.* In Reference 113 it was mentioned that in contrast to monolithic materials, where the plastic zone size follows an incremental and predictable behavior with crack length, composite materials exhibit a steep increase in the plastic zone size at a particular crack length. Such behavior was attributed to

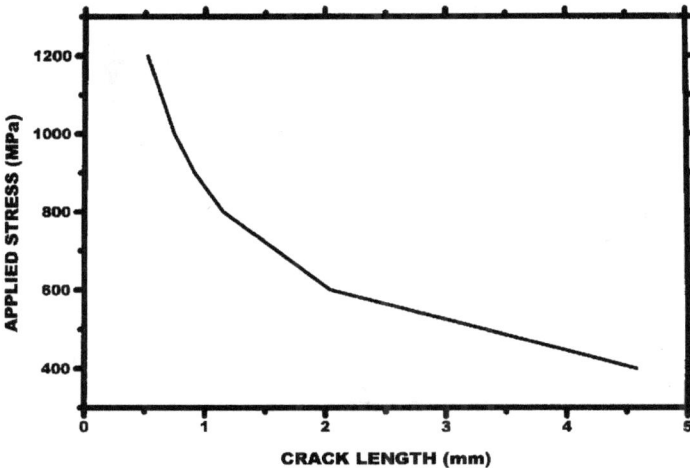

Figure 7.46 Fiber bridging degradation curve for a 32% SCS-6/Ti-15-3.

the inability of the undebonded fibers to provide any further constraint. In Reference 113 it is also mentioned that since the amount of plasticity ahead of the crack tip is controlled more by the constraint effect provided by the undebonded fibers and less by the matrix yield stress (which mainly provides a constant resistance), it is rational to assume that there is a particular crack length that signifies conditions of constraint effect degradation. In other words, there is a particular crack length at which the rate, d(plastic zone size)$/dN$ cannot be further predicted by a modified Dugdale's plastic strip model.[114] In terms of the proposed model, this critical state starts when the crack tip stresses are of sufficient magnitude to instantaneously debond two fiber-rows. Mathematically, such case can be represented by Eq. (7.27)[8]

$$\sigma_{cons} = \frac{2\sqrt{2}}{\pi}\left(\sigma_{3d}\sqrt{(1 - n_2^2)} - \sigma_2 n_1 + \sigma_2 n_2\right) \qquad (7.54)$$

with

$$n_1 = \frac{a}{a + 2D} \quad \text{and} \quad n_2 = \frac{a + 2D - d}{a + 2D}$$

Data derived from Eq. (7.45) are shown graphically in Fig. 7.47.

In Reference 111 it was proposed that crack instability (final crack length) will occur when a large number of fibers ahead of the crack tip have been debonded and therefore the fiber constraint effect on crack tip plasticity is minimum. As a result, crack tip plasticity becomes

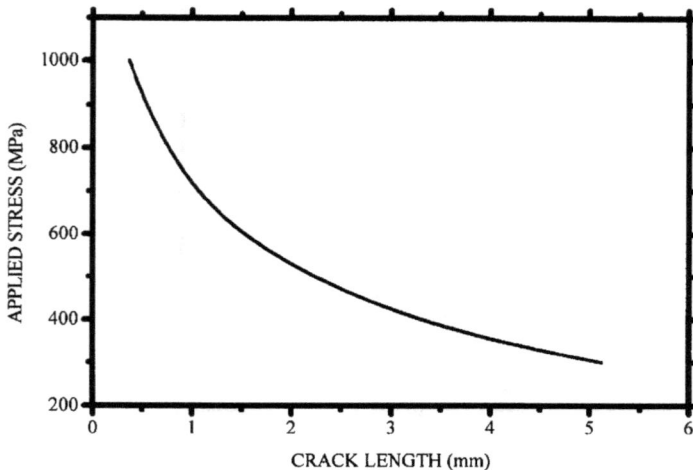

Figure 7.47   Constraint effect degradation curve for a 32% SCS-6/Ti-15-3.

maximum and hence CTOD → maximum. Consequently, all the fibers ahead of the crack tip begin to fail and the flow resistance of the material decreases to a value controlled by the matrix yield stress and the fiber volume fraction (Eq. 7.51). The conditions for crack instability also derives from Eq. (7.27), by substituting $\sin^{-1} n_2 = \pi/2$ (since $n_2 \to 1$) and $\cos^{-1} n_2 \approx \sqrt{2}\sqrt{d/(a+d+a_{in})}$

$$(\sigma_2 - \sigma_1)\sin^{-1} n_1 - \frac{\pi}{2}\sigma_2 - \sigma_{3d}\sqrt{2}\sqrt{\frac{d}{a+d+a_{in}}} + \frac{\pi}{2}\sigma = 0 \qquad (7.55)$$

Solving for n_1, Eq. (7.55) yields

$$n_1 = \left(\frac{\frac{\pi}{2}\sigma_2 + \sigma_{3d}\sqrt{2}\sqrt{\frac{d}{a+d+a_{in}}} - \frac{\pi}{2}\sigma}{\sigma_2 - \sigma_1}\right) \qquad (7.56)$$

Since $iD/2 + d/2 \to \infty$, $n_1 \to 0$, and thus the stress for crack instability for a given crack length is given by rearranging Eq. (7.56)

$$\sigma_{ins} = \frac{2\sqrt{2}}{\pi}\sigma_{3d}\sqrt{\frac{d}{a+d+a_{in}}} + \sigma_2 \qquad (7.57)$$

Or, by employing Eq. (7.51)

$$\sigma_{ins} = \frac{2\sqrt{2}}{\pi}\sigma_{3d}\sqrt{\frac{d}{a+d+a_{in}}} + \sigma_{ym}(1 - V_f) \qquad (7.58)$$

A typical crack instability curve has been drawn in Fig. 7.48. From Eq. (7.58) it is clear that if all fibers ahead of the crack tip are fractured, then $\sigma_{3d} \to 0$ and instability will occur due to $\sigma_{ins} = \sigma_{ym}(1 - V_f)$ which signifies conditions of general yielding.[115]

7.4.5 The fatigue damage map

By collecting the information provided by the above methods, a design engineer is able to visualize in terms of applied stress (data taken from design or FEA) versus crack length (data taken from material quality control process or NDI) the progressive damage of a component during fatigue. A typical example of such an assembly is the so called fatigue damage map[41,111,113] (Fig. 7.49).

In Fig. 7.49, area A represents a theoretical Kitagawa–Takahashi curve for MMC composites. The stress level at minimum expected

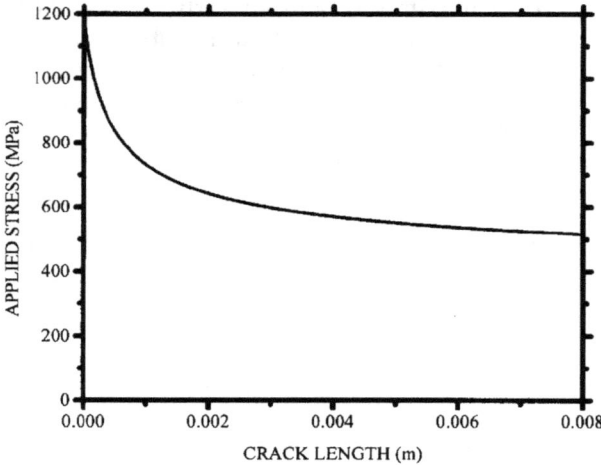

Figure 7.48 Crack instability curve for a 32% SCS-6/Ti-15-3.

crack length represents an upper safe boundary for safe life. Although crack arrest is controlled by both the constraint effect and the closure stress, it is rational to accept that at short crack lengths no significant σ_1 can be expected. Additionally, since the development of closure stress would lead to an earlier crack arrest, the application of the model without modeling σ_1 in relation to the crack length can still be regarded as reliable. In area B, the distinction of the steady-sate crack growth considering the severity of the constraint effect could provide

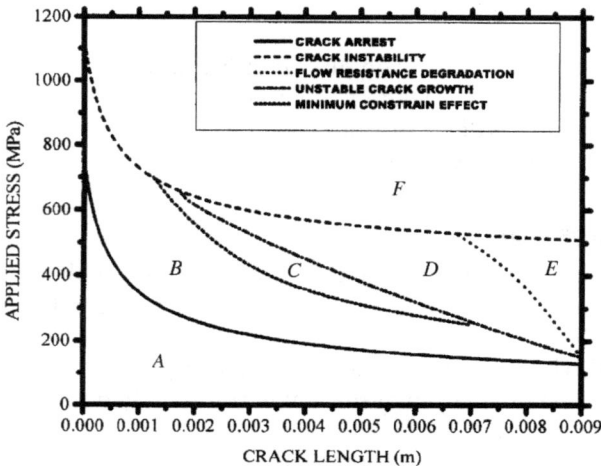

Figure 7.49 The fatigue damage map for a 32% SCS-6/Ti-15-3.

useful information, especially in cases where damage tolerant design is combined with nondestructive inspection. This area is considered to be extremely tolerable to flight stress environments since it is able to withstand overloads, ground loads, etc., by taking full advantage of the two basic crack-resisting mechanisms (constraint effect and fiber bridging). For many, area B is considered to define the limits of damage tolerant design. The transition line between area B and area C represents the beginning of degradation behavior in terms of the constraint effect. In general, area C reflects stress concentration ahead of the crack tip of such density that a single fiber row is incapable of providing full constraint. However, it should be noted that since area C represents almost 40% of the total life of the material, it could be used to define the maximum limit of the operational life of a component, especially in costly replacement procedures. Area D represents the initiation of unstable crack growth since no effective bridging could be developed and the constraint effect is already minimum. Consequently, the material in the area has lost all the crack-resisting features. Area D is to be regarded as a lower bound prior to crack instability. Area E, represents degradation of the flow resistance of the material in terms of fiber failure ahead of the crack tip. Although it is of no fatigue resistance interest, from an engineering point of view, it represents conditions where the static mechanical properties of the material are characterized by deterioration. For many this area represents an artificial value of fracture toughness. Finally, in area F, the material is characterized by a steep increase of the crack plasticity due to extensive fiber debonding and conditions of general yielding.

References

1. J. E. Schoutens (1989) Metal matrix composites, in *Reference Book for Composites Technology*, (S. M. Lee, ed.), Technomic, Lancaster.
2. W. S. Johnson (1987) *Fatigue Damage Accumulation in Various Metal Matrix Composites*, NASA Technical Memorandum 89116, Virginia.
3. W. Thomson (1980) *Mathematical and Physical Papers*, vol. 3, Cambridge University Press, Cambridge.
4. R. Talreja (1990) Damage characterization, in *Fatigue of Composite Materials*, (K. L. Reifsnider, ed.) Elsevier Science, New York, 79–103.
5. G. C. Sih (1987) Microstructure and damage dependence of advanced composite material behavior, in *Proceedings, Workshop on Composite Materials Response: Constitutive Relations and Damage Mechanisms*, Glasgow, (G. C. Sih, G. F. Smith, I. H. Marshall, and J. J. Wu, eds.), Elsevier Applied Science, Oxford, 1–23.
6. J. Aveston, G. A. Cooper, and A. Kelly (1971) Single and multiple fracture, in *Proceeding of a Conference on the Properties of Fibre Composite*, National Physical Laboratory, IPC Science and Technology Press, 15–26.
7. M. D. Senmeier and P. K. Wright (1989) The effect of fiber bridging on fatigue crack growth in titanium matrix composites, in *Fundamental Relationships between*

Microstructure and Mechanical Properties of Metal Matrix Composites, (M. N. Gungor and P. K. Liaw, eds.), Proceedings TMS Fall Meeting, Indiana, 441–457.

8. M. Y. He and A. G. Evans (1991) The strength and fracture of metal/ceramic bonds, *Acta Metall. Mater.*, **39**, No. 7, 1587–1593.

9. W. D. Brewer and J. Unman (1982) *Interface Control and Mechanical Property Improvements in Silicon Carbide Titanium Composites*, NASA Technical Paper 2066.

10. I. Greaves, J. R. Yates and H. V. Atkinson (1994) The role of the interface in the initiation of fatigue cracks in SCS-6/titanium MMCs, *Composites*, **25(7)**, 692–697.

11. M. Y. He and A. G. Evans (1991) The strength and fracture of metal/ceramic bonds, *Acta Metall. Mater.*, **39**, No. 7, 1587–1593.

12. J. Cook and J. E. Gordon (1964) A method for the control of crack propagation in all-brittle systems, *Proc. Royal Soc. London*, **A282**, 508–520.

13. S. S. Wang and I. Choi (1983) The interface crack between dissimilar anisotropic composite materials, *J. Appl. Mech.*, **50**, 169–178.

14. J. R. Rice and G. C. Sih (1965) Plane problems of cracks in dissimilar media, *Trans. ASME*, June, 418–423.

15. M. Comninou (1977) The interface crack, *J. Appl. Mech.*, Dec., 631–636.

16. H. F. Wang, W. W. Gerberich and C. J. Skowronek (1993) Fracture mechanics of Ti/Al2O3 interfaces, *Act. Metall. Mater.*, **41**, No. 8, 2425–2432.

17. P. Ehrburger and J. B. Donnet (1980) Interface in composite materials, *Phil. Trans. Royal Soc. London*, **A2894**, 495–505.

18. P. W. Erickson and E. P. Plueddemann (1974) Historical background of the interface, in *Composite Materials*, vol. 6, (E. P. Plueddemann, ed.), Academic Press, New York, 1–29.

19. M. Y. He and J. W. Hutchinson (1991) Kinking of a crack out of an interface, *J. Appl. Mech.*, **56**, 270–278.

20. S. M. Jeng, J.-M. Yang and C. J. Yang (1991) Fracture mechanisms of fiber-reinforced titanium alloy matrix composites—Part III: Toughening behavior, *Mater. Sci. Eng.*, **A138**, 181–190.

21. B. N. Cox and D. B. Marshall (1991) Crack Bridging in the fatigue of fibrous composites, *Fatigue Fract. Eng. Mater. Struct.*, **14**, No. 8, 847–861.

22. J. P. Lucas (1992) Delamination fracture: effect of fiber orientation on fracture of a continuous fiber composite laminate, *Eng. Fract. Mech.*, **42**, No. 3, 543–561.

23. J. J. Masson, K. Weber, M. Miketta, and K. Schulte (1991) Optimizing of processing parameters and mechanical behavior of a carbon fibre reinforced aluminium matrix composite, in *Metal Matrix Composites, Processing, Microstructure and Properties*, Proc. 12th Riso International Symposium on Metallurgy and Materials Science, Roskidle, Denmark, 509–514.

24. C. A. Rodopoulos (1996) Fatigue studies under constant and variable amplitude loading in MMCs, Ph.D thesis, University of Sheffield, U.K.

25. M. F. Kanninen and C. H. Popelar (1985) *Advanced Fracture Mechanics*, Oxford University Press, New York.

26. J. Bacuckas, Jr. and W. S. Johnson (1992) *Application of Fiber Bridging Models to Fatigue Crack Growth in Unidirectional Titanium Matrix Composites*, NASA Technical Memorandum, 107588.

27. S. Jansson, H. E. Déve, and A. G. Evans (1991) The anisotropic mechanical properties of a Ti matrix composite reinforced with SiC fibers, *Metall. Trans.*, **22A**, 2975–2984.

28. Y. S. Lee, M. N. Gungor, and P. K. Liaw (1991) Modelling of transverse mechanical behavior of continuous fiber reinforced metal–matrix composites, *J. Compos. Mater.*, **25**, 536–556.

29. K. Schulte and K. Minoshima (1993) Damage mechanisms under tensile and fatigue loading of continuous fibre-reinforced metal matrix composite, *Composites*, **24**, No. 3, 197–208.

30. S. M. Jeng, P. Alassoeur, and J. M. Yang (1992) Fracture mechanisms of fibre reinforced titanium alloy matrix composites V: Fatigue crack propagation, *Mater. Sci. Eng.* **A154**, 11–19.

31. W. S. Johnson (1988) *Fatigue Testing and Damage Development in Continuous Fiber Reinforced Metal Matrix Composites*, NASA Technical Memorandum 100628, Langley.

32. W. S. Johnson (1989) Mechanisms controlling fatigue damage development in continuous fiber reinforced metal matrix compositges, *Adv. Fract. Res.*, **ICF 7**, 897–905.

33. W. S. Johnson (1993) Damage development in titanium metal-matrix composites subjected to cyclic loading, *Composites*, **24**, No. 3, 187–196.

34. W. S. Johnson, S. J. Lubowinski, and A. L. Highsmith (1990) Mechanical characterisation of unnotched SCS6/Ti-15-3 metal matrix composites at room temperature, (J. M. Kennedy, H. H. Moeller, and W. S. Johnson, eds.) ASTM STP 1080, American Society for Testing and Materials, Philadelpha, 175–191.

35. R. S. Haaland 1990) Mechanical behavior of 7091 aluminium/SiC interfaces, in *Fundamental Relationships between Microstructure and Mechanical Properties of Metal Matrix Composites*, (M. N. Gungor and P. K. Liaw, eds.), The Metals Society, 779–791.

36. S. M. Jeng, P. Alassoeur, J.-M. Yang, and S. Aksoy (1991) Fracture mechanisms of fiber-reinforced titanium alloy matrix composites, Part-IV: Low cycle fatigue, *Mater. Sci. Eng.*, **A148**, 67–77.

37. D. P. Walls, G. Bao, and F. Zok (1993) Mode I fatigue cracking in fiber reinforced metal matrix composite, *Acta Metall. Mater.*, **41**, 2061–2071.

38. A. R. Ibbotson, C. J. Beevers, and P. Bowen (1991) Stable and unstable crack growth transitions under cyclic loading in a continuous fibre reinforced composite, *Scripta Metall. Mater.*, **25**, 1781–1789.

39. K. R. Bain and M. L. Cambone (1990) Fatigue crack growth of SCS-6/Ti-6-4 metal matrix composite, in *Fundamental Relationships between Microstructures and Mechanical Properties of Metal Matrix Composites*, (M. N. Gungor and P. K. Liaw, eds.), The Metals Society, 459–469.

40. E. R. de los Rios, C. a. Rodopoulos, and J. R. Yates (1996) A model to predict the fatigue life of fibre-reinforced titanium matrix composites under constant amplitude loading, *Fatigue Fract. Eng. Mater. Struct.*, **19**, No. 5, 539–550.

41. C. A. Rodopoulos (1997) Fatigue damage map for metal matrix compositges—A useful tool for design against fatigue, in *Proceedings of 1st Hellenic Conference of Composite Materials*, (S. A. Paipetis, E. E. Gdoutos, eds.), Kyriakidis Bros., 545–569.

42. D. Walls, G. Bao and F. Zok (1991) Effects of fiber failure on fatigue cracking in a Ti/SiC composite, *Scripta Metall.*, **25**, 911–916.

43. D. M. Harmon, C. R. Saff, and D. L. Graves (1989) Strength prediction for metal matrix composites, in *Metal Matrix Composites: Testing, Analysis and Failure Modes*, (W. S. Johnson, ed.), ASTM STP 1032 American Society for Testing and Materials, Philadelpha, 222–236.

44. D. M. Harmon, C. R. Saff, and D. L. Graves (1989) Damage initiation and growth in fiber reinforced metal matrix composites, in *Metal Matrix Composites: Testing, Analysis and Failure Modes*, (W. S. Johnson, ed.), ASTM STP 1032, American Society for Testing and Materials, Philadelpha, 237–250.

45. P. C. Paris, M. P. Gomez, and W. E. Anderson (1961) A rational analytic theory of fatigue, *Trend Eng.*, **13**, 9–14.

46. J. K. Shang and R. O. Ritchie (1989) Crack bridging by uncracked ligaments during faqtigue-crack growth in SiC reinforced aluminum alloy composites, *Metall. Trans. A.*, **20**, 897–908.

47. W. S. Johnson (1979) Characterization of fatigue damage mechanisms in continuous fiber reinforced metal matrix composites, Ph.D. thesis, Duke University.

48. G. J. Dvorak and W. S. Johnson (1981) Fatigue mechanisms in metal matrix composite laminates, in *Advances in Aerospace Structures and Materials*, ASME AD-01, American Society of Mechanical Engineers, New York, 21–34.

49. B. N. Cox (1991) Extrinsic factors in the mechanics of bridged cracks, *Acta Metall. Mater.*, **39**, No. 6, 1189–1201.

50. L. Ghosn, P. Kantzos, and J. Telesman (1990) *Modeling of Crack Bridging in a Unidirectional Metal Matrix Composite*, NASA Technical Memorandum 1044355.
51. D. B. Marshall, B. N. Cox, and A. G. Evans (1985) The mechanics of matrix cracking in brittle-matrix fiber composites, *Acta Metall*, **33**, No. 11, 2013–2021.
52. B. Budiansky, J. W. Hutchinson, and A. G. Evans (1986) Matrix fracture in fiber-reinforced ceramics, *J. Mech. Phys. Solids*, **34**, No. 2, 167–189.
53. L. N. McCartney (1987) Mechanics of matrix cracking in brittle-matrix fiber composites, *Proc. Royal Soc. London A*, **409**, 329–350.
54. R. M. McMeeking and A. G. Evans (1990) Matrix fatigue cracking in fiber composites, *Mech. Mater.*, **9**, 217–227.
55. Y. -C. Chiang, A. S. D. Wang, and T.-W. Chou (1993) On matrix cracking in fiber reinforced ceramics, *J. Mech. Phys. Solids*, **41**, 1137–1154.
56. X. R. Wu and A. J. Carlsson (1991) *Weight Functions and Stress Intensity Factor Solutions*, Pergamon Press, New York.
57. B. N. Cox and C. S. Lo (1992) Load ratio and notch effects for bridged fatigue cracks in fibrous composites, *Acta Metall. Mater.*, **40**, 69–80
58. Y.-C. Chiang, A. S. D. Wang, and T.-W. Chou (1993) On Matrix cracking in fiber reinforced ceramics, *J. Mech. Phys. Solids*, **41**, 1137–1154.
59. X. R. Wu and A. J. Carlsson (1991) *Weight Functions and Stress Intensity Factor Solutioins*, Pergamon Press, New York.
60. T.-H. B. Nguyen and J.-M. Yang (1994) Elastic bridging for modelling fatigue crack growth propagation in a fiber-reinforced titanium matrix composite, *Fatigue Fract. Eng. Mater. Struct.*, **17**, No. 2, 119–131.
61. G. P. Cherepanov (1979) *Mechanics of Brittle Fracture*, McGraw-Hill, New York.
62. H. F. Bueckner (1971) Weight functions for the notched bar, *Z. Angew. Math. Mech.*, **51**, 97–109.
63. J. Llorca and M. Elices (1992) A cohesive crack model to study the fracture behavior of fiber-reinforced brittle-matrix composites, *Int. J. Fract.*, **54**, 251–267.
64. P. Kantzos (1991) Fatigue crack growth in T-based metal matrix composites, M.S. thesis, Pennsylvania State University.
65. E. Y. Luh, R. H. Dauskardt, and R. O. Ritchie (1990) Cyclic fatigue–crack growth behavior of short fatigue cracks in SiC-reinforced lithium aluminosilicate glass–ceramic composite, *J. Mater. Sci. Lett.*, **9**, 719–725.
66. G. R. Irwin (1964) Structural aspects of brittle fracture, *Appl. Mech. Res.*, **3**, 65–81.
67. L. J. Chosn, J. Telesman, and P. Kantzos (1993) Specimen qeometry effects on fiber bridging in composites, in *Fatigue '93*, (J.-P. Bailon and J. J. Dickson, eds.), 1231–1238.
68. B. Budiansky (1986) Micromechanics II, *Proceedings of the Tenth US National Congress of Applied Mechanics*, ASME Publications, 25–31.
69. Y. C. Gao, Y. M. Mai, and B. Cotterell (1988) Fracture of fiber-reinforced materials, *Z. Angew. Math. Phys.*, **39**, 550–572.
70. D. L. Davidson, K. S. Chan, A. McMinn, and G. R. Leverant (1989) Micromechanics and fatigue crack growth in an alumina-fiber-reinforced magnesium alloy composite, *Metall. Trans.*, **20A**, 2369–2378.
71. R. M. McMeeking and A. G. Evans (1990) Matrix fatigue cracking in fiber composites, *Mech. Mater.*, **9**, 217–227.
72. E. R. de los Rios, C. A. Rodopoulos, and J. R. Yates (1995) Micro-mechanical crack growth modelling of fibre-reinforced composites, in *Experimental Techniques and Design in Composite Materials*, (M. S. Found, ed.), Sheffield University Press, 304–320.
73. K. Schulte and K. Minoshima (1993) Damage mechanisms under tensile and fatigue loading of fibre-reinforced metal matrix composites, *Composites*, **24**, No. 3, 197–208.
74. A. Navarro and E. R. de los Rios (1988) Compact solution for a multizone BCS crack model with bounded or unbounded end conditions, *Phil. Mag. A*, **57**, 43–50.
75. B. A. Bilby, A. H. Cottrell, and K. H. Swinden (1963) The spread of plastic yield from a notch, *Proc. Royal Soc. London A*, **272**, 304–314.

76. A. Navarro and E. R. de los Rios (1992) Fatigue crack growth modelling of successive blocking of dislocations, *Proc. Royal Soc. London A*, **437**, 375–390.
77. E. R. los Rios, C. A. Rodopoulos, and J. R. Yates (1995) The effect of fibre, matrix mechanical properties on the fatigue crack propagation in a fibre-reinforced titanium matrix composite, in *High Technology Composites in Modern Applications*, (S. A. Paipetis and A. G. Youtsos, eds.), University of Patras, Applied Mechanics Laboratory, 316–327.
78. J. M. Yang, S. M. Jeng, and C. J. Yang (1991) Fracture mechanisms of fibre-reinforced titanium alloy matrix-composites, Part I: Interfacial behavior, *Mater. Sci. Eng.*, **A138**, 155–167.
79. T. J. Mackin, P. D. Warren, and A. G. Evans (1992) Effects of fibre roughness on interface sliding in composites, *Acta Metall. Mater.*, **40**, 1251–1257.
80. D. Walls, G. Bao, and F. Zok (1991) Effects of fibre roughness on interface sliding in composites, *Acta Metall. Mater.*, **25**, 911–916.
81. C. J. Yang, S. M. Jeng, and J.-M. Yang (1990) Interfacial properties measurement for SIC fiber-reinforced titanium alloy composites, *Scr. Metall. Mater*, **24**, 469–474.
82. W. S. Johnson, S. J. Lubowinski, and A. L. Highsmith (1992) Mechanical characterisation of unnotched SCS-6/Ti-15-3 metal matrix composites at room temperatures, (J. M. Kennedy, H. H. Moeller, and W. S. Johnson, eds.), ASTM STP 1080, American Society for Testing and Materials, Philadelpha, 175–191.
83. D. L. Davidson (1992) The micromechanics of fatigue crack growth at 25°C in Ti-6Al-4V reinforced with SCS-6 fibres, *Metall. Trans.*, **23A**, 865–879.
84. K. S. Chan (1993) Effects of interface degradation on fibre bridging of composite fatigue cracks, *Acta Metall. Mater*, **41**, 761–768.
85. D. M. Kotchick, R. C. Hink, and R. E. Tressler (1975) Gauge length and surface damage effects on the strength distributions of silicon carbide and sapphire filaments, *J. Compos. Mater.*, **9**, 327–336.
86. J. J. Masson and E. Bourgain (1992) Some guidelines for a consistent use of the Weibull statistics with ceramic fibres, *Int. J. Fract.*, **55**, 303–319.
87. R. A. Naik, W. D. Pullock, and W. S. Johnson (1991) Effect of a high-temperature cuycle on the mechanical properties of silicon carbide/titanium metal matrix composites, *J. Mater. Sci.*, **26**, 2913–2920.
88. Royal Society (1984) *Report of Working Group on Subcritical Crack Growth*, Royal Society, London.
89. B. A. Lerch, D. R. Hull, and T. A. Leonhardt (1988) *As-received Microstructure of a Sic/Ti-15-3 Composite*, NASA Technical Memorandum 100938.
90. A. C. Pickard (1986) *The application of 3-Dimensioinal Finite Element Methods to Fracture Mechanics and Fatigue Life Prediction*, Engineering Materials Advisory Services Ltd.
91. J. C. Newman Jr. and I. S. Raju (1981) *Stress-Intensity Factor Equations for cracks in Three-Dimensional Finite Bodies*, NASA Technical Memorandum 83200.
92. N. Tsangarakis, J. M. Slepetz and J. Nunes (1985) Fatigue behavior of alumina fiber reinforced aluminum composites, (J. R. Vinson and M. Taya, eds.) ASTM STP-864, American Society for Testing and Materials, Philadelpha, 131–152.
93. T. W. Clyne and P. J. Withers (1993) *An Introduction to Metal Matrix Composites*, Cambridge University Press, Cambridge.
94. H. Kitagawa and S. Takahashi (1976) Applicability of fracture mechanics to very small cracks in the early stage, in *Proceedings of the 2nd International Conference on the Behavior of Materials*, ASM, 627–631.
95. K. J. Miller (1982) The short crack problem, *Fatigue Fract. Eng. Mater. Struct.*, **5**, 223–232.
96. D. P. Rooke and D. J. Cartwright (1974) *Compendium of Stress Intensity Factors*, HMSO, London.
97. J. Schapery (1968) Thermal expansion coefficients of composite materials based on energy principles, *J. Comp. Mater.*, **2**, 380–404.
98. M. R. James (1989) Residual stresses in metal matrix composites, in *Proceedings of International Conferences on Residual Stresses (IRCS-2)*, (G. Beck, S. Denis, and A. Simon, eds.), Elsevier, 429–435.

99. J. F. Durodola and B. Derby (1994) An analysis of thermal residual stress in Ti-6-4 alloy reinforced with SiC and Al2O3 fibres, *Acta Metall. Mater.*, **42**, No. 5, 1525–1534.

100. C. T. Sun, J. L. Chen, G. T. Sha, and W. E. Koop (1990) Mechanical characterization of SCS-6/Ti-6-4 metal matrix composites, *J. Compos, Mater.*, **24**, 1029–1059.

101. S. Mall and P. G. Ermer (1991) Thermal fatigue behavior of a unidirectional SCS6/Ti-15-3 metal matrix composite, *J. Compos. Mater.*, **25**, 1668–1686.

102. E. E. Gdoutos, D. Karalekas, and I. M. Daniel (1991) Thermal stress analysis of a silicon carbide/aluminium composite, *Exp. Mech.*, **18**, 202–208.

103. E. Zywicz and D. M. Parks (1988) Thermo-viscoplastic residual stresses in metal matrix composites, *Compos. Sci. Technol.*, **3**, 295–315.

104. U. Ramamurty, F.-C. Dary, and F. W. Zok (1996) A method for measuring residual strains in fiber-reinforced titanium matrix composites, *Acta Mater.*, **44**, No. 8, 3397–3406.

105. K. M. Brown, R. W. Hendricks, and W. D. Brewer (1990) X-ray diffraction measurements of residual stresses in SiC/Ti composites, in *Fundamental Relationships between Microstructures and Mechanical Properties of Metal Matrix Composites,* (M. N. Gungor and P. Liaw, eds.), The Metals Society, 269–286.

106. E. R. de los Rios (1998) Dislocation modelling of fatigue crack growth in polycrystals, *Eng. Mech.*, **5**, No. 6, 363–368.

107. A. Wöhler (1860) Versuche über die festigeit der Eisenbahnwagenaschsen, Zeitschrift für Bauwesen, **10** [English summary (1867), *Engineering* 4, 160–161].

108. F. A. McClintock (1963) On the plasticity of the growth of fatigue cracks, in *Fracture of Solids*, Vol. 20 (D. C. Drucker and J. J. Gilman, eds.), Wiley NY, 65–102.

109. P. Bowen (1992) Characterisation of crack growth from an unbridged defect in continuous fibre reinforced titanium matrix composite, in *Test Techniques for Metal Matrix Composites II*, (N. D. R. Goddard and P. Bowen, eds.), IERA Technology 107–126.

110. L. J. Ghosn, J. Telesman, and P. Kantzos (1990) Fatigue crack growth in unidirectional metal matrix compositges, in *Fatigue '90*, (H. Kitagawa and T. Tanaka, eds.), Engineering Materials Advisory Service, 893–898.

111. E. R. de los Rios, C. A. Rodipoulos, and J. R. Yates (1996) Damage tolerant fatigue design in metal matrix composites, *Int. J. Fatigue*, **19**, No. 5, 379–387.

112. H. Tada, P. C. Paris, and G. R. Irwin (1985) *The Stress Analysis of Cracks Handbook*, 2nd ed., Del Research, St. Louis, MO.

113. E. R. de los Rios, C. A. Rodipoulos, and J. R. Yates (1998), The fatigue behaviour of metal matrix composites under single overloads, *Fatigue Fract. Eng. Mater. Struct.*, **21**, 1503–1511.

114. D. S Dugdale (1960) Yielding of steel sheets containing slits, *J. Mech. Phys. Solids*, **8**, 100–108.

115. B. A. Bilby, A. H. Cottrell, and K. H. Swinden (1963) The spread of plastic yield from a notch, *Proc. Royal Soc. London A* **272**, 304–314.

Case Studies:
Aerospace Engineering
Construction Engineering
Medical Engineering

Behavior of Composite Structures under Impact Loading

Th. Kermanidis and G. Labeas

Laboratory of Technology and Strength of Materials,
Mechanical Engineering & Aeronautics Department,
University of Patras, Greece

8.1 Introduction

This chapter discusses some research topics concerning the crash simulation of composite systems with energy absorption elements. The current research areas related to crashworthiness are reviewed first, including crashworthiness requirements and characteristics, failure behavior of composite structures in crash conditions and modeling strategies using nonlinear numerical codes. Various energy absorbing composite elements are considered in detail and their failure behavior is numerically simulated. One of the case studies that is considered in detail is the "tensor skin concept," an energy absorbing composite system that was originally developed to improve the crashworthiness of helicopters under water impact. Other composite components analyzed include sine-wave beams, box-beams, and sandwich panels, which are used extensively in the construction of the lower part of aircraft composite fuselages. An important issue, which is covered here, is the verification of the numerical simulation procedure by experimental work. The experimental results utilized to assess and validate the numerical procedure and verify the numerical results were derived within European Research Projects in this field.

During recent decades, remarkable efforts have been made in replacing advanced aluminum alloys by polymer-based composite

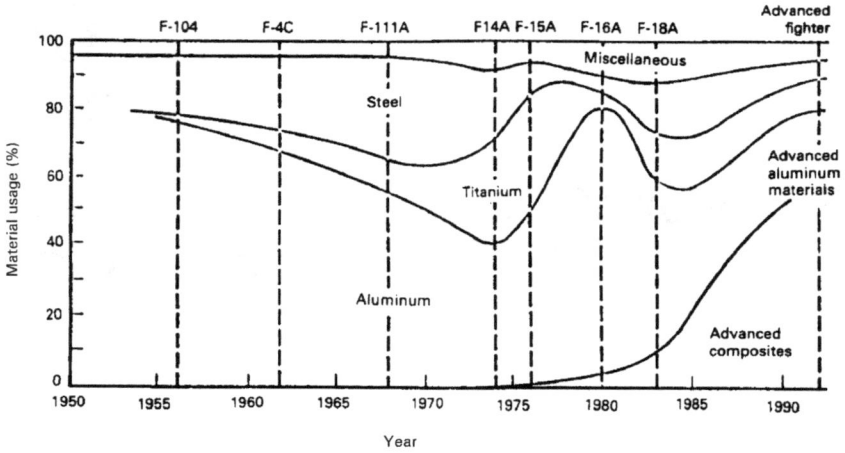

Figure 8.1   The utilization of composite materials in military aircraft during 1950–1990.[1]

materials in the construction of civil and military aircraft. The main driving force for this substitution is the weight reduction effort in order to increase payload and fuel economy in civil aircraft. Moreover, in military aircraft, improved performance and radar cross-section reduction (stealth designs) can be achieved. Figures 8.1 and 8.2 show the trend of the aircraft industry toward the direction of composite aircraft and rotorcraft.

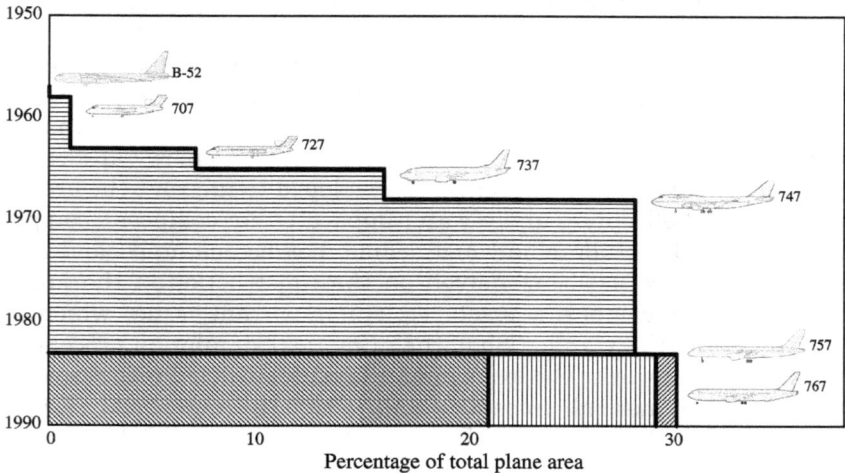

Figure 8.2   Materials weight distribution in advanced transport aircraft.[1]

One of the major problems that should be solved before the use of composite materials in entire airframe designs is the understanding of how composites behave as load-carrying materials. Some of these loads are service loads that are expected to be experienced in day-to-day operations, e.g., aerodynamic, engine, and landing loads. Other types of loading are less frequently experienced but are still critical and have to be considered. Crash loads fall under this category and are discussed in the present study.

The recent trends in the evolution of composite airframes tend to accept that crashworthiness should be incorporated into the fuselage design. The main goal of this development is to maintain the structural integrity at the passenger cabin and provide the highest probability of occupant survival with minimal injuries in a prescribed crash environment, at least in the milder accidents. When composites are used in the fuselage construction, the major problem associated to the crashworthiness design is the understanding of how composites behave as load-carrying materials, and specifically the assessment of their ability to withstand impact damage. It is well known that carbon reinforced thermosets are brittle in nature having low energy absorption characteristics. Novel composite systems with potential improvement of energy absorption capabilities are needed to overcome these drawbacks. Parallel to this development, to provide higher degree of protection to the occupants it seems to be necessary to make major changes in the design philosophy by introducing energy-absorbing structural subcomponents in the structural areas that are directly affected by the crash force.

The purpose of this chapter is to summarize some of the principles for providing composite design concepts working toward the maximum crash survivability of the occupants, as developed within the European Research Project "Design for Crash Survivability" (CRASURV). After a short description of the crash environment, we present the crashworthiness requirements, the philosophy of how to proceed to increase the structural crashworthiness of a composite fuselage, as well as the role of an effective numerical simulation procedure of the response of impacted composite substructures.

8.2 Crashworthiness of Aircraft Structures

8.2.1 The crash environment

In the past, aircraft have been designed with high structural integrity with the aim of avoiding crash situations, which means that no crash tolerance could be resisted and no crashworthiness issues have been taken into account. While crash avoidance has been and will continue to be the main theme in aircraft safety, crashes of any type have been proven to be inevitable, which resulted in increasing concern by the airworthiness authorities for the survivability of commercial aircraft accidents. Hence, design for crash survivability has become of major importance.

The study of crashworthiness includes, among other main activities, the collection of background data on crash accidents, in order to justify the survivable crash scenarios. A survivable accident can be defined as one in which the forces transmitted to the occupant do not exceed the limits of human tolerance to abrupt accelerations. The aircraft crash environment ranges from the insignificant hard landing to the nonsurvivable event. In any vehicle crash, the decelerating forces depend upon the kinetic energy of the vehicle and the distance through which it moves during deceleration. In a crash with pure vertical velocity component, the movement is limited by the deformation of the terrain, which may vary from soft soil to rigid ground, as well as by the structure itself. If the vehicle crashes with a high longitudinal velocity component, the loading depends on friction, the structure's crush capability, as well as the local crush strength of the impacting obstacles (trees, barriers, rocks, etc.). In real situations, the most common case is that a crash occurs with combinations of longitudinal and vertical velocity components. The crash environment for helicopters additionally provides a high potential for rollover because of the vertical location of the center of gravity and the turning rotor.

During a typical crash, the aircraft structure progressively collapses at its crush strength and the deceleration forces are transmitted to the occupied sections. Those accidents are survivable in which the forces transmitted to the occupant through the seat and restraint system do not exceed the human tolerance and in addition the fuselage area remains sufficiently intact to provide a protective shell within which an occupant can survive.[2] Crashes on take-off and landing around airfields have been shown to be the most common survivable crash scenarios. It should become clear that only in relatively low-speed impacts may the design considerations be effective. In general, if the vertical velocity component is above 12 m/s or the horizontal component above 18 m/s, then the occupants' probability of survival is marginal.[3] Therefore, most efforts are now devoted to the study of the low kinetic energy aircraft impacts. The first and most obvious goal is to reduce the number of injuries and deaths caused by aircraft impacts. Second, by a careful design there may be a reduction of the repair costs following a low-speed impact.

Before the structural integrity of any crashworthiness vehicle can be evaluated, the crash loads acting on the structure must be known. Extensive studies have been performed to identify the crash environ-

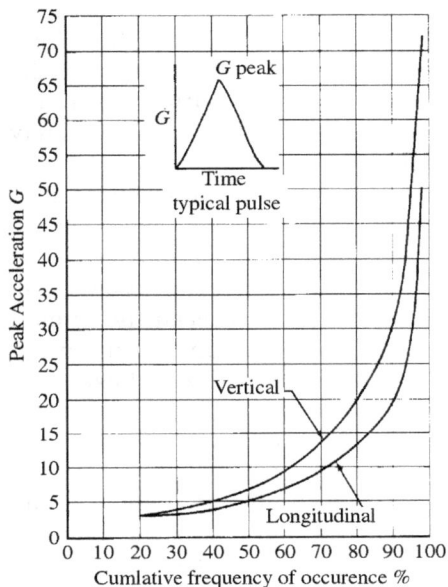

Figure 8.3 Longitudinal and vertical peak accelerations.[3]

ment of a survivable accident. From these studies, the velocity and acceleration data were estimated and documented.[3] In Fig. 8.3, the longitudinal and vertical peak accelerations versus the percentage frequency of occurrence are plotted.[3] The 95th percentile survivable crash conditions were selected as a reasonable design environment and are recommended for use in the design of structural subcomponents, seats, restraint systems, and other attached parts in aircraft fuselages.

8.2.2 Structural crashworthiness requirements

Crash survival requires that the vehicle and the occupant both survive the impact, so that the occupant is able to evacuate the vehicle before any post-crash environmental hazards, such as fire or water, become intolerable.[3] The maintenance of the occupant living space has the first priority among the successive requirements that should be fulfilled, which implies a crashworthiness design of the structure. Additional important requirements are the minimization of the crash pulse in conjunction with a proper occupant restraint system to avoid severe occupant injury, as well as the prevention of post-crash outbreak of fire. The capability of a vehicle to provide a protective space during a crash depends on its strength, mass distribution, and total kinetic energy at the time of impact, and the distance over which the kinetic energy is dissipated. The crashworthiness requirements are different for the various types of aircraft. The main concern in the present study applies to the fixed-wing civil aircraft with pressurized, cylindrically shaped fuselage, which usually crashes with a relatively low vertical and high longitudinal velocity.

As mentioned above, the most important design requirement in a crashworthiness composite vehicle is that the passenger cabin should maintain its structural integrity in order to minimize injuries to the occupants. At the same time, it should be insured that the deceleration forces that might occur during the impact will not exceed the human body limits. Thus, the design objective is to determine a minimum-weight system at an acceptable cost that provides the highest probability of occupant survival in a crash environment.[2] To enable such a crashworthiness design, it is necessary that the performance of the aircraft structure under crash conditions be predictable. An understanding of how the local structure reacts under dynamic loading conditions and also of how the local structure influences the dynamic behavior of the global structure is another important aspect in establishing design guidelines for crashworthiness.

Fulfillment of the crashworthiness requirements implies that the accurate behavior of airframes can be known. Much research activity has been undertaken in this field during recent years. Moreover, prediction of the crash performance is the subject of two European Community-funded research projects, namely the project "Crashworthiness for Commercial Aircraft" (CRASH) (1993–1995), which has focused on metallic structures, and the project "Design for Crash Survivability" (CRASURV) (1996–1999), which focuses on composite aircraft structures.

In general the prediction of the response of materials and structures under impact loading is quite difficult. To facilitate the study of the behavior of the impacted structure, impacts can generally be classified into three main categories, according to the phenomena that follow the impact. In the first category, impacts of low load intensity leading to stresses below the material yield point are considered. In this case, the major design aspect of the entire structure is focused on the resistance against the external forces. In the second category, impacts with higher load intensity can be considered, where the behavior of the structure involves high material nonlinearity, large deformations, local heating and failure initiation. In this impact category the response is highly localized and is affected rather by the material behavior in the vicinity of the loading area than by the geometry of the global structure. The third category includes impacts with very high load intensity. In this category, additionally to the major phenomena of the previous category, various types of failure take place and propagate within the structure, causing local failure of structural components and subsequently partial or total collapse of the entire structure. This type of impact is the most difficult in the analysis, as a large number of interacting failure modes occur. It therefore becomes necessary that both the material behavior and failure mechanisms, as well as the connectivity of the components and the corresponding load paths in the entire structure, are well understood in a very detailed manner.

Although the design and analysis techniques for static, buckling and vibration loading are fairly well established, techniques for design of composite structures under impact loading are still a major research activity. Apart from the crash of a vehicular structure, a related in-service situation that requires an impact analysis to prove the resistance of the structure is the foreign object impact. The foreign object impact is usually treated as a low- or medium-velocity impact and can be classified in the first or second category described above, while a crash is classified in the third category. However, methodologies and techniques employed in both problems are similar.

Energy absorption characteristics of composite structures under various types of loading are of importance for the development of techniques for the prediction of the structural behavior under crash conditions. A major problem where composite materials are concerned is their limited ability to absorb impact energy. It is well known that carbon fiber reinforced matrices of thermoset type are quite brittle and their delamination and cracking is common even in relatively low levels of impact. To overcome this problem of limited energy absorption capability of composite materials, two paths should be followed in parallel. First, the investigation of novel composite systems with increased energy absorption capabilities should be undertaken, and second, the design concept of the structural areas that are subjected to impact loads should be changed by introducing energy-absorbing structural subcomponents. For the first issue, thermoplastic resins (e.g., PEEK and PEI) that have higher impact toughness are under examination. Where a more effective energy-absorbing system is required, laminates made of polyethylene (PE) fibers embedded in epoxy matrices can be considered, because their "strain to failure" limits are much higher than the common 1–2% that is the usual case of ordinary composite systems.[4] To deal with the second issue, special design techniques for composite components are required to enable crashworthy structures. Typical subcomponents of this type are box-beams, sine-wave beams, and "tensor skin" strips, as shown in Fig. 8.4. To identify the suitability of these substructures for a crashworthy design, preliminary static crush and dynamic mass drop tests should

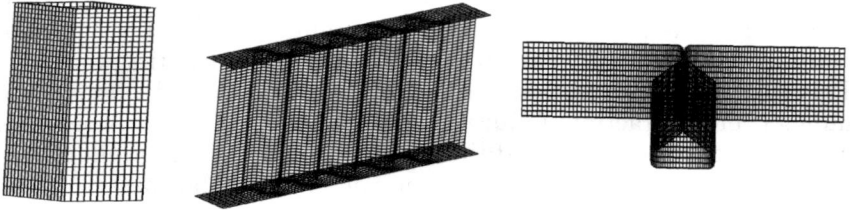

Figure 8.4 Typical aircraft crashworthiness subcomponents.

be performed. However, tooling, manufacturing, and testing of components is expensive. Suitable simulation techniques for the prediction of crash behavior of composite structures are required to allow parameter studies and minimize the tests.

The two major objectives that should be accomplished to enable the prediction of impact behavior of composite structures are (a) understanding of the material behavior under impact conditions, through materials testing and study of failure mechanisms, and (b) the development of numerical simulation tools and the verification of them by applying them to the simulation of full-scale airframe composite structures. In the present study, theoretical aspects concerning the simulation techniques and the modeling of the material behavior are described and applied in the case of the crash simulation of specific composite substructures. As the complete structure is not considered here, the subject of prediction of the impact loading transferred in the structure under various crash scenarios is not discussed here.

8.3   Tools for the Simulation of Impacted Composite Structures

8.3.1   The finite element technique

Vehicular structures such as cars, trains, planes, and fast boats are partly produced using composite materials. Their structural components often comprise many elements such as beams, rings, plates, and shells that are assembled to form the structure. The analytical simulation treatment is only applicable for elementary structures of simple geometry and boundary conditions, which is not the case for crashes of complex vehicular structures. Accordingly, numerical procedures should be utilized in the analysis. An adequate numerical code should consider large deflections, local buckling and post-buckling, nonlinear material behavior, progressive failure, and damage propagation. A numerical technique that has these characteristics is the widely applied finite element (FE) method. The development of successful FE models of aircraft structures requires accurate geometrical representation of the structure, accurate modeling of the material behavior, knowledge of the actual loads that will be applied to the FE model, and finally a real structure to test and verify the model results. Using the FE technique, the real structure is simulated with a large number of small elements. The structure's response to the impulsive loading comprises the deformed shape, displacement histories, stress and strain distributions in the time domain, as well as acceleration and velocity time histories. The FE formulation of a time-depended problem has the general form:

$$[M]\{\ddot{u}\} + [C]\{\dot{u}\} + [K]\{u\} = \{F^a\} \tag{8.1}$$

where $[M]$, $[C]$, and $[K]$ are the structural mass, damping, and stiffness matrices, respectively, $\{\ddot{u}\}$, $\{\dot{u}\}$, and $\{u\}$ are the nodal acceleration, velocity, and displacement vectors, respectively and $\{F^a\}$ is the applied load vector. Equation (8.1) can be considered as a set of "static" equilibrium equations that includes also inertia ($[M]\{\ddot{u}\}$) and damping ($[C]\{\dot{u}\}$) terms. The stiffness, mass, and damping matrices, as well as the load vector can be functions of $\{u\}$, $\{\dot{u}\}$ or $\{\ddot{u}\}$, which means that Eq. (8.1) is in general nonlinear. A typical solution algorithm that is utilized in the general-purpose commercial FE codes for the solution of the nonlinear equation system at discrete time points is the Newmark integration method coupled with the Newton–Raphson iteration scheme.[5] The Newmark method is an implicit, unconditionally stable, direct integration scheme for solving equations in the time domain. An implicit integration method determines the solution at time $(t + \Delta t)$ in terms of the problem variables at time $(t + \Delta t)$, while an explicit algorithm determines the solution at $(t + \Delta t)$ in terms of the variables in time (t). The unconditional stability implies that, in linear analysis, for any given time step size, the error introduced at a specific time point will not grow without bound in the solution of the subsequent time points. The Newmark method operates directly on the equations of motion. The equations are integrated step by step using finite difference expansions between the time points n and $n + 1$, i.e., the time interval Δt:

$$\{U_{n+1}\} = \{U_n\} + [(1 - \delta)\{\ddot{U}_n\} + \delta\{\ddot{U}_{n+1}\}]\,\Delta t \tag{8.2}$$

$$\{U_{n+1}\} = \{U_n\} + \{U_n\}\,\Delta t + [(\tfrac{1}{2} - a)\{\ddot{U}_n\} + a\{\ddot{U}_{n+1}\}]\,\Delta t \tag{8.3}$$

where α and δ are the Newmark integration parameters, $\Delta t = t_{n+1} - t_n$ is the time step and $\{\ddot{u}\}$, $\{\dot{u}\}$ and $\{u\}$ are the nodal displacement, velocity and acceleration vectors, respectively.

Inserting Eqs. (8.2) and (8.3) into Eq. (8.1), the following relation is obtained:

$$(\alpha_0[M] + \alpha_1[C] + [K])\{U_{n+1}\} = \{F^a\} + [M](\alpha_0\{u_n\} + \alpha_2\{U_n\} + \alpha_3\{\ddot{U}_n\})$$
$$+ [C](\alpha_1\{u_n\} + \alpha_4\{U_n\} + \alpha_5\{\ddot{U}_n\}) \quad (8.4)$$

To solve the nonlinear equation system (8.4), the iterative Newton–Raphson method is employed, in the form

$$[K_i^T]\{\Delta u_i\} = \{F^a\} - \{F_i^{nr}\} \qquad \{u_{i+1}\} = \{u_i\} + \{\Delta u_k\} \quad (8.5)$$

where $[K_i^T]$ is the efficient coefficient matrix and $\{F_i^{nr}\}$ is the vector of restoring loads.

The right-hand side of Eq. (8.5) is the residual or out-of-balance vector, i.e., the amount of the system that is out of equilibrium. To ensure the solution convergence, the residual ($F^a - F^{nr}$) is required to be sufficiently smaller than a specified tolerance (e.g., 1/1000 of the average forces). Therefore, in each time interval (n), several iterations (i) are required for solution convergence. One drawback of the implicit procedure occurs in the presence of high nonlinearity, which is always encountered in the case of impact. In this case it is possible that the iterative process diverges, leading either to a rapid growth of the solution error or to a repetitive nonconvergent pattern. The fulfillment of the contact conditions is an additional drawback. A complete stabilization of the contact conditions is required, which means that if one contact condition changes from closed to open, another iteration will be made. The only way to overcome these problems is to decrease the incremental time step size $\Delta t = \{u_{n+1}\} - \{u_n\}$. However, this is not always successful; furthermore, it is not possible to predict the total required computing time in advance.

To overcome these drawbacks of the implicit solution procedure described above, the solution of the equation system (8.1) for the case of nonlinear dynamic problems is more effectively done by the explicit solution procedure. The explicit solution procedure is based on an explicit central difference integration rule:

$$\{U_{n+1}\} = \{u_n\} + \Delta t \cdot \{U_{n+1/2}\} \quad (8.6)$$
$$\{U_{n+1/2}\} = \{U_{n-1/2}\} + \Delta t\{\ddot{U}_n\} \quad (8.7)$$

The central difference integration operator is explicit in the sense that the kinematics state may be computed using the known values of $u_{(n-1/2)}$ from the previous iteration. The acceleration at the beginning of the increment may be computed by

$$\{\ddot{U}_n\} = M^{-1}(F^{\{a\}} - I^{(int)}) \quad (8.8)$$

which means that the explicit procedure does not require either iterations or calculation of the tangent stiffness matrix. The explicit integration operator is stable, provided that Δt is smaller than the smallest time required for an elastic wave to travel along any element of the FE mesh, which means that the following inequality holds:

$$\Delta t \leq \frac{L}{\lambda_d} \tag{8.9}$$

where λ_d is the dilatational wave speed of the material. If, for example, the mesh contains almost uniform elements and multiple materials, the smallest element with the highest wave speed will determine the stable time increment. Obviously, the explicit procedure is best suited for analyzing high-speed dynamic effects. The use of small increments, which is dictated by the stability limits, also simplifies the implementation of the contact conditions. On the other hand, the application of the explicit solution scheme in semistatic simulations, which requires more time, will lead to enormous solution time. A method for speeding up the analysis is to scale up the mass density. As the wave speed of the material is inversely proportional to the square root of the density, an increase of the material density by a factor of 100, for example, will decrease the material wave speed by 10, which means 10 times faster solution. However, to avoid introduction of unrealistic dynamic effects, it should be ensured that the increase in inertia force, induced by the mass scaling, will remain limited so that its effect on the overall structure's response will be negligible. Another advantage of the explicit procedure is that the run time increases proportionally to the mesh size. If, for example, a mesh refinement by a factor of 2 in all directions is performed in a 2-D or a 3-D problem, an increase of the run time by 8 and 18, respectively, will occur. This is because the number of elements will increase by 4 and 9, respectively, and the time step will decrease by a factor of 2 in both cases. Some of the nonlinear dynamic commercial codes with explicit solution algorithms implemented are LS-DYNA3D by CAD-FEM, PAM-CRASH by ESI, RADIOSS by MECALOG, and DYCAST by Grumman Aerospace. In the present case study, the nonlinear dynamic FE code utilized for the simulations of the impact of composite structures is the PAM-CRASH code.

8.3.2 Modeling aspects of composite structures in PAM-CRASH FE code

One of the sensitive parts of a successful numerical simulation of an impacted structure, using nonlinear dynamic FE codes, is the development and calibration of suitable material damage models. Such models must properly represent the material response, as well as the stiffness and strength degradation at high deformation rates. The material behavior models should suitably describe the elastic properties degradation and the damage initiation and propagation.

The PAM-CRASH code enables the modeling of composite layered structures, using four node shell elements with one integration point per layer, combined with a material type, coded "type 130," that represents the anisotropic material behavior. Elastic fiber-matrix damaging behavior, or elastoplastic behavior with damage, can be modeled by material "type 130." Different material properties can be defined for each layer, each requiring stiffness, strength, and damage progression data. For each layer the initial undamaged in-plane stiffness properties E_{11}, E_{22}, G_{11}, and v_{12} should be provided, for the calculation of the initial modulus matrix $\mathbf{C}_0$. A damage function d,[6] enables the representation of the degradation of the initial modulus matrix $\mathbf{C}_0$, when an initial undamaged phase is exceeded. The modulus matrix behaves according to:

$$\mathbf{C}(d) = \mathbf{C}_0(1 - d) \tag{8.10}$$

The damage function d is a scalar parameter that depends upon strain as

$$d(\varepsilon) = d_v(\varepsilon_v) + d_s(\varepsilon_s) \tag{8.11}$$

where d_v is the volumetric damage due to a volumetric equivalent strain ε_v, and d_s is the shear damage parameter due to a shear equivalent strain ε_s. For the simple case of one uniaxial test, the equivalent volumetric and shear damage values are defined as

$$\varepsilon_v = \varepsilon_{kk} = (1 - v_{12} - v_{13})\varepsilon_{11} \tag{8.12}$$

$$\varepsilon_s = [\tfrac{1}{2} e_{ij} e_{ij}]^{1/2} = (\varepsilon_{11}/\sqrt{3})(1 + v_{12} + v_{13} + v_{12}v_{13} + v_{12}^2 + v_{13}^2)^{1/2} \tag{8.13}$$

In Eqs. (8.12) and (8.13) ε_{kk} is the trace of the total strain tensor and e_{ij} are the components of the deviatoric strain tensor respectively. The scalar parameter ε_v represents the first invariant of the volumetric strain tensor, while the scalar ε_s represents the second invariant of the deviatoric strain tensor. The implemented damage law in PAM-CRASH assumes that the fracturing damage parameter d is zero for an equivalent strain between zero and ε_i (Fig. 8.5a). After the value ε_i is reached, the fracturing damage factor d grows linearly between the values ε_i and ε_1. Between ε_1 and ε_u the damage factor d grows linearly again, with a different slope. The damage parameters that correspond to the strains ε_i, ε_1, and ε_u are d_i, d_1, and d_u respectively, where d_u is the stage at which the ultimate damage is reached. The elasticity

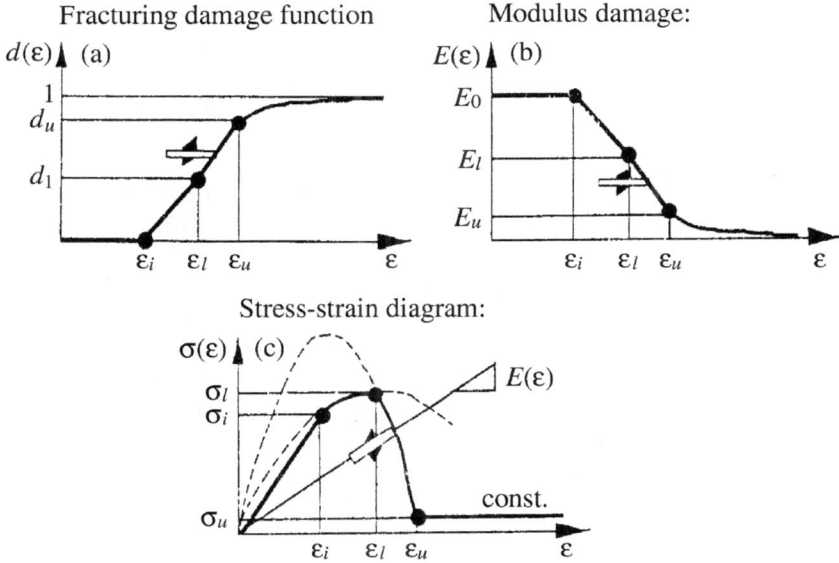

Figure 8.5 Fracturing damage function, modulus degradation and stress–strain diagram.[6]

modulus is assumed to degrade according to Fig. 8.5b and is related to uniaxial data according to Fig. 8.5c.

The calibration of the material damage models can be performed either using the tension–compression stress–strain curves to introduce only volumetric damage, or using the shear stress–strain curves to introduce only shear damage. In the former case, the slopes can be calculated from the tension–compression uniaxial data as

$$E_{v0} = \frac{\sigma_i}{\varepsilon_{vi}} \qquad E_{vl} = \frac{\sigma_l}{\varepsilon_{vl}} \qquad E_{vu} = \frac{\sigma_u}{\varepsilon_{vu}} \tag{8.14}$$

The volumetric damage values are then

$$d_{vl} = 1 - \frac{E_{vl}}{E_{v0}} \qquad d_{vu} = 1 - \frac{E_{vu}}{E_{v0}} \tag{8.15}$$

The use of either only volumetric or only shear damage can lead to the successful representation of the experimental tension and compression data. However, this is not the case for the shear behavior, which is overestimated or underestimated if only volumetric or only shear damage is used. To address this problem, the damage models are calibrated starting with the calculation of the shear damage from the shear coupon tests. After selecting a set of values for the strains ε_i, ε_1, and ε_u, the slopes G_i, G_1 and G_u and the corresponding shear damage parameters d_1 and d_u are calculated. Different combinations of ε_i, ε_1

and ε_u should be tried to achieve a good prediction of the τ–γ curves for each material. The results obtained, using shear damage only, have shown an enormous overestimation of the tension–compression strengths, when composite fabrics are modeled. This happens because the composite fabrics have a completely different behavior in tension–compression (ε_{max} is between 1% and 3%) and shear (γ_{max} is between 10% and 18%). For this reason, volumetric damage parameters should be introduced afterwards, to cut off the final strength and match the tension–compression data. Although the use of this volumetric damage enables better representation of material behavior, sometimes it can be unsuccessful in the simulation of fabric composites. In the simulations that follow, the structures made from composite fabrics were modeled using material damage models based on shear tests and shear damage only, which theoretically overestimates the tension–compression behavior but has led to more reasonable numerical results. The developed material models are checked by a one-element FE model in pure tension and pure shear loading.

In sandwich structures the bonded interfaces should be modeled using a tied contact algorithm. The algorithm coded in PAM-CRASH as contact "type 2" is used in the simulations that follow. This algorithm requires the input of normal N_s and shear T_s strengths of the bond and enables the contact failure, after the contact force of the tied nodes is exceeded. The failure occurs[6] when

$$\left(\frac{N}{N_s}\right)^2 + \left(\frac{T}{T_s}\right)^2 \geq 1 \qquad (8.16)$$

For the modeling of the contact between the structure interfaces, the self-contact algorithm, coded in PAM-CRASH as contact "type 26" is applied. This contact algorithm automatically searches for elements contacting each other and introduces an internal contact force between these elements.

The simulation of tools can be represented via rigid walls. The rigid walls can have various shapes to sample the tool geometry. Finite or infinite mass can be assigned to the tools, while a velocity curve can be introduced.

8.4 Case Studies on the Crash Failure Behavior of Novel Composite Aircraft Structures

The experience developed in impact dynamics research on metallic aircraft structures tends to expand into the field of composites. In the United States, major efforts have been made since the early 1970s by

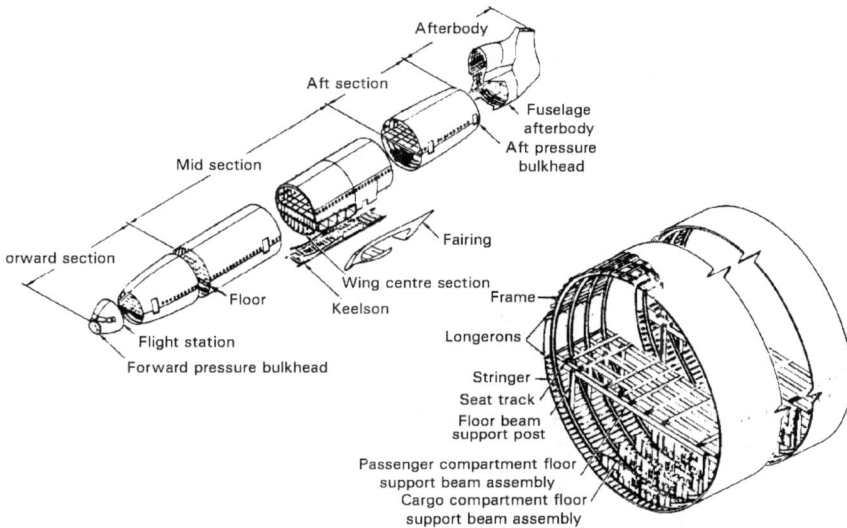

Figure 8.6 Internal structure of a typical civil aircraft fuselage.[1]

NASA as well as aircraft companies and research institutes. The research has been focused on the understanding of behavior and failure mechanisms, prediction of impact loads under different crash types, investigation of the energy-absorption mechanisms, and simulation of the impact structural response. A typical composite impact dynamics research program should comprise the following elements:

1. The development of a database for understanding the behavior, responses, failure mechanisms, and general loads associated with both conventional and innovative design concepts.

2. Analytical studies and development of numerical simulation tools relating to composite structures.

3. Studies in scaling of composite structures under static and dynamic loads.

4. Full-scale tests of composite structures to verify the performance of structural concepts.

The target of the research activities is to obtain improved crashworthy designs and meet the performance and energy absorption requirements. From a structural point of view, crashworthiness improvements should be performed both in the fuselage structure and in the subfloor structure. A typical transport fuselage structure, shown in Fig. 8.6,[1] is composed of frame components. The design of energy-absorbing composite frames is the fundamental issue for the development of the fuselage structure.

In parallel, various crashworthiness composite components, i.e. beams, panels, and cruciforms, should be developed for implementation in subfloor boxes of helicopter and aircraft fuselages as energy-absorbing elements. In the following, typical structural elements for utilization in subfloor designs are studied. The results of the simulations are compared with the corresponding experimental results of the structural subcomponents. However, the first example presented comes from the car industry and is a typical box-beam composite structure.

8.4.1 Simulation of the mass drop test of a composite box beam

The component analyzed in this section is a composite box-beam used in the front end structure of a prototype passenger car. The geometrical and material data are taken from reference 9. The beam is 1 m long and has a honeycomb construction. The geometry and lay-up of

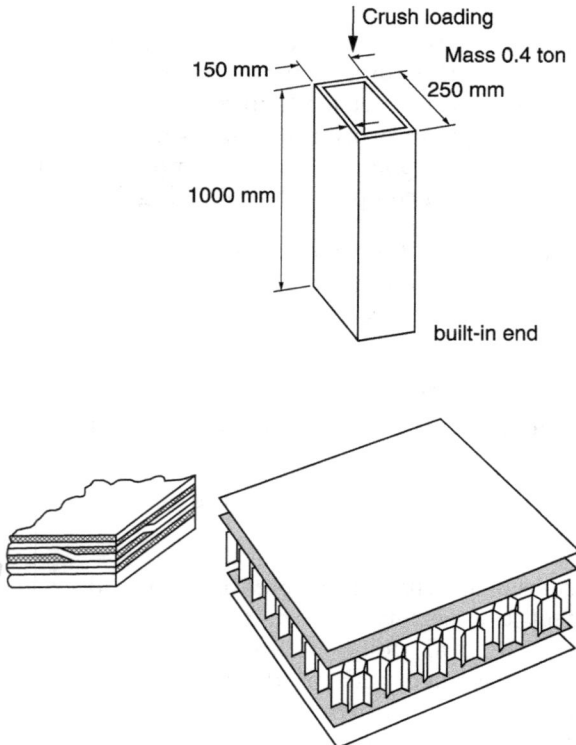

Figure 8.7 Geometry and material data of the composite box-beam.

the composite laminate are shown in Fig. 8.7. The beam is clamped at one end vertically and a mass of 0.4 ton drops on the other end, centrally and parallel to the longitudinal axis of the column.

The materials used for the construction of the sandwich facings of the beam are high-stiffness and high-strength unidirectional Tonen FT500 carbon/epoxy layers, a layer of DuPont aramid/epoxy cloth at (0/90) lamination, and a PAN-based T300 carbon/epoxy fabric of (45/−45) lamination. The honeycomb core consists of aramid honeycomb material. For the numerical simulation of the mass drop test, a finite element box-beam model is developed, comprising 1680 solid elements combined with PAM-CRASH material "type 31," which represent the aramid honeycomb core, while the inner and outer honeycomb facings are modelled using 3360 layered shell elements combined with material "type 130." Material damage models are developed for all the involved laminates according to the procedures described in Section 8.2. The critical points in the present simulation are (a) exact material representation by correctly calibrated damage models, which is always a critical point in impact numerical simulations, and (b) the slight random eccentricity that needs to be introduced to the coordinate values of all the nodes of the FE mesh, because the load is exactly axial. This latter point can influence the value of the peak force, as will be described later.

The simulation is performed for the first 6 milliseconds of the experiment, which is the most important period for the deformation of the column. In Fig. 8.8 the calculated deformed shapes of the box-beam at different times are shown. The rigid wall force versus time

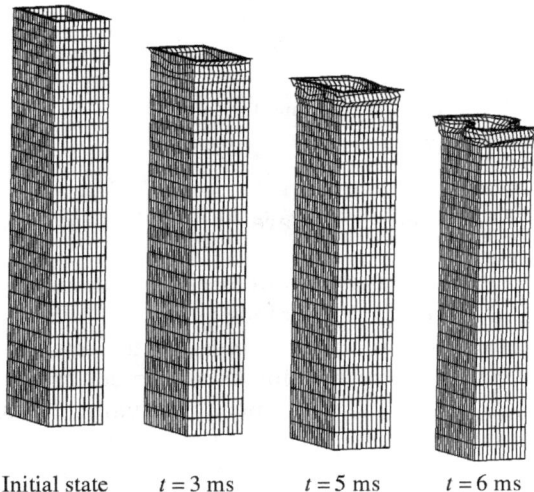

Initial state $t = 3$ ms $t = 5$ ms $t = 6$ ms

Figure 8.8 Calculated deformed meshes of the box-beam at different time intervals.

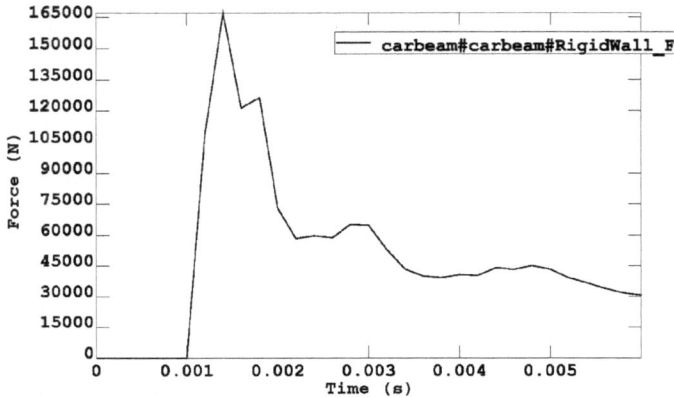

Figure 8.9 Calculated rigid wall force versus time for the box-beam mass drop test.

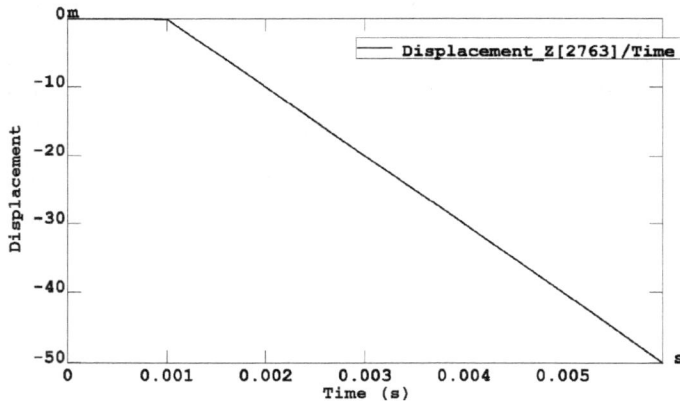

Figure 8.10 Calculated rigid wall displacement versus time for the box-beam mass drop test.

and the rigid wall displacement versus time are plotted in Figs. 8.9 and 8.10, respectively.

A comparison between the calculated rigid wall forces and displacement of Figs. 8.9 and 8.10 and the measured values of Fig. 8.11 shows generally good agreement. However, at the beginning of the simulation (at $t = 1.2$ ms), an initial peak force value is observed in the calculated curve that is much higher than the measured value. This is most likely because of the imperfections of the real structure, which are to an extent introduced in the simulation concept by inducing random eccentricities to the coordinate values the FE nodes, as

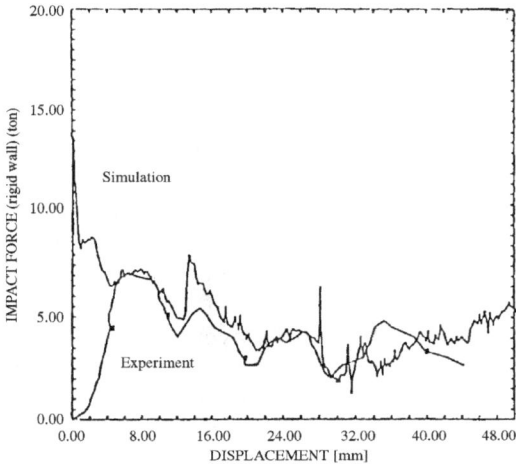

Figure 8.11 Measured rigid wall force versus time of the box-beam mass drop test in comparison to calculated values.[9]

discussed above. The same problem seems to occur in the numerical results of reference.[9]

8.4.2 Simulation of a static crush test of a sine-wave beam

Originally, composites were considered as poor energy-absorbing materials, as mentioned earlier, because of their low strain to failure characteristics. The ability to absorb energy could be improved partly through material improvements, in order to enable activation of progressive failure via fiber pull-out, matrix cracking, or delamination, However, the most significant improvements could only be achieved through innovative design of the structural components. According to References 10 and 11, the sine-wave beam concept is the most efficient structural design concept yet developed when using hybrid lamination techniques, as it combines high load-carrying capacity in shear and high energy absorption capability in compression. The only alternative shape that has higher ability to absorb energy is cylindrical, but often this cannot be used as structural element.[11]

The sine-wave beam takes its name from the shape of its web, which actually is a combination of circular sections in a corrugated or sine-wave-like shape. To prevent shear buckling of the web, a deep sine-wave angle of about 120° is chosen.[12] The advantage of sine-wave beams over other configurations is that they have some hoop constraint, although it is limited due to the cyclic inversion of the radius. The most common application of the sine-wave beam is in the composite underfloor structures of aircraft and helicopters.

The materials used in the construction of sine-wave beams are hybrid laminates consisting of fabrics at ±45° and unidirectional layers at 0°. In Fig. 8.12 the geometric and material data of the subfloor sine-wave beam investigated in the present section is shown.[13] The web of the beam has a length of 480 mm and a width of 224 mm, while its flange width is 60 mm. The outer plies of the laminate are made of carbon-aramid/epoxy hybrid fabric and the inner plies of carbon/epoxy UD tape. The ±45° layers are for shear stiffness, while the 0° layers are needed for energy absorption in a crash. A trigger mechanism at 35 mm from the bottom of the web is applied to reduce the peak loads without affecting the energy absorption. The most suitable trigger mechanism for these hybrid beams is an eccentricity at the outside of the web laminate, created by the inclusion a strip or foil and a 5 mm gap in the unidirectional core.[11]

The sine-wave beam described above is tested in vertical compression.[12] All the data for the static crush test are obtained.[13] For the

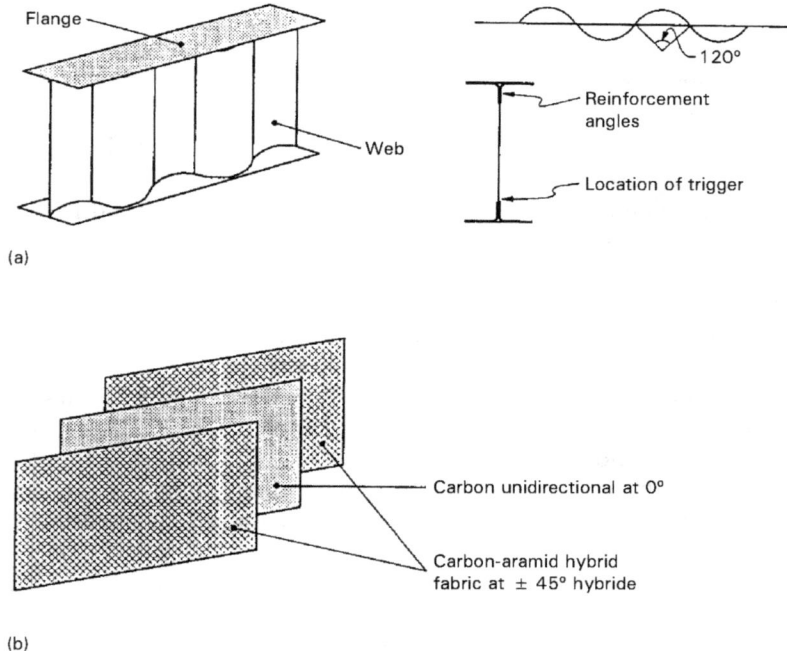

(a)

(b)

Figure 8.12 Geometry and materials of the sine-wave beam: (a) sine-wave beam concept; (b) web lay-up.[13]

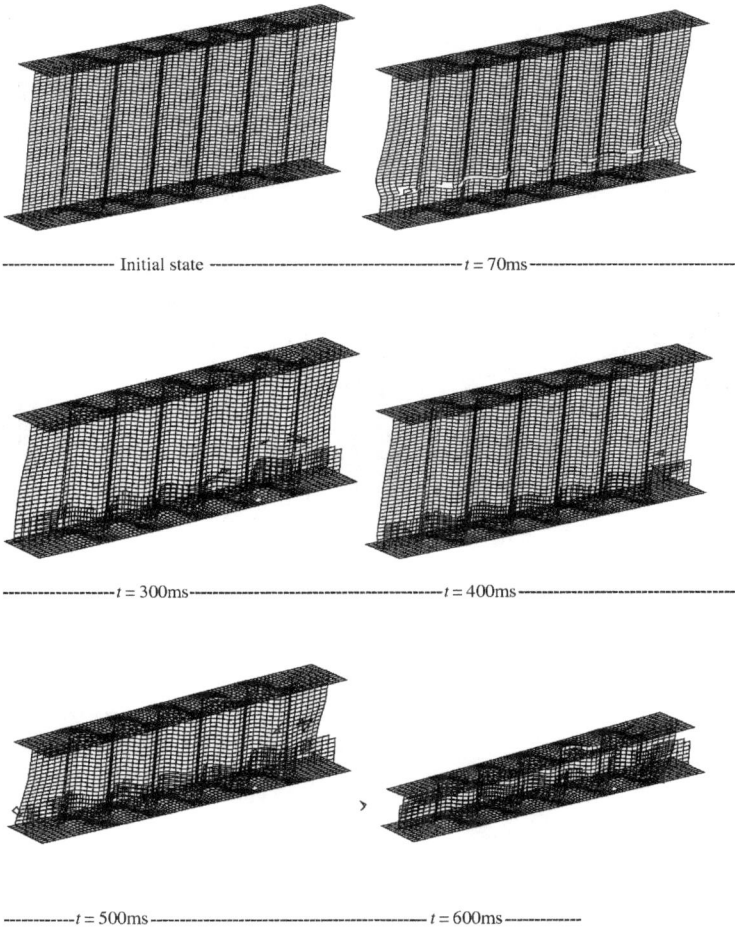

-------------------- Initial state --$t = 70\text{ms}$--------------------------------

--------------------$t = 300\text{ms}$--$t = 400\text{ms}$--------------------------------

------------$t = 500\text{ms}$--- $t = 600\text{ms}$-------------

Figure 8.13 Deformed shapes of the sine-wave beam at various time intervals.

numerical simulation of this panel, the FE mesh developed comprises 3018 layered shell elements (5183 nodes), combined with material type 130. The trigger mechanism, which is located 35 mm from the bottom flange, is modeled by introducing dummy plies of reduced elasticity modulus in the lay-up of the corresponding elements. The role of the trigger is to introduce an eccentricity in this area and cause failure initiation at an earlier state and from this location, reducing the peak force. It is very important to model the trigger area properly, otherwise global buckling of the web occurs and the simulation cannot predict the real failure modes. The simulation of the folding process is shown

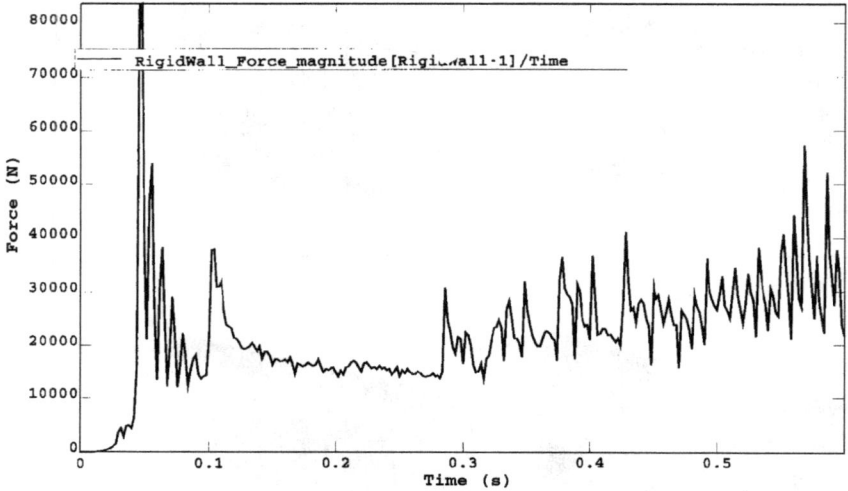

Figure 8.14 Calculated rigid wall force of the sine-wave beam test.

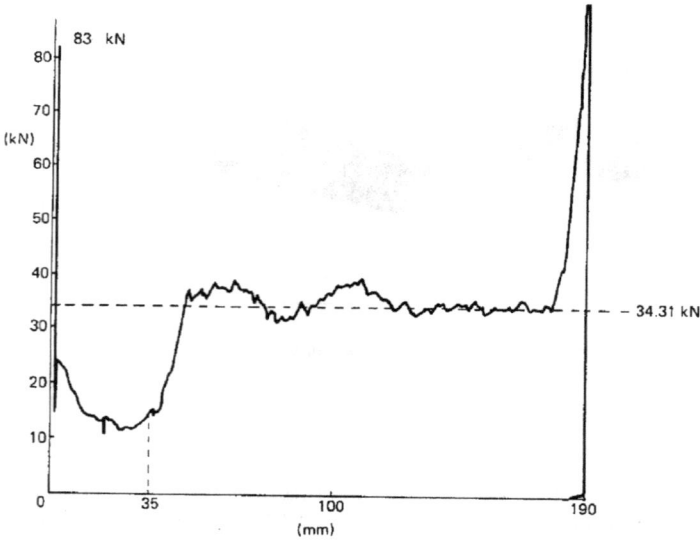

Figure 8.15 Measured impactor force versus impactor displacement.

in Fig. 8.13, where the deformed shapes of the beam are plotted at various time intervals. The calculated rigid wall force versus time is plotted in Fig. 8.14, while the measured tool force versus tool displacement is shown in Fig. 8.15. The predicted peak force of 83 kN corresponds to failure initiation at the trigger area and is equal to the measured value. After that point, the broken web touches the bottom flange and starts to fold until its final collapse. During this stage the calculated value of the mean force is found to be between 22 kN and 30 kN, which is close to the measured mean force value of 34.31 kN. The oscillation of the calculated force value is due to the loss of contact between the web and the lower flange just after the element's elimination.

8.5   Studies on the Impact Behavior of the "Tensor Skin" Concept

The tensor skin concept, as stated in Reference 14, was developed to improve the crashworthiness of helicopters with fuselage structures made of composite materials for case of impacts on water. The traditional composite skins may fail early by transverse pressure loads, but the tensor skin concept, when applied to the bottom skin and sandwich panels of a helicopter, is able to transfer the water pressure loads to the substructure and force it to absorb energy in the crushing mode for which the substructure is designed.

The initial development of the tensor skins started with the work,[15] in which one-dimensional tensors consisting of a folded sheet were loaded with a transverse force. The folded sheet had different geometries, as shown in Fig. 8.16.

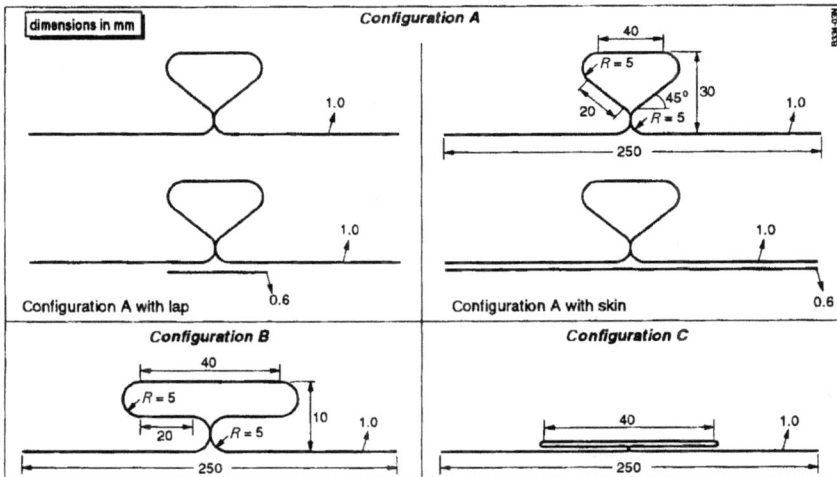

Figure 8.16   Configurations of 1-D tensor elements.[5]

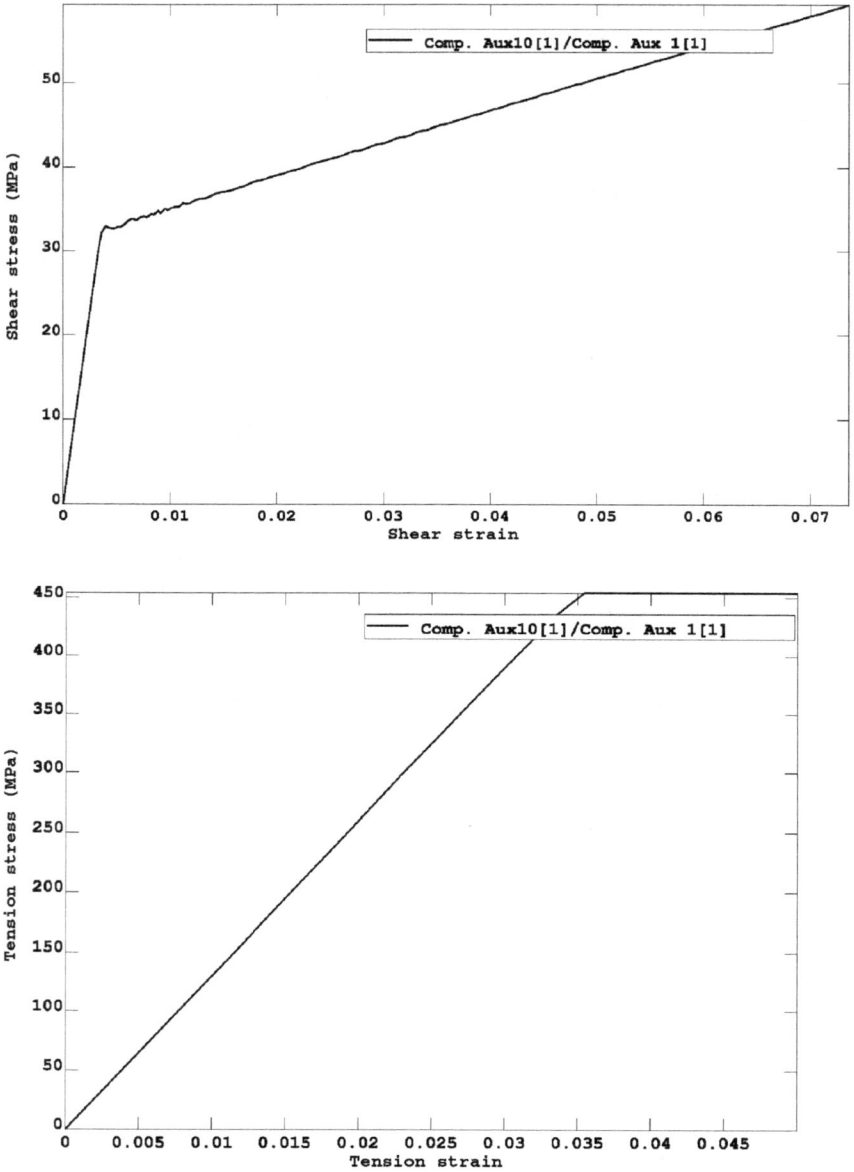

Figure 8.17   Normal and shear stress–strain response of Dyneema fabric material.

At a certain force the sheet unfolds and the load increases again, until the final failure occurs, which is dictated by the strength of fibers. In Reference 15, different materials are tested and it is shown that the ordinary composite material systems have an unsatisfactory behavior. For this reason, 1-D "tensor skin" strips made of polyethylene (PE) fibers embedded in epoxy matrix (Dyneema layers) have been developed.[15] The advantage of the Dyneema layers it that they allow "strain to failure" limits much higher than the common 1–2% of the ordinary composite systems. Therefore, when the strip is loaded in tension or bending, the beam unfolds and deflects by forming "plastic hinges" before it stretches and fails in tension. This unfolding mechanism allows the absorption of impact energy.

8.5.1 Crushing behavior of the one-dimensional "tensor skin" strip

In Reference 15, static crush bending tests are performed for 1-D tensor strips of length 250 mm, width 50 mm, thickness of 1 mm and lay-ups $[0/90]_3$ and $[\pm 45/90/\pm 45]$. These tests are simulated in the present section, using the finite element method. The PAM-CRASH code is used to develop a FE mesh of the 1-D strip, consisting of 3300 layered shell elements combined with material "type 130." The damage models developed for the Dyneema layers are developed according to the procedure described in the previous sections and are based on material properties.[17] To check the developed damage model, a FE model of the Dyneema layer, consisting of only one element, is imposed in pure tension and pure shear loading. The calculated normal and shear stress–strain curves from the PAM-CRASH simulation are shown in Fig. 8.17.

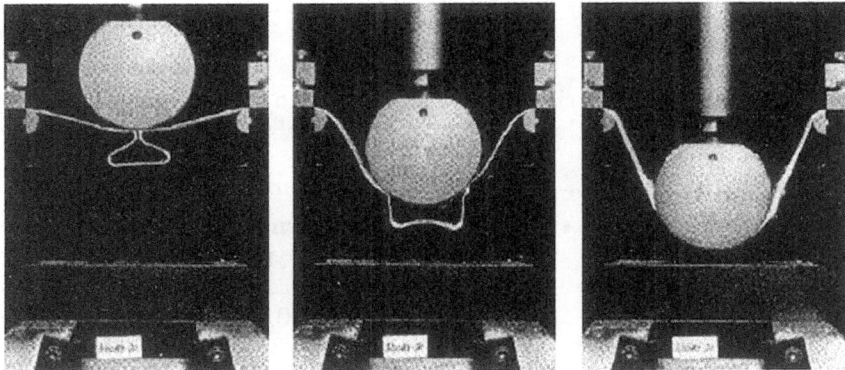

Figure 8.18 Deformed configurations of the 1-D tensor strip.[15]

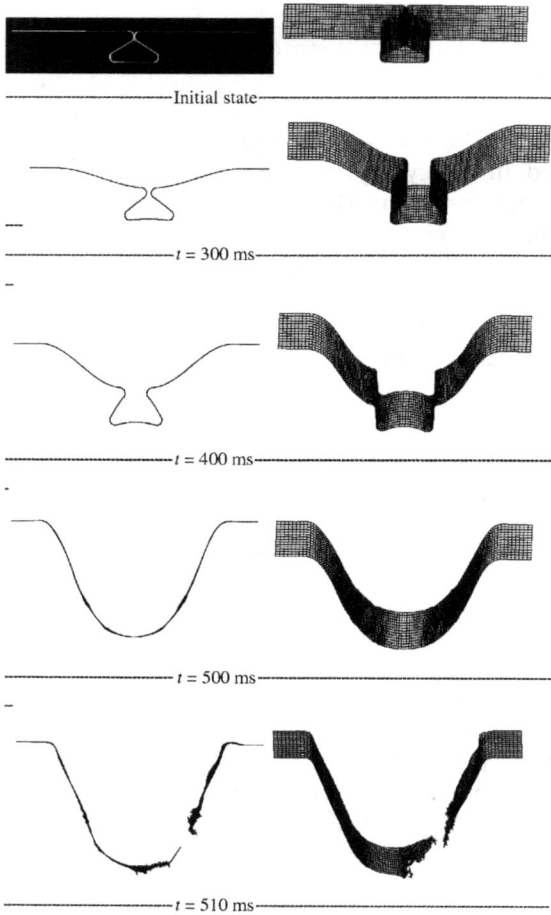

Figure 8.19 Calculated deformed shapes of the 1-D tensor strip.

Two semi-spherical rigid walls are used to model the clamping system. In Fig. 8.18 the unfolding process that occurs during the test is shown.

The calculated deformed shapes for the $[0/90]_3$ lay-up at various time intervals, are plotted in Fig. 8.19. Comparing Figs. 8.18 and 8.19, it is observed that the unfolding process of Dyneema is successfully simulated. The calculated rigid wall force versus time is plotted in Figs. 8.20 and 8.21 for the $[0/90]_3$ and $[\pm45/90/\pm45]$ lay-ups, respectively. One important observation is that for both lay-ups the rigid wall force is less than 500 N during the whole unfolding process and increases sharply only before the final failure. Therefore, most of the

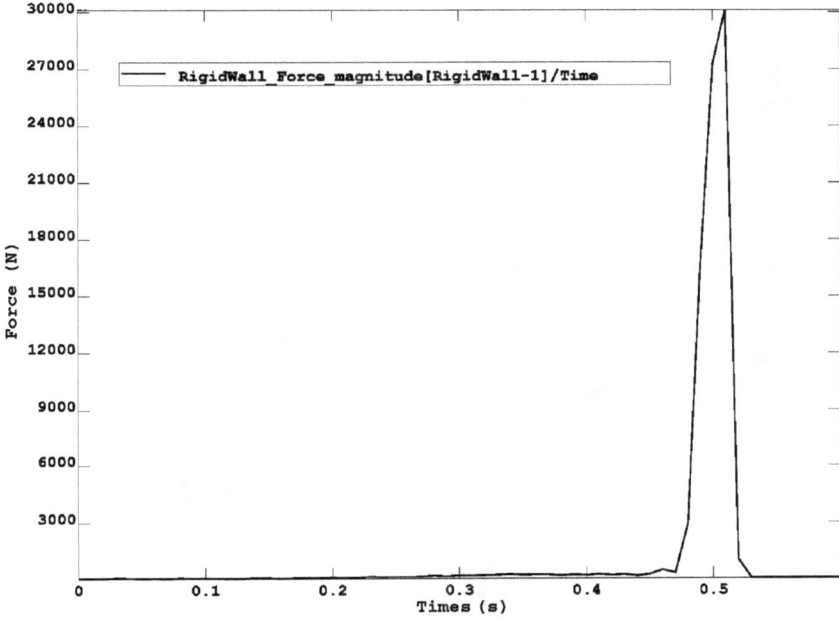

Figure 8.20 Calculated rigid wall force versus time of the [0/90]$_3$ lay-up.

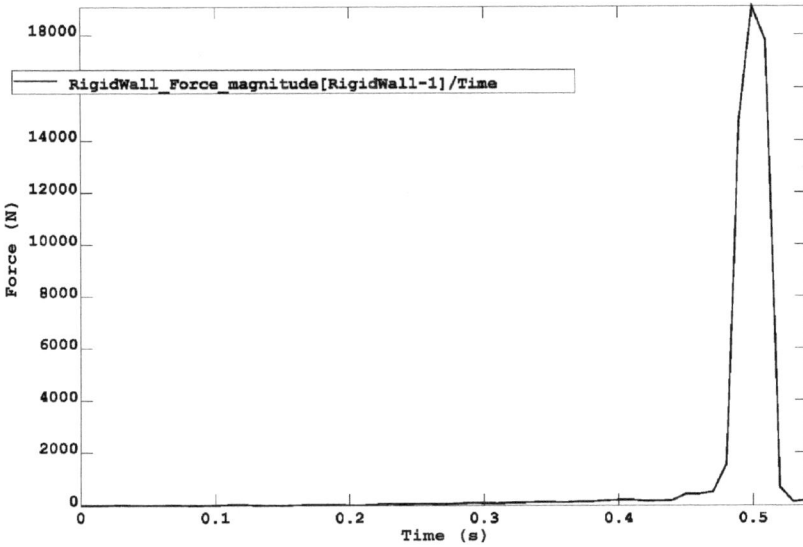

Figure 8.21 Calculated rigid wall force versus time of the [±45/90/±45] strip.

energy absorption occurs in the final deformation stage. The predicted peak forces were measured at 30.83 kN for the $[0/90]_3$ and 19.25 kN for the $[\pm 45/90/\pm 45]$ lay-up, respectively.[15] The calculated peak forces have values within 10% of the measured ones.

8.5.2 Simulation of the two-dimensional tensor skin strip static crush test

The two-dimensional "tensor skin" strip considered in this section was developed[15] to provide a more realistic structure for practical applications that could fail progressively and enable absorption of higher amounts of energy. The two-dimensional "tensor skin" strip is a sandwich construction with the cross section shown in Fig. 8.22 and

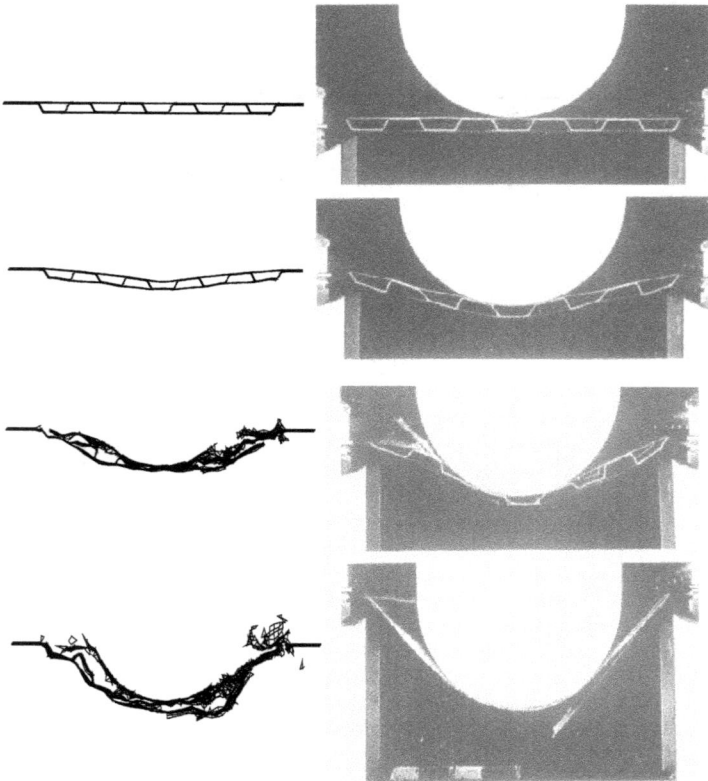

Figure 8.22 Calculated (left) and actual (right) deformed shapes of the 2-D tensor skin strip static crush test, at various time intervals.

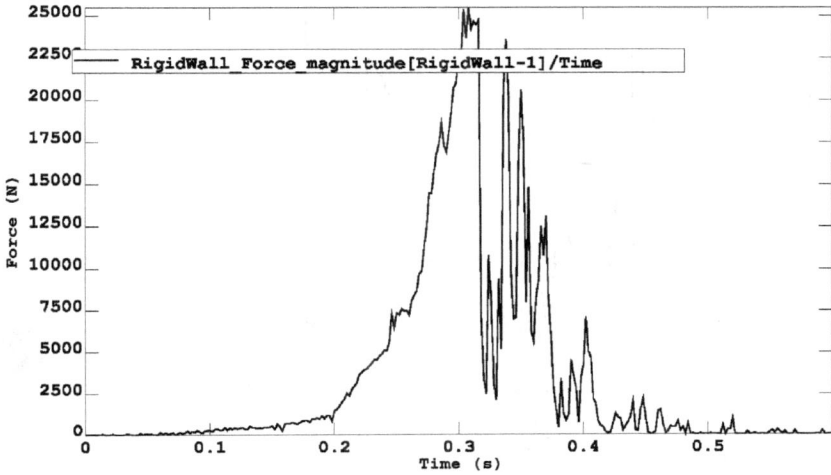

Figure 8.23 Rigid wall force versus time of the 2-D tensor strip static crush test.

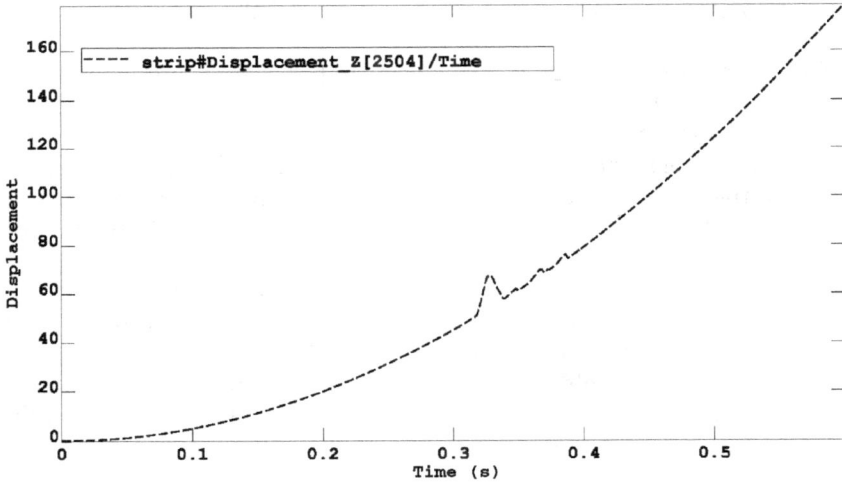

Figure 8.24 Rigid wall force versus time of the 2-D tensor strip static crush test.

width of 135 mm. The static crush test of the 2-D tensor strip is reported.[15] In Fig. 8.22 the calculated (left) and the actual (right) deformed shapes of the 2-D "tensor skin" strip static crush test are presented at various time intervals.

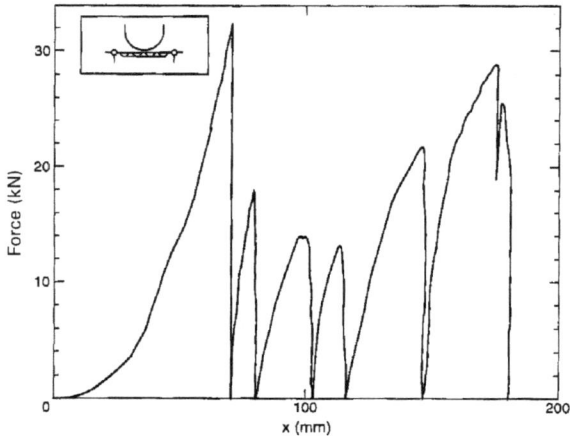

Figure 8.25 Measured tool force versus time of the 2-D tensor strip static crush test.[15]

The rigid wall force and displacement versus time of the 2-D tensor skin strip static crush test are shown in Fig. 8.23 and 8.24, respectively.

From Figs. 8.23 and 8.24 it is deduced that the outer and inner faces fail at 0.4 s, which corresponds to a rigid wall force of 25.2 kN and rigid wall displacement of 62 mm. The time for the final panel failure is 0.49 s, which corresponds to a rigid wall displacement of 124 mm. In Fig. 8.25 the measured tool force versus tool displacement is plotted,[15] where the outer and inner faces start to fail at 31 kN and 70 mm tool displacement. Comparing Figs. 8.23 and 8.24 with Fig. 8.25, a good correlation is observed between the calculated and measured tool force and displacement from the time of the failure initiation up to $t = 0.49$ ms. However, in the simulation, the Dyneema layers fail before their total unfolding (see also Fig. 8.22), at a maximum tool displacement of 124 mm, which is lower than the experimentally measured tool displacement of about 180 mm.

8.5.3 Simulation of the static crush tests of the three-dimensional tensor skin panels

A more practical realization of the "tensor skin" concept is the three-dimensional tensor panel, which is a 540 × 540 mm square panel with the cross section shown in Fig. 8.22. A photograph of the panel is shown in Fig. 8.26. The outer face is made from three aramid/epoxy layers with [±45/0/±45] lay-up and the inner "hat" shaped face is

Figure 8.26   The 3-D "tensor skin" panel.

made of three hybrid carbon-aramid/epoxy layers with the same lay-up. The corrugated core consists of three Dyneema layers with different orientation. To estimate the energy absorption capability of the structure, static crush experiments on two panels, one with $[0/90]_3$ Dyneema lamination, coded "Mid0St," and one with $[\pm45]_3$ Dyneema lamination, coded "Mid45St," were performed and reported.[15]

For the simulation of these experiments, a FE model of the 3-D panel is developed. All the data for the modeling have been taken.[18] The FE mesh comprises 3300 elements (3837 nodes). Both the normal N_s and shear T_s strengths of the tied interfaces are assumed to be 5 kN. These values are calculated from the mechanical properties of the bond (Argomet F13) and the average area of the elements of the FE mesh. The calculation of these strength values is a very critical point of the analysis. The strength values determine the introduction of one of the basic failure modes of the system, i.e., the debonding of the three different layers of the panel. If this debonding occurs early, then all the load is carried by the layers individually. This means that the total bending stiffness is seriously reduced, which results in premature collapse of the whole system. The element elimination strains are assumed to be 0.6 for Dyneema and 0.2 for the aramid and carbon/aramid fabric material systems.

8.5.3.1   Results of the simulation of the "Mid0St" static crash test.   The deformed shapes of the outer (left), middle (center), and inner (right) faces of the Mid0St static crash test, at various time intervals are

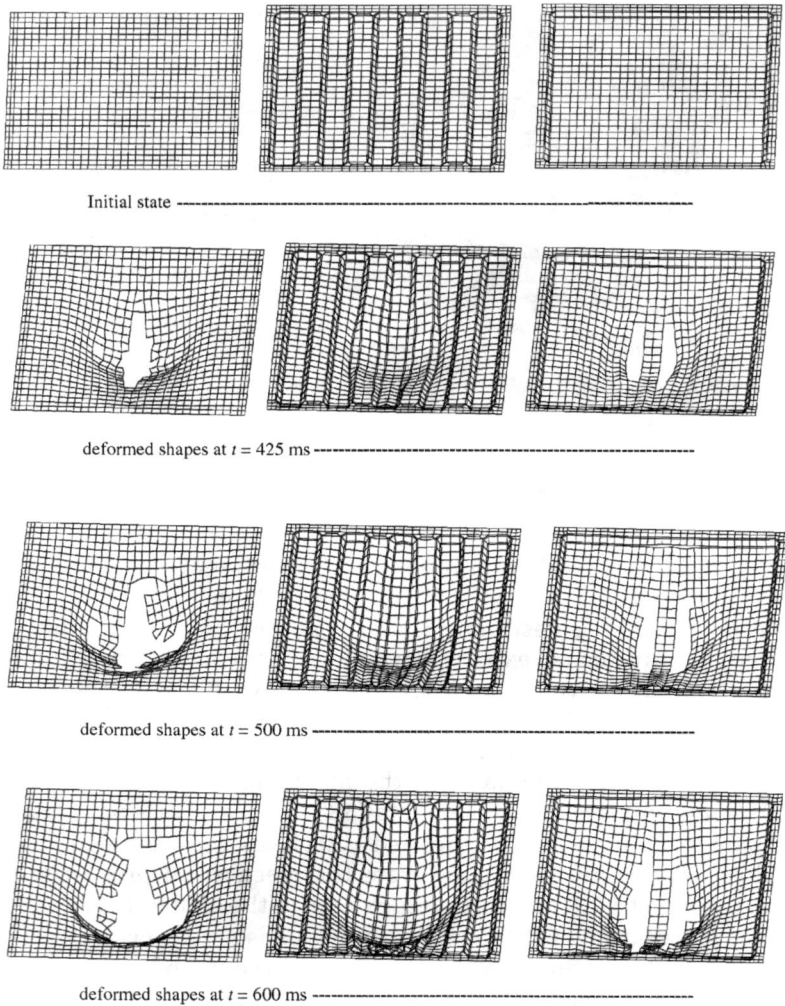

Initial state --

deformed shapes at $t = 425$ ms --

deformed shapes at $t = 500$ ms --

deformed shapes at $t = 600$ ms --

Figure 8.27   Deformed shapes of the outer (left), middle (center), and inner (right) faces, at various time intervals, of the "Mid0St" static crush test.

shown in Fig. 8.27. The calculated rigid wall force and displacement versus time are shown in Figs. 8.28 and 8.29, respectively. From these figures, the outer and inner faces fail at 0.38 s, which corresponds to a rigid wall force of 32 kN and rigid wall displacement of 38 mm. The time for the final Dyneema failure is 0.76 s, which corresponds to a rigid wall force of 69.8 kN and rigid wall displacement 148 mm.

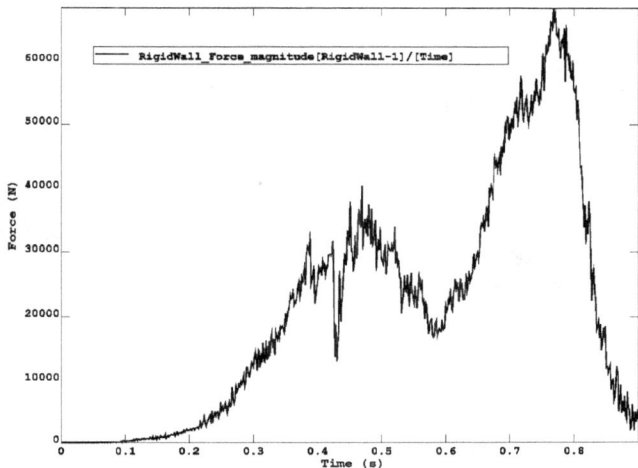

Figure 8.28 Rigid wall force versus time of the "Mid0St" static crush test.

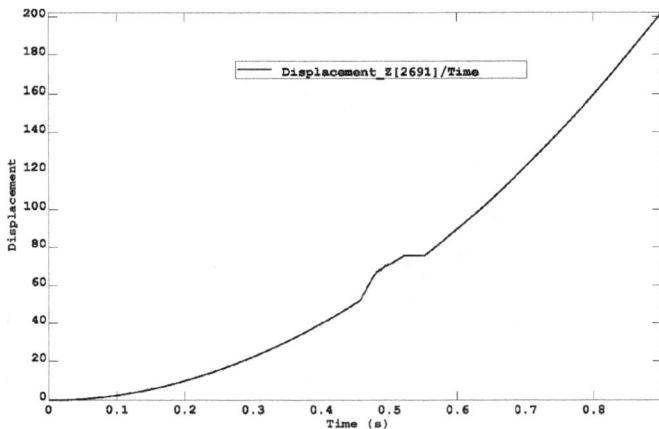

Figure 8.29 Rigid wall force versus time of the "Mid0St" static crush test.

In Fig. 8.30 the measured tool force versus tool displacement is plotted,[19] from which it is seen that the outer and inner faces fail at a tool force of 30.89 kN and rigid wall displacement of 40 mm, while the Dyneema layer fails at a tool force of 74.7 kN and rigid wall displacement of 140 mm. Comparing Figs. 8.28 and 8.29 with Fig. 8.30, a very

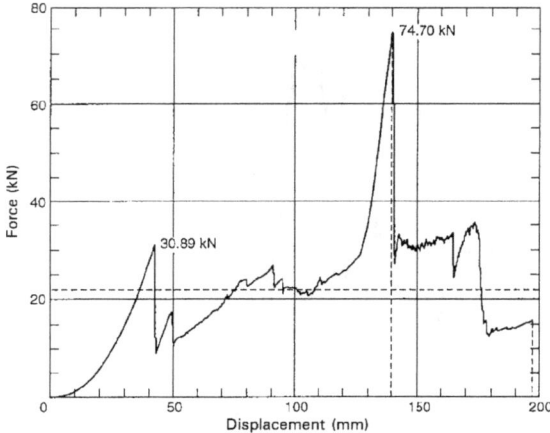

Figure 8.30 Measured tool force versus tool displacement of the "Mid0St" test.

good agreement is observed between the calculated and measured tool force and displacement.

8.5.3.2 Results of the simulation of the "Mid45St" static crash test. The calculated deformed shapes of the outer (left), middle (center), and inner (right) faces, of the "Mid45St" static crash test at various time intervals are shown in Fig. 8.31.

The calculated rigid wall force and displacement versus time are shown in Figs. 8.32 and 8.33, respectively.

From Figs. 8.31 and 8.32, the outer and inner faces fail at $t = 0.39$ s, which corresponds to a rigid wall force of 36 kN and rigid wall displacement of 39 mm. The time for the final Dyneema failure is 0.84 s, which corresponds to a rigid wall force of 149 kN and rigid wall displacement of 178 mm. In Fig. 8.34, the measured tool force versus tool displacement is plotted,[19] from which it is seen that the outer and inner faces fail at a tool force of 31.8 kN and rigid wall displacement of 42 mm, while the Dyneema layer fails at a tool force of 171 kN and rigid wall displacement of 149 mm. Comparing Figs. 8.31 and 8.32 with Fig. 8.34, a very good agreement is observed between the calculated and measured tool force and displacement, especially concerning the peak loads.

Finally, in Fig. 8.35 the failure of the outer and inner faces of the panel is shown. A comparison between Figs. 8.35 and 8.31 shows that the failure modes of the three panel layers are well simulated.

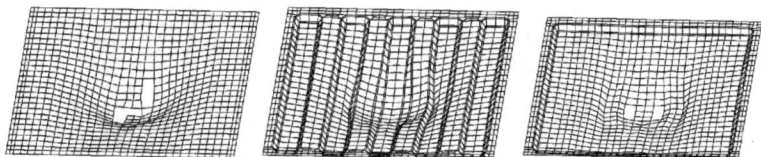

Deformed shapes at $t = 380$ ms

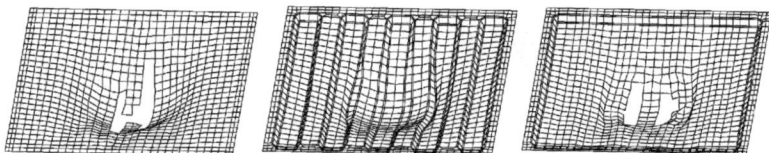

Deformed shapes at $t = 500$ ms

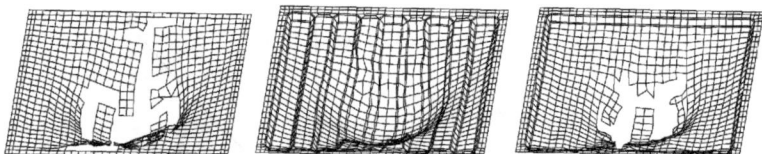

Deformed shapes at $t = 700$ ms

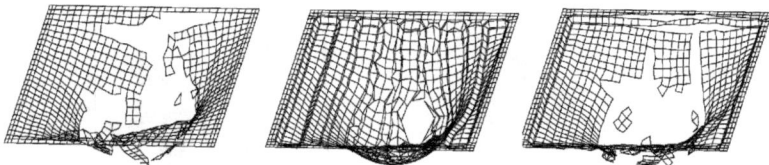

Deformed shapes at $t = 830$ ms

Figure 8.31 Deformed shapes of the outer (left), middle (center), and inner (right) faces, at various time intervals, of the "Mid45St" static crush test.

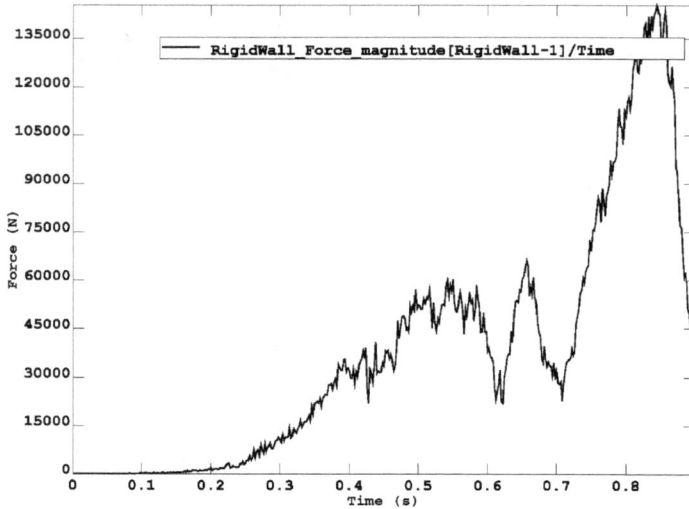

Figure 8.32 Rigid wall force versus time of the "Mid45St" static crush test.

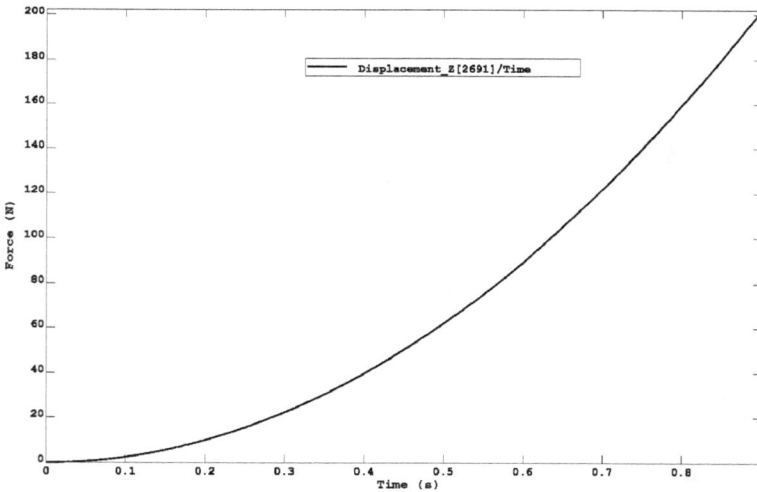

Figure 8.33 Rigid wall force versus time of the "Mid45St" static crush test.

8.5.4 Simulation of the mass drop tests of the tensor skin panel

A better and more representative understanding of the behavior of the square "tensor skin" panels described in previous sections, can be obtained from dynamic tests. For this reason, the panels are clamped at all edges and an impactor of specified mass drops on the panels

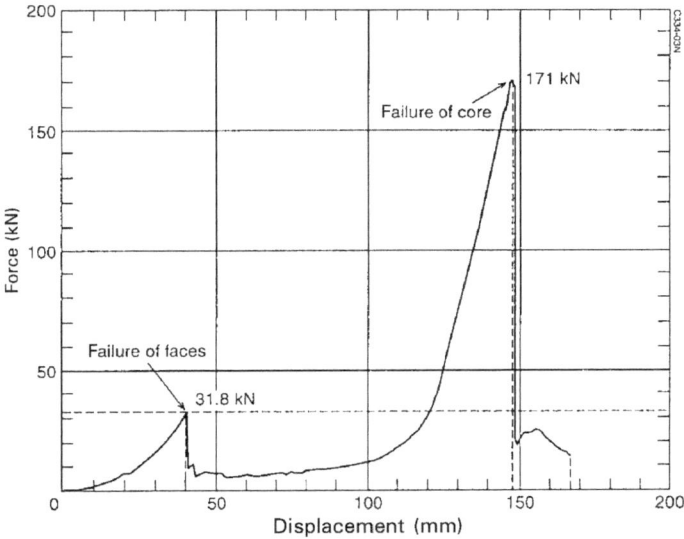

Figure 8.34 Measured tool force versus tool displacement of the "Mid45St" test.

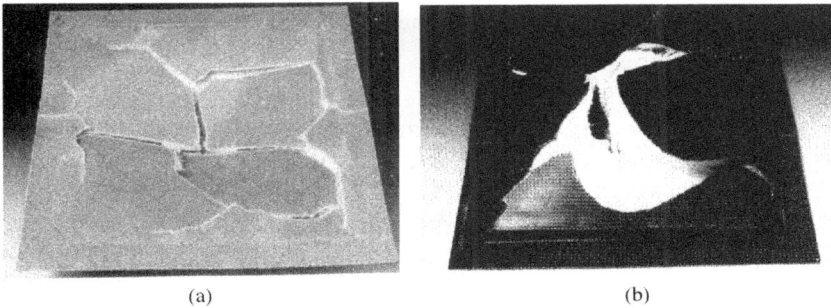

(a)   (b)

Figure 8.35 Failure of the outer (a) and inner (b) faces from the "Mid45St" test.

centrally, at a predefined initial velocity. The two dynamic tests are coded "TSDy2D" and "TSDy3D." These tests are simulated numerically in this section. All the experimental data required for the simulations are taken from.[8] The "TSDy2D" panel has two corrugated Dyneema layers of a total thickness of 0.74 mm and is impacted by a 59.55 kg mass at an initial velocity of 7.244 m/s. The "TSDy3D" panel has three corrugated Dyneema layers of a total thickness of 1.11 mm and is impacted by a 78.95 kg mass at a velocity of 7.534 m/s. The experimental set-up is shown in Fig. 8.36.

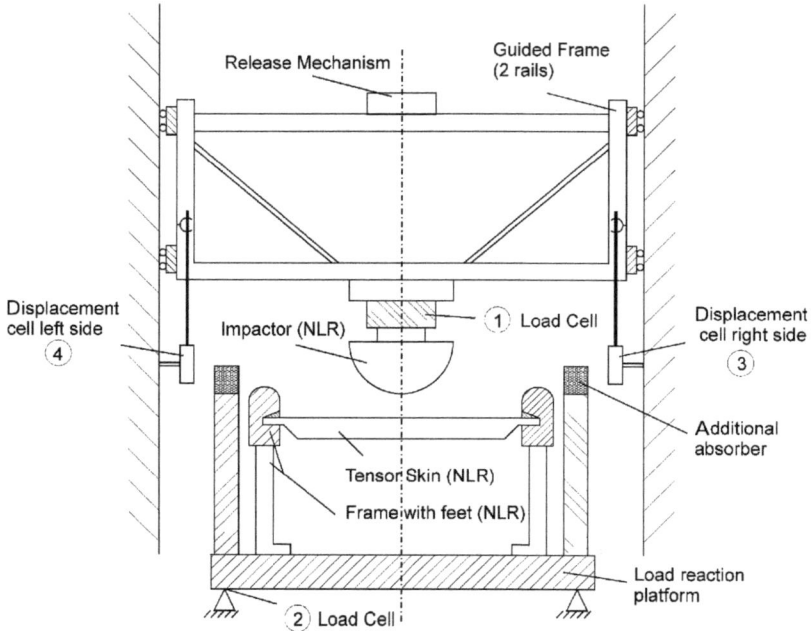

Figure 8.36 Experimental set-up for the dynamic mass drop tests.[8]

For the simulation of the two dynamically tested corrugated panels, the same FE mesh as developed for the simulations of the static crush tests is used. The impactors are simulated as spherical rigid walls, taking into account their corresponding masses and initial velocities. The connection between the three layers of the panel are considered as tied interfaces with failure. The normal N_s and the shear T_s strengths of the tied interface are assumed to be 3 kN, which is significantly lower than the values used in the simulations of the static tests, as it was observed that in the dynamic tests the interlayer bonding fails almost completely and quite early. As stated earlier, these values are very critical for the simulation, as one of the basic failure modes of the panel is directly related to these values. The element elimination strains are assumed to be 0.6 for Dyneema and 0.1 for the aramid and carbon/aramid fabric material systems. In the present dynamic simulations, the selection of the elimination strains is also very critical. If the elimination strains are selected at lower values than the real ones, elements that should be eliminated in a specific time interval will remain in the FE mesh, inducing unrealistic forces to other areas of the model. Self-contact (type 26) is considered for the whole structure.

The results of the simulations described in the following include calculated deformed shapes at various time intervals, calculated rigid wall (rw) forces versus time, and some of the corresponding experimental results, from.[8] The deformed shapes of the three panel faces in the "TSDy3D" dynamic test at various time intervals are shown in Fig.

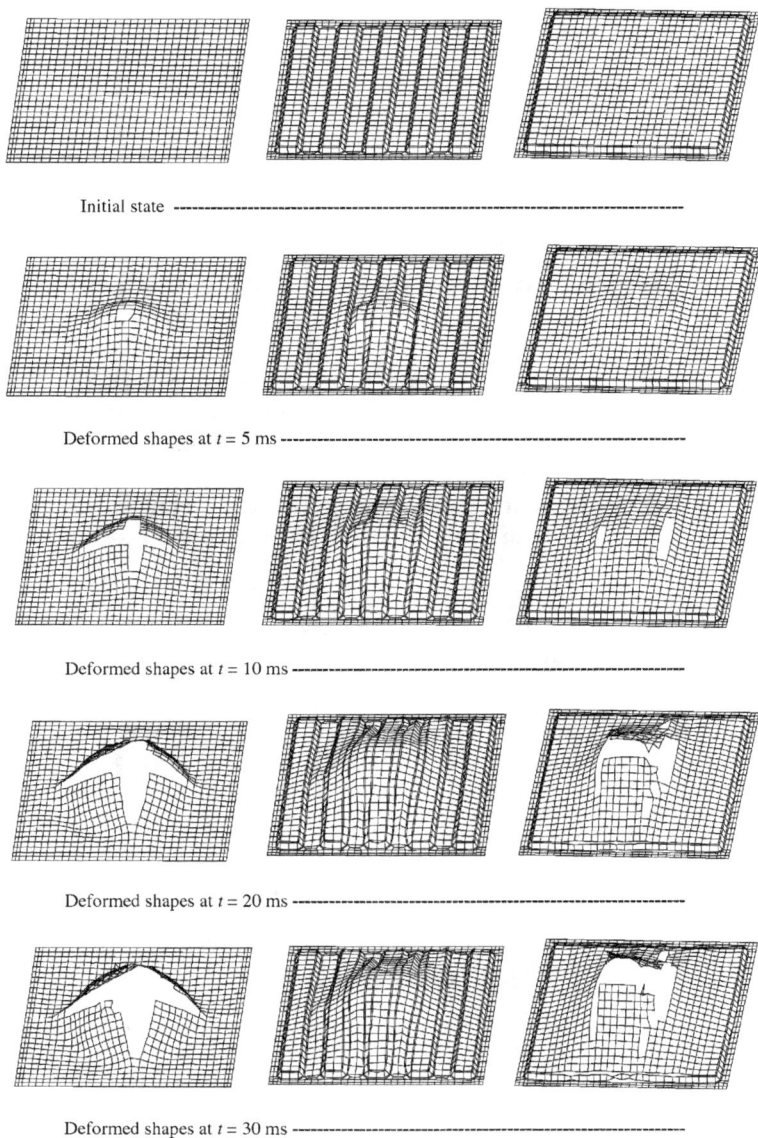

Initial state --

Deformed shapes at $t = 5$ ms --

Deformed shapes at $t = 10$ ms --

Deformed shapes at $t = 20$ ms --

Deformed shapes at $t = 30$ ms --

Figure 8.37 Deformed shapes of the outer (left), middle (center), and inner (right) faces, at various time intervals, of the "TSDy3D" dynamic test.

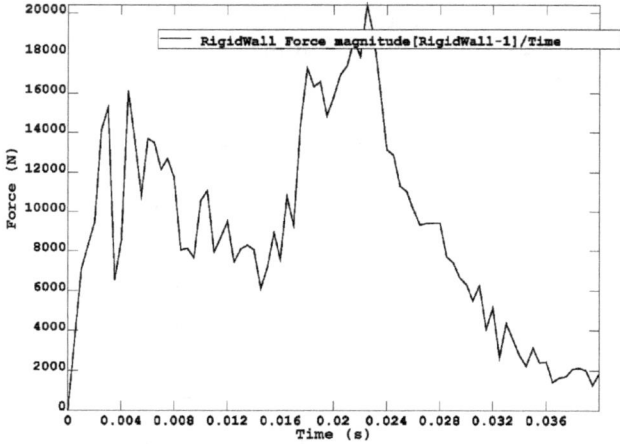

Figure 8.38 Rigid wall force versus time of the "TSDy3D" dynamic test.

8.37 and the rigid wall force versus time curve in Fig. 8.38. The experimentally measured tool force versus tool displacement for both the "TSDy2D" and "TSDy3D" tests are shown in Fig. 8.39. From a comparison between experimental and calculated results of the "TSDy3D" test, the following considerations arise: (1) The failure modes of the three layers of the "TSDy3D" test are well simulated.

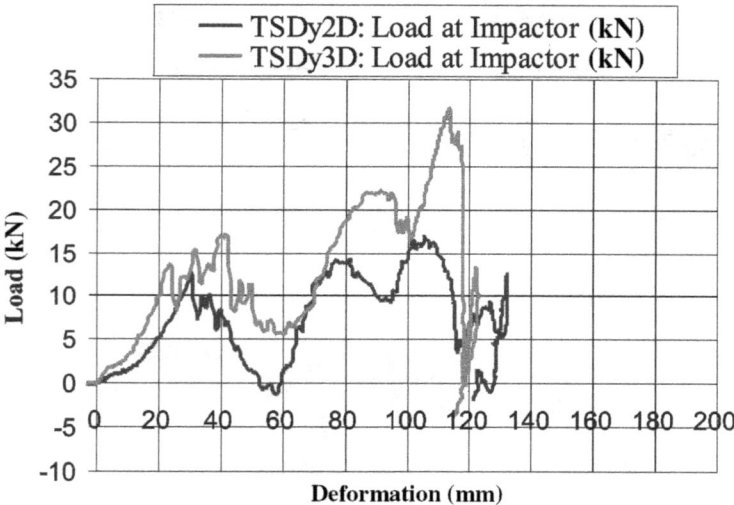

Figure 8.39 Experimentally measured rigid wall forces of the dynamic tests.[8]

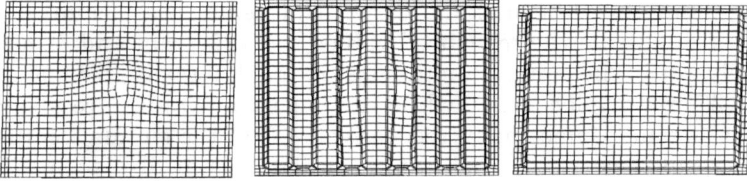

Deformed shapes at $t = 5$ ms

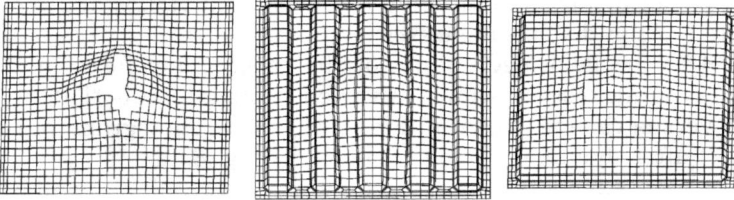

Deformed shapes at $t = 10$ ms

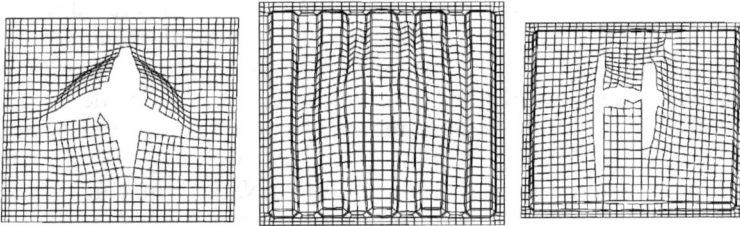

Deformed shapes at t = 20 ms

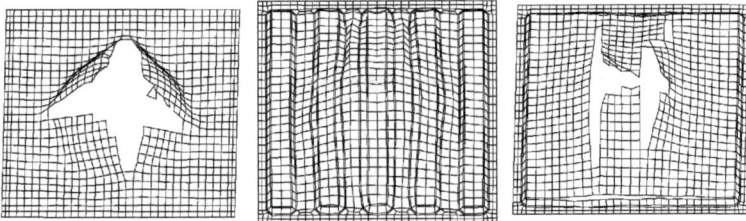

Deformed shapes at $t = 30$ ms

Figure 8.40 Deformed shapes of the outer (left), middle (center), and inner (right) faces, at various time intervals, of the "TSDy2D" dynamic test.

(2) In the simulation the maximum rw displacement is 96 mm and occurs at 25 ms, while the measured values are 122.6 mm and 30 ms, respectively. (3) In the simulation the first rw peak force is 15.3 kN, at 5.5 ms, at 37 mm displacement, while the measured values are about 17.5 kN, at 40 mm displacement, respectively. (4) In the simulation the second rw peak force is 17.5 kN, at 18 ms, at 82 mm displacement, while the measured values are about 22.5 kN, at 88 mm displacement, respectively. (5) In the simulation the third rw peak force is 20.5 kN, at 23 ms, at 92 mm displacement, while the measured values are about 31.7 kN, at 115 mm displacement, respectively.

In Fig. 8.40 the deformed shapes of the three panel faces in the "TSDy2D" dynamic test are shown at various time intervals. The rigid wall force versus time is shown in Fig. 8.41, and a photograph of the failed panel after dynamic testing is shown in Fig. 8.42.

Comparing the experimental with the calculated results, the following considerations arise: (1) the failure modes of the inner and outer layers of the "TSDy2D" test are generally well simulated. (2) The Dyneema layer fails partially in the test, which is not predicted in the simulation. (3) In the simulation the maximum rw displacement is 100 mm and occurs at 26 ms, while the measured values are 132.3 mm and 32 ms, respectively. (4) In the simulation the first rw peak force is 13 kN, at 2.5 ms, at 25 mm displacement, while the measured values are about 13 kN, at 30 mm displacement, respectively. (5) In the simulation the second rw peak force is 16.8 kN, at 6 ms, at 50 mm displacement, while the measured values are about 14.5 kN, at 78 mm displacement, respectively. (6) In the simulation the third rw peak

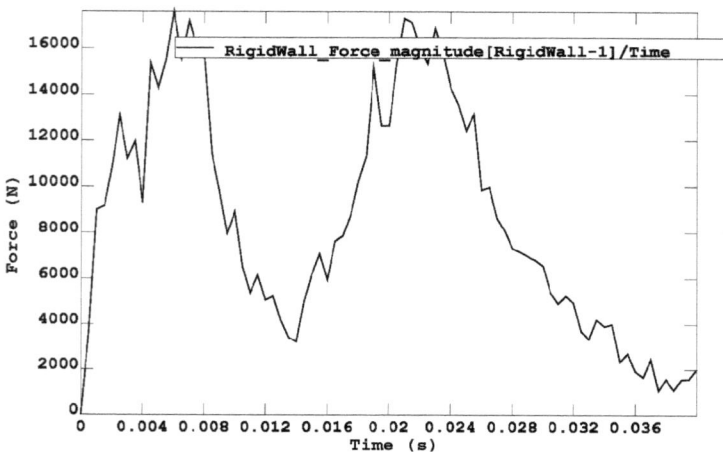

Figure 8.41 Rigid wall force versus time of the "TSDy2D" dynamic test.

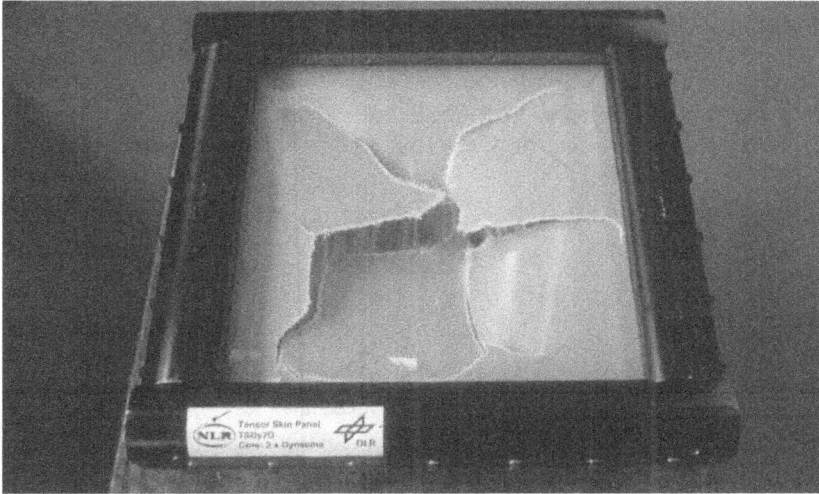

Figure 8.42 Failure mode of the outer face of the "TSDy2D" panel.

force is 16.5 kN, at 22 ms, at 98 mm displacement, while the measured values are about 17 kN, at 108 mm displacement, respectively.

8.6 Conclusions

It is very important to understand the structural behavior of composites under crash loads. To achieve this, a large experimental database is required that includes experimental static and dynamic tests of conventional and innovative composite structures. The innovative systems should meet the crashworthiness requirements, as well as demands of performance, integrity, and effectiveness. Numerical analysis tools are essential; they are used to study the energy-absorption characteristics and the scaling effects between full-scale behavior and tested models, and contribute to the verification procedure.

In the present work the numerical analysis is based in the finite element method. The PAM-CRASH FE code has been successfully applied for the simulation of experimental testing of various innovative crashworthiness components, such as box-beams, sine-wave beams and various practical realizations of the "tensor skin" concept. In general, the failure process of the simulated structures is successfully predicted. Good agreement is observed between the calculated and measured tool forces and displacements in most cases. It is important that in all cases the prediction of the real crushing modes is accurate, although it requires careful selection of the modeling parameters and sophisticated utilization of the modeling capabilities of the numerical codes. However, the exact representation of the behavior of the composite material is difficult using the available composite material damage models. In conclusion, the simulation approach followed can be considered as a supporting tool that could be helpful in the design and testing processes of innovative crashworthiness concepts.

References

1. Niu M.C.-Y. (1988) *Airframe Structural Design*, Conmilit Press Ltd.
2. Hansen J. S. and Tennyson R. C. (1988) Study of the dynamic behavior of stiffened composite fuselage shell structures, *Proceedings of the Energy Absorption of Aircraft Structures as an Aspect of Crashworthiness*, Luxembourg, AGARD.
3. Fenves S., Perrone N., Robinson A., and Schnobrich W. (1973) *Numerical and Computer Methods in Structural Mechanics*, Academic Press, New York, 558–563.
4. Thuis H., Vries H., and Wiggenraad J. (19) *Subfloor Skin Panels for Improved Crashworthiness of Helicopters*, NLR-TP 95082 U.
5. Swanson Analysis Systems (1996) *ANSYS Users Manual*, Rev. 5.1.
6. Engineering Systems International (1997) PAM-CRASH *Users Manual*.
7. Ubels L.C. (1997) *Behavior of Shear Panels*, NLR-TR 97534.
8. Kohlgrueber D. and Weissinger H. (1997) *D.4.1.3. Results of Dynamic Tests of Tensor Skin Panels*, DLR-IB 435-97/31.
9. Haug E. and De Rouvray A. (1993) Crash response of composite structures, in *Structural Crashworthiness and Failure*, (Jones N and Wierzbicki T., eds.), Elsevier Science.
10. Kindervater, C. M. (ed.) (1990) *Energy Absorption of Composites as an Aspect of Aircraft Structural Crash Resistance*, 4th European Conference on Composite Material, ECCM IV, Stuttgart.
11. Thuis H. and Wiggenraad J. (1992) *The Influence of Trigger Mechanisms and Geometry on the Crush Characteristics of Sine Wave Beams*, NLR contract report CR 92133 C.
12. Bark L. W., Cronkhite, J. D., Burrows L. T., and Neri L. M. (1988) Crash testing of advanced composite energy absorbing repairable cabin subfloor structure, American Helicopter Society Meeting, Virginia, October 23–27.
13. Lestari W. (1994) *Crashworthiness Study of a Generic Composite Helicopter Subfloor Structure*, NLR-TR 93590 L.
14. Michielsen A. L. P. J. and Wiggenraad J. F. M. (1997) *Review of Crashworthiness Research of Composite Structure*, NLR-CR 97046.
15. Obdam A. (1993) *Composiet Tensors, Concept Evaluatie, Report Intership*, NLR (in Dutch),
16. Ubels L. C. (1997) *Initial Material Data, Crasurv Task 1.1*, NLR-TR 97308L.
17. Li Q. M., Mines R., and Birch R. S. (1997) *Initial Material Data for Composite Materials*, University of Liverpool, IRC/151/97.
18. Michielsen A. L. P. J. (1997) *Specification of Sub-components and Box Structures*, NLR-CR 97315.

Composites as Strengthening Materials of Concrete Structures

Thanasis C. Triantafillou

Department of Civil Engineering, University of Patras, Patras, Greece

9.1 Introduction

The issue of upgrading the civil engineering infrastructure has been one of great importance for over a decade. Deterioration of bridge decks, beams, girders and columns, buildings, parking structures, and others may be attributed to aging, to environmentally induced degradation, to poor initial design and/or construction, to lack of maintenance, and to natural events such as earthquakes. The infrastructure's increasing decay is frequently combined with the need for upgrading so that structures can meet more stringent design requirements (e.g., increased traffic volumes in bridges exceeding the initial design loads), and hence the aspect of civil engineering infrastructure renewal has received considerable attention over the past few years throughout the world. At the same time, seismic retrofit has become at least equally important, especially in areas of high seismic risk.

Recent developments related to materials, methods, and techniques for structural strengthening have been enormous. One of today's state-of-the-art techniques bears the stamp of fiber reinforced polymer (FRP) composites, which are currently viewed by structural engineers as "new" and highly promising materials in the construction industry. Composite materials for strengthening of civil engineering structures are available today mainly in the form of either thin unidirectional *strips* (with thickness in the order of 1 mm) made by pultrusion, or flexible *fabrics*, made of unidirectional or bidirectional fibers (and sometimes preimpregnated with resin). Composites have found their way as strengthening materials in reinforced concrete (rc) elements (beams, slabs, columns, etc.) in thousands of applications worldwide where conventional strengthening techniques may be problematic. For instance, one of the popular techniques for upgrading rc elements has traditionally involved the use of steel plates epoxy-bonded to the external surfaces (e.g., tension zones) of beams and slabs. This technique is simple and effective as far as both cost and mechanical performance are concerned, but suffers from several disadvantages[1]: corrosion of the steel plates resulting in bond deterioration; difficulty in manipulating heavy steel plates in tight construction sites; need for scaffolding; and limitation in available plate lengths for flexural strengthening of long girders, resulting in the need for joints. Replacing the steel plates with FRP strips provides satisfactory solutions to these problems. Another common technique for the strengthening of rc structures involves the construction of reinforced concrete (either cast in-place or shotcrete) jackets (shells) around existing elements. Jacketing is clearly quite effective as far as strength, stiffness, and ductility is concerned, but it is labor intensive, it often causes disruption of occupancy, and it provides rc elements, in many cases, with an undesirable increase in stiffness. Another disadvantage of steel jackets is the requirement for corrosion protection. Here, too, the conventional jackets may be replaced with FRP fabrics or strips wrapped around rc elements, thus providing substantial increase in strength (axial, flexural, shear, torsional) and ductility without much affecting the stiffness. Similar arguments apply equally to civil engineering structures made of materials other than concrete; masonry (including historic monuments) and timber are typical examples (e.g. Reference [2]). The discussion in this chapter will be limited to the application of composites as strengthening materials for concrete structures only, primarily due to space limitations but also because such structures are met much more frequently than others in structural intervention projects. But the reader should bear in mind that many of the concepts applied to concrete are transferable to masonry and even timber construction with little additional effort.

The reasons why composites are increasingly used as strengthening materials of reinforced concrete elements may be summarized as follows[2]: immunity to corrosion; low weight (about $\frac{1}{4}$ of that of steel), resulting in easier application in confined space, elimination of the need for scaffolding, and reduction in labor costs; very high tensile strength (both static and long-term, for certain types of FRP materials); stiffness that may be tailored to the design requirements; large deformation capacity; and practically unlimited availability in FRP sizes or FRP geometry and dimensions. Composites suffer from certain disadvantages too, which are not to be neglected by engineers: in contrast to steel, which behaves in an elastoplastic manner, composites in general are linear elastic to failure (which, however, occurs at large strains) without any significant yielding or plastic deformation-induced ductility. Additionally, the cost of materials on a weight basis is several times higher than that for steel (though when cost comparisons are made on a strength basis, they become less unfavorable). Moreover, some fibers, e.g., carbon or aramid, are anisotropic, resulting in incompatible thermal expansion coefficients between composites and concrete. Finally, their exposure to high temperatures (e.g. in the case of fire) may cause premature degradation and collapse (some epoxy resins start softening at about 80°C). It is my firm opinion that FRP materials should not be thought of blindly as a replacement for steel (or other materials) in structural intervention applications. Instead, the advantages offered by them should be evaluated against potential drawbacks, and final decisions regarding their use should be based on consideration of several factors, including not only mechanical performance aspects but also constructibility and long-term durability.

In this chapter I wish to give an overview of the application of composites as externally bonded reinforcement of concrete structures. Following a general description of materials and techniques related to the application of composites as external reinforcement of concrete elements, the chapter is divided into three main sections, with each of them devoted to one particular aspect of strengthening. The first section deals with composites as *flexural* strengthening reinforcement, the second covers *shear* strengthening, and the third is devoted to the application of composites for *confinement*. These sections are followed by a closing summary of the main topics covered in the chapter and the key conclusions. The material presented herein emphasizes aspects related primarily to mechanical behavior and failure analyses. Where appropriate, design guidelines are also provided, so that the reader may obtain a better feeling for both analysis and design aspects. Last, it should be emphasized that several of the topics presented in this chapter are subjects of ongoing research and development, and the details of various modeling approaches may be subject to future revisions. Nevertheless, a considerable effort has been made here to present material that is state-of-the-art in the area of composites as strengthening materials in concrete construction.

Long-term properties and behavior of FRP-strengthened concrete are not covered in this chapter. This should not be perceived as an incompleteness: it is well established today that the creep, relaxation, fatigue strength, and durability behavior of rc elements strengthened with advanced composites (such as CFRP, which is the most common type of external reinforcement) is excellent and does not lead to restrictions on application and dimensioning.

9.2   Materials and Techniques

This section provides general information on FRP materials used in concrete strengthening, on concepts and techniques of their application, and on recently developed advanced applications of FRP as externally bonded reinforcement of concrete.

9.2.1   Materials

Composite material strengthening systems consist of a wide range of combinations of resins and fibers, which are briefly described below. Typical resins belong in the general class of epoxies, polyesters, or vinyl esters, while fibers are made of carbon, aramid, or glass, resulting in CFRP (carbon FRP), AFRP (aramid FRP) or GFRP (glass FRP) composites, respectively.

Resins

- *Primer:*   The purpose of the primer is to impregnate the surface of the concrete in order to improve the adhesive bond for the saturating resin or adhesive.

- *Putty:*   The putty is occasionally used to fill concrete surface holes and pits in order to prevent bubbles from forming during curing and also to provide a smooth surface to which the FRP can bond.

- *Saturating resins:*   The saturating resin is used to impregnate the reinforcing fibers, fix them in place, and provide a shear medium between fibers to transfer load effectively.

- *Adhesives:*   The purpose of the adhesive is to provide a shear load path between the concrete surface and the composite material.

- *Protective coatings:*   When used, the protective coatings provide external protection of the bonded FRP reinforcement from potentially damaging environmental effects.

Fibers.   Continuous fibers, such as carbon, aramid, and glass, are typical reinforcements for commercially available FRP strengthening systems, with carbon fibers being the most commonly used. Typical fiber properties are summarized in Table 9.1.[3] Note that values in this table are only indicative of static strength of unexposed fibers; design values of the FRP composite systems accounting for reductions due to

TABLE 9.1   Typical Properties of Fibers

Material	Elastic Modulus (GPa)	Tensile Strength (MPa)	Ultimate Tensile strain (%)
Carbon			
High strength	215–235	3800–4800	> 1.4
Ultra high strength	215–235	3800–6200	> 1.5
High modulus	350–500	> 3100	> 0.5
Ultra high modulus	500–700	> 2400	> 0.2
Glass			
E	70	1850–2700	> 4.5
S	85–90	3500–4800	> 5.4
Aramid			
Low modulus	70–80	3500–4100	> 2.5
High modulus	110–125	3500–4100	> 1.6

long-term loading, environmental exposure, etc., are normally supplied by the manufacturer.

9.2.2   FRP products and methods of application

The most common FRP strengthening systems are summarized below.[4]

- Dry unidirectional fiber sheet and unidirectional fabric, or knitted fabric, where fibers run predominantly in one direction. Installation on the concrete surface requires saturating resin.

- Dry multidirectional fabric or knitted fabric, where fibers run in at least two directions. Installation requires saturating resin.

- Resin-preimpregnated uncured unidirectional sheet or fabric, where fibers run predominantly in one direction. Installation may be done with or without additional resin.

- Resin-preimpregnated uncured multidirectional sheet or fabric, where fibers run predominantly in two directions. Installation may be done with or without additional resin.

- Dry fiber tows that are wound or otherwise mechanically placed onto the concrete surface. Resin is applied to the fiber during winding.

- Preimpregnated fiber tows that are wound or otherwise mechanically placed onto the concrete surface. Product installation may be executed with or without additional resin.

- Premanufactured cured laminate sheets or shaped shells, which are installed through the use of adhesives. The former are typically in the form of thin ribbon strips or grids that may be delivered in a rolled coil, while the latter are factory-made split shells that can be fitted around columns or other elements and be bonded in multiple layers.

9.2.3 Special techniques

Application of prestressed FRPs.   In some cases it may be advantageous to bond the external FRP reinforcement onto the concrete surface in a prestressed state. Both laboratory and analytical research, e.g. References [5, 6], shows that prestressing represents a significant contribution to the advancement of the FRP strengthening technique, and methods have been developed to prestress the FRP composites under real-life conditions.[7]

Prestressing the strips prior to bonding provides delay of formation of cracks and finer distribution of cracks in the concrete; this improves the serviceability and durability of concrete elements and increases the concrete strength. The procedure for applying a prestressed FRP strip is shown schematically in Fig. 9.1. When the prestressing force is too high, failure of the beam due to release of the prestressing force will occur at the two ends, as a result of the development of high shear stresses in the concrete just above the FRP. Hence the design and construction of the end zones requires special attention. Tests and analysis have shown that if no special anchorages are provided at the ends, FRP strips shear off (from the ends) with prestress levels in the order of only 5–6% of their tensile strength (for CFRP). But a

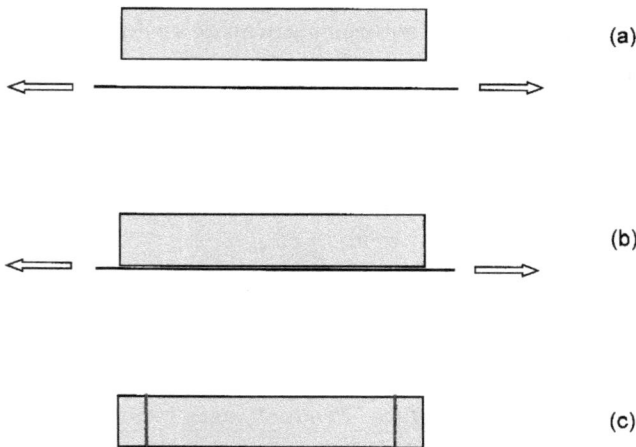

Figure 9.1 Strengthening with prestressed FRP strips: (a) prestressing; (b) bonding; (c) end anchorage and FRP release upon hardening of the adhesive.

technically and economically rational prestress would require a considerably higher degree of prestressing, in the range of 50% of the FRP tensile strength, which may only be achieved through the use of special anchorages applying vertical confinement (see Fig. 9.1c). Such systems have been developed by the EMPA (Swiss Federal Laboratories for Materials Testing and Research) and recently by the German firm of Leonhard, Andrä und Partner,[8] in conjunction with SIKA Chemie.

Automated wrapping. The technique of FRP strengthening through automated winding of tow or tape was first developed in Japan in the early 1990s and a little later in the United States. The technique, shown schematically in Fig. 9.2, involves continuous winding of wet fibers around columns or other structures (e.g., chimneys, as has been done in Japan) by means of a robot. A key advantage of the technique, apart from good quality control, is the rapidity of installation.

Fusion-bonded pin-loaded straps. Another interesting development of the FRP strengthening technique involves replacing solid and relatively thick laminates (Fig. 9.3a) by the system shown in Fig. 9.3b, known as pin-loaded strap.[9] The strap comprises a number of nonlaminated layers formed from a single, continuous, thin, *thermoplastic* matrix tape. The outside, final layer of the tape is fixed by a fusion bonding process to the previous layer, making contact onto the surrounding concrete using an end fixture. Such a system enables the individual layers to move relative to each other, thus reducing the unwanted secondary bending stresses. Careful control of the initial tensioning process allows interlaminar shear stress concentrations to

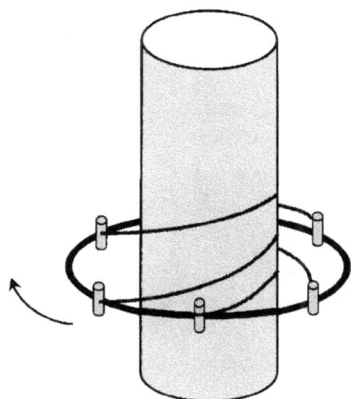

Figure 9.2 Automated rc column wrapping.

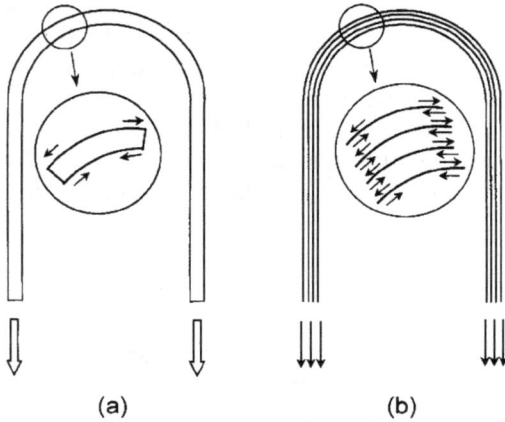

Figure 9.3 Wrapping with (a) thick laminates and (b) nonlaminated straps.

(a)   (b)

be reduced, so that a uniform strain distribution in all layers is achieved.

9.3 Flexural Strengthening

9.3.1 General

Reinforced concrete elements, such as beams and columns, may be strengthened in flexure through the use of FRP composites epoxy-bonded to their tension zones, with the direction of fibers parallel to that of high tensile stresses (member axis). The concept is illustrated in Fig. 9.4, where unidirectional FRP laminates or fabrics are used to strengthen a reinforced concrete element in flexure, and in Fig. 9.5, which shows a practical application. The failure analysis of such elements may follow well-established procedures for reinforced concrete structures, provided that: (a) the contribution of external FRP reinforcement is taken into account properly; and (b) special

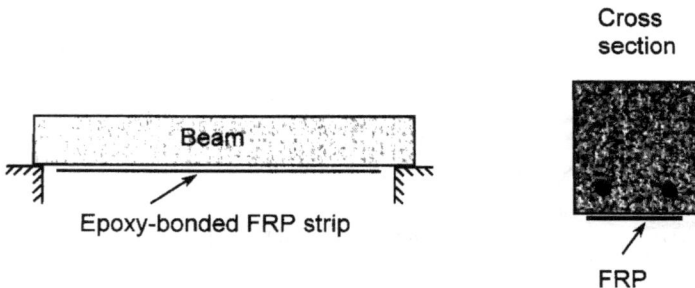

Cross section

Beam

Epoxy-bonded FRP strip

FRP

Figure 9.4   (a) Flexural strengthening of rc beam; (b) cross section.

Figure 9.5   Flexural strengthening of rc beams with CFRP strips. (*Courtesy SIKA HELLAS.*)

consideration is given to the issue of bond between the concrete and the FRP. A basic understanding of the failure analysis and design of reinforced concrete elements (e.g., beams, slabs, columns, shear walls) strengthened in flexure with externally bonded FRP is provided in the following.

9.3.2   Materials modeling

Figure 9.6 illustrates idealized stress–strain curves for concrete in compression (a), steel in either tension or compression (b) and FRP in tension (c). These curves, along with the assumption that the slip at the concrete–FRP interface may be ignored (an assumption that is justified for most structural adhesives applied at thicknesses in the order of 1.0–1.5 mm, in which case viscoelastic phenomena, such as axial and interlaminar shear creep as well as relaxation, are negligible), form the basis for the failure analysis of concrete elements strengthened in flexure. Central to the analysis of these elements is the identification of all the possible *failure modes*; this is presented next.

9.3.3   Failure modes

A reinforced concrete element strengthened in flexure with externally epoxy-bonded FRP reinforcement may fail according to one of the failure modes described below.

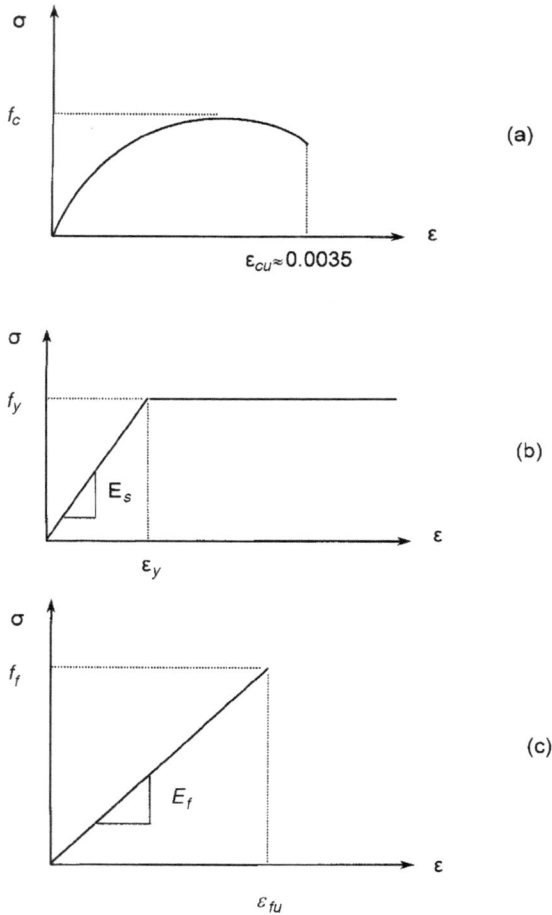

Figure 9.6 Idealized stress–strain curves for (a) concrete in compression; (b) steel in tension or compression; (c) FRP in tension.

- *Steel yielding followed by concrete crushing.* The flexural strength may be reached with yielding of the tensile steel reinforcement followed by crushing of the concrete in the compression zone, while the FRP is intact.

- *Steel yielding followed by FRP fracture.* For relatively low ratios of both steel and FRP, flexural failure may occur with yielding of the tensile steel reinforcement followed by tensile fracture of the FRP.

- *Concrete crushing.* For relatively high reinforcement ratios, failure of the rc element may be caused by compressive crushing of the concrete while both the steel and the FRP are intact.

- *FRP peeling-off at the outermost crack in the anchorage zone.*   The FRP may peel off in the anchorage zone as a result of shear fracture through the concrete (Fig. 9.7a).

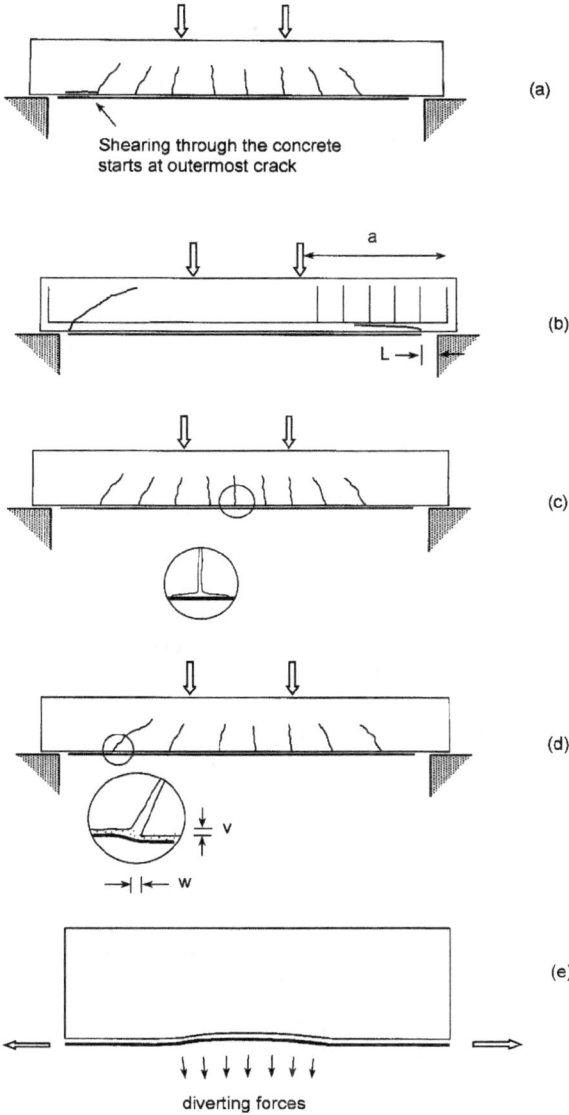

Figure 9.7   (a) FRP peeling-off at the outermost crack in the anchorage zone; (b) FRP plate-end shear failure; (c) FRP peeling-off at flexural cracks; (d) FRP peeling-off at shear cracks; (e) FRP peeling-off due to unevenness of concrete surface.

- *FRP plate-end shear failure.*   FRP plate-end shear failure will occur when the shear capacity of the concrete at the end of the FRP is exceeded (Fig. 9.7b).

- *FRP peeling-off at flexural cracks.*   Flexural (vertical) cracks in the concrete may propagate horizontally and thus cause peeling-off of the FRP in regions far from the anchorage (Fig. 9.7c).

- *FRP peeling-off caused by shear cracks.*   Shear cracking in the concrete generally results in both horizontal and vertical opening (Fig. 9.7d), which may lead to FRP peeling-off.

- *FRP peeling-off caused by unevenness of the concrete surface.*   The unevenness or roughness of the concrete surface may result in localized debonding of the FRP, which may propagate and cause peeling-off (Fig. 9.7e).

- *Bond failure in the adhesive.*   In the rather extreme case where the strength of the adhesive is lower than the strength of concrete, the debonding or peeling-off mechanisms described above could involve failure of the adhesive.

- *Bond failure in the interfaces between FRP, adhesive and concrete.* Debonding through the FRP-adhesive or adhesive-concrete interface is very rare and might occur only when the surface conditions during the FRP application are insufficient.

- *Shear failure in the FRP (interlaminar shear).*   Shear failure in the FRP might occur as a result of tensile stress transfer in the fibers. This failure mechanism will initiate when the maximum shear stress in the FRP reaches its shear strength.

More detailed treatment of the failure mechanisms listed above is given next.

9.3.4   Failure analysis of rc elements strengthened in flexure

9.3.4.1   Steel yielding followed by concrete crushing.   The strain and stress distribution in the critical cross section of a doubly reinforced rc element when the flexural capacity is reached due to steel yielding followed by concrete crushing is shown in Fig. 9.8. Crucial in the analysis for the ultimate limit state in flexure (flexural capacity) is consideration of the fact that during strengthening the rc element may not be fully unloaded. This implies the existence of initial strains (ε_0 at the extreme tensile concrete fiber), which should be added to those caused by additional loads after strengthening. The moment capacity of the strengthened element is calculated from equilibrium of internal

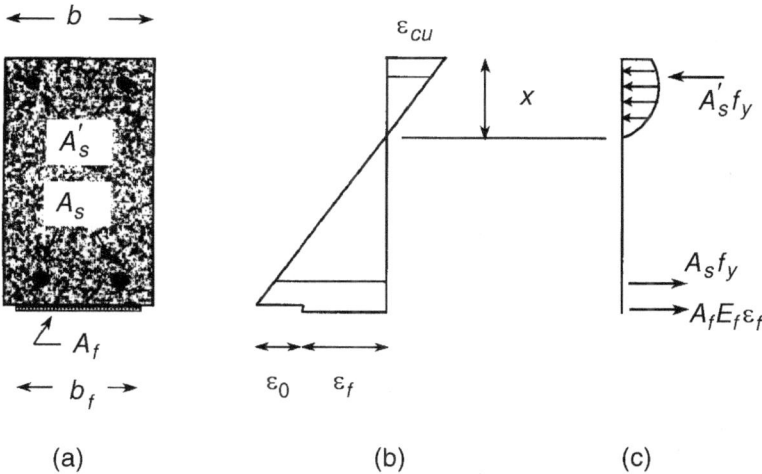

Figure 9.8 Flexure analysis for failure due to steel yielding followed by concrete crushing: (a) cross section; (b) strain distribution; (c) stress distribution at failure.

moments, after the distance x of the neutral axis from the extreme compressive concrete fiber is obtained from internal force equilibrium. Having obtained x, a check should be made to ensure that: (a) the tensile steel strain has exceeded the yield limit, $\varepsilon_y = f_y/E_s$, but has not reached the ultimate strain ε_{su}; (b) the strain in the FRP, ε_f, has not reached the ultimate strain ε_{fu}.

This failure mode is ideal to design a strengthened rc element for, as it is associated with significant deformations resulting from yielding of the tensile steel reinforcement.

9.3.4.2 Steel yielding followed by FRP fracture. Figure 9.9 shows the strain and stress distribution at flexural failure of a critical rc element section when the FRP fractures in tension before the concrete crushes in compression. Here, too, strain compatibility and internal force and moment equilibria result in determining the location of the neutral axis as well as the moment capacity. The checks to be made regarding strains are: (a) the maximum concrete strain ε_c should be below the ultimate strain ε_{cu}; and the tensile steel strain, which will certainly exceed its yield limit, $\varepsilon_y = f_y/E_s$, should be less than the ultimate steel strain ε_{su}.

As in the previous case, this failure mechanism is also desirable, because steel yielding is associated with high deformability, which is of crucial importance in the design of rc elements. Previous studies[10]

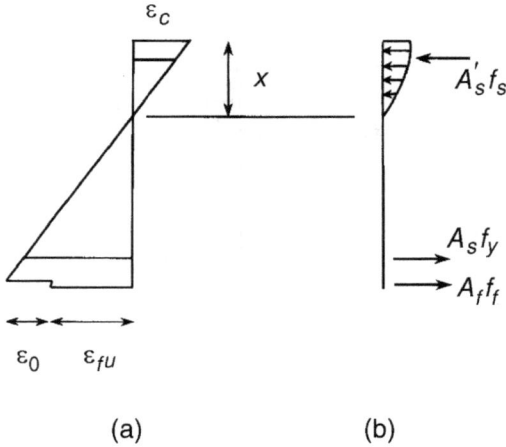

Figure 9.9 Flexure analysis for failure due to steel yielding followed by FRP fracture: (a) strain distribution; (b) stress distribution at failure.

have shown that this mechanism is possible only if both the tensile steel area A_s and the FRP area A_f are very low.

9.3.4.3 Concrete crushing. In case A_s and/or A_f is large, the concrete will crush in compression before any other material (steel, FRP) reaches its tensile capacity. The relevant strain and stress distribution at the cross section level is illustrated in Fig. 9.10. Strain compatibility

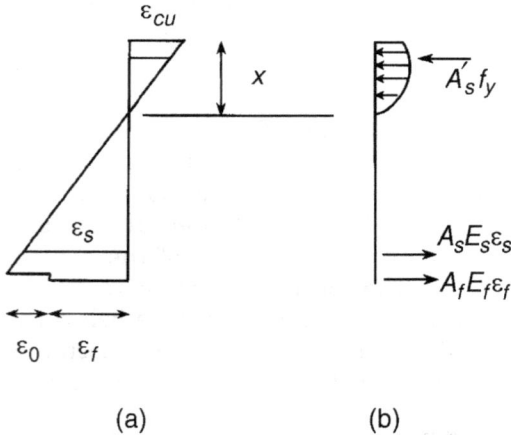

Figure 9.10 Flexure analysis for failure due to concrete crushing: (a) strain distribution; (b) stress distribution at failure.

and internal force and moment compatibility will provide the neutral axis depth and the moment capacity as above, and the only check to be made is that the tensile steel strain has not reached the yield limit, $\varepsilon_y = f_y/E_s$.

This failure mode is brittle (as deformations, and hence warning, at imminent collapse are minimal) and should be avoided by proper dimensioning of the strengthening system.

9.3.4.4 FRP peeling-off at the outermost crack in the anchorage zone.

Debonding and peeling-off at the outermost crack in the anchorage zone may be best characterized by using simple *fracture mechanics* modeling. The failure mechanism illustrated in Fig. 9.7a may be analyzed with the simple model of the anchorage zone shown in Fig. 9.11a, combined with concepts of fracture mechanics. According to this model, debonding may be treated as a mode II (or, better, mixed mode I and II) fracture process that takes place through shearing (and tension, if mode I is also considered) in the weakest material near the FRP–concrete interface. Unless very high-strength concrete or very low-strength adhesives are used (both cases are rare), fracture will propagate through the concrete, which is the weakest material

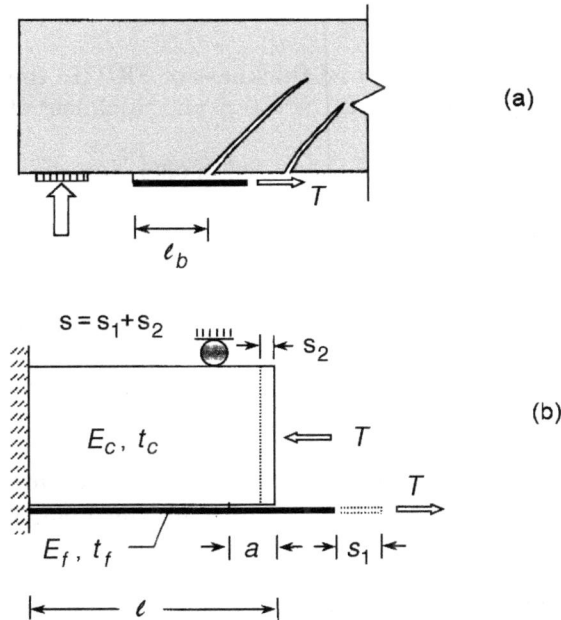

Figure 9.11   (a) FRP anchorage zone; (b) simple model for analysis of bond failure.

near the interface. Fracture propagation and initiation may be treated by either linear or nonlinear fracture mechanics, as described below.

Linear elastic fracture mechanics (LEFM) model. Based on the assumptions that (a) all materials are homogeneous, isotropic and linear elastic; (b) the adhesive is only exposed to shear forces; and (c) the thickness of the adherents and the adhesive and the width b_f of the bonded plate are all constant throughout the bond line, one may apply the *compliance method* (e.g., Reference [11]) to calculate the maximum tensile force in the FRP, T_{max}, corresponding to crack propagation in the bond line of the model shown in Fig. 9.11b as follows (e.g., Reference [12]):

$$T_{max} = \sqrt{2 b_f G_{IIc} \Big/ \frac{\partial C}{\partial a}} \tag{9.1}$$

where G_{IIc} = mode II critical strain energy release rate. Considering small (axial) deformations and ignoring the development of bending moments and the deformations in the bond layer, simple beam theory[12] results in the following expression for the derivative of the compliance:

$$\frac{\partial C}{\partial d} = \frac{1}{E_f t_f} + \frac{1}{E_c t_c} \tag{9.2}$$

where E_f and t_f are the elastic modulus and thickness of FRP (in the fiber direction), and E_c and t_c are the elastic modulus and thickness of concrete.

The apparent disadvantage of the LEFM-based model described above is the assumption of linear elastic materials, an assumption which is, in general, not justified with concrete materials. Naturally, this has led to adoption of nonlinear fracture mechanics (NLFM) modeling, described next.

Nonlinear fracture mechanics (NLFM) model. The problem of FRP bond failure in the anchorage zone may be approached more realistically by assuming that the bond line is characterized by a nonlinear relationship between the local bond shear stress, τ, and the local slip, s, as shown in Fig. 9.12a, which also shows the shear strength of the bond zone, τ_{max}. Also note that the area below the τ–s curve represents the mode II fracture energy, G_{IIf}, which is defined as the energy required to bring a local bond element to shear fracture (debonding). It is important to note here that, as clearly suggested by experimental evidence, the constitutive material law of Fig. 9.12a and consequently G_{IIf} do not depend on the bond length, l_b, provided that this length does not fall below a certain value, $l_{b,max}$. In addition to the above constitutive law, a crack model for shear, the physical meaning of

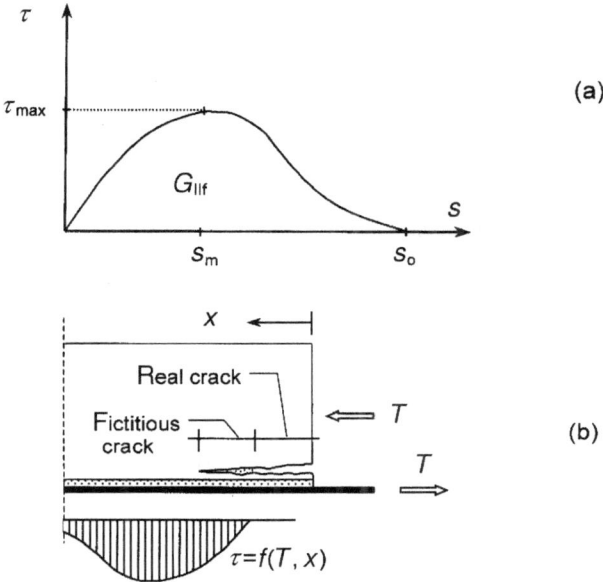

Figure 9.12 (a) Local bond constitutive law for NLFM analysis; (b) fictitious crack model for bond failure in the anchorage zone.

which is shown in Fig. 9.12b, may be used to derive the maximum tensile force in the FRP, T_{max}. Following this approach, Täljsten[12] derived the expression for T_{max} corresponding to the geometry of Fig. 9.11b as given next:

$$T_{max} = \tau_{max} b_f (l-a) F \left(\frac{\tau_{max}^2 (l-a)^2}{E_f t_f G_{IIf}}, \frac{E_f t_f}{E_c t_c}, g \right) \tag{9.3}$$

where g represents the shape function of the τ–s curve. Simple shapes (e.g., linear ascending branch, linear ascending and descending branches) of the τ–s curves result in analytical solutions for Eq. (9.3), while for more realistic shapes of the τ–s curve numerical stepwise calculations may be needed. For the simplest case of linear elastic bond behavior (τ–s curve with linear ascending branch), Täljsten[12] gives the following expression for F in Eq. (9.3):

$$F = \frac{\sqrt{2}}{\omega_1} \tanh\left(\frac{\omega_1}{\sqrt{2}} \right) \tag{9.4a}$$

$$\omega_1 = \left(\frac{\tau_{max}^2 (l-a)^2 \left(1 + \frac{E_f t_f}{E_c t_c} \right)}{E_f t_f G_{IIf}} \right)^{1/2} \tag{9.4b}$$

The validity of Eqs. (9.4) has been checked and verified successfully for concrete strengthened with either steel plates[12] or CFRP.[13]

A similar NLFM approach was also employed by Holzenkämpfer,[14] who characterized bond failure of steel plates bonded to concrete using the bond law of Fig. 9.12 with linear ascending and descending branches. Adopting the Mohr–Coulomb failure criterion, the value of τ_{max} is expressed as a function of the concrete surface's tensile strength, f_{ctm}; the slips s_m and s_0 are derived from the deformation of a representative volume of the bond zone; and the mode II fracture energy can be approximated as

$$G_{IIf} \approx C_f f_{ctm} \tag{9.5}$$

where C_f is a constant that may be determined by linear regression analysis of the results corresponding to fracture energy testing (e.g. double shear tests). Finally, for bond lengths exceeding the limiting value, $l_{b,max}$, the maximum bond force is given as

$$T_{max} = c_1 k_b b_f \sqrt{E_f t_f f_{ctm}} \tag{9.6}$$

where k_b is a geometric factor:

$$k_b = 1.06 \sqrt{\frac{2 - (b_f/b_c)}{1 + (b_f/400)}} \geq 1 \tag{9.7}$$

Note that the width of FRP, b_f, and concrete, b_c, in Eq. (9.7) is given in millimeters. $l_{b,max}$ is given as

$$l_{b,max} = \sqrt{\frac{E_f t_f}{c_2 f_{ctm}}} \tag{9.8}$$

c_1 and c_2 in Eqs. (9.6) and (9.8) may be obtained through calibration with test results. For bond lengths $l_b < l_{b,max}$, the ultimate bond force is calculated as follows[14]:

$$T = T_{max} \frac{l_b}{l_{b,max}} \left(2 - \frac{l_b}{l_{b,max}} \right) \tag{9.9}$$

Neubauer and Rostásy[15] have performed over 50 bond tests (similar configuration as in Fig. 9.11b) with the width and thickness of CFRP strips, the bond length, and the concrete compressive strength as variables. In these tests, bond failure was characterized by sudden crack propagation either through the concrete only (for concrete with compressive strength 25 MPa) or through the CFRP (interlaminar failure) after a few centimeters of concrete failure (for concrete strength equal to 55 MPa), which started from the loaded end of the bond length. Despite the different failure modes, Rostásy and Neubauer[16] concluded that the same fracture mechanism, dependent on the concrete fracture energy, was responsible for the initiation of bond failure, while the interlaminar failure was considered as a secondary effect, caused by the high local tensile stresses. Calibration of Holzenkämpfer's model using the data[15] resulted in the following values $c_1 = 0.64$ and $c_2 = 2$ in Eqs. (9.6) and (9.8), corresponding to a mean value of $C_f = 0.204$ and a standard deviation of 0.0527.

Anchorage failure check.   Failure or not of the FRP in the anchorage zone according to the mechanism described above may be checked by comparing the tensile force that develops in the FRP due to bending of the strengthened element with the maximum force that may be resisted for any given anchorage length (T or T_{max}, depending on the anchorage length, properly reduced by a strength reduction factor). This check has to be made (at least theoretically) at every cross section along the length of the rc element.

9.3.4.5 FRP plate-end shear failure.   Tests by several investigators,[17,18] have indicated that when externally bonded plates stop at a certain distance from the supports (as is typically the case in strengthening applications) a nearly vertical crack can initiate at the plate end (plate end crack) and then grow as an inclined shear crack (left of Fig. 9.7b). However, by virtue of internal stirrups, the shear crack may be arrested and the bonded-on plate separated from the concrete at the level of the longitudinal reinforcement (right of Fig. 9.7b). This failure mechanism will be activated when the maximum shear stress near the plate end reaches a critical value. Jansze[18] employed this modeling approach along with the *fictitious shear span* concept (illustrated in Fig. 9.13) to compute the shear resistance of plated beams along the lines of an expression provided by the CEB-FIP Model Code.[19] The resulting equations are summarized below:

Fictitious shear span:

$$a_L = \sqrt[4]{\frac{(1 - \sqrt{\rho_s})^2}{\rho_s} \, dL^3} \qquad (9.10)$$

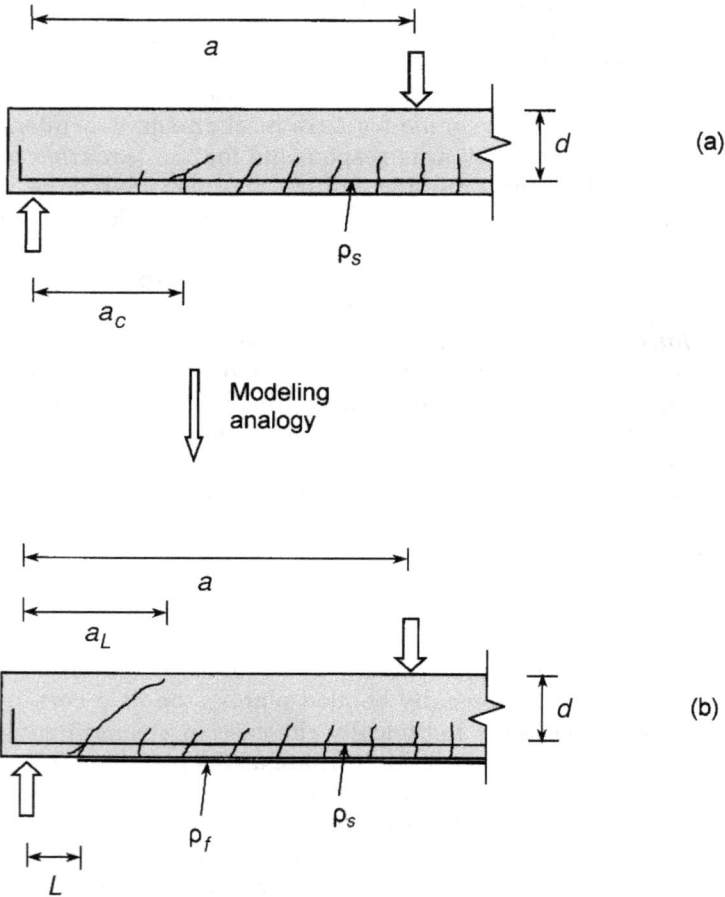

Figure 9.13 (a) Concept of fictitious shear span; (b) modeling analogy for the analysis of plate-end shear failure.

Shear strength:

$$\tau_{cu} = 0.18 \sqrt[3]{3 \frac{d}{a_L}} \left(1 + \sqrt{\frac{200}{d}}\right) \sqrt[3]{100 \rho_s f_c} \qquad (9.11)$$

Maximum shear force: $V_{cu} = \tau_{cu} b d$ (9.12)

Restrictions: $a > L + d, \qquad a_L < a$ (9.13)

In the above equations $\rho_s = A_s / bd$, and f_c = compressive strength of concrete. All other variables are defined in Fig. 9.13.

9.3.4.6 FRP peeling-off at flexural cracks in the constant moment region.
Another failure mechanism of FRP-plated elements may initiate from
near vertical (flexural) cracks in high-moment regions. When such
cracks become wide, the tensile stress in the FRP must be redistributed, and this leads to local debonding (Fig. 9.7c); final failure is
reached by propagation of debonding towards the end of the plate. The
current practice toward suppressing this failure mechanism requires
that the maximum strain in the FRP, $\varepsilon_{f,\max}$, be limited below a certain
value, which may be derived based on testing (and supported by
analysis). Neubauer and Rostásy[15] have suggested that this value
should be taken as the minimum of five times the yield strain of
internal tensile steel reinforcement, ε_{ys}, and 50% of the FRP strain
capacity, ε_{fu}.

9.3.4.7 FRP peeling-off caused by shear cracks. Shear cracks in
concrete elements are inclined, and are associated with both horizontal, w, and vertical, v, opening displacements, primarily due to the
aggregate interlock and dowel action mechanisms.[20] Similarly to what
was described in the previous section, the horizontal crack opening
displacement may initiate debonding. But the vertical crack opening
displacement may also cause debonding, as it induces direct tension in
the concrete layer between the FRP and the embedded longitudinal
steel reinforcement (Fig. 9.7d). Whether peeling-off will initiate or not
in this case (for a given horizontal crack opening displacement)
depends on a number of parameters, including the vertical crack
opening displacement, the flexural and shear rigidity of the FRP,
and the tensile strength of concrete. The author's opinion is that
this failure mechanism has not yet been quantified properly by the
research community (although some attempts have been made),[10,21,22]
and hence no attempt at detailed analytical treatment will be made
here.

9.3.4.8 FRP peeling-off caused by unevenness of the concrete surface.
Debonding of the FRP due to possible unevenness of the concrete
surface is another failure mechanism that has not been studied
thoroughly. Experimental evidence suggests that this mechanism
may be avoided by adopting certain practical execution rules and
limitations on concrete surface roughness. Most of these limitations
refer to a maximum concrete roughness over a given length, and
depend on FRP type and dimensions (e.g., thickness). Specific details
should be looked for in specifications normally provided by suppliers of
FRP strengthening systems.

9.3.4.9 Bond failure in the adhesive. Debonding of the FRP through the adhesive is a very rare failure mechanism that will not be activated under normal circumstances, that is, when typical high quality structural epoxies are employed. Such epoxies have shear and tensile strengths that exceed those of concrete by far, and hence this mechanism will always be preceded by one that involves failure through the concrete. Possible exceptions might be encountered at relatively high-temperature applications (where the strength of adhesives drops) or when the concrete substrate has unusually high strength.

The failure analysis for the adhesive debonding mechanism could be either strength-based or fracture energy-based. The strength-based approach would involve calculation of the maximum shear stress in the adhesive and comparison of this with the adhesive's shear strength. Similar arguments hold for more complex states of stress, say a combination of shear and tensile stresses, in which case the use of an appropriate failure criterion for the adhesive would be imperative. The fracture energy-based approach would involve determination of the mode II (in combination with mode I, if tensile stresses are important too) stress intensity factor and comparison with the critical stress intensity factor. As mentioned earlier, debonding through the adhesive is highly unlikely to occur, and thus the elaborate treatment of either of the two approaches mentioned above is not justified in the framework of this book.

9.3.4.10 Bond failure in the interfaces between FRP, adhesive and concrete. Debonding through the interfaces between FRP and adhesive or adhesive and concrete is another failure mechanism that is rather unlikely to occur, unless the concrete or FRP surface conditions are insufficient during the adhesive application process. This may be avoided through proper quality control, which would call for a number of surface preparation steps (normally specified by the FRP strengthening system supplier). In the extreme case where poor quality control results in insufficient interfaces, debonding through these interfaces may be tackled as described above, that is, by adopting either strength-based of fracture energy-based theories (the latter may also be combined with advanced interface mechanics theories). Since this failure mode may easily be avoided by means of proper surface preparation (e.g., cleaning of FRP with acetone and sandblasting of concrete), no detailed analysis will be attempted in this work.

9.3.4.11 Shear failure in the FRP (interlaminar shear). Tensile stresses in unidirectional fiber reinforced polymers are carried by the fibers through shearing of the polymer matrix. If the shear stresses between fibers reach the shear strength of the polymer or the shear strength of the polymer–fiber interface, interlaminar shear in the FRP will occur. This failure mode is (theoretically) possible in pultruded FRP strips as well as between different layers of FRP sheets. But typical polymer matrix materials (and structural adhesives) have shear strengths that are several (6–8) times higher than that of concrete, so this failure mechanism is very rare and hence no further elaboration will be given here.

9.3.5   FRP material selection and structural design

Proper design of flexural strengthening should include a long list of actions, of which the most important are given below (in a rather simplified manner).

- Selection of a high-stiffness FRP material, preferably CFRP, in order to minimize the thickness of the external reinforcement and maximize its performance (e.g., long-term resistance).

- Determination of the FRP cross section for a given strength increase so that failure will be governed by steel yielding followed by concrete crushing (this step is performed in analogy with conventional reinforced concrete dimensioning). The failure mechanism of steel yielding followed by FRP fracture would also be acceptable, but limitations on maximum FRP strains (to avoid premature debonding) are likely to suppress it. Of crucial importance here is to ensure that the *degree of strengthening* (flexural capacity of strengthened element divided by that before strengthening) is limited below a certain value, of the order of 2, so that in the extreme case of FRP loss (e.g., during a fire) the element will maintain a safety factor under service loads in the order of at least 1.1–1.2.

- Verification (check) of the premature FRP debonding failure mechanisms.

- Provision of special mechanical anchors (especially near the FRP ends) if premature anchorage failure is anticipated.

- Proper execution (preparation of surfaces, bonding, etc.) by meeting stringent quality control requirements.

9.4   Shear Strengthening

9.4.1   General

Shear strengthening of concrete elements using FRP may be provided by bonding the external reinforcement with the principal fiber direction as parallel as practically possible to that of maximum principal tensile stresses. For the most common case of structural members subjected to lateral loads, that is, loads perpendicular to the member axis (e.g., beams under gravity loads or columns under seismic forces), the maximum principal stress trajectories in the shear-critical zones form an angle with the member axis that may be taken roughly equal to 45°. However, it is normally more practical to attach the external FRP reinforcement with the principal fiber direction perpendicular to the member axis. Concepts for the use of FRPs as externally bonded reinforcement to enhance the shear capacity of concrete elements of rectangular cross section are illustrated in Fig. 9.14, and a practical application is shown in Fig. 9.15.

9.4.2 Failure mechanisms

When the concrete element reaches its shear capacity (that is just before it fails in shear), the external FRP is stretched in the principal fiber direction up to a strain level that is, in general, less than the tensile fracture strain $\varepsilon_{f,u}$. This strain is defined as *effective strain*, $\varepsilon_{f,e}$ to reflect the fact that if it were multiplied by the FRP elastic modulus in the principal fiber direction, E_f, and the available FRP cross-sectional area, it would provide the total force carried by the FRP at shear failure of the element. The effective FRP strain is extremely difficult, if not impossible, to calculate based on rigorous analysis. But it may be estimated based on simple modeling and through proper analysis of experimental data (described later).

Central to the estimation of $\varepsilon_{f,e}$ is the understanding of the mechanism governing the behavior of the FRP when the concrete element's shear capacity is reached. As shear cracks open, the FRP will either fail in *tensile fracture* (in the vicinity of the cracks or near corners of the cross section, due to stress concentrations) or *debond*, depending on a number of parameters. Experimental evidence suggests that a certain degree of debonding will develop (especially near cracks on the concrete surface) even if FRP failure is due to tensile fracture, and hence it may be argued that the parameters affecting the FRP failure mode are all associated, to some degree, with debonding. Therefore, it may be stated that $\varepsilon_{f,e}$ depends heavily on the FRP bonded length, its relation to the "effective bond length" (through which FRP–concrete interface shear bond stresses develop), and the relation of the latter to the "development length" (defined as that necessary to reach FRP tensile fracture before debonding). Apart from the bond conditions (surface preparation, execution, etc.), the development length depends on the FRP axial rigidity, which may be expressed by the product of E_f times the FRP area fraction ρ_f (FRP cross-sectional area divided by

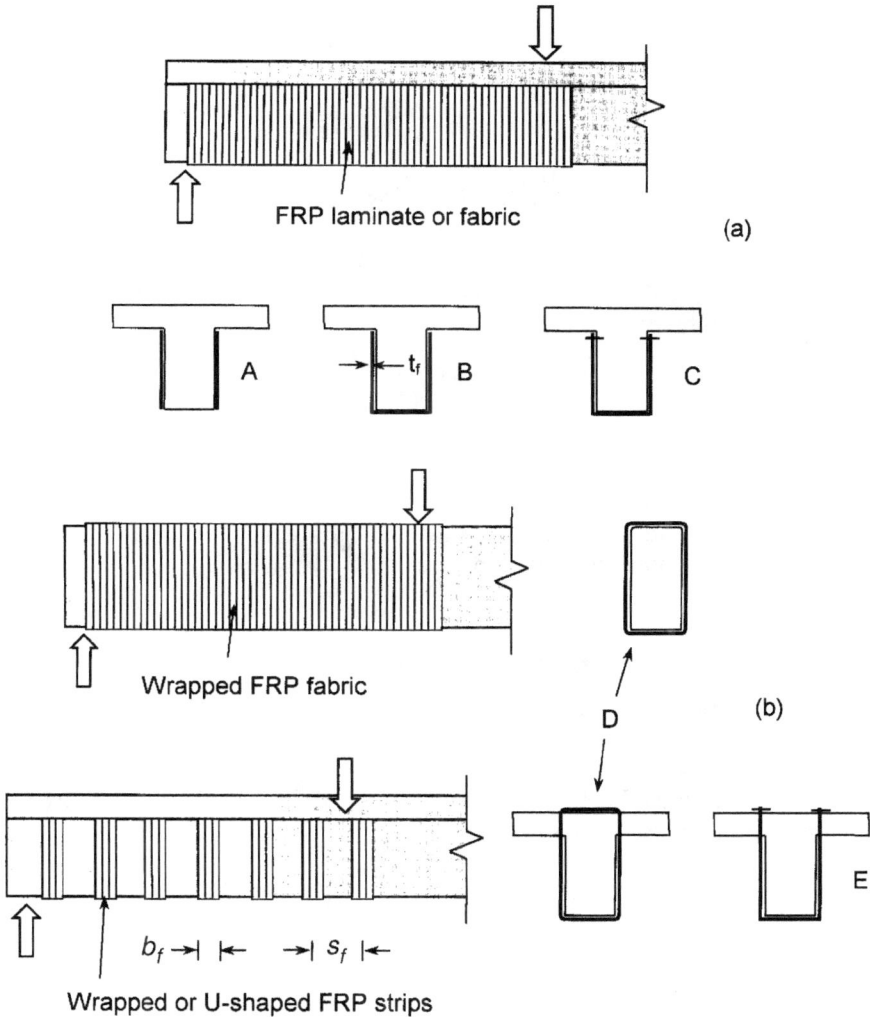

Figure 9.14 Schematic illustration of reinforced concrete beam strengthened in shear with FRP: (a) FRP laminates or fabrics bonded to the web; (b) wrapped or U-shaped fabrics or strips (the concept shown as D is also applicable to columns).

area of concrete), and is inversely proportional to the shear, that is to the tensile, strength of concrete. It is well established today that the tensile strength of concrete is proportional to $f_c^{2/3}$,[19] where f_c is the compressive strength, so that one may finally argue that $\varepsilon_{f,e}$ depends heavily on the quantity $E_{f\rho f}/f_c^{2/3}$.[23] In other words, thick and/or stiff FRP laminates or fabrics will debond "more easily" than thin and/or

Figure 9.15 FRP shear strengthening application in Greece with CFRP fabrics. (Courtesy SIKA HELLAS)

flexible ones; and the weaker the base material (concrete) is, the "earlier" will debonding develop.

9.4.3 FRP contribution to shear capacity

In terms of modeling, it is quite likely that engineers will find it quite convenient to treat externally bonded FRP shear reinforcement in a manner that is identical to the case of internal (steel) reinforcement (stirrups). As may be found in any standard textbook on reinforced concrete analysis and design, internal steel reinforcement is treated as part of an idealized truss (truss model), and its contribution to the capacity of the reinforced concrete element at shear failure (that is the shear force V_s carried by the stirrups) may be calculated from the following simple equation[24,25]:

$$V_s = k f_{yw} \rho_w b_w d (1 + \cot \beta) \sin \beta \qquad (9.14)$$

where d = effective depth; b_w = minimum width of cross section over the effective depth; f_{yw} = yield stress of steel stirrups; ρ_w = area fraction of steel, equal to cross-sectional area of stirrups, A_{sw}, divided by $b_w s$, where s = spacing of stirrups; β = angle between principal fiber direction and longitudinal axis of member; and k = code-dependent constant (e.g., equal to 0.9 according to the Eurocode; 1.0 according to the American code, ACI; and 1/1.15 according to the Japanese code, JCI).

In analogy to the above equation, the contribution of FRP to shear capacity (that is the shear force V_f carried by the external strips or fabrics) may be calculated from[26]

$$V_f = kE_f \varepsilon_{f,e} \rho_f b_w d (1 + \cot \beta) \sin \beta \qquad (9.15)$$

In the case of continuous fabrics (e.g., Fig. 9.14a) $\rho_f = 2t_f/b_w$, while in the case of strips of width b_f uniformly spaced at distance s_f (Fig. 9.14b) $\rho_f = (2t_f/b_w)(b_f/s_f)$.

9.4.4 Effective FRP strain

The effective FRP strain, $\varepsilon_{f,e}$ in the above equation may be obtained from tests. Imagine two identical concrete beams properly dimensioned so that failure will be controlled by shear. If one of the beams is strengthened with externally bonded reinforcement, it may be assumed that the difference in shear-carrying capacity is attributed to the external reinforcement. Hence Eq. (9.15) may be applied and $\varepsilon_{f,e}$ obtained. In a recent study, Triantafillou and Antonopoulos[23] studied carefully more than 80 test data published in over 15 research papers and applied the above procedure. The outcome of this study is shown in Figs. 9.16 and 9.17 (with k taken equal to 0.9), for specimens that failed by debonding and by FRP tensile fracture (which may occur when shear cracking develops or a little latter), respectively. In agreement with the qualitative arguments made above, it can be seen that $\varepsilon_{f,e}$ clearly decreases as $E_f \rho_f / f_c^{2/3}$ increases. The results in Fig. 9.16 are useful in characterizing CFRP-strengthened elements, as there are no data points for AFRP and those points for GFRP are very limited. However, the author's view is that the type of fibers should not affect $\varepsilon_{f,e}$ considerably in the case of debonding. If debonding is not dominant, the effective strain appears to depend on the type of FRP: different trends are shown in Fig. 9.17 for CFRP and AFRP strengthening materials, due to the very different fracture strains of these materials. If debonding is not dominant, the effective strain appears to depend on the type of FRP, due to the difference in the fracture strain of various FRPs. Moreover, the experimental results suggest that fully wrapped FRP (for instance, in the case of columns or rectangular beams) is not likely to fail due to debonding; fracture of FRP combined with or following concrete shear cracking is expected to dominate in this case.

The best fit power-type expressions to the test data shown in Figs. 9.16 and 9.17 (dashed lines in these figures) are summarized below.

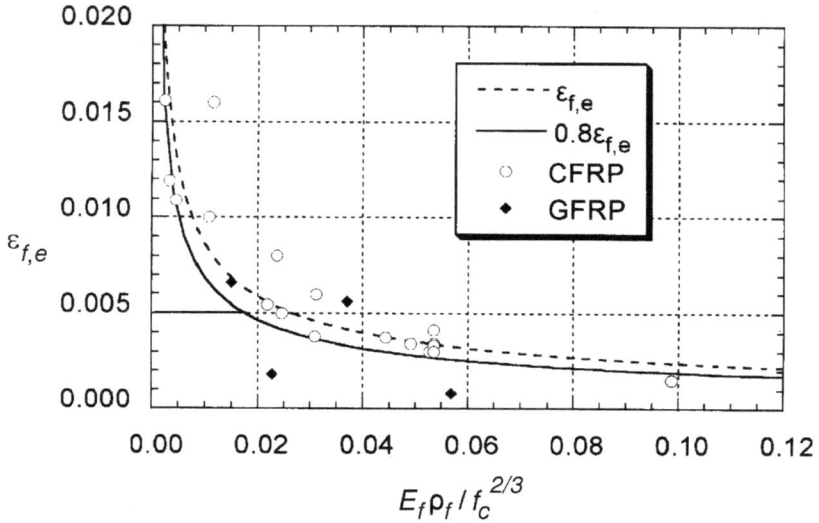

Figure 9.16   Effective FRP strain in terms of $E_f\rho_f/f_c^{2/3}$—shear failure combined with FRP debonding.

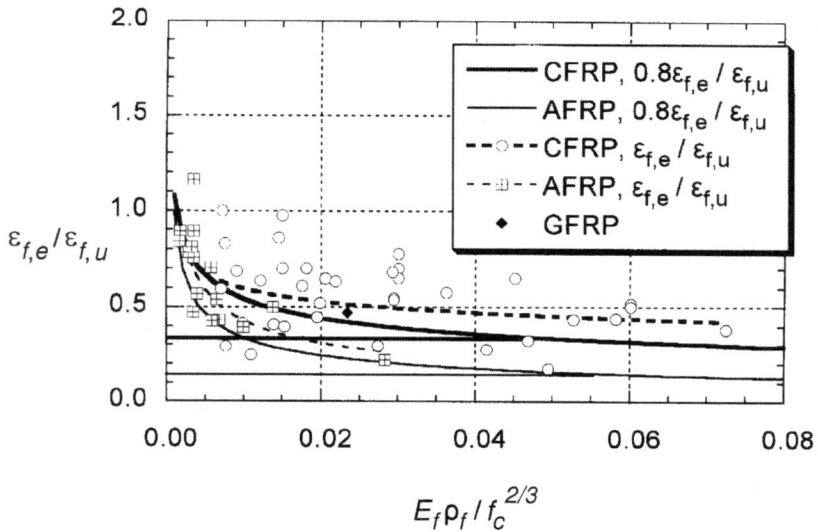

Figure 9.17   Normalized FRP strain in terms of $E_f\rho_f/f_c^{2/3}$—shear failure combined with or followed by FRP fracture.

Premature shear failure due to debonding (for CFRP only, to be used with caution for other FRPs):

$$\varepsilon_{f,e} = 0.65 \left(\frac{f_c^{2/3}}{E_f \rho_f}\right)^{0.56} \times 10^{-3} \qquad (9.16)$$

Shear failure combined with or followed by CFRP fracture:

$$\frac{\varepsilon_{f,e}}{\varepsilon_{f,u}} = 0.17 \left(\frac{f_c^{2/3}}{E_f \rho_f}\right)^{0.30} \qquad (9.17)$$

Shear failure combined with or followed by AFRP fracture:

$$\frac{\varepsilon_{f,e}}{\varepsilon_{f,u}} = 0.048 \left(\frac{f_c^{2/3}}{E_f \rho_f}\right)^{0.47} \qquad (9.18)$$

It is important to note that in these equations (and in Figs. 9.16 and 9.17) f_c is in MPa and E_f is in GPa.

Given the above, dimensioning of the external FRP shear reinforcement for a certain increase in the element's shear capacity becomes a straightforward task throught the use of Eq. (9.15), in which $\varepsilon_{f,e}$ is to be taken as the minimum of Eq. (9.16) and (9.17) for CFRP or Eq. (9.16) and (9.18) for AFRP (unless the FRP is fully wrapped around the element, so that Eq. (9.17) or (9.18) is always applicable). A point of crucial importance here is to provide an upper limit to $\varepsilon_{f,e}$ (see horizontal cut-offs in Figs. 9.16 and 9.17), in order to ensure that the shear integrity of concrete is maintained sufficiently, so that other mechanisms, such as the aggregate interlock, may be activated too. This limit should be in the order of 0.004–0.005.[26–28]

The last comment regarding the use of Eq. (9.15) applies to the use of safety factors. As the introduction of these factors is code-dependent (and hence outside the scope of this book), this topic will not be treated in detail here. A summary of code-dependent formulations for the contribution of FRP to shear capacity is given below, but more details may be found.[23]

Eurocode format:

$$V_{fd} = 0.9 \frac{\alpha \varepsilon_{f,e}}{\gamma_f} E_f \rho_f b_w d (1 + \cot \beta) \sin \beta \qquad (9.19)$$

where γ_f = partial safety factor for FRP and α = mean strain reduction factor.

ACI format:

$$\phi_f V_f = \phi_f \varepsilon_{f,e,A} E_f \rho_f bd(\sin \beta + \cos \beta) \tag{9.20}$$

where $\varepsilon_{f,e,A} = 0.9\varepsilon_{f,e}$ and ϕ_f = strength reduction factor.

JCI format:

$$V_{fd} = \frac{1}{1.15} \frac{\alpha \varepsilon_{f,e,J}}{\gamma_f} E_f \rho_f b_w d(1 + \cot \beta) \sin \beta \tag{9.21}$$

where $\varepsilon_{f,e,J} = 1.035\varepsilon_{f,e}$.

9.4.5 FRP material selection

To obtain a better insight into the contribution of FRP to shear capacity, the above equations are plotted in Fig. 9.18 for two concrete strengths, $f_c = 20$ MPa and 40 MPa, assuming a CFRP strengthening material with $\varepsilon_{f,u} = 0.015$, $\beta = 90°$ (fibers perpendicular to axis of member), $\alpha = 0.8$ and $\gamma_f = 1.20$ (unless CFRP fracture controls, where $\gamma_f = 1.15$). A careful examination of Fig. 9.18 leads to the conclusion that for values of $E_f \rho_f$ below a limiting value $(E_f \rho_f)_{\text{lim}}$, the design is

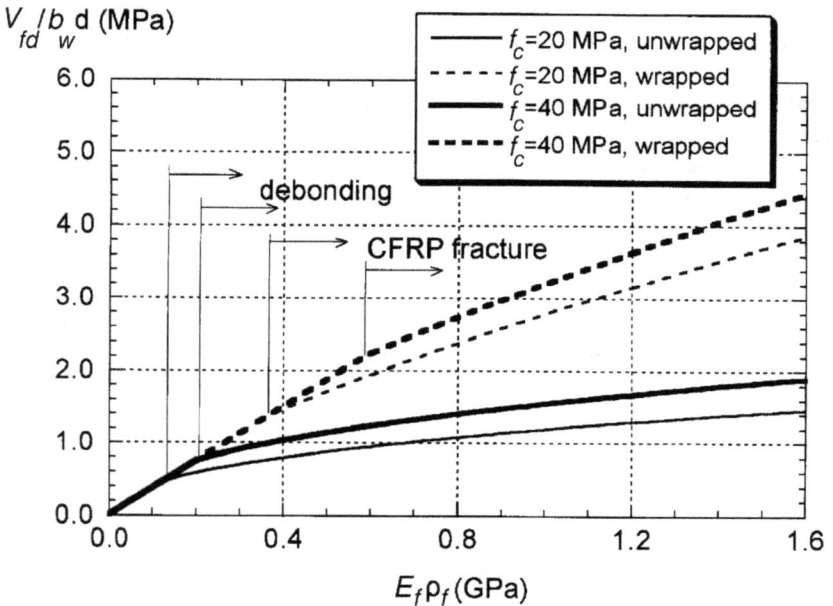

Figure 9.18 FRP contribution to shear capacity in terms of $E_f \rho_f$ for two concrete strengths and wrapped versus unwrapped elements.

governed by the upper limit to $\varepsilon_{f,e}$, that is, no FRP failure mechanism is activated, and hence the contribution of FRP to shear capacity is proportional to $E_f\rho_f$. For values of $E_f\rho_f$ exceeding $(E_f\rho_f)_{\text{lim}}$, failure is governed: (a) by debonding combined with shear fracture, if the FRP is not properly anchored (e.g. side or U jackets); or (b) by shear fracture combined with or followed by CFRP fracture, if the composite material is anchored properly (e.g. fully wrapped). In the first case the increase in shear capacity with $E_f\rho_f$ is relatively small but the concrete strength plays an important role, whereas in the second case the increase in shear capacity with $E_f\rho_f$ becomes quite substantial, but the role of concrete is of secondary impotance. It may therefore be reasonable to suggest that $E_f\rho_f$ should not exceed $(E_f\rho_f)_{\text{lim}}$ (shown by Triantafillou and Antonopoulos[23] to be equal to $0.018f_c^{2/3}$), unless debonding may be prevented or the attachment of FRP to concrete may be improved by specially designed mechanical anchorages.

Another point of interest is the orientation of fibers. As Eq. (9.15) suggests, this should be chosen as close as possible to 45° (or ±45° in case of reversed loading, e.g. seismic), but convenience of execution will normally call for FRP materials placed with the principle fiber orientation at 90° to the member axis.

As an additional recommendation towards the proper design of rc elements strengthened in shear with FRP, it should be pointed out that the spacing s_f of strips (if such strips are used) should be limited below a certain maximum value, so that no diagonal crack may be formed without intercepting a strip. Moreover, full wrapping appears to be far more effective than partial jacketing (U-shaped or side jackets). When full wrapping is not feasible (for instance, when there is no access to the top side of T-beams), it is recommended that FRP strips be attached to the compressive zone of the rc member through the use of simple mechanical anchors. Finally, the degree of strengthening (defined in analogy to flexural strengthening) should be limited here too, so that the safety factor under service load is of the order of 1.2.

9.4.6 Elements with circular cross section

The material presented above refers mainly to rc elements of rectangular (or nearly rectangular) cross sections. If the cross section is circular (e.g., some columns), the contribution of FRP (wrapped around the column) to shear capacity is controlled by the tensile strength of the FRP jacket, but is limited to a maximum value corresponding to excessive dilation of the concrete due to aggregate interlock (one of the key shear force transfer mechanisms) at inclined cracks. By limiting the concrete dilation, that is, the radial strain (which is equal to the FRP hoop strain), to a maximum value, say, ε_{max}, one may easily show that for inclined cracks forming an angle θ with the column axis, the FRP contribution to shear capacity is as given below:

$$V_f = \varepsilon_{max} E_f \rho_f \frac{1}{2} \frac{\pi D^2}{4} \cot \theta \tag{9.22}$$

where D = column diameter and ρ_f = volumetric ratio of FRP. The derivation of Eq. (9.22) is easily understood if one assumes that at shear failure all the FRP material crossing an inclined crack is strained uniformly at ε_{max}. Experimental evidence suggests that ε_{max} is in the order of 0.004.[27]

9.5 Confinement

9.5.1 General

In existing reinforced concrete columns or column-type elements (e.g., bridge piers, chimneys) where insufficient transverse reinforcement and/or seismic detailing is provided, three types of failure modes can be observed under seismic loading.[29]

The first (and most critical) failure mode is the shear failure, where inclined cracking, cover-concrete spalling, and rupture or opening of the transverse reinforcement may cause brittle failure. If the shear capacity of columns is found below a desired level, shear strengthening should be applied according to the concepts presented in the previous section.

The second column failure mode consists of a *confinement* failure (lack of adequate transverse reinforcement) of a flexural plastic hinge region (column end), where flexural cracking may be followed by cover-concrete crushing and spalling, buckling of the longitudinal reinforcement, or compressive crushing of the concrete. Plastic hinge failures are typically limited to short column regions and are associated with considerable inelastic deformations; thus they are less destructive and much more desirable than shear failures. This desired ductile plastic hinging at the column ends can be achieved through adequate confinement, which will prevent cover-concrete spalling, longitudinal reinforcement buckling, and concrete crushing.

Finally, several columns (especially in bridges) feature lap splices at the lower end. While the confinement concepts discussed previously for plastic hinges also apply to lap-spliced column ends, the flexural strength of the column can only be developed and maintained when debonding of the reinforcement lap splice is prevented. Such debonding occurs once vertical cracks develop in the cover concrete and progresses with increased dilation and cover spalling. The associated rapid flexural strength degradation can be prevented or limited with increased lap length and confinement.

The discussion provided above emphasizes that the enhancement of confinement in structurally deficient reinforced concrete columns in seismic regions is of outmost importance. Confinement may also be beneficial in nonseismic zones, where, for instance, the axial load capacity of a column must be increased due to higher vertical loads (e.g., increased traffic in a bridge). In any case, confinement may be provided by wrapping rc columns with FRPs, in which the principal fiber direction coincides with the column circumferential direction (see Fig. 9.19 for a real application of the concept). Details of the behavior of concrete confined with FRP jackets are given next.

9.5.2 Behavior of concrete confined with FRP

Concrete is a restraint-sensitive material, which means that its response under axial loading depends heavily on the lateral restraint provided through confinement.[30] Let us examine, as an example, the case of a cylindrical concrete column with diameter D wrapped with an FRP jacket having thickness t_f and modulus of elasticity E_f in the circumferential direction, which will typically coincide with the principal fiber direction. Axial loading will cause shortening of the column and lateral expansion, which, in turn, will cause the FRP jacket to

Figure 9.19 Confinement of rc columns through wrapping of resin-impregnated carbon fibers.

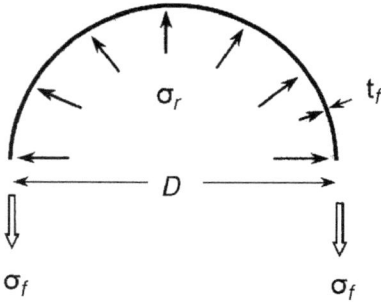

Figure 9.20 Confining action of circumferentially applied FRP.

extend in the circumferential direction and develop confinement pressure onto the concrete (Fig. 9.20). Assuming that the stress in the FRP is σ_f, one may easily show that the confinement stress, σ_r, exerted on the concrete in the radial direction equals

$$\sigma_r = \frac{2t_f}{D}\sigma_f = \frac{1}{2}E_f\rho_f\varepsilon_f \qquad (9.23)$$

where ρ_f is the volumetric ratio of confining FRP and ε_f is the circumferential FRP strain. The confinement stress reduces the shear stress components of the triaxial stress tensor in the concrete and suppresses microcracking and crack initiation/propagation.

One approach to examining the behavior of FRP-confined concrete is to look at its volumetric response, which is shown in Fig. 9.21.[31] The volumetric strain in this figure is equal to the sum of the three principal strains, tensile strains and dilation are considered negative, and the axial stress is normalized with respect to the strength of unconfined concrete, f_{c0}. As the figure shows, the FRP jacket effectively curtails the lateral expansion of concrete shortly after the unconfined strength is reached. It then reverses the direction of volumetric response, and the concrete responds through large and stable volume contraction (this is not the case with steel confinement jackets, where yielding is associated with unstable volumetric expansion, see Fig. 9.21).

The stress–strain response of FRP-confined concrete is illustrated schematically in Fig. 9.22. The figure displays a distinct bilinear response with a sharp softening and a transition zone at a stress level that is near the strength of unconfined concrete. After this stress the tangent stiffness stabilizes at a constant value, until the concrete reaches its ultimate strength f_{cc} when the jacket reaches tensile failure at a stress $f_{f,u,e}$ (and a corresponding strain $\varepsilon_{f,u,e}$) which is, in general, less than the uniaxial tensile strength $f_{f,u}$. This reduction is

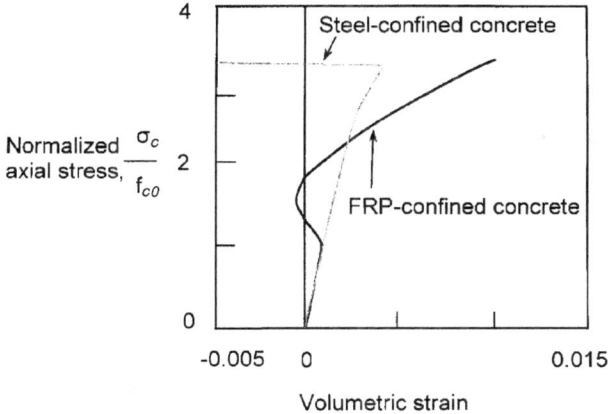

Figure 9.21 Volumetric response of FRP-confined concrete.

attributed to several factors, including (a) the triaxial state of stress in the FRP (due to axial loading and confining action, but also due to bending, e.g., at corners of low radius), and (b) the quality of execution (potential local ineffectiveness of some fibers due to misalignment, and overstressing of others; damaged fibers at sharp corners or local protrusions, etc).

Confinement of rc columns using FRP may also be provided in an *active* way, by applying the composite material through automated or manual winding with the fibers in a prestressed state. This concept does not change any of the behavior characteristics described above, except that an additional confining stress, given by Eq. (9.23) with σ_f

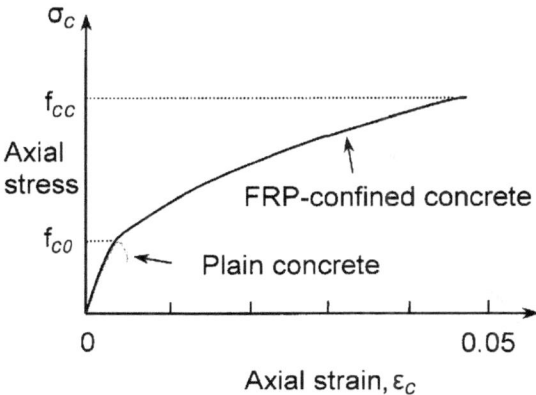

Figure 9.22 Axial stress–strain response of FRP-confined concrete versus plain concrete.

replaced by the prestress level, must be added to that caused by column shortening.

9.5.3 Analytical modeling for circular columns

From arguments discussed above, it is realized that reliable models for FRP-confined concrete have to account for a number of parameters, including (a) the circumferential stiffness of the FRP; (b) the continuous effect of the restraint provided by the FRP on the dilation tendency of the concrete; and (c) the composite action of the FRP-concrete column and the FRP-concrete interaction, based on micromechanics. Despite the fact that several models have been reported in the literature, the author's view is that proper modeling of FRP-confined concrete is just a little beyond its infancy. Nevertheless, one of the most recently proposed models, which appears to be in good agreement with test data (but suffers from being almost purely phenomenological), is briefly described next. The model was proposed by Samaan et al.,[32] and is based on the four-parameter relationship of Richard and Abbott,[33] which is also shown in Fig. 9.23.

$$\sigma_c = \frac{(E_1 - E_2)\varepsilon_c}{\left[1 + \left(\frac{(E_1 - E_2)\varepsilon_c}{f_0}\right)^n\right]^{1/n}} + E_2\varepsilon_c \tag{9.24}$$

$$f_{cc} = f_{c0} + 6f_r^{0.7} \quad \text{(MPa)} \tag{9.25a}$$

$$E_1 = 3950\sqrt{f_{c0}} \quad \text{(MPa)} \tag{9.25b}$$

$$E_2 = 245.61f_{c0}^{0.2} + 0.336E_f\rho_f \quad \text{(MPa)} \tag{9.25c}$$

$$f_0 = 0.872f_{c0} + 0.371f_r + 6.258 \quad \text{(MPa)} \tag{9.25d}$$

$$\varepsilon_{cu} = \frac{f_{cc} - f_0}{E_2} \tag{9.25e}$$

where n is taken equal to 1.5 and f_r is the maximum value of the confining stress, given by Eq. (9.23) with $\sigma_f = f_{f,u,e}$.

The constitutive law of Eq. (9.24) for FRP-confined concrete is plotted in Fig. 9.24 for several values of the product $E_f\rho_f$, and assuming that the ultimate FRP circumferential strain is $\varepsilon_{f,u,e} = 0.01$ (typical value for CFRP). The figure illustrates the role of FRP confinement in terms of both strength and deformability, and proves that confinement is increasingly beneficial as $E_f\rho_f$ increases.

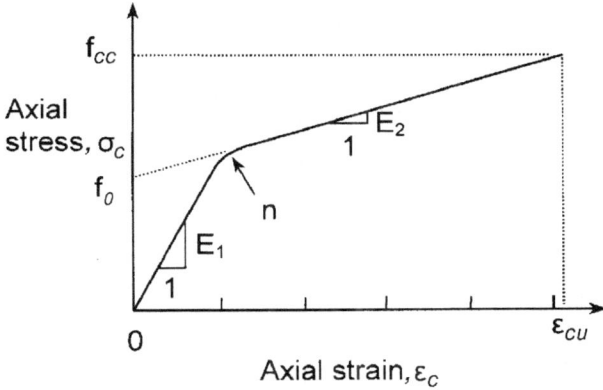

Figure 9.23 Parameters of bilinear confinement model.

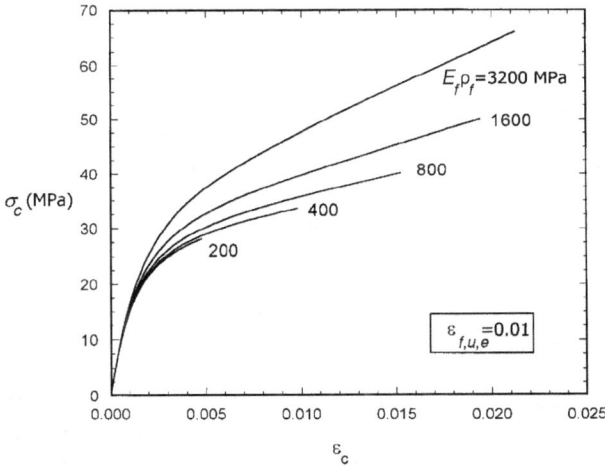

Figure 9.24 Axial stress–strain curves for various values of $E_f \rho_f$.

9.5.4 Rectangular columns

Confinement of rc columns is less effective if the cross section is rectangular. In this case, the confining stress is transmitted to the concrete through the four corners of the cross section, and increases with the corner radius R. Based on statistical analysis of test data, the following relationship for the strength of FRP-confined concrete columns of *square* cross section has been proposed[34]:

$$f_{cc} = f_{c0}\left[0.169 \ \ln\left(\frac{2R}{B}\frac{f_r}{f_{c0}}\right) + 1.32\right] \qquad (9.26)$$

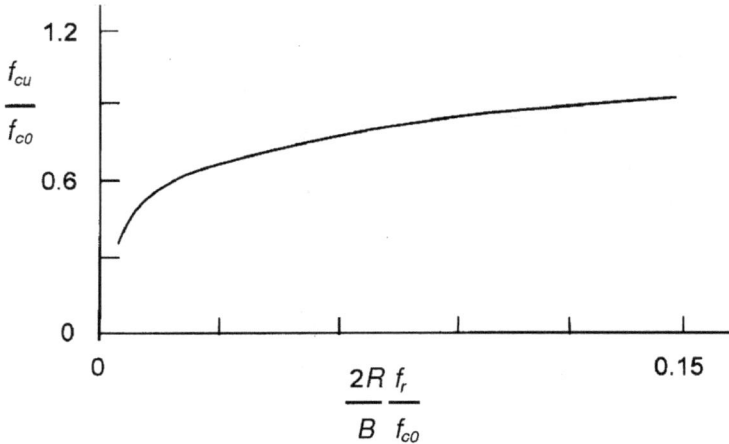

Figure 9.25 Normalized strength of FRP-confined concrete versus $(2R/B)(f_r/f_{c0})$.

where $B =$ inside dimension of FRP tube. It may be noted that the above equation, plotted in Fig. 9.25, does not render a value for cross sections with sharp edges. On the other hand, such sharp corners should always be avoided in real applications (through rounding-off with a grinder), and hence this singularity is not of concern in practical confinement analysis.

9.5.5 FRP material selection for confinement

The analyst/designer of a reinforced concrete column-type element (of circular cross section) to be confined by FRP will normally have to decide upon the desired (target) maximum strain level in the concrete, ε_{cu}. This strain will be determined based on both strength and (if so desired) *ductility* requirements. Ductility is typically quantified by the ratio of curvature at failure of a cross section to curvature at yielding of the internal steel reinforcement. With ε_{cu} known, the FRP material properties may be selected through combination of Eqs. (9.23) and (9.25a,c–e) as follows:

$$\varepsilon_{cu} = \frac{f_{c0} + 6\left(\frac{1}{2}E_f\rho_f\varepsilon_{f,u,e}\right)^{0.7} - \left[0.872f_{c0} + 0.371\left(\frac{1}{2}E_f\rho_f\varepsilon_{f,u,e}\right) + 6.258\right]}{245.61f_{c0}^{0.2} + 0.336E_f\rho_f}$$

(9.27)

The above formula is intentionally not simplified, because the form of Eqs. (9.24) and (9.25) (constitutive law for FRP-confined concrete) may change in the future should another confinement model be adopted. For a given FRP material (that is E_f and $\varepsilon_{f,u,e}$), Eq. (9.27) may be solved for the required FRP quantity ρ_f, through which the shell thickness may be obtained. Finally, the above equations suggest that the FRP quantity required to provide a certain confinement level increases as both E_f and $\varepsilon_{f,u,e}$ increase: stiff and highly deformable FRPs (in the circumferential direction) maximize the effectiveness of the confining shell.

9.6 Summary and Conclusions

A considerable volume of both analytical and experimental work has been done to date on the use of fiber reinforced polymer composites as strengthening materials of reinforced concrete structures. FRP materials offer engineers a variety of outstanding properties, including corrosion resistance, low weight, high strength, high deformability, and availability at practically unlimited sizes, with tailor-made properties; hence they provide unique solutions in several cases where conventional materials may prove inadequate. On the negative side, the cost of advanced FRP materials (e.g., CFRP) is relatively high and the sensitivity of some of their long-term properties to certain effects calls for special protective measures (e.g., AFRP may be particularly sensitive to ultraviolet irradiation and moisture absorption).

Practical worldwide applications of FRP in the form of epoxy-bonded strips or fabrics number a few thousands up to date. These applications are typically related to *flexural strengthening* of beams, slabs or columns, *shear strengthening* of beams or columns, and *confinement* of columns or column-type elements (e.g., chimneys).

Flexural strengthening is provided by bonding FRP strips or fabrics in the tension zones of reinforced concrete elements. The failure analysis of such elements follows well-established procedures for reinforced concrete structures, provided that the contribution of external FRP reinforcement is taken into account properly, and that special consideration is given to the failure modes associated with bond failure at the FRP–concrete interface. The latter requires analysis techniques that are based on classical strength of materials concepts but also on advanced fracture mechanics theories (which have been applied successfully, yielding simple design equations).

Shear strengthening is achieved by applying FRP strips or fabrics either covering partially or wrapped all around reinforced concrete elements. The failure analysis of shear-strengthened elements is based on the well-known truss model for reinforced concrete, properly modified to account for premature FRP failures due to debonding.

Finally, FRPs may be applied in the form of external jackets wrapped around columns to provide confinement and hence increase the strength and ductility. The confinement effect may be taken into account through proper modeling of the constitutive laws for concrete.

As a final remark, I emphasize that the use of advanced composites as strengthening reinforcement of concrete (and other structures) is only just out of its infancy. However, there are clear indications that the FRP strengthening technique will increasingly continue to be the perfect choice for many structural intervention projects, involving buildings, bridges, historic monuments, and other structures. Hence, education and training of structural engineers in this direction (which has, hopefully, been offered to some degree in this chapter) is, and will be, of great value. But it should also be emphasized that, even in the future, when further decreases in the price of high-performance fibers is expected, FRP composites will not replace conventional materials; instead, they will be used selectively to supplement them.

References

1. Meier, U. (1987) Bridge repair with high performance composite materials, *Mat. Technik*, **4**, 125–128 (in German).
2. Triantafillou, T. C. (1998) Strengthening of structures with advanced FRPs, *Prog. Struct. Eng. Mater.*, **1**, No. 2, 126–134.
3. American Concrete Institute (1999) *Guidelines for Selection, Design and Installation of Fiber Reinforced Polymer (FRP) Systems for Externally Strengthening Concrete Structures*. Detroit, MI.
4. American Concrete Institute (1996) *State-of-the-Art Report on Fiber Reinforced Plastic Reinforcement for Concrete Structures*, ACI Report 440R-96, Detroit, MI.
5. Triantafillou, T. C., Deskovic, N., and Deuring, M. (1992) Strengthening of concrete structures with prestressed fiber reinforced plastic sheets, *ACI Struct. J.*, **89**, No. 3, 235–244.
6. Deuring, M. (1993) *Strengthening of RC with Prestressed Fiber Reinforced Plastic Sheets*, EMPA Research Report 224, Dübendorf, Switzerland (in German).
7. Luke, P. S., Leeming, M. B., and Skwarski, A. J. (1998) ROBUST results for carbon fibre, *Concrete Eng. Int.*, **2**, No. 2, 19–21.
8. Maier, M. (1999) Personal communication.
9. Winistoerfer, A. and Mottram, T. (1997) The future of pin-loaded straps in civil engineering applications, in *Recent Advances in Bridge Engineering, Proceedings of the US-Canada-Europe Workshop on Bridge Engineering*, (U. Meier and R. Betti, eds.), EMPA, Switzerland, 115–120.
10. Triantafillou, T. C. and Plevris, N. (1992) Strengthening of RC beams with epoxy-bonded fibre-composite materials, *Mater. Struct.*, **25**, 201–211.
11. Knott, J. F. (1973) *Fundamentals of Fracture Mechanics*, Butterworths, London.
12. Täljsten, B. (1996) Strengthening of concrete prisms using the plate bonding technique, *Int. J. Fract.*, **82**, 253–266.
13. Brosens, K. and Van Gemert, D. (1997) Anchoring stresses between concrete and carbon fibre reinforced laminates, in *Non-Metallic (FRP) Reinforcement for Concrete Structures, Proceedings of the 3rd International Symposium*, Japan, Concrete Institute, **1**, 271–278.
14. Holzenkämpfer, P. (1994) Ingenieurmodelle des verbundes geklebter bewehrung für betonbauteile, Ph.D. dissertation, TU Braunschweig, Germany (in German).

15. Neubauer, U. and Rostásy, F. S. (1997) Design aspects of concrete structures strengthened with externally bonded CFRP-plates, in *Concrete+Composites, Proceedings of the 7th International Conference on Structural Faults and Repair*, **2**, 109–118.
16. Rostásy, F. S. and Neubauer, U. (1997) Bond behaviour of CFRP-laminates for the strengthening of concrete members, in *Composite Construction—Conventional and Innovative*, IABSE Conference Report, Innsbruck, 717–722.
17. Oehlers, D. J. and Moran, J. P. (1990) Premature failure of externally plated reinforced concrete beams, *ASCE J. Struct. Eng.*, **116**, No. 4, 978–995.
18. Jansze, W. (1997) Strengthening of reinforced concrete members in bending by externally bonded steel plates, Ph.D. dissertation, TU Delft, The Netherlands.
19. CEB-FIP Model Code (1990) *CEB Bulletin d'Information* No. 203, Comité Euro-International du Béton, EPF Lausanne, Switzerland.
20. Fardis, M. N. and Buyukozturk, O. (1979) Shear transfer model for reinforced concrete, *J. Eng. Mech. Div., Proc. Am. Soc. Civil Eng.*, **105**(EM2), 255–275.
21. Kaiser, H. (1989) Strengthening of reinforced concrete with epoxy-bonded carbon fiber plastics, Ph.D. dissertation, ETH, Zürich, Switzerland (in German).
22. Blaschko, M. (1997) *Strengthening with CFRP*, Münchner Massivbau Seminar, TU München (in German).
23. Triantafillou, T. C. and Antonopoulos, C. P. (2000) Design approach for concrete members strengthened in shear with FRP, *ASCE J. Compos. Construct.*, accepted for publication.
24. Park, R. and Paulay, T. (1975) *Reinforced Concrete Structures*, Wiley, New York.
25. Eurocode No. 2 (1991) *Design of Concrete Structures*, European prestandard ENV 1992-1-1.
26. Triantafillou, T. C. (1998) Shear strengthening of reinforced concrete beams using epoxy-bonded FRP composites, *ACI Struct. J.*, **95**, No. 2, 107–115.
27. Priestley, M. J. N. and Seible, F. (1995) Design of seismic retrofit measures for concrete and masonry structures, *Construct. Build. Mater.*, **9**, No. 6, 365–377.
28. Khalifa, A., Gold, W. J., Nanni, A., and Aziz, A. M. I. (1998) Contribution of externally bonded FRP to shear capacity of rc flexural members, *ASCE J. Compos. Construct.*, **2**, No. 4, 195–202.
29. Paulay, T. and Priestley, M. J. N. (1992) *Seismic Design of Reinforced Concrete and Masonry Buildings*, Wiley, New York.
30. Pantazopoulou, S. J. (1995) Role of expansion on mechanical behavior of concrete, *ASCE J. Struct. Eng.*, **121**, No. 12, 1795–1805.
31. Mirmiran, A. and Shahawy, M. (1997) Behavior of concrete columns confined by fiber composites, *ASCE J. Struct. Eng.*, **123**, No. 5, 583–590.
32. Samaan, M., Mirmiran, A., and Shahawy, M. (1998) Model of concrete confined by fiber composites, *ASCE J. Struct. Eng.*, **124**, No. 9, 1025–1031.
33. Richart, R. M. and Abbott, B. J. (1975) Versatile elastic-plastic stress strain formula, *ASCE J. Eng. Mech.*, **101**, No. 4, 511–515.
34. Mirmiran, A., Shahawy, M., Samaan, M., El Echary, H., Mastrapa, J. C., and Pico, O. (1998) Effect of concrete parameters on FRP-confined concrete, *ASCE J. Compos. Construct.*, **2**, No. 4, 175–185.

Composites in Concrete Construction

K. Pilakoutas

Department of Civil & Structural Engineering,
Sir Frederick Mappin Building, Mappin Street,
Sheffield S1 3JD

10.1 Introduction

It is estimated that if just one concrete offshore platform for oil extraction were to be reinforced with CFRP, it would absorb more than the world's current annual production of carbon fibres. This example highlights the massive amounts of structural reinforcement required in construction, which amount to two to three orders of magnitude more than the entire production of resin matrix composites. Hence, it is reasonable to conclude that, as the construction industry develops the appetite for composites, new supply solutions will have to be found and a drop in prices will become inevitable with the increased volume of production.

The construction industry has used composites since the early 1950s,[1] but it was only in the 1970s that composites were used as concrete prestressing materials and in the late 1980s that they started being used as reinforcement in concrete.[2] This chapter deals with the special case of using composites as concrete reinforcement.

There are many reasons why civil and structural engineers may need to use resin matrix–continuous fibre (fibre reinforced polymer, FRP) reinforcement in concrete. The primary reason is durability; other reasons include electromagnetic neutrality, high strength, and low weight. Each of these reasons will be briefly examined in the following and likely applications will be identified.

10.1.1 Durability of reinforced concrete

It is currently estimated that the infrastructure repair and maintenance bill in the European Union is around US$30 billion and a similar figure is estimated for the United States. A large proportion of this expense is due to problems with concrete structures. Hence, solutions enhancing the longevity of concrete structures, such as the use of FRP reinforcement, always attract much research and development effort.

Reinforced concrete constructions have revolutionized the construction industry for over a century, due to its versatility and low cost. For many years, the public perception was that concrete is permanent. After all, the Romans used it in constructions and there are many examples where it is still serving its purpose, two thousand years on.

Unlike Roman concrete, however, concrete in its modern use is reinforced with steel, since inherently it has low tensile strength. Reinforcement in concrete is provided not only to add tensile strength, but also to enhance the element's shear strength and ductility and to reduce crack widths, structural deflections, and vibrations. Hence, when reinforcement is used the product is a "composite" material (concrete and tensile reinforcement), which constitutes reinforced concrete (RC). Like all composites, the characteristics of RC depend on its constituent materials.

When it comes to durability, the alkaline environment of concrete provides the necessary protection to conventional steel reinforcement from the elements. Nonetheless, when exposed to the environment, conventional steel has the irresistible urge to return to its original oxide state and corrodes (see Fig. 10.1). Hence, civil engineers have adopted thick concrete covers to the steel reinforcement together with measures to reduce concrete crack widths and permeability, while cement scientists have developed highly alkaline cements. However, the environmental attack is relentless and occurs mostly in the form of carbonation or sulphate, or chloride attack. Carbonation is natural, taking place due to the diffusion of CO_2 from the atmosphere, and though it does not harm the concrete structurally, it reduces its alkalinity. Chlorides or sulphates may be present in the water, in additives (chemicals added to improve the concrete properties) or in the aggregates of fresh concrete (in places such as the Middle East) and can cause the rapid deterioration of concrete. However, these chemicals can also penetrate the concrete surfaces exposed in marine environments, in the ground, and, in recent decades, in colder climates as a direct result of the use of de-icing salts.[4]

Once the alkalinity of concrete is reduced, in the presence of oxygen and water, steel will have the tendency to corrode. Corrosion is a

Figure 10.1 Heavy deterioration of a concrete structure due to corrosion of reinforcement. (*Courtesy Dr. C. J. Lynsdale.*)

complicated process in which steel becomes part of an electrochemical cell. Due to the excellent electrical conductivity of steel, free electrons can be transported from remote parts of the element, hence the anode and cathode can be anywhere along its length. The iron in steel reduces to iron oxides commonly known as rust. These oxides have a much larger volume than the iron they consume and this creates a mechanism for the build-up of internal stresses. These stresses are capable of eventually splitting the concrete near the surface. The resulting crack allows the much faster ingress of chemicals and the reaction speeds up until the concrete cover is fully damaged and spalls off. Complete failures in structures due to corrosion are rare, since the signs are not in general easy to miss. However, in prestressed concrete structures, the signs of corrosion are not so easy to detect. The sudden and unexpected failure of a post-tensioned highway concrete bridge in Wales in 1985 led to the ban of prestressing steel strands in bridge post-tensioning by the UK Highways Agency (this ban was only lifted in recent years).[5]

10.1.1.1 Conventional solutions. Attempts to reduce steel reinforcement corrosion have led over the years to the development of higher-alkalinity cements and to many other solutions, some of which even led to the development of new industries! The conventional approach to preventing corrosion is the reduction in the permeability of concrete, primarily through the provision of a good concrete cover. In addition, strict quality control is necessary to ensure that the structure has a cover that is 100% fault-proof. However, even when the conventional wisdom is fully applied, the risk of corrosion still exists in highly aggressive environments. To reduce the risk of corrosion in such highly aggressive environments, new industries are providing new solutions such as concrete surface protective coatings to stop the ingress of CO_2 and other water-soluble chemicals, corrosion inhibitor admixtures at the wet stage, epoxy coating of reinforcement, and galvanising of reinforcement. A more innovative approach adopted in recent decades is cathodic protection. This technique, which was initially developed as a rehabilitation measure, utilises an electric current or a sacrificial anode to protect the main reinforcement.[6]

10.1.1.2 New solutions. Most of the above solutions have either had failures or are expensive; hence, in certain projects (especially in the Middle East), conventional steel reinforcement is still not allowed. Stainless steel reinforcement is one of the more robust anticorrosion solutions. Such reinforcement only needs to be applied near the surface and, hence, the additional cost in a major project is minimized. However, engineers still express a concern in using different steel alloys side by side due to fears of bimetallic reaction leading to corrosion.

Resin matrix–continuous fibre composites in the form of FRP (fibre reinforced polymer) reinforcement appeared in the market in the early 1990s as another solution to the corrosion problem,[2,7] even though the ability of such composites to resist the alkaline environment of concrete was not thoroughly investigated. However, durable FRP reinforcement now exists in the market that has been designed to resist the alkaline concrete environment.

10.1.1.3 Applications. The use of FRP in concrete for anticorrosion purposes is expected to find applications in structures in or near marine environments, in or near the ground, in chemical and other industrial plants, in places where good-quality concrete cannot be achieved, and in thin structural elements. Currently, most applications of FRP reinforcement in concrete are found in Japan, where many demonstration projects were developed in the early 1990s. Research and development is now actively taking place in many countries, most prominently in North America and Europe. In Europe, the EUROCRETE project, which was a project involving

Figure 10.2 The first concrete footbridge in Europe with only FRP reinforcement. (*Courtesy EUROCRETE project.*)

many industrial partners and the University of Sheffield, installed the first completely FRP reinforced footbridge in 1996 (Fig. 10.2). This was followed by a number of other demonstration projects including another bridge in a golf course in Norway and a fender in Qatar. Figure 10.3 shows the reinforcement of the fender unit before casting, and Fig. 10.4 shows the fender in position.

In thin structural elements, there is the possibility of reducing the concrete cover needed to protect the reinforcement, and, this might lead to reduced sections and new designs. Examples of such elements include cladding panels, parapets, and manhole covers.

10.1.2 Electromagnetic neutrality

Steel reinforcement can interfere with magnetic currents and, hence, it is usually avoided in applications where magnetic neutrality is required. Examples of these include the bases of large motors, magnetic scanning equipment, and magnetic levitation systems such as the MAGLEV. In Japan, much of the research on the use of composites in concrete was driven by the research on the MAGLEV.

Electromagnetic interference is progressively a nuisance especially to the mobile telecommunications industry and to the defence indus-

Figure 10.3 FRP reinforcement for a fender unit. (*Courtesy EUROCRETE project.*)

try. Applications in these industries are likely to increase with time, in the vicinity of both transmitting stations and receiving devices. Concrete fence posts and fence panels, reinforced with FRP reinforcement, were used around such a station during the EUROCRETE project (Fig. 10.5).

10.1.3 High strength

FRP reinforcement can develop higher strengths than conventional reinforcement and, hence, can be utilised to reduce congestion of reinforcement in certain applications. However, as will be seen later, the strength is developed at a high strain, and this has other struc-

Figure 10.4 FRP reinforced concrete fender in position. (*Courtesy EURO-CRETE project.*)

Figure 10.5 Concrete fence reinforced with FRP around a transmitter. (*Courtesy EURO-CRETE project.*)

tural implications. Hence, it is not anticipated that the high strength of FRP will be a major advantage in many RC applications. However, if FRP is prestressed, not only is the high strength utilized, but also the lower elastic modulus will imply lower losses in the longer term. Hence, a major market for FRP is likely to be found in the concrete prestressing industry. In addition to concrete prestressing, FRP cables can be used in cable-stay bridges and in other anchoring applications such as ground anchors or rock bolts.

However, one aspect of FRP behavior that should not be overlooked is stress corrosion, particularly of glass FRP, which means that carbon FRP is likely to dominate most of the above-mentioned applications.

10.1.4   Light weight

The light weight of FRP may have some practical advantages in construction, but again it is not anticipated to be the driving force behind its application as reinforcement in concrete. Normally, the weight of the concrete is high, and hence, small savings in reinforcement weight will not be significant. However, in some exceptional circumstances, the use of lightweight reinforcement may speed up construction, especially in inaccessible or confined spaces, where it is difficult to have many workers side by side.

The light weight of FRP becomes a real advantage when dealing with externally bonded reinforcement for repair purposes. This application of FRP is dealt with in Chapter 9.

10.1.5   Other applications

It is worth pointing out that the aim of using FRP reinforcement is not to totally replace steel reinforcement from RC structures, but to identify areas in which steel is totally unsuitable for use or the structural advantages of steel are not required. Examples of such applications include:

- Nonstructural elements, such as partition walls and cladding panels
- Deep structural elements, such as above openings in buildings or bridges with no headroom restrictions
- Elements in which larger deformations are an advantage, or not a problem, such as in highway separators or parapets, or where the higher deflections allow the member's inertia to be better mobilized,
- Elements in which high damping is required, such as in large suspended floors or in bridges

- Elements in which good fatigue characteristics are required, such as in bridges and machine bases

- Elements in which arching action can be developed, such as in vaults, domes, and other three-dimentional continuous structures

- Elements in which extremely heavy loading is to be carried over short spans, such as in warehouses or short rail bridges and tunnels.

It is conceivable that most concrete structures could be designed with FRP rather than steel reinforcement if the supply and costs were to change considerably. However, it should be appreciated that, historically, the structural form of constructions depends on the material characteristics used at the time. Hence, due to the different mechanical properties of FRP reinforcement, concrete constructions reinforced with FRP may appear different in structural terms from current structures.

10.2   Types of FRP Reinforcement

Currently, many manufacturers are developing nonferrous alternatives to conventional steel reinforcement and, consequently, a variation in the nature of end products is expected. The commonly used nonferrous fibers include carbon, glass, and aramid, which are all extremely strong. However, because they are very thin, they suffer from imperfections over long lengths. To avoid this problem, these fibers are impregnated with a resin matrix and the result is commonly called fiber reinforced polymer (FRP). Since resins are very weak and very flexible as well as expensive, a minimum resin volume ratio, enough to bind the fibers together, is always desirable. A fiber volume ratio of around 70% is about the maximum that can be achieved for producing effective FRP rods or plates.

10.2.1   Manufacture

FRP materials can be manufactured using many techniques, including manual lay-up, moulding, tube rolling, filament winding, braiding, and pultrusion.[7] Pultrusion is the most common method used for manufacturing FRP rods in Europe and America. The pultrusion process is explained schematically in Fig. 10.6.[8] The process starts with several spools of fibers in the form of strands. The fibers are pulled through a series of guides, where they are brought together in the desired arrangement. Next, they are passed through a resin path for impregnation. Finally, the wet fibers are passed through heated dies, where the excess resin is squeezed out of the fibers and the curing

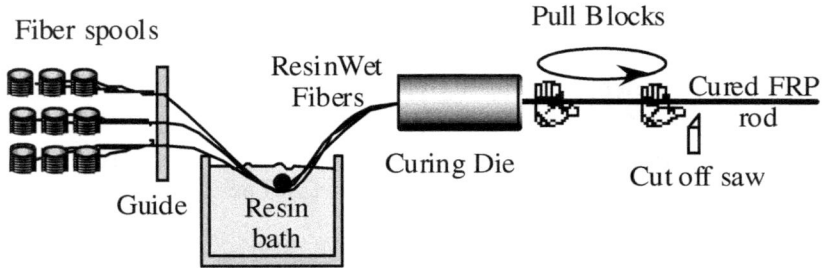

Figure 10.6 Schematic representation of the pultrusion process.

process initiates. The entire process is kept moving by pulling the cured bar.

FRP materials are available in a variety of shapes and sizes: rods, plates, grids, spirals, and links are some common examples used in the construction industry. Some of the different types of FRP reinforcement currently available in the market are shown in Figs. 10.7 to 10.16.

FRP can easily be manufactured in any shape while the resin is not cured. After that stage, the shape cannot be changed and this creates a potential problem in use. Hence, unlike steel, FRP materials produced by thermosetting resins cannot be bent on site. Attempts to bend a straight bar will result in stress concentration leading to the fracture of the bar. To overcome this problem, engineers must adopt different

Figure 10.7 FRP Grid NEFMAC. (*NEFCOM Co., Ltd.*)

Figure 10.8 Aramid fiber composite material ARAPREE. (*Nippon Aramid Co., Ltd.*)

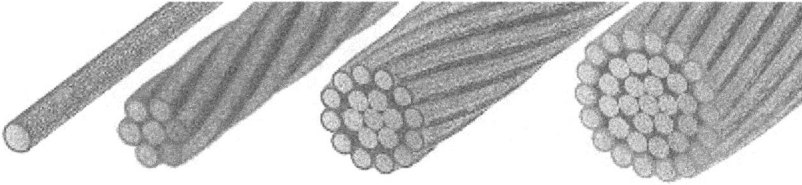

Figure 10.9   Carbon fiber composite cables CFCC. (*Tokyo Robe MFG. Co., Ltd.*)

Φ6mm round rod

Φ6mm deformed rod

Φ8mm round rod

Φ8mm deformed rod

Flat plate rod

Φ12.5mm strand
(under development)

Figure 10.10   Aramid FRP TECHNORA ROD. (*Teijin Ltd.*)

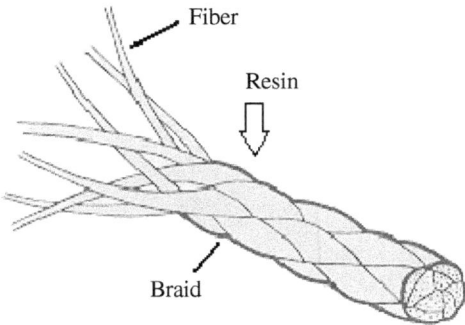

Fiber

Resin

Braid

Figure 10.11   High-performance fiber composite material FIBRA. (*Shinko Wire Co., Ltd.*)

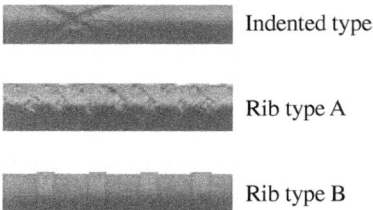

Indented type

Rib type A

Rib type B

Figure 10.12   CFRP rods LEADLINE. (*Mitsubishi Kasei Corporation.*)

Figure 10.13 High-performance fiber composites (thermoplastic resin).

Figure 10.14 C-bars. (*Marsall Industries Composites, Inc.*)

Figure 10.15 FRP rebar. (*Corrosion Proof/Hughes Brothers, Inc.*)

Figure 10.16 EUROCRETE rebar. (*Eurocrete Ltd.*)

design approaches by taking into account the different material properties of FRP bars. New FRP materials are currently being developed using thermoplastic materials, which could be bent locally after the application of heat.

Several techniques are used to improve the rebar bond characteristics, including cutting indentations in the resin surface, overmolding a new surface on the bar, and coating the surface with sand. The EUROCRETE rebars are produced by introducing a suitably designed peel-ply on the surface of the bar prior to curing. After curing, the peel-ply is removed, leaving the rebar with a rough surface.

10.2.2 Properties

FRP rods are fundamentally different from conventional steel rebars, having different color and texture, and lower weight and flexibility. As a result, they project a new, cleaner, and "high-tech" image for reinforcement in concrete structures. A study of their properties reveals their advantages and drawbacks compared to conventional steel reinforcement, as shown in Table 10.1.

FRP materials, unlike steel, are strongly anisotropic. Their mechanical properties are different in the two transverse directions, having the longitudinal axis as the stronger one. In addition, their mechanical properties vary significantly from one product to another depending, mainly, on the nature and volume of fibers, the shape of the cross section, the mechanical properties of resin, and the fiber orientation. For these reasons, it is very difficult to specify universal values for the mechanical properties of all FRP materials; thus, only indicative values can be given.

Generally, FRP bars develop much greater tensile strength than conventional high-strength steel, depending on the nature of the fibers. Glass FRP bars can develop more than twice the tensile strength of steel, whereas carbon and aramid bars can develop more than three times the tensile strength of steel. A general comparison of the ranges of properties of FRP and steel rebars is shown in Fig. 10.17.

The ultimate tensile strength of FRP bars, and specially that of glass FRP bars, is sensitive to the rebar diameter and decreases with an increase in diameter. This is because fibers located near the center of the bar cross- section are not subjected to as much stress as those fibers near the outer surface of the bar, due to the resin-dependent shear-lag phenomenon.[9] The compressive strength of FRP is not as great as their tensile strength due to buckling of the fibers, but in RC

TABLE 10.1   Main Advantages and Disadvantages of FRP Materials

Advantages	Disadvantages
Higher ratio of strength to self-weight (10–15 times greater than steel)	Higher cost (more than 10 times by weight)
Carbon and aramid fibre reinforcement have excellent fatigue characteristics	Lower elastic modulus (except some carbon FRPs)
Excellent corrosion resistance and electromagnetic neutrality	Glass FRP reinforcement suffers from stress corrosion
Low axial coefficient of thermal expansion	Lack of ductility

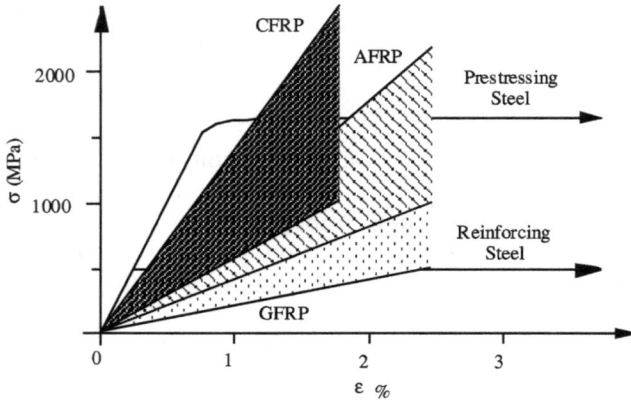

Figure 10.17   Stress-versus-strain relationship for FRP and steel bars.

this is not a major problem, since the rebars are essentially used in tension.

The elastic modulus of FRP bars is generally lower than that of steel. Usually, GFRP bars have the lowest elastic modulus, around 20–30% of that of steel, and carbon FRP the highest, between 50 and 75% of that of steel, as shown in Fig. 10.17. This lower value of modulus of elasticity is expected to play an important role in the deformability of FRP RC members. FRP bars remain practically elastic up to failure, unlike steel rebars that exhibit ductile behavior. This aspect of FRP behavior raises the fear of brittle failure and will be dealt with later.

During the EUROCRETE project, several types of FRP rebars were developed specifically to resist the concrete alkaline environment. Glass and carbon fibers were used. Aramid fibers were also tried but were not optimized in performance. Circular sections of diameter 8.5 mm and 13.5 mm and 8 mm-square rebars with surface indentations were developed. The typical mechanical characteristics for the GFRP bars showed an elastic modulus of 45 GPa and strength (in concrete) of around 800 MPa. Similarly, the CFRP bars showed an elastic modulus of 115 GPa and strength (in concrete) of around 1400 MPa. The determination of strength of FRP rods is a subject of debate.

Different manufacturers adopt different testing procedures and report different types of strength, such as theoretical, guaranteed, characteristic, average, etc. Due to their directional properties, FRP will show different strengths depending on the amount of lateral pressure applied in the grips during testing. Researchers have developed tests that minimize the lateral pressure in the development region and these tests normally lead to higher strengths. However,

FRP rebars in concrete are bonded in the concrete and the bond demand can be very intense near cracks. As a result, the strengths obtained in rebars that rupture in tension due to flexure in concrete beams are lower than theoretically calculated, or than strengths obtained in tests with minimized lateral pressure.

The directional properties of FRP also affect their shear strength, which is very low compared to their tensile strength. FRP bars can be cut easily with a simple saw in a direction transverse to the longitudinal axis. The lower shear strength of FRP rebars can also influence their bonding behavior to concrete.

Since the material characteristics are so different from those of steel, it can be concluded that, to develop an understanding of the structural behavior of FRP RC elements and components, a review of all the fundamental assumptions made in RC design and analysis is required.

10.3   Structural Considerations and Failure Modes

The successful replacement of conventional steel with FRP as concrete reinforcement requires the examination of many aspects and likely modes of failure. This section will deal with the various modes of failure or fracture expected from composite elements made of FRP RC, such as

1. Flexural failure due to concrete crushing or FRP tensile fracture

2. Serviceability failures due to deflections or cracking

3. Shear failure of concrete or FRP reinforcement

4. Bond failure

5. Time-dependent failures

Before examining these failures, it is necessary to assess whether the conventional approach to RC analysis is still valid. Therefore, the assumption that plane sections remain plane, often called "Bernoulli's assumption," needs to be verified. Figure 10.18[10] shows experimental results that verify that provided bond failures can be avoided, the plane sections assumption is valid. This enables us to adopt the conventional techniques for analysis of RC sections.

10.3.1   Flexural failures

Due to its high tensile strength, FRP reinforcement of the same cross-sectional area is expected to lead to stronger structural elements as

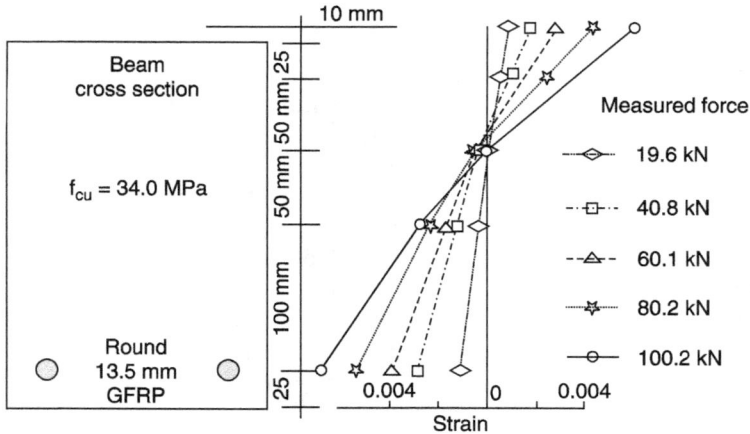

Figure 10.18 Strain distribution along the height of a FRP reinforced section.

long as that strength is fully mobilized. In RC design, when the strength of reinforcement is fully utilized, the section is considered to be "under-reinforced." Nevertheless, the effectiveness of flexural reinforcement is heavily undermined when the cross section becomes "over-reinforced." A RC section becomes over-reinforced when the reinforcement does not reach its full potential and concrete crushes in compression. When the failure in concrete and reinforcement is simultaneous, then that section is called "balanced."

For conventional steel reinforcement, the strength to stiffness ratio is similar to that of normal concrete and, hence, the neutral axis depth for the balanced rectangular section is around the middle of the overall effective depth. For FRP reinforcement, the strength to stiffness ratio is an order of magnitude greater than that of concrete and, hence, the neutral axis depth for the balanced section is very close to the compressive end, as shown in Fig. 10.19.

The above considerations imply that:

1. A large proportion of the cross-section is subjected to tensile strains.

2. Much larger flexural deflections are expected for similar cross sections.

3. A greater strain gradient exists in the compressive zone.

4. Due to (1), shear deformations are expected to increase and shear strength to decrease.

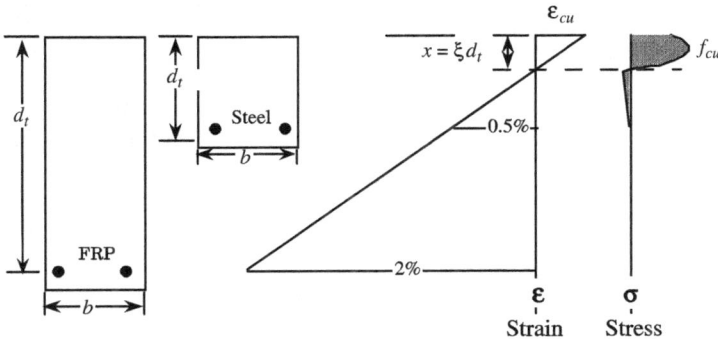

Figure 10.19 Strain distribution for a GFRP reinforced concrete section.

5. Due to (1) and (4), anchoring of the FRP rebars may be more difficult.

Pre-stressing or post-tensioning the FRP reinforcement will eliminate most of the above problems; however, it makes the construction process much more difficult and expensive.

10.3.1.1 Types of flexural failure. If all the other modes of failure are avoided, flexural failure will be reached either by crushing of the concrete in compression (Fig. 10.20) or by rupturing of the FRP reinforcement in tension. Unconfined concrete normally crushes in

Figure 10.20 Concrete crushing.

compression at a strain of around 0.0035. Its failure is brittle and in steel RC elements it can be avoided by appropriate design, which can lead to steel yielding. The tensile rupture of FRP reinforcement depends on its type, but also on its bond characteristics. By its nature, RC cracks in tension and the FRP reinforcement is there to prevent the opening crack. However, due to the very large difference in stiffness between the cracked and uncracked section, the stress in the reinforcement is expected to vary substantially from one section to the other. This puts a very high demand for bond or surface shear stresses and this can lead to FRP rupture at lower strength than the material strength.

To predict the mode of failure of RC sections, it is necessary to examine the stress that is developed in the reinforcement when concrete failure takes place. This is demonstrated in Fig. 10.21 versus percentage amount of reinforcement. The diagram shows that at concrete failure the yield strength of steel is exceeded for reinforcement ratios below 3.5%. This value depends on the strength of reinforcement and concrete. This means that for lower ratios of reinforcement, the section is under-reinforced and, given that steel has ductility, ductile failures will be obtained. Since engineers prefer ductile to brittle failure, the entire basis of conventional RC design is based on this type of failure occurring first. For that reason, sections that are over-reinforced are not allowed in design and, for optimum design, balanced sections are required.

It is clear from Fig. 10.21 that very small amounts of FRP reinforcement (lower than 0.5%) are necessary to lead to under-reinforced

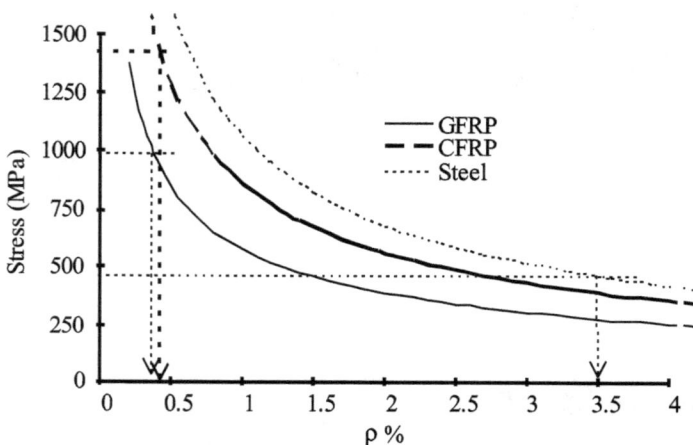

Figure 10.21 Stress in reinforcement at concrete failure versus percentage amount of reinforcement.

sections. Such small amounts of reinforcement are undesirable, because they offer little additional resistance to the concrete section after concrete cracking and will lead to very large cracks and large deflections.

Consequently, it has to be accepted that FRP over-reinforced concrete sections will be inevitable in most structural applications. To overcome this problem, other sources of ductility may be utilized, if necessary, as will be described later.

It is obvious from Fig. 10.21 that, as the reinforcement ratio increases, the stress developed in the FRP bar decreases. Eventually, this stress reduces below the strength of steel and, as a result, larger areas of reinforcement are required to achieve the same moment. This has cost implications and it can be seen that for the lower ratios of reinforcement, the use of FRP reinforcement is more efficient.

Although it is not advocated that steel reinforcement should be replaced with FRP in RC sections without due consideration to all design implications, it is acknowledged that, initially, engineers may be asked to do just that. Therefore, a brief guide to the available flexural design approaches is given in the next section.

10.3.1.2 Flexural design approaches. In this section, three different approaches are examined when substituting one type of reinforcement with another: equal stiffness; equal strength; and equal area. Three steel RC beams are designed according to the above approaches, and comparisons are made with two GFRP reinforced beams for which experimental results are available. The steel reinforced beams are designed and analyzed without the use of safety factors in order to facilitate comparisons. The concrete strength used in the beams is C40 (40 MPa) and a concrete cover of 25 mm is assumed. The strength and elastic modulus were taken as 460 MPa and 200 GPa for steel and 766 MPa and 45 GPa for the GFRP reinforcement. The beam cross section is 250 mm deep and 150 mm wide. The beams are subjected to four-point bending with shear span 766 mm. The results obtained are shown in Table 10.2 together with results of the two GFRP beams tested during the EUROCRETE project.[11]

Stiffness approach. In the stiffness approach, the GFRP and reference steel beam are designed to have the same reinforcement equivalent stiffness, that is,

$$E_{FRP}A_{FRP} = E_S A_S \qquad (10.1)$$

where E_{FRP} and E_S are the elastic moduli of FRP and steel reinforcement, respectively, and A_{FRP}, A_S are the cross-sectional areas of FRP and steel reinforcement, respectively.

TABLE 10.2 Capacity of Beams Based on Different Design Approaches

Material	Design Approach	Reinforcement Area (mm²)	Reinforcement Percent (%)	Reinforcement Capacity (kN)	Section Capacity (kN)	Failure Pattern	Neutral Axis (mm)
GFRP	GB10	429.4	1.31	329	100.2	Crushing	56.0
Steel	Stiffness	96.6	0.29	44	32.1	Yield	56.0
Steel	Strength	717.7	2.17	329	183.5	Crushing	112.3
Steel	Area	429.4	1.31	198	126.0	Yield	100.6
GFRP	GB13	286.3	0.87	220	85.5	Crushing	47.0
Steel	Stiffness	64.4	0.20	30	21.6	Yield	47.0
Steel	Strength	477.1	1.45	219	137.4	Yield	104.4
Steel	Area	286.3	0.87	132	88.8	Yield	86.8

For beams reinforced according to this approach, serviceability limit state conditions such as deflections and crack width are automatically satisfied. The area of GFRP reinforcement is significantly larger than the area of steel, by E_S/E_{FRP} (the modular ratio). Since a much larger area of stronger reinforcement is used, the FRP cross section will inevitably become over-reinforced. This method does not lead to any significant change in the position of the neutral axis between the reference and the GFRP beam. However, a massive increase in the ultimate capacity of the GFRP reinforced beam is achieved through the change of the failure pattern, i.e., reinforcement yield changes to concrete compressive failure. This capacity is shown to be 3 and 4 times the steel RC capacity for beams GB10 and GB13,[11] respectively.

Strength approach. In the strength approach, the GFRP and the reference beam are designed to have the reinforcement with the same equivalent strength, that is,

$$f_{FRP}A_{FRP} = f_S A_S \qquad (10.2)$$

where f_{FRP} and f_S are the design strengths of FRP and steel reinforcement, respectively.

The equal strength approach adopts a smaller amount of GFRP reinforcement and, hence, after initial concrete cracking, this results in a significant reduction of the beam stiffness. The serviceability limit state conditions that apply to steel reinforced sections will be exceeded. Despite the significant reduction in the area of reinforcement, this method again produces over-reinforced sections due to the lower elastic modulus of GFRP bars. This results in a change of the failure mode and leads to a reduction in the ultimate capacity of the section. In both GFRP reinforced beams, shown in Table 10.2, a reduction in capacity of around 40% is observed, which means that the design strength of the glass is not reached.

Area approach. The equal area approach is very simple since it adopts the same amount of reinforcement irrespective of its type. This approach leads to a decrease in the stiffness of the beam, as well as to a change in the beam mode of failure. This is as a direct consequence of the strength to stiffness ratio and, consequently, of the different level at which sections reinforced with GFRP and steel become balanced. Reinforced concrete sections with low reinforcement ratios will lead to GFRP RC beams with similar ultimate capacities as steel RC beams, such as GB13. However, for higher ratios of reinforcement a small reduction in capacity is expected, as in the case of beams GB10.

A summary of the three approaches is given in Table 10.3. Obviously the choice of design approach depends on the design constraints. When stiffness is of paramount importance, such as when dealing with deflections, then more FRP reinforcement will be required than steel. However, in many applications, when strength is important, savings in the amount of reinforcement might be possible.

10.3.1.3 Deflection serviceability. From Table 10.3, it is clear that when deflection serviceability limit states govern, then only the similar rebar stiffness approach is appropriate. However, since the elastic modulus of the GFRP rebar, for example, is 4–5 times less than that of steel, to achieve an equal stiffness, 4–5 times more area of reinforcement needs to be provided. Clearly, in this case, using FRP reinforcement will lead to an expensive solution, which may under-utilize the rebar strength by 10–15 times.

Alternative ways of controlling deflections include prestressing the rebar, reducing the span, and increasing the effective depth of the element.

10.3.1.4 Crack control. The control of crack widths is another design criterion that requires the use of reinforcement. Controlling cracks at the early stages of concrete depends more on the location and distribu-

TABLE 10.3 Flexural Design Approaches

FRP Beams with				Service	
Similar Rebar	A_{FRP}/A_S	Section Capacity		Deflection	Stiffness
Stiffness	E_S/E_{FRP}	Higher		Same	Same
Strength	f_S/f_{FRP}	Same or lower		Much higher	Much lower
Area	1	Higher or lower		Higher	Lower

tion of the reinforcement rather than its stiffness. Consequently, light FRP mesh reinforcement is ideal for this task.

Controlling cracks arising as a result of large tensile strains due to flexure in concrete requires not only a dense distribution of reinforcement with good bond characteristics, but also high stiffness. FRP reinforcement is clearly at a disadvantage in this respect, compared to steel. Nonetheless, crack control is a design constraint that is applied for durability as well as for aesthetic considerations. Due to the corrosion resistance of FRP, engineers working on design recommendations are proposing the relaxation of crack width limitations, and a figure of around 0.5 mm appears to be emerging.

10.3.1.5 Flexural capacity. All three design approaches described above are capable of achieving a flexural capacity similar to or higher than that of steel reinforced elements. Clearly, the most economical approach will be the one utilizing the full strength of the FRP material, since that means less area of flexural reinforcement and possibly less initial material cost. The question that needs to be answered, nevertheless, is whether, at low ratios of reinforcement, the FRP reinforced sections will have a reasonable level of capacity. This is examined in Fig. 10.22, which shows the normalized moment capacity versus the reinforcement percentage ratio. This figure demonstrates that GFRP and CFRP RC sections have a higher capacity than steel RC sections for values of ρ below 1.5% and 3%,

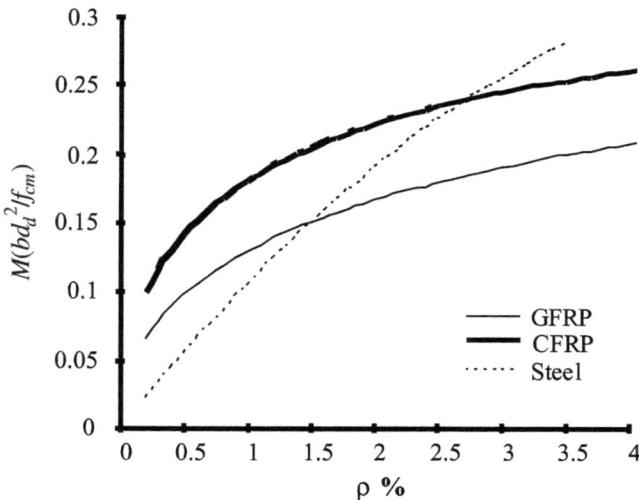

Figure 10.22 Moment capacity versus reinforcement percentage ratio, ρ (%).

respectively. Hence, it is possible to achieve the same flexural capacities for FRP and steel RC elements, but with smaller amounts of FRP reinforcement. However, larger ultimate deflections will prevail in FRP reinforced sections.

10.3.2 Deformability and ductility

The issue of ductility is of considerable concern to engineers, who prefer to see warnings of impending collapse of structures rather than unpredictable brittle failures. To some extent, engineers have a false sense of security in ductility, since in statically determinate structures, large deformations tend to develop very quickly with a small increase in load. In such cases, the warnings of failure may come too late for positive action. Redistribution due to ductility, however, is very important, especially in seismic regions, even though it is not an easily quantifiable asset. Hence, it is not advisable to use the currently available FRP reinforcement in structures that rely on plastic redistribution and high levels of energy dissipation through the reinforcement.

In many structural situations, buckling of compressed steel elements or compressive failure of concrete or masonry is the expected ultimate failure mode. Such structural elements are designed with safety levels similar to those of elements that are more ductile. Hence, the abrupt failure of FRP reinforced elements should be seen in the same context and should be understood, rather than worry the engineer unduly. It is also worth pointing out that large deformations and wide cracking normally accompany flexural failure of FRP RC elements. In addition, FRP RC elements carry higher loads as the deformations increase and warning of impending failure will be given soon after the design load is applied. In fact, failure of FRP reinforced elements is only likely to occur at deflections much greater than those expected from steel reinforced sections at yield of the steel reinforcement.

The likely curvatures expected when a RC section crushes in compression are shown in Fig. 10.23. It can be seen that at the critical cross section, the steel reinforced section is expected to have a higher curvature than the FRP equivalent sections for the same amounts of reinforcement, up to the point when steel sections become over-reinforced. In the figure, the vertical arrows show the strain expected in concrete when steel starts yielding.

Although steel RC sections can exhibit large curvatures on the sectional level, on the structural level these curvatures are concentrated in the regions of reinforcement yield. The likely curvature distributions for steel and FRP reinforced beams are shown in Fig.

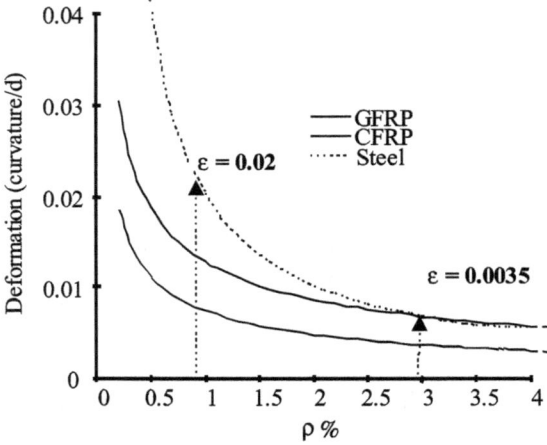

Figure 10.23 Sectional deformation at failure versus reinforcement percentage ratio, ρ (%).

10.24 (for fully cracked concrete). It can be seen that the curvature distribution in the FRP reinforced beam is more uniform and, as a result, the beam can deflect more than the steel RC beam, even though its maximum curvature is lower.

It can be concluded that, for practical purposes, FRP and steel reinforced structural elements in flexure will fail at similar maximum displacements. This was confirmed in numerous experiments conducted by the EUROCRETE project.

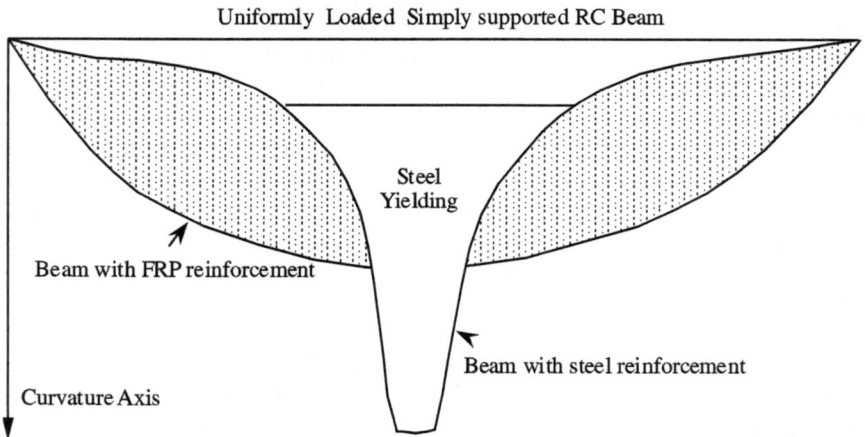

Figure 10.24 Expected curvature distribution for a beam at failure.

The relatively high deformability of FRP RC beams can serve as a warning of failure for structural purposes. It can also be utilized in situations where deformability is not a disadvantage, such as in many ground structures and elements that are likely to be subjected to impact or explosive loading.

When structural ductility is of highest importance, alternatives techniques can be used to enable FRP reinforced elements to develop ductile behavior, such as:

- Lateral confinement of the concrete compressed zone to provide concrete ductility
- FRP hybrid rods or combinations of FRP rods with different material characteristics, failing or being mobilized at different strains, to provide pseudo-ductility
- Plastic bond failure to develop pseudo-plastic behavior
- Enhanced structural redundancy through the addition of sacrificial elements that do not lead to collapse once they fail.

None of these mentioned techniques is straightforward and they require the use of unconventional procedures and lateral thinking. The following example is given to demonstrate that such alternatives are viable.

10.3.2.1 Example of FRP beam behaving in a pseudo-ductile manner. Frangou and Pilakoutas[12] undertook a series of tests at the University of Sheffield using a patented technique of external lateral post-tensioning (PCT/EP94/01222, University of Sheffield). The tests comprised of beams, 2.5 m long with a 150 × 250 mm cross section, each reinforced with two 20 mm diameter GFRP (similar to EURO-CRETE) rods. The beams were tested in bending by applying a two-point load at a distance of 700 mm from each support. The results obtained from one such beam, BT1, before and after repair, are shown in Fig. 10.25. The original beam BT1 failed, as expected, due to concrete crushing in a brittle manner, since it was over-reinforced. After failure, the beam was repaired by replacing the crushed concrete and was strengthened by applying externally tensioned metal strips. Steel strips 12.7 mm width, 0.5 mm thickness and of 950 N/mm^2 ultimate stress were used.

As shown in Fig. 10.25, the strengthening of the beam not only increased the load-carrying capacity by almost 20%, but also allowed the beam to behave in a more ductile manner. This example demonstrates that it is possible to design over-reinforced FRP RC members, which exhibit pseudo-ductile behavior prior to failure. The ductility

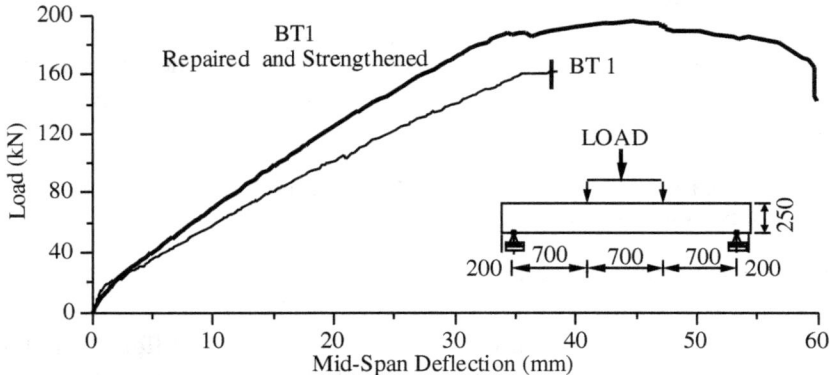

Figure 10.25 Load–deflection curve for beam reinforced with GFRP, before and after repair and strengthening.

developed in this way is not due to inelastic behavior of the reinforcement, but is a result of the nonlinear behavior of confined concrete. Such ductility is called "pseudo-ductility" because the nonlinear characteristics of confined concrete are a result of internal damage and, hence, the potential for energy dissipation under load reversal is limited.

10.3.3 Shear failures

The shear resistance of RC concrete elements is normally considered to be a result of the combination of resistances due to concrete and shear reinforcement. When shear demand exceeds the inherent shear capacity of concrete, shear reinforcement is provided. Shear failure, in elements without shear reinforcement, follows diagonal cracks, which start in the tensile region and eventually penetrate the compression zone. Once that happens, failure is abrupt and catastrophic and, hence, has to be avoided. This is normally done with the provision of shear reinforcement in the form of links.

The concrete shear resistance is thought to be the result of a combination of mechanisms arising from concrete in compression, aggregate interlock in the tensile region, and dowel action of flexural reinforcement. All these mechanisms are expected to decrease in FRP reinforced elements. First, the concrete area under compression is reduced, as explained in the previous section. Then, the lower stiffness and anisotropy of FRP materials means that dowel strength and stiffness are lower than for steel. Aggregate interlock is also known to decrease with increased crack widths. However, since there is not much agreement among engineers as to the precise proportions of the above mechanisms, an empirical approach is normally adopted to resolve this problem.

To investigate the shear strength and deformation characteristics of FRP reinforced elements, extensive testing is currently being undertaken in many parts of the world and was also part of the EUROCRETE investigations. Initial design recommendations proposed by Japanese,[13,14] European,[15] and North American researchers[16,17] all adopted conservative assumptions based on conventional design equations. These approaches and newly developed approaches are described in the following.

10.3.3.1 Strain approach. The premise in this approach is that the concrete section does not actually recognize what it is reinforced with, but only experiences forces and strains. Hence, if the design using FRP reinforcement maintains the same strain when the design forces are developed, then that design will by definition lead to the same results, i.e., a safe design. For this reason, Clarke et al.[15] recommended the use of an equivalent area of steel in Table 9 of the British Standard for RC design,[18] for determining the concrete shear resistance. Hence, the area of steel is modified by

$$A_S = A_{FRP}\left(\frac{E_{FRP}}{E_S}\right) \tag{10.3}$$

where E_S, the elastic modulus of steel reinforcement, is taken as 200 MPa.

Initial results from the EUROCRETE project validated this assumption,[10] as shown in Table 10.4. However, results from punching shear slab testing[19,20] show that the strain approach is conservative (Table 10.5). All the slabs reported in Table 10.5 failed in punching shear. The first two slabs had GFRP reinforcement, the third had CFRP rein-

TABLE 10.4 Comparison of Analytical and Experimental Shear Capacities in Beams

Beam No.	Shear Reinforcement	Concrete f_{cu} (MPa)	P_{ult} (kN) (Flexure)	V_{ult} (kN) (Factored)	(strain approach) (Unfactored)	$P_{failure}$ (kN)	Failure Mode
GB2	No	38.1	95.58	**39.0**	48.8	52.9	Shear
GB5	Yes (100%)	31.2	**84.9**	155		105.1	Flexure
GB6	No	32.9	97.1	**37.2**	46.5	43.9	Shear
GB10	Yes (50%)	39.8	**98.1**	98.4		103	Flexure
GB11	Yes (25%)	39.8	98.1	**69.9**	78.9	98	Shear
GB12	Yes (25%)	39.8	146.9	**69.9**	78.9	133.1	Shear

TABLE 10.5 Experimental and Predicted Punching Shear Capacities

| | Experimental Results | | | | Predicted Results | | |
Slab	σ_{max} (MPa)	σ_{avg} (MPa)	Load (kN)	Strain Approach Load (kN)	σ (MPa)	Stress Approach Load (kN)	Modified Approach Load (kN)
SG2	418	318	271	224	650	415	273
SG3	360	332	237	194	533	340	237
SC2	627	508	317	260	616	355	316
SGS2	649	446	270	286			272

forcement, and the last slab had GFRP flexural reinforcement and CFRP shear reinforcement.

The strain approach is conservative since the FRP reinforcement does not yield at the same strain as steel and a higher stress can be developed in the rebars before punching shear failure can occur. Hence, the above strain based assumption results in a lower-bound solution, since the FRP reinforcement is capable of carrying higher forces after the limiting (yield) strain of steel.

10.3.3.2 Stress approach. The equal stress approach assumes that the strain in the longitudinal reinforcement does not have an effect on concrete shear resistance and, hence, considers just the design force (or stress) in the reinforcement. Such a stress-based assumption is an upper-bound solution, since when the same force is developed, the strain in the concrete and FRP flexural reinforcement is much higher than that of the corresponding strain in steel rebar. The modified value for the equivalent area of steel can only be obtained by finding the point of intersection of the shear capacity curve with the flexural capacity curve. Hence, the area of steel is modified by

$$A_S = A_{FRP}\left(\frac{\sigma_{FRP}}{f_S}\right) \tag{10.4}$$

where σ_{FRP} is the stress in the FRP bar (obtained using an iterative procedure to determine the stress at which the flexural capacity exceeds the punching shear capacity).

Results using the stress approach are shown in Table 10.5 for FRP RC slabs and clearly demonstrate that this approach provides an upper limit of the capacity.

10.3.3.3 Modified approach. Having established upper and lower limits, improved solutions can be found very quickly. Recently,

El-Ghandour[19] has proposed a new approach based on experimental work carried out at the University of Sheffield using the EUROCRETE bar. This approach takes partial advantage of the force that can be developed by FRP reinforcement beyond the lower-bound strain limit to a new value of 0.0045. According to this approach, the equivalent area of steel is obtained as in the case of the strain approach by multiplying by a strain correction factor $\phi = 1.8$. Hence, the area of steel is modified by

$$A_S = A_{FRP}(E_{FRP}/E_{\text{steel}})(\phi) \tag{10.5}$$

where
$$\phi = \varepsilon_{FRP}/\varepsilon_{\text{yield steel}} \tag{10.6}$$

The correction factor for strain, $\phi = 1.8$, was obtained by allowing the tensile strain in the failure region to reach a value of 0.0045 when punching shear failure occurs, instead of limiting it to around 0.0025 (yield strain of steel) as in the case of the strain approach. Results from slab testing proved that this modified approach gives accurate punching shear capacity predictions for both GFRP and CFRP RC flat slabs, as shown in Table 10.5.

Figure 10.26 shows the experimental punching shear capacities of a GFRP RC slab.[20] The maximum and average experimental stress values are shown in this graph. In addition, the punching shear capacity predictions according to the above three approaches are shown against the experimental results. It is obvious that the modified approach predicts most accurately the punching shear capacity, with

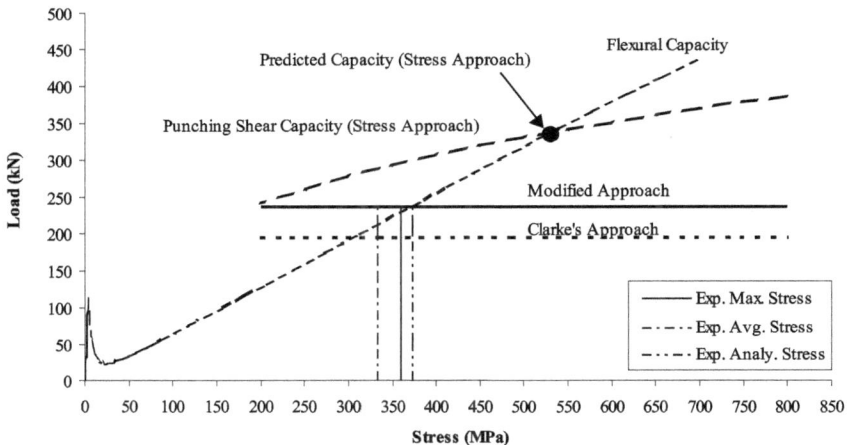

Figure 10.26 Predictions for the punching shear capacity of a GFRP reinforced slab.

the two other approaches providing lower and upper bound solutions. "Analy. Stress" represents the stress predicted by section analysis. This value appears to be very close to the experimental maximum value.

The validity of the modified approach was finally checked by comparisons with test results of another study by Matthys and Taerwe.[21] Series C, CS and H represent slabs reinforced with CFRP NEFMAC grids, CFRP mesh with sanded surface bars, and hybrid type of FRP NEFMAC grids comprising both glass and carbon fibres. Figure 10.27 shows comparisons of the test results with the modified approach predictions. The results were modified to account for the higher concrete strength expected at the time of testing. The graph provides further evidence of the validity of the modified approach in predicting the punching shear capacity of FRP RC slabs.

10.3.3.4 Contribution of shear reinforcement. When the applied shear stress exceeds the design concrete shear strength, shear reinforcement is normally provided. The provision of such reinforcement is again subject to strain restrictions when using FRP reinforcement. The strain approach maintains the strain limit of 0.0025, at which steel rebars normally yield.

This approach leads to very conservative results, since in the case of GFRP reinforcement, this limits the stress to 112.5 MPa. As result, the amount of GFRP reinforcement required for shear becomes very large. During the EUROCRETE project, the amount of shear reinforcement required to reinforce beams according to the above approach was

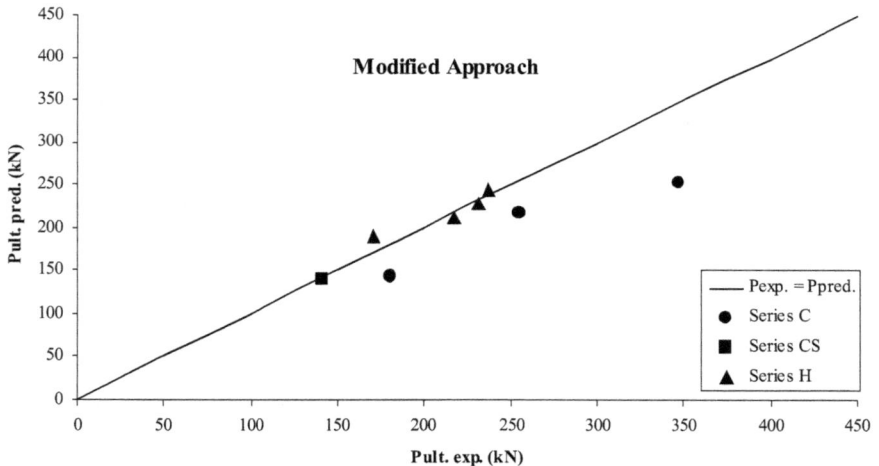

Figure 10.27 Modified approach predictions for slabs tested by Matthys and Taerwe.[21]

halved in one case (GB10), with no severe consequences, as shown in Table 10.4. The reinforcement was further reduced to 25% of that required in beams GB11 and GB12; although shear failure took place, it occurred at much higher loads than predicted.

This approach was again examined by El-Ghandour et al.[20] Examination of the results in Table 10.5 for SGS2, with CFRP shear reinforcement, shows that punching shear failure occurred at an average strain level in the vertical legs of 0.0041. This corresponds to an average stress level of 446 MPa in the CFRP shear reinforcement. The experimental punching shear failure load was 270 kN. El-Ghandour et al.[20] proposed that the above strain limit be relaxed by using a multiplication factor of 1.8.

Using the strain approach assumptions for both concrete shear resistance and strain in shear reinforcement (0.0025), it is found that the predicted capacity for this slab is 286 kN. Although this prediction is only slightly above the experimental one, both major assumptions appear to be violated. First, the strain in the experimental value of strain in the shear reinforcement of 0.0041 exceeds the design value of 0.0025, which means that the strain approach underestimates the contribution of shear reinforcement. Second, the assumption that concrete contributes fully to the punching shear resistance once a punching shear crack is formed (i.e., adding the full concrete shear resistance to the reinforcement contribution), appears to overestimate the contribution of the concrete resistance. This is obvious since by adding two conservative estimations for the concrete shear resistance and shear reinforcement contribution, the result still overestimates the slab capacity.

It is accepted that the concrete contribution to the punching shear resistance is substantially reduced after the initiation of the major shear crack, and it was proposed by El-Ghandour[19] that only 50% of the concrete shear resistance can be relied upon at this stage. This is in line with the ideas of the American code ACI 318.[22]

It was also considered that the FRP shear reinforcement can be mobilized at a higher strain level than previously accepted and the modification factor of 1.8 times the proof strain of steel was proposed (i.e., 0.0045).

Additionally, it was also noted from the experimental work at Sheffield University[19] that only one layer of shear reinforcement was fully activated at failure of the slabs. This means that the maximum spacing for shear reinforcement of 0.75d in BS 8110[18] should be reduced to 0.5d, in line with ACI 318.[22]

Using all the above proposals, the predicted capacity of slab SGS2 is 272 kN (see Table 10.5), which is in perfect agreement with the experimental value.

10.3.3.5 Effectiveness of shear reinforcement. Shear reinforcement is normally provided in the shape of shear links that bend at the corners around flexural rebars. The shape of FRP shear links is not ideal for carrying the stresses due to the development of lateral compression stresses. Continuous FRP elements are inherently weaker in the direction perpendicular to the fiber than along the fibers. At the corners of links, the geometry imposes a 90° change of direction of the force and stress. This can only be achieved through stresses perpendicular to the fiber direction. The magnitude of these stresses depends on the radius of curvature at the corner and the thickness of the link. For small radii of curvature, the lateral strain can be very high, leading to a massive loss of uniaxial tensile strength.

10.3.4 Bond of FRP Rebars

Bonding between concrete and FRP rebars is one of the fundamental aspects of structural behavior that needs to be understood before design guidelines for FRP materials can be developed for use in the construction industry. Adequate levels of bond strength and stiffness are required between reinforcement and concrete to transmit forces effectively from one material to the other. Locations of particular concern are the ends of rebars, since bending of FRP bars is not always possible, and splices. In flexural structural elements, splitting of concrete in the tension zone is the most likely mode of bond failure. This type of failure is substantially different from and more dangerous than the pull-out mode, since it happens at a much lower bond stress level and the residual bond stress of the rebar decreases rapidly to zero.

The bond splitting behavior of FRP rebars in concrete is expected to be different from that of conventional steel rebars since key parameters influencing bond performance are different, such as

- Lower FRP modulus of elasticity
- Much lower shear strength and stiffness in the longitudinal and transverse direction
- High normal strains expected before failure.

The bond characteristics of FRP pultruded rods are governed by the surface shear strength of the resin matrix. Suitably designed resins have shear strengths higher than concrete and, hence, bond failure in surface textured rods is expected at the concrete interface. Deformations on the surface of FRP rebars cannot enhance the bond characteristics in the same way as for steel rebars, since shear failure can occur through the composite and not just through the concrete.

However, despite the fact that a lower maximum bond strength is expected from FRP rebars, the more ductile nature of the bond can lead to a better distribution of the bond stresses and, hence, lead to reduced anchorage lengths.

Two major experimental series of tests were undertaken at the University of Sheffield, as part of the EUROCRETE project,[23] to evaluate the bond behavior of FRP rebars in concrete. In the first series more than 100 specimens were tested in direct pull-out,[23,24] whereas in the second series the bond development of the FRP rebars was examined in over 35 RC beams tested in four-point bending. The following two subsections discuss the main parameters that influence bond failures in pull-out and beam tests. Pull-out tests are representative of the way testing is undertaken to determine bond strength, while beam tests offer a more realistic view of bond in structures.

10.3.4.1 Pull-out failures. Several parameters that influence bond performance were examined, such as the nature of fibers, the bar diameter, shape and deformations, the embedment length, and concrete strength. The adopted pull-out test arrangement is shown in Fig. 10.28.

The rebars were embedded in 150 mm concrete cubes and the embedment length was fixed in multiples of the bar diameter to facilitate comparisons. The embedment lengths were designed so that no splitting failures would occur. The two ends of the rebar in the concrete cube were debonded to minimize the end effects. Three displacement transducers were used to measure the slip of the rebar at the loaded end so as to eliminate the effects of accidental bending.

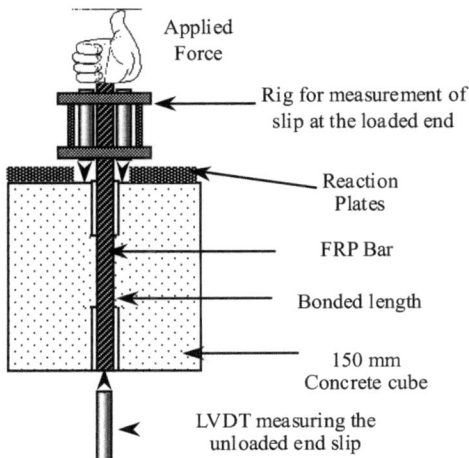

Applied Force

Rig for measurement of slip at the loaded end

Reaction Plates

FRP Bar

Bonded length

150 mm Concrete cube

LVDT measuring the unloaded end slip

Figure 10.28 Pull-out test arrangement.

Average bond stress, τ, is calculated by dividing the maximum load by the bonded rebar perimeter.

Characteristic bond-slip curves for carbon and glass EUROCRETE FRP rebars, for a typical embedment length of 8 times the 8 mm diameter of round rebars, are illustrated in Figs. 10.29 and 10.30. The concrete compressive cube strength for both specimen was 41 MPa. It can be observed that both rebars behaved in a similar manner, only starting to slip at relatively large bond stresses. Due to their higher stiffness, the carbon rebars appear to fail at a lower slip displacement.

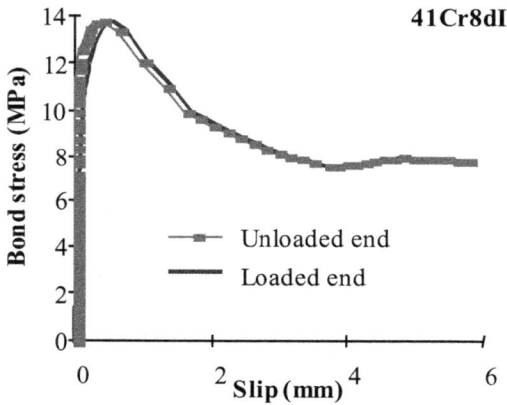

Figure 10.29   Stress–slip relationship for a CFRP bar.

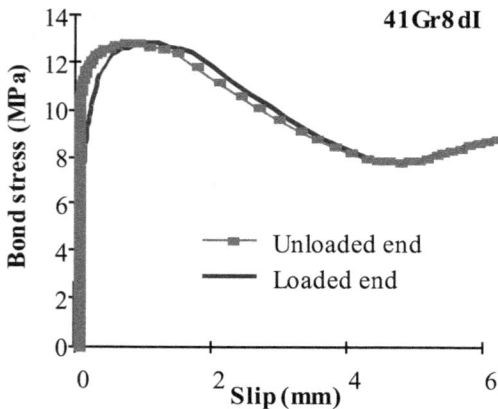

Figure 10.30   Stress–slip relationship for a GFRP bar.

The effect of various parameters on the pull-out bond strength is discussed in the following.

Rebar material.   Both carbon and glass FRP rods exhibit similar bond behavior and their maximum average bond stress at the characteristic embedment length of 5 diameters (according to RILEM/CEB test[25]) was around 12 MPa. According to CEB Bulletin 151,[26] the maximum average bond stress for deformed steel rebars for the same embedment length is around 15 MPa for concrete C40. The fact that the FRP rebars develop more than 80% of the bond strength expected from steel rebars is remarkable considering their lower elastic moduli and different nature of the bonded surface. It should also be noted that the bond strength of epoxy-coated rebars, which are mainly used as anticorrosive reinforcement, varies from 67% to 95% of that of deformed steel rebars, which is comparable to the bond strength of FRP rebars.

The above results are in agreement with results published by other researchers[27–30] despite the fact that different types of FRP rods are used. This fact may lead to the conclusion that bond strength is independent of the rebar material of its elastic modulus, but this issue will be examined later in more detail.

Embedment length.   The maximum average bond stresses developed by carbon and glass FRP rebars versus the embedment length are presented in Fig. 10.31. Linear regression was used to determine the best-fit line through the data points. As expected, the average bond strength is seen to decrease as the embedment length increases, but this does not happen at a slower rate than expected from steel reinforcement. This could be a result of the nonlinear distribution of bond stress on the bar.

Bar diameter and shape.   Figure 10.32 shows that the rate of decrease of bond strength with embedment length is also influenced by the bar diameter. This rate of bond strength decrease is lower for

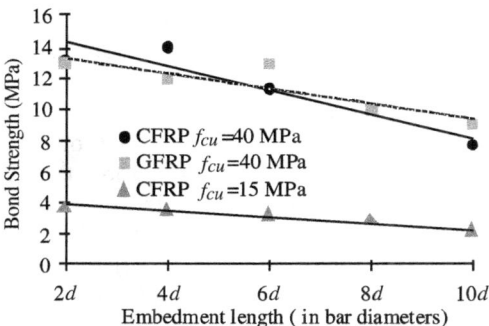

Figure 10.31 Effect of embedment length on bond strength.

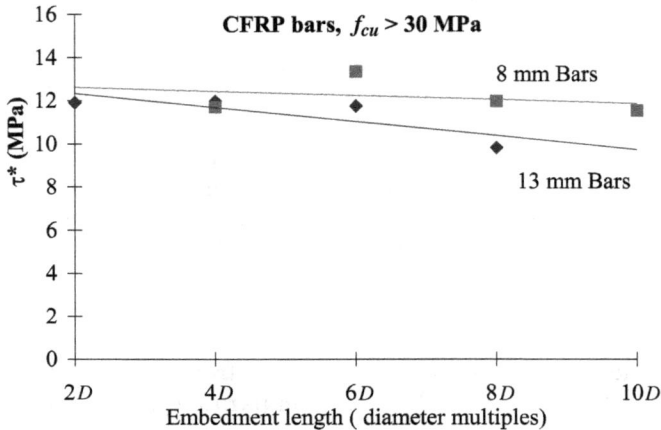

Figure 10.32   Effect of embedment length and bar diameter on bond strength.

smaller-diameter bars (8–8.5 mm) than for larger-diameter bars (13.5 mm). The rate of decrease of the bond strength with the embedment length is also thought to be a function of the ductility of the bond. A perfectly plastic bond characteristic will lead to a decrease in average bond strength with embedment length, while a brittle bond will result in the "unzipping" of larger embedment lengths.

Bond development in FRP rebars is affected by their diameter due to the resin's low shear stiffness. In a large-diameter rebar pulled in tension through the surface, there can be a differential movement between the core and the surface fibers. This results in a nonuniform distribution of normal stresses through the cross section of the bar. This effect, known as "shear lag," is an important factor that imposes a limit on the maximum size of FRP bar diameter that can be used as rebar.

During the EUROCRETE project, square and round cross sections were examined. The results indicate that an increase of bond strength up to 20% over circular cross sections can be achieved by square cross sections. This was attributed mainly to the wedging effect of the corners.

The rebar surface deformations also have an important effect on the bond strength. Malvar[27,30,31] suggested that surface deformations of about 5.4% of the bar diameter are sufficient to provide adequate bond behavior to concrete. Achillides[23] demonstrated that EUROCRETE rebars tested with reduced deformations (less than 0.5 mm) reduced the bond strength to 25% of that expected of the normal deformations. Tests have shown that smooth bars in general pulled out at very low bond strengths, but the results were very variable, sometimes reaching relatively high strengths.

Concrete strength. For steel rebars, concrete strength plays an important role in the development of bond strength. This is because the pull-out failure in well-confined specimens takes place due to concrete shear failure. The shear strength of concrete is a function of its tensile and compressive strength.

In contrast, in the scientific literature it is reported that the role of concrete strength is not as significant in determining the FRP rebar bond strength, probably due to a different failure mode. In FRP rebars, failure occurs near the surface of the rebar by peeling of part of the surface layer. Consequently, the bond strength of FRP rebars is not controlled by the concrete strength but by the shear strength of the bar resin.

During the EUROCRETE project, it was observed from scanning electron microscope (SEM) images that interface failure occurred only when the concrete strength was greater than a value of 20–30 MPa. Figures 10.33 to 10.35 show the SEM images of the CFRP rebar surface before and after pull-out failure as well as the surface of the concrete. Figure 10.33 shows the resin-rich CFRP rebar surface formed by peel ply. In Fig. 10.34, the rebar surface after pull-out is shown to have concrete attached to it. However, Fig. 10.35 shows that the concrete surface also contains resin and fiber residue. This means

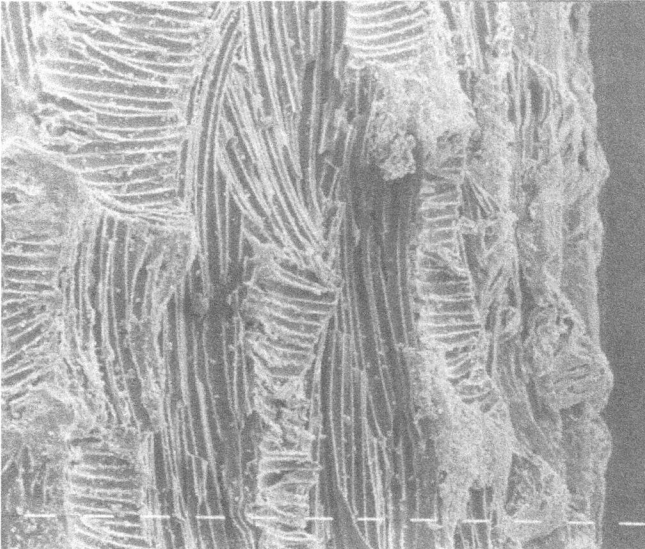

Figure 10.33 SEM image of CFRP bar before testing.

Figure 10.34 SEM image of CFRP bar after testing.

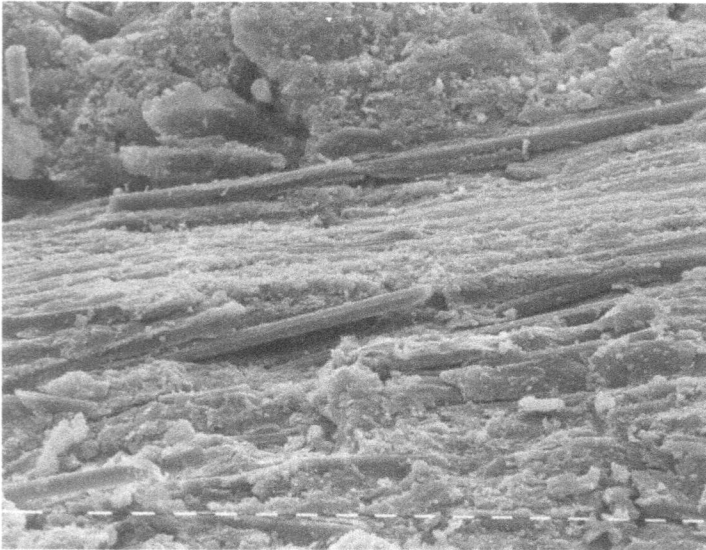

Figure 10.35 SEM image of concrete failure interface with CFRP bar.

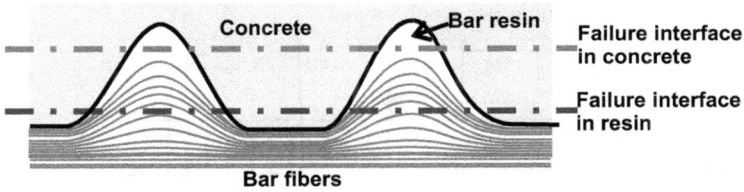

Figure 10.36 Shear failure at the surface layer of FRP bars. (*After Reference 23*).

that the failure, though mostly in the concrete, also affects the rebar surface. Hence, it can be concluded that the shear strength between fibers and resin controls the bond capacity of FRP. The height of the failure surface from the bar axis (see Fig. 10.36) is assumed to depend on the relative strength of resin and concrete. In the EUROCRETE experiments, the height of the failure interface of GFRP rebars was found to be lower than that of CFRP rebars.

For low-strength concrete (15 MPa), concrete was found crushed between the ribs of the FRP rebar and bond strength was mainly controlled by concrete strength. The effect of different concrete strengths on the maximum developed bond stress, for the same diameter bars, is shown clearly in Fig. 10.37.

10.3.4.2 Bond-splitting in beam tests. Boundary conditions in conventional pull-out tests are rarely encountered in practice, but such tests are undertaken because they are easy to carry out and offer an economic way of bond strength relative assessment. Concrete structural elements are normally exposed to combinations of flexure and

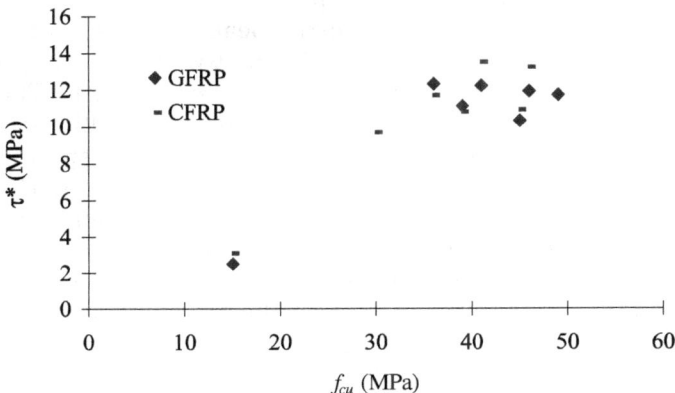

Figure 10.37 Influence of concrete cube strength on bond strength for FRP bars.

Single bar Anchorages

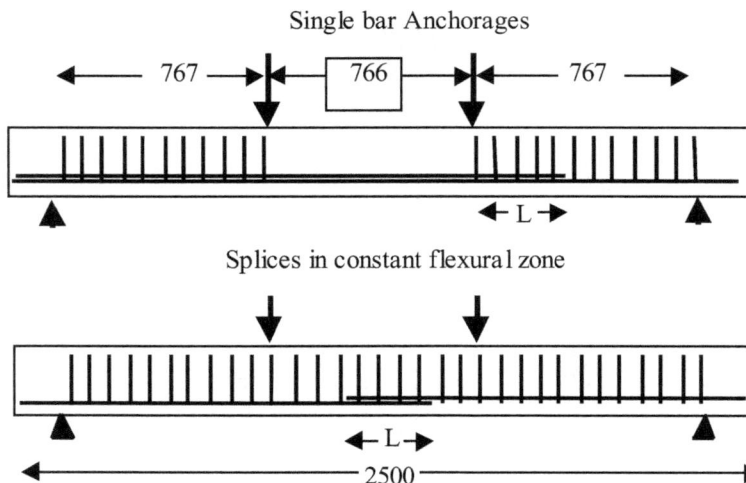

Splices in constant flexural zone

Figure 10.38   Beam testing arrangement (dimensions in mm).

shear loads. Under flexure conditions, splitting of concrete in the tension zone is the most common mode of bond failure, which is substantially different from the pull-out mode. Hence, beam testing is necessary to understand the splitting mode of failure.

More than 35 medium-scale rectangular beams (2500 × 250 × 150 m) were prepared and tested under a four point bending arrangement[23] (Fig. 10.38) for the EUROCRETE project. Different types (bar size, nature of fibers) and configurations of FRP reinforcement were used in the tests.

Material properties. EUROCRETE rebars were used in the tests. The links used as shear reinforcement in the beams of this experimental series were specially manufactured filament-wound rectangular (200 × 100 mm) GFRP links with 10 × 5 mm cross section. The tensile strength of the used links was measured to be around 425 MPa and their elastic modulus around 47 MPa. They were fixed at a spacing of 75 mm center-to-center either only in the shear span or along the whole beam, depending on the arrangement of the main reinforcement. Although, the amount of shear reinforcement provided to the beams was well below the minimum requirements proposed by the strain approach, this spacing was adopted in the previous phases of testing for similar beams and successfully prevented shear failures.

Compression reinforcement was used mainly to enhance the concrete compressive strength at the top of the beam. The average concrete strength measured in this series of tests was in the range 35–45 MPa.

Test specimens and procedure.   Nine beams were prepared with the arrangement shown in Fig. 10.38, each having slightly different reinforcement arrangement. A summary of the reinforcement arrangement in each beam is shown in Table 10.6, together with the typical loading arrangement.

All the beams were instrumented with strain gauges attached on the main rebars and links. External displacement transducers were used to measure vertical and horizontal deflections, as well as surface

TABLE 10.6   Reinforcement Arrangements for the Beams Designed to Fail in Bond

Beam	Anchorage Length, L (mm)	Bottom Cover to Diameter Ratio	Main (bottom) Shear Reinforcement	Arrangement of Main (bottom) Reinforcement
GB29	250	1.85	3 GFRP bars (13.5 mm) GFRP links 75 mm c/c in the shear span	
GB30	300	1.85	3 GFRP bars (13.5 mm) GFRP links 75 mm c/c in the shear span	
GB31	300	1.85	4 GFRP bars (13.5 mm) GFRP links 75 mm c/c in all the way	
GB32	300	3.13	3 CFRP bars (8 mm) GFRP links 75 mm c/c in the shear span	
CB33	300	3.13	4 CFRP bars (8 mm) GFRP links 75 mm c/c in all the way	
GB34	370	2.94	3 GFRP bars (8.5 mm) GFRP links 75 mm c/c in the shear span	
GB35	300	2.94	3 GFRP bars (8.5 mm) GFRP links 75 mm c/c in the shear span	
GB36	300	2.94	4 GFRP bars (8.5 mm) GFRP links 75 mm c/c in all the way	
CB37	580	3.13	3 CFRP bars (8 mm) GFRP links 75 mm c/c in the shear span	

extensions. The difference in strain values along the longitudinal bars was used to calculate the bond stress distribution in bars.

Experimental results. Various types of bond splitting cracks were developed in the beams depending on the arrangement of reinforcement in the cross section. Table 10.7 shows the types of splitting cracks observed together with the developed bond values in each case. In this table, the starting letter of the beam code indicates whether the beam was reinforced with carbon, (C) or glass (G) FRP EUROCRETE bars. In all cases, the crack initially developed at the very end of the rebar and, as the load increased, the crack extended along the whole length of the anchorage. The underside of one of the beams after failure is shown in Fig. 10.39.

The maximum average bond strength, τ^*, developed over the anchorage or splice, is given in Table 10.7. The maximum strain value of the strain gauge attached at the loaded end of the rebar was used in the calculation of τ^* over the whole embedment length.

Due to the difference in mode of failure, the bond splitting strength in the case of beams is lower than that developed by the same rebars in pull-out tests. In pull-out tests the pull-through failure develops by shearing the surface deformations of the rebar, whereas in the case of beams the concrete cover split and the rebar slipped through the concrete without any disturbance of the rebar surface. This significant difference in bond strength noted here casts a shadow on the reliability of pull-out tests for assessing the bond behavior of FRP rebars in concrete members. This is because, for most practical applications, bond failure of rebars is likely to be due to splitting.

Another important observation is that the bond splitting behavior of CFRP and GFRP rebars appears to be significantly different. By comparing the bond strengths developed in the single anchorage of beam CB32 with the respective strengths of the anchorages in beams GB34 and GB35, it is obvious that CFRP rebars develop higher bond splitting strengths than GFRP rebars under similar conditions. The

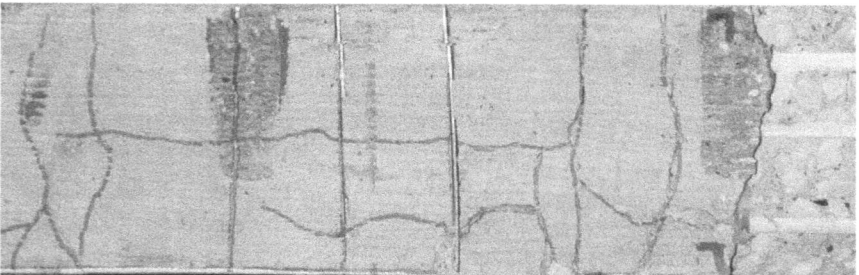

Figure 10.39   Splitting crack developed under the single bar anchorage in beam GB34.

TABLE 10.7   Bond Splitting Behaviour and Average Bond Strength Developed

Beam	Type of Failure		Embedment Length (mm)	Maximum τ^* (MPa)
GB29	Bond splitting crack under the middle bar		250	3.2
GB30	Bond splitting crack under the middle bar		300	2.7
GB31	Face and side bond splitting failure		300	3.8
CB32	Bond splitting crack under the middle bar		300	4.6
CB33	Side bond splitting failure		300	5.7
GB34	Bond splitting crack under the middle bar		370	3.2
GB35	Bond splitting crack at the side face of beam		300	3.0
GB36	Face and side bond splitting failure		300	4.1
CB37	Bond splitting crack under the middle bar		580	3.7

same observation can be also made in the case of spliced rebars in beams CB33 and GB36. It is estimated that CFRP rebars develop around 30% higher bond splitting strength than similar diameter GFRP rebars.

These results are in contrast with the results of the pull-out tests, which showed that CFRP and GFRP rebars developed similar bond behavior. GFRP rebars induce splitting at a lower normal and bond stress, but having a lower modulus of elasticity, at a higher normal strain. Hence, it appears that the deformability of FRP rebars in the axial and transverse directions significantly influences their bond splitting behavior. It can be concluded that pull-out tests cannot be used reliably to compare the bond behavior of FRP rebars when bond splitting failure is expected in a concrete member and, hence, a different test will have to be used.

From this phase of testing it can also be seen that the maximum average bond strength developed in the case of spliced rebars, for example, in beam GB31, was much greater than in the single bar anchorage in beam GB30. To determine τ^* in the case of splices, the strain values of the gauges attached at the loaded end of each pair of spliced rebars were considered. The value of τ^* calculated over the spliced length was 3.8 MPa at ultimate load level, which was much higher than the value of 2.7 MPa developed in the single anchorage in beam GB30. Similar observations were also made in the case of beams CB32, CB33, and GB35, GB36 where spliced rebars developed significantly higher bond splitting values. A possible explanation for this is that the plane of failure in the case of splices is different from that in the case of single bars, as shown in Table 10.7. As a result, the influence of stirrups has a different effect on the bond splitting strength of the rebars in each case and requires further examination. However, it is considered that reduced bond strengths normally used for the design of splice lengths are not needed in the case of the FRP rebars.

Strain and bond stress distribution along a single anchorage. The distribution of normal strains along a typical single anchorage rebar of beam GB30 is shown in Fig. 10.40, for successive load steps. The slope of each curve is proportional to the bond stress developed on the bar. The bond stress profiles are shown in the middle graph of Fig. 10.40. It is clear that for low load levels the peak bond stress develops at the end of the rebar, up to the load level when the rebar starts slipping (around 50 kN) and the first splitting crack initiates under the rebar.

After the development of the first splitting crack (see load level 60 kN of the third graph of Fig. 10.40), the peak bond stress migrates from the end of the rebar toward the middle of the beam as load increases. The reason for this movement can be the propagation of the splitting cracks along the rebar. To verify this, specially arranged displacement transducers were positioned at the bottom face of the beam along the length of the middle bar to monitor the widths of the splitting crack during the test. The crack width propagation along the rebar is shown in the third graph of Fig. 10.40. It is clear that the splitting crack developed at the rebar end at a load just below 60 kN and propagated eventually all the way toward the direction of the loading point. The negative values shown for crack widths are due to the Poisson effect on the concrete, which is in tension in the perpendicular direction. The nonzero values of strains and bond stresses at zero load level can be attributed to the fact that these values were recorded at the end of a load cycle, which may have locked in some strains.

An important observation is associated with the movement of the peak bond stress value along the anchorage length. In beams tested

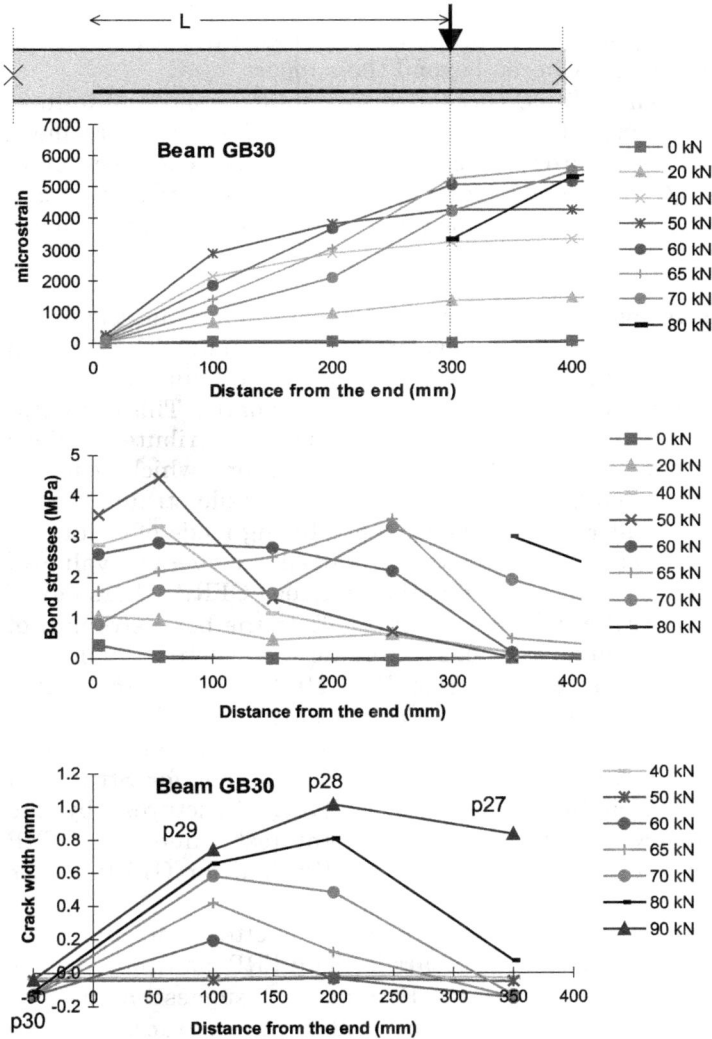

Figure 10.40 Normal strains, bond stresses, and crack widths along the anchorage length.

with rebars anchored under the supports, the peak bond value migrates from the loaded end in the middle of the beam toward the free end of the anchorage at the support as the load increases.[23,24] Here, this is not the case since the initial peak bond stress appears at the end of the bar. This difference is due to the different location of the anchorage length with respect to the flexural cracked zone of the beam. In the current case, the whole anchorage length lies within the cracked region of the beam, and hence the bond stresses are less influenced by the propagation of the cracks toward the support, as is the case with rebars anchored beyond the support.

All beams shown in Table 10.7 failed in a splitting mode of failure, despite the presence of sufficient amount of shear reinforcement (GFRP links, 10×4 mm cross section, and spacing 76 mm center to center). The concrete cover to the rebars was 25 mm, which was 2–3 times the diameter of the bar, depending on whether the bar diameter was 8 mm or 13.5 mm.

10.3.4.3 Conclusions on bond behavior. It is concluded that the in the beam tests the CFRP rebars develop a substantially higher bond strength than the GFRP rebars, whereas in the pull-out tests carbon and glass FRP rebars developed similar bond strengths. This comparatively lower bond strength of GFRP rebars may be attributed to their higher deformability in the longitudinal direction, which leads to higher tensile strains at failure. These higher tensile strains appear to play an important role in inducing the splitting mode of failure.

In the beam tests, GFRP rebars developed an average value of maximum bond strength around 3 MPa, whereas CFRP rebars developed more than 4 MPa, which is comparable to the bond strength of deformed steel rebars. The characteristic bond stress value of deformed steel rebars, according to BS 8110[18] (without the safety factor) for concrete with $f_{cu} = 35$ MPa is 4.1 MPa. This supports the assumption that CFRP rebars develop bond stresses similar to those in steel. In addition, the recently published "Guidelines for Structural Design of FRP Reinforced Concrete Building Structures" by the Japanese Ministry of Construction[13] present test values for CFRP rebars (13 mm diameter) around 4.5–4.7 MPa and for GFRP rebars around 3.2 MPa.

Splices with EUROCRETE rebars developed better bond characteristics than the single anchorage rebars for both FRP materials. This is unexpected since most Codes of Practice[13-18] suggest a 40–50% increase in the anchorage length in the case of splices, to accommodate a reduction in the expected bond strength. This may be explained by the fact that the plane of failure in the case of spliced rebars is different from that for anchored rebars. In the case of spliced rebars a horizontal crack running through the plane of the spliced rebars was observed, whereas for anchored rebars a vertical splitting crack was formed at the bottom side of the beam, directly under the rebar. As a result, shear links are expected to have a different effect on the formation of the splitting crack in each case.

Another interesting is that the peak bond stress on the FRP rebar initially develops near the section of load application. This peak gradually propagates towards the end of the beam, ahead of flexural cracking. The area under the curves can be used to estimate the average bond stress developed at each load step. However, when a splitting crack initiates at the very end of the rebar, the local value of bond strength immediately drops. As the load increases, the splitting crack develops along the rebar toward the mid-span and the peak bond strength propagates toward the mid-span ahead of the splitting crack. This propagation of the peak bond demand is in the reverse direction from that in well-anchored beams.

10.3.5   Other types of failure

The basic types of failure were examined in the above. Other types of failure are also possible depending on the materials used.

10.3.5.1 Stress corrosion. Fracture under low levels of sustained stress is a problem in glass fibers and, hence, is also expected in GFRP rebars. Preliminary design guidelines[15] limit the long-term stress on GFRP to about 30% of the material strength. This has obvious implications in design of structural elements that are to sustain significant permanent stress.

10.3.5.2 Creep. Creep is a problem of increased deflections under sustained loads. Even though the fibers of FRP may be resistant to creep, the resin is susceptible to creep and a certain degree of creep is expected from FRP composites. However, having in mind that concrete creeps considerably with time when subjected to sustained stress, especially in dry environments, the internal redistribution in stress required to account for concrete creep is much greater than would be expected from FRP. Hence, long-term deflections in FRP RC can still be determined in the conventional manner.

10.3.5.3 Fire. Fire or elevated temperatures are problems that always concern civil engineers, especially when dealing with buildings. Again, although the fibers may be resistant to elevated temperatures, the resin that binds them will soften or deteriorate at relatively low temperatures, depending on its composition. The loss of the resin stiffness will result in bond softening, but may not lead to serious problems if parts of the rebar are anchored in cooler regions. Since not much research has been carried out in this field, future research is necessary to clarify the temperatures under which FRP reinforcement can be relied upon in concrete.

10.3.5.4 Long term durability. The issue of the long-term durability is the one that drives the development of FRP reinforcement. However, it does not mean that the durability of FRP reinforcement is always guaranteed. What can be said is that FRP reinforcement can be designed to resist specific environments such as the environment in concrete and that only such reinforcement can be used in concrete with certainty.

References

1. L. Holloway (1994) *Handbook of Polymer Composites for Engineers*. Woodhead Publishing Limited.
2. J. L. Clarke (1993) *Alternative Materials for the Reinforcement and Prestressing of Concrete*, Chapman and Hall.
3. CEB Bulletin 151 (1982) *Bond Action and Bond Behaviour of Reinforcement*, State-of-the-art Report, Committee Euro-International du Breton.
4. R. Holland (1997) *Appraisal and Repair of Reinforced Concrete*, Thomas Telford, London.
5. E. A. Byars and T. McNulty (eds.) (1997) Management of concrete structures for long-term serviceability, Thomas Telford, London.
6. R. T. L. Allen and S. C. Edwards (eds.) (1987) *Repair of Concrete Structures*, Blackie, Glasgow.
7. G. E. Bakis (1993) FRP reinforcement: materials and manufacturing, in *Fiber-Reinforced-Plastic (FRP) Reinforcement for Concrete Structures: Properties and Applications*, (A. Nanni, ed.), Elsevier, 167–58.
8. K. Pilakoutas, Z. Achillides, and P. Waldron (1997) Non-ferrous reinforcement in concrete structures, in *Innovation in Composite Materials and Structures* (M. B. Leeming and B. H. V. Topping, eds.), Civil-Comp Ltd., Edinburgh, 47–58.
9. S. Faza and H. GangaRao (1993) Glass FRP reinforcing bars for concrete, in *Fiber-Reinforced-Plastic (FRP) Reinforcement for Concrete Structures: Properties and Applications* (A. Nanni, ed.), Elsevier, 167–188.
10. N. Duranovic, K. Pilakoutas, and P. Waldron (1997) FRP reinforcement for concrete structures: design considerations, in *Proceedings of the Third International Symposium on Non-metallic (FRP) Reinforcement for Concrete Structures*, vol. 2, Japan Concrete Institute, 527–534.
11. N. Duranovic, K. Pilakoutas, and P. Waldron (1997) Tests on concrete beams reinforced with glass fibre reinforced plastic bars, in *Proceedings of the Third International Symposium on Non-metallic (FRP) Reinforcement for Concrete Structures*, vol. 2, Japan Concrete Institute, 479–485.
12. M. Frangou and K. Pilakoutas (1994) Novel technique for the repair and strengthening of RC columns, in *Proceedings of the Fifth U.S. National Conference on Earthquake Engineering*, vol. III, 637–646.
13. Japanese Ministry of Construction (July 1995) *Guidelines for Structural Design of FRP Reinforced Concrete Building Structures* (Draft), FRP Reinforced Concrete Research Croup, Building Research Institute.
14. JSCE (1996) *Code for Design and Construction of Concrete Structures with Continuous Fiber Reinforcement* (Draft), Japan Society of Civil Engineers.
15. J. Clarke, D. O'Regan, and C. Thirugnanendran (1997) *EUROCRETE PROJECT: Modification of Design Rules to Incorporate Non-Ferrous Reinforcement*, Sir William Halcrow and Partners, London.
16. Canadian Highway Bridge Design Code (1996) *Fibre Reinforced Structures*, Section 16 (Draft).
17. American Concrete Institute (1996) *State-of-the-Art Report in Fiber Reinforced Plastic (FRP) Reinforcement for Concrete Structures*, ACI 440R- 96.

18. BS 8110 (1985), *The Structural Use of Concrete*, British Standards Institution, London.
19. A. W. El-Ghandour (1999) Behaviour and design of FRP RC slabs, Ph.D. thesis, Sheffield University, U.K.
20. A. W. El-Ghandour, K. Pilakoutas, and P. Waldron (1998) Behaviour of FRP RC flat slabs with CFRP shear reinforcement, in *Proceedings of the European Conference on Composite Materials*, ECCM-8, Naples.
21. S. Matthys and L. Taerwe (1997) Punching tests on concrete slabs reinforced with FRP grids, *Proceedings of the Third International Symposium on Non-Metallic (FRP) Reinforcement for Concrete Structures*, vol. 2, Japan Concrete Institute, 559–566.
22. American Concrete Institute 318-95 (1995) *Building Code Requirements for Reinforced Concrete*, ACI Committee 318, Detroit.
23. Z. Achillides (1998) Bond behaviour of FRP bars in concrete, Ph.D. thesis, Sheffield University, U.K.
24. Z. Achillides, K. Pilakoutas, and P. Waldron (1997) Bond behaviour of FRP to concrete, in *Proceedings of the Third International Symposium on Non-metallic (FRP) Reinforcement for Concrete Structures*, vol. 2, Japan Concrete Institute, 341–348.
25. RILEM/CEB/FIP (1978) *Bond Test for Reinforcing Steel 2. Pullout test*, Recommendation RC6, Committee Euro-International du Beton.
26. CEB Bulletin 151 (1982) *Bond Action and Bond Behaviour of Reinforcement*. State-of-the-Art Report, Committee Euro-International du Beton.
27. J. L. Malvar (1995) Tensile and bond properties of GFRP reinforcing bars, *ACI Mater. J.*, **92**, No. 3, 276–285.
28. J. Larralde and R. Silva-Rodriqez (1993) Bond and slip of FRP repairs in concrete. *J. Mater. Civil Eng.*, **5**, No. 1, 30–39.
29. B. Benmokrane et al. (1994) Investigation on bond performance of FRP rebars, presented at ACI Convention, San Francisco, Session on Bond of FRP Rebars.
30. A. Nanni, M. Al-Zaharani, S. Al-Dulaijan, C. Bakis, and T. Boothby (1995) Bond of FRP reinforcement to concrete—experimental results, in *Non- Metallic (FRP) Reinforcement for Concrete Structures* (L. Taerwe, ed.), ESFN Spon, 137–145.
31. J. L. Malvar (1992) Bond of reinforcement under controlled confinement, *ACI Mater. J.*, **89**, No. 6, 593–601.
32. J. L. Malvar (1994) Bond stress-slip characteristics of FRP rebars, presented at ACI Convention, San Francisco, Session on Bond of FRP Rebars.

Composites in Biomedical Engineering

Yannis F. Missirlis

*University of Patras, Department of Mechanical
Engineering, Rion, Patras 26500, Greece*

11.1 Introduction

Nature has utilized the concept of composite materials and structures almost invariably. Biological systems, both animals and plants, exhibit a remarkable, and apparently optimized with regard to their function, combination of structural elements to construct composite structures at all levels: from cell components to whole organs.

As human tissue and organ replacement has advanced since the 1970s and the needs and prospects for further advancement are pressing, the search for novel materials and in particular composites is obvious.

In this chapter we shall present representative examples of biological structures at various levels of structural hierarchy along with some of their key properties that make them suitable for their function. Then some examples of artificial material structures will be given with emphasis on their properties that make them suitable for tissue/organ replacement or assistance.

11.2 Biomembranes

In order to create the necessary degree of functional order among the various biomolecules that are continuously being produced and transported and/or are reacting to maintain "life," several "organelles" are formed within the cytoplasm of animal cells. These organelles are separated from the surroundings by their membranes. Each cell also is individualized by its membrane.

The cell plasma membrane is generally considered as a three-layered composite system.[1] Different types of cells have specialized composition of their membranes, but a general structure is as follows: (a) The central part of the membrane is composed of a lipid bilayer in which protein molecules are embedded. The lipids in the two mono-layers are asymmetrically distributed, but the proteins are surrounded by clouds of specific lipid. (b) The outer part is the glycocalyx, i.e., a macromolecular film formed by the oligosaccharides of the glycolipid heads and the polypeptide heads of the glycoproteins. (c) At the intracellular side one finds a macromolecular network termed the cytoskeleton, which couples to the lipid bilayer. The combination of the protein/lipid bilayer with the cytoskeleton forms the basis for the mechanical properties of the cells.

The human erythrocyte, or red blood cell (RBC), provides an excellent model for studying the mechanical and rheological properties of the intact cell membrane because the RBC lacks the nucleus and other organelles and practically it can be thought of as a flexible thin-shell vessel containing a quasi-Newtonian fluid (hemoglobin solution). At rest, the RBC looks like a biconcave disk, as shown in Fig. 11.1. Typical sizes are cell diameter $8\,\mu m$, cell height $2.5\,\mu m$, cell surface $140\,\mu m^2$, cell volume $110\,\mu m^3$, cell membrane thickness $80\,\text{Å} = 8 \times 10^{-3}\,\mu m$. A simplified schematic of the cell's membrane is shown in Fig. 11.2. Each cell has about 1×10^5 spectrin tetramers, and ankyrin molecules, $3-4 \times 10^4$ actin molecules, and 2×10^5 Band III and Band 4.1 molecules. The spectrin–actin network is beautifully organized in a triangular complex (Fig. 11.3).

It has been shown that both the structural organization of the cytoskeleton network and its coupling to the lipid/protein bilayer is modulated by a continuous biochemical energy consumption in the form of phosphorylation and dephosphorylation reactions. In other words, the mechanical properties of the cell membrane depend on the biochemical state of the membrane components, which is manifested partly by the polymerization or depolymerization of spectrin. As a simplified hypothesis, it is suggested that the cytoskeleton (i.e., the spectrin–actin proteinous network) confers on the membrane its "solid" mechanical characteristics while the lipid bilayer confers its "liquid" viscous characteristics.

Several techniques have been utilized to measure the mechanical properties of cells and particularly of cell membranes. They employ such concepts as the application of specific shear strain rates on the cell membrane (by placing cells in defined flow fields) or the subjection of cell membranes to stretching, shearing, bending, or to a combina-

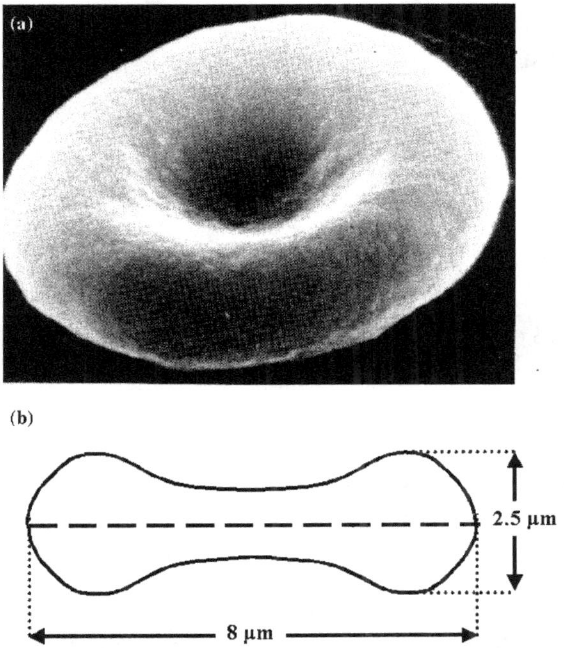

Figure 11.1 (a) A scanning electron micrograph of a human red blood cell (RBC), ×14000. (b) Schematic cross-sectional shape of an RBC at rest.

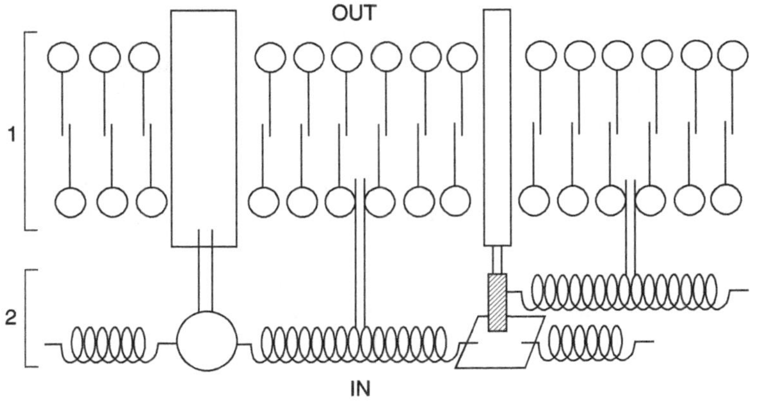

Figure 11.2 Highly schematic view of a portion of the RBC membrane. 1 is the lipid bilayer (represents a molecule of a lipid) with integral proteins transversing it (□ is termed band III and is a glycophorin). 2 is the cytoskeleton composed mainly of the proteins spectrin (-∿∿∿-) and actin (▱). The cytoskeleton is attached to the lipid/integral protein bilayer via the proteins ankyrin (○) and band 4.1 (▤) or directly with the lipids.

Spectrin Tetramer

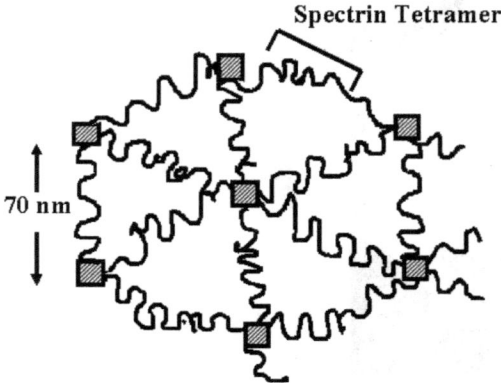

70 nm

Figure 11.3 A simplified schematic of the RBC membrane cytoskeleton: the lines connecting the corners are spectrin tetramers (∿), the corner heads are actin molecules (▨). Halfway along the triangle side are found ankyrin molecules (not shown).

tion of such stresses and measuring the respective moduli. The micropipette aspiration technique,[2] being one that has been used on many occasions, will be described in brief. A schematic is shown in Fig. 11.4.

A dilute cell suspension is introduced in a glass chamber of approximately $10 \times 10 \times 1$ mm that contains a buffered saline solution and is open in the one of the 1×10 mm sides. Through this side a glass micropipette, with a tip diameter of the order of $1\,\mu$m (depending on the cell type and/or the short of experiment) and filled with the same buffered saline solution is introduced. The micropipette is attached to a sensitive aspiration pressure measuring system and can be moved in three dimensions with a micromanipulator. The chamber is placed on the stage of an inverted microscope. By manipulation, the pipette tip can be made to touch a particular cell and, by the application of negative (aspiration) pressure, a portion of the cell can be moved inside the pipette. The whole process can be viewed and recorded for later examination.

Depending on the experimental conditions, various elastic and viscous moduli can be calculated. Furthermore, by manipulating the cell membrane structural elements, either with crosslinking agents or with lysing enzymes, the contribution of the various lipids or proteins to the whole membrane mechanics can be estimated.

11.3 Soft Tissues

Specific cells, in particular the fibroblasts, produce and export the two major structural proteins of mammals, namely collagen and elastin, along with a soft matrix consisting of mucopolysaccharides (also called

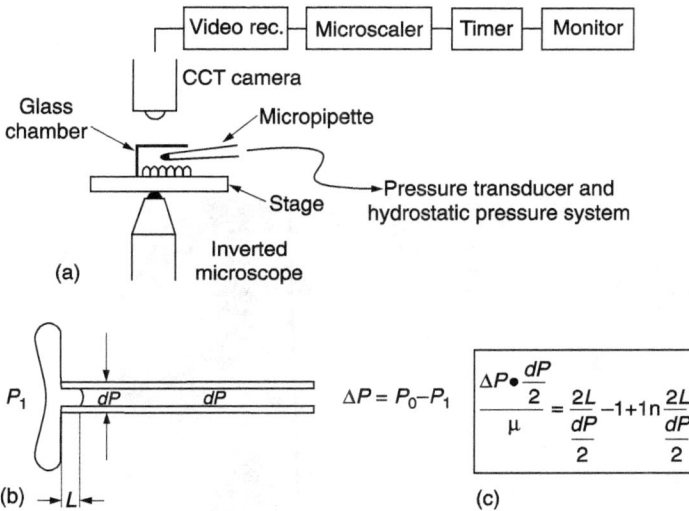

Figure 11.4 (a) A schematic representation of the equipment necessary for the micropipette aspiration experiment (see text for details). (b) Schematic of a RBC partly aspirated inside a glass micropipet of internal diameter dP, with aspirating pressure $\Delta P = P_0 - P_1$, where P_0 is the hydrostatic pressure applied at the pipette (negative) and P_1 is the atmospheric pressure. (c) The equation relating the length of the aspirated RBC inside the micropipette, L, to the applied pressure, ΔP. μ is the elastic shear modulus of the RBC membrane.[2]

proteoglycans or glycosaminoglycans) in which the collagenous fibers and elastinous bands are embedded to form the multitude of soft tissue structures of the body. Selected examples will be presented to illustrate various aspects of their composite nature.

11.3.1 Vascular vessels and heart valves

The classical anatomical description of the aorta (the major blood vessel) is of a trilaminate structure, the intima (inside, in contact with the flowing blood), the media, and the adventitia.[3] The aortic system has many branches that eventually end up as capillary tubes. The internal diameter (in humans) of these vessels starts at 25–30 mm and diminishes down to 5–10 μm. Furthermore, the system is never at rest; it is constantly under unsteady pressure.

The heart valves are also described as composed of four layers, namely, the arterialis (facing the aorta), the fibrosa, the spongiosa, and the ventricularis.[4] A more detailed examination of this structure, however, reveals a multilayered arrangement, as shown in Fig. 11.5. The exact architecture and composition of each layer, their interconnecting mechanisms, and the role of each one in the mechanical function of the valve leaflet have not been thoroughly investigated so far. The picture in Fig. 11.5 suggests also that this structure is so complex that any attempt to copy it exactly using man-made materials would be extremely difficult.

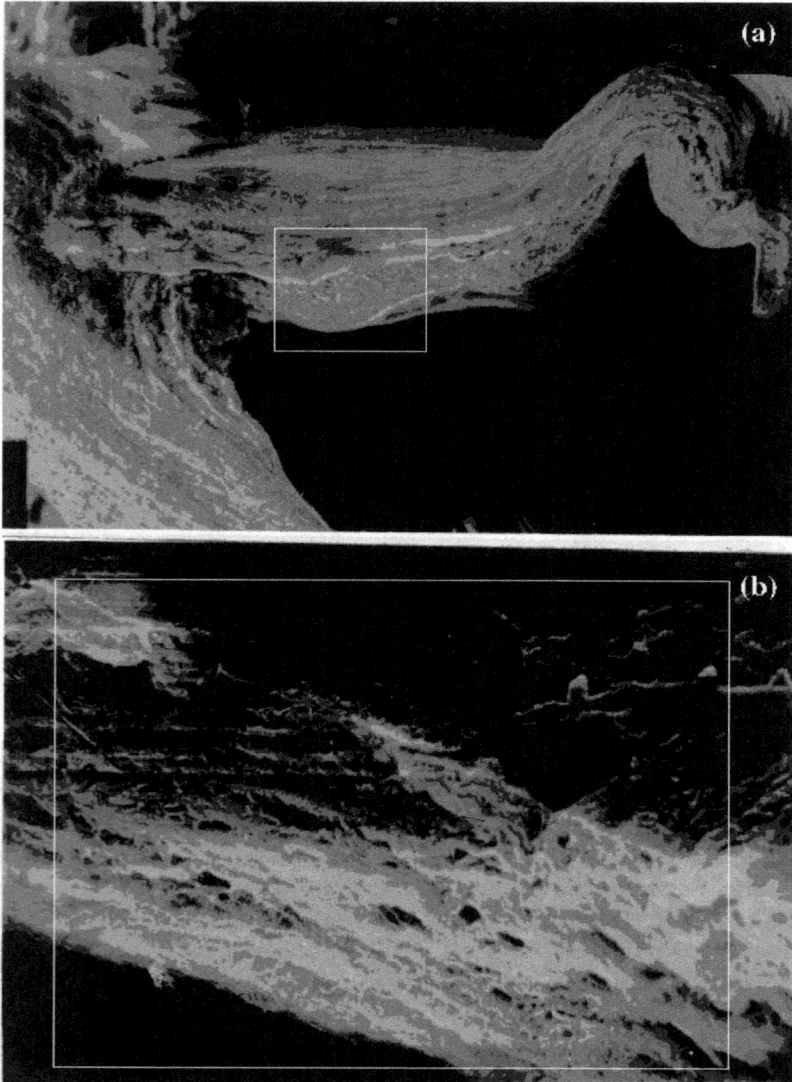

Figure 11.5 SEM photographs of human aortic valve. (a) Cross section of valve cusp and aortic ring, ×26. (b) Cross section and ventricularis, ×340.

In both vessels and valve leaflets the structural composition is that of collagen, elastin, and proteoglycans. The relative amount both in form (size of fibrils, orientation, etc.) and in mass of these components, as well as their association and bonding among themselves, varies greatly according to the anatomical position.

Investigatation of the mechanical behavior of the cardiovascular tissues is complicated for the following reasons:

- Each vessel is continuously under muscular and neural control.
- The properties change with time (aging).
- The vessel is never at rest (always under stress).
- The whole system is attached externally to the surrounding tissues, whereby its mechanical performance is orchestrated.

The various approaches used for such studies are as follows:

1. "In-vivo," i.e., in living humans or animals using noninvasive techniques (for example, with ultrasound).
2. "In-vivo," using invasive techniques (for example, strain gauges).
3. "In-vitro," i.e., the vessel or the valve has been taken outside the body, the living cells have died, and therefore only the passive mechanical properties are measured either
   (a) in whole vessels (or segments), or
   (b) on rectangular strips cut from the vessel wall.

Each method obviously has its own limitations. To get the best possible picture of vessel mechanics it is necessary to carry out a battery of associated tests, as described previously.

The influence of age, post-mortem time of experiment, etc., on the mechanical properties of biological materials are described with a wealth of information in the classic reference by Yamada.[5] A few points, however, need to be mentioned additionally, and are illustrated in Fig. 11.6.

The first point to be made is that when a strip of bovine aortic tissue is subjected to a tensile experiment, nonlinear behavior is observed that depends strongly on the anatomical position (direction) of the strip (see the two curves designated as A in Fig. 11.6). A histological examination of this composite structure reveals that there is a preferential orientation of the collagen fibers in the circumferential direction. Therefore, the whole process of harvesting the vessel from the animal, its preservation, the opening and cutting away of the strip might influence the resulting measurements. For example, one may have cut load-bearing fibers that run obliquely with respect to the

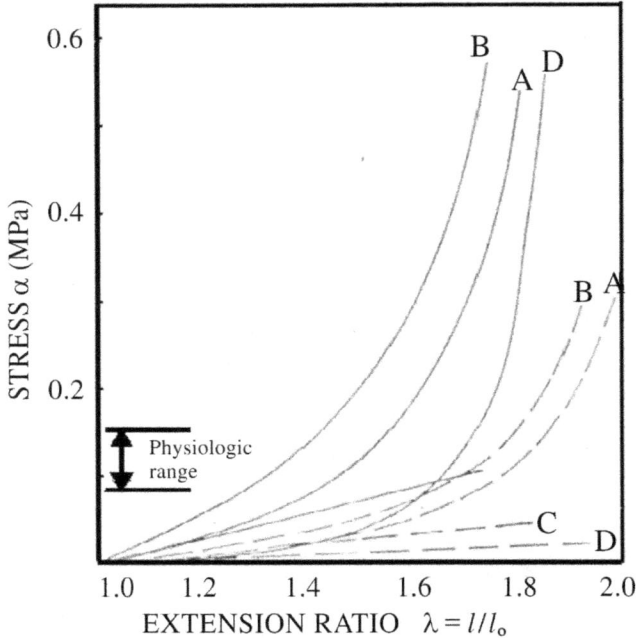

Figure 11.6 Typical stress–strain curves for bovine aortic tissue. Observe the strong anisotropy. See text for details. A, native tissue; B, defatted; C, decoagulated; D, deelasticated. Solid curves, circumferential sample; dashed curves, longitudinal sample.

parallel edges of the strip so that some of them do not contribute to bearing the applied load. In such a case the calculated stress is lower than the real stress.

Another feature of Fig. 11.6 is an attempt to calculate the contribution of the different components (collagen, elastin, lipids, proteoglycans) to the overall mechanical response of the aortic tissue. In a series of experiments, similar strips were subjected to successive physicochemical or biochemical treatments by which the tissue was defatted or deelastinated, and so on.[4,6] In general it seems that the existence of lipids and proteoglycans does not strongly affect the stress–strain curve; elastin is important in the initial, low-stress region, while collagen manifests itself later at higher stresses. However, the importance of intercomponent bonding and its role in the transfer of forces among them has not yet been studied thoroughly.

11.3.2 Skin

The human skin is an important regulatory (for heat and moisture) and immunological (prevention of certain diseases) organ. It performs other physiological functions as well, and protects the internal organs from mechanical, chemical, or electrical injury. As with most biological tissues, skin is a multilayered composite structure, with continuous turnover of its cells. Anatomically it can be thought as being made of two parts: the outer, called the *epidermis*, and the inner, called the *dermis*. These two parts are separated by a thin collagenous layer, termed the *basement membrane*. Below dermis one finds muscle, vessels, and adipose tissue.

Epidermis consists of many layers of different types of cells, which have a variety of functions (the outermost layer consisting mostly of dead cells that easily leave the surface). Dermis also has cells, glands, the hair follicles, nerve endings, and various types and degrees of organization of collagen fibers, elastin bundles, and oxytalan fibers (similar to elastin).[3]

The mechanical properties of skin, therefore, are largely due to collagen and elastin networks (as discussed in the previous section on vascular vessels) and depend on many other factors such as location in the body, age, hydration, and others. A typical stress–strain curve of a uniaxially loaded skin strip is shown in Fig. 11.7.

11.4 Hard Tissues: Bone

Bone and tooth are considered as "hard" (or mineralized, or calcified) tissues in mammals, in contradistinction to the rest of the tissues, termed "soft." The major components of all bony structures are inorganic apatite crystals (mainly hydroxyapatite: $Ca_{10}(PO_4)_6(OH)_2$) and organic collagen. However, many factors including species, age, sex, the individual bone, and position within each individual bone among others, determine the composition of a particular bone sample. It is important to note that as much as one-third of the bone mass (in the long bones) may be water in the form of bone cells, blood, and neural networks.

An important structural feature of the bone is that it is continuously being remodeled. Not only during the growth of an individual but also during the "steady-state" of the individual's development. At least three different types of cells are involved in the process of bone generation and bone resorption, i.e., in bone remodeling: (1) the *osteoblasts*, which produce and export outside collagen and associated polysaccharides; (2) the *osteoclasts*, large multinucleated cells, which erode and resorb bony material; and (3) the *osteocytes*, which have

N   Native
DL   Deelastinated
DC   Decollagenated

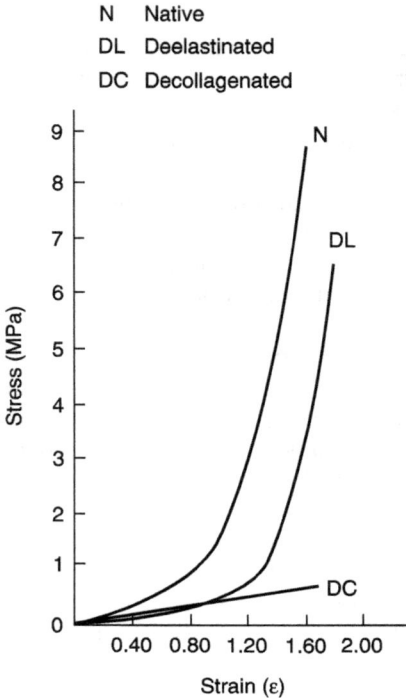

Figure 11.7 Stress–strain curves of abdominal canine skin: N, native; DL, deelasticated; DC, decoagulated. (*Missirlis and Ahood, unpublished data.*) See text for details.

many long, branched processes and occupy the greater part of the *lacunae* and *canaliculi*, i.e., the open spaces in the midst of the bony structures. The role of osteocytes is possibly to maintain the local architecture and to facilitate the transport of materials along the canals. The dynamic remodeling of bone is a response to the application of stresses. Unstressed or overstressed situations lead to weakening of the bone by stimulating the osteoclastic activity.

As the different bones serve different functions (scull, ear, tooth, finger, etc.) so their structure varies. An example will be given here of the major support bones of the adult human skeleton, such as the *femur*. Starting at the *molecular level* of organization of the noncellular structure, one finds the *tropocollagen* molecule, approximately 1.5×280 nm in size, in close association with apatite crystallites of about $4 \times 20 \times 60$ nm size.[7] At the next hierarchical level, the *ultrastructural* level, a microfibrinal composite structure made up of collagen and apatite crystals is formed into fibers about 3–5 μm thick. These fibers, usually having a preferred orientation, make up *lamellae* which in one *microstructural* case (there are others) are organized concentrically to make up *osteons* (see also Fig. 11.8). In

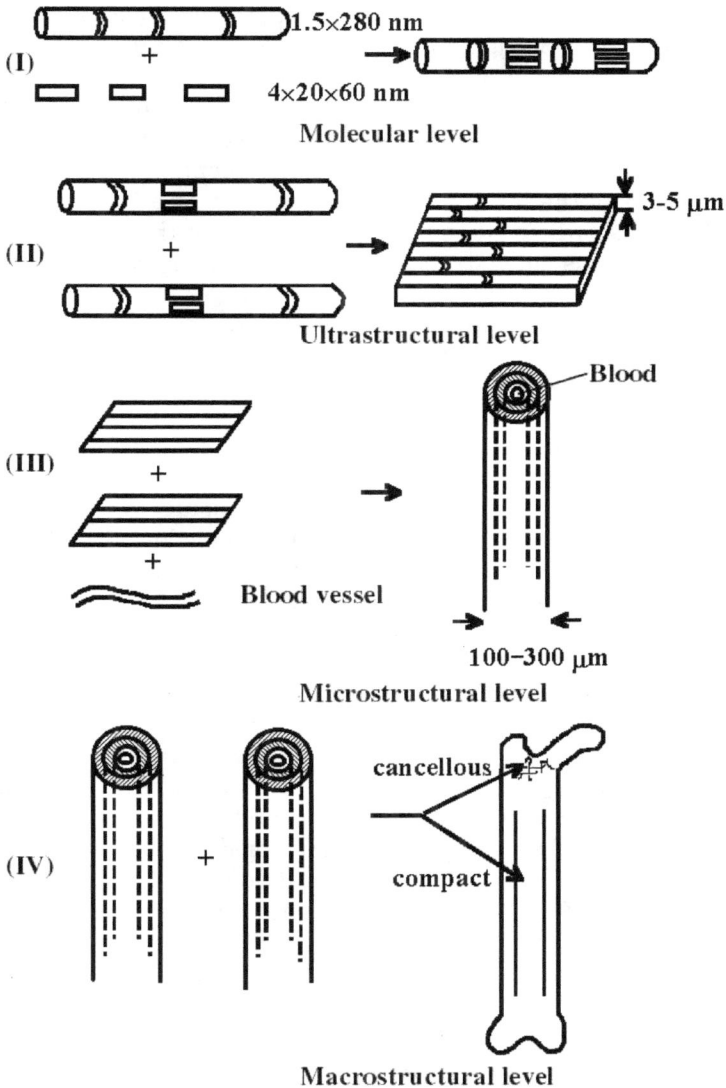

Figure 11.8 A highly schematic drawing showing the four levels (I to IV) of hierarchical organization of bone. At the first (molecular) level, tropocollagen molecules and apatite crystals make up fibrils. Next at the ultrastructural level the microfibrils are organized into lamellae, which in turn may form osteons at the microstructural level. The osteons (or haversian systems) may form either compact or cancellous bone structures.

the central canal of the osteons one finds one or two blood vessels. Osteons are about 100–300 μm in diameter.[8] At the next level, the *macrostructural*, the groups of osteons can form two different architectural designs. One is called *compact* or *cortical* bone: a dense structure, found mainly in the outer shell of the shafts of long bones (the load bearing bones), the only pores being the central (*haversian*) canals of the osteons, the places where the steocytes are located and the sites of bone resorption. At the articulated ends of the long bones one finds a more porous bony structure called *spongy* or *trabecular* or *cancellous* bone.

Both types—compact or cancellous—bone are anisotropic and heterogeneous and their mechanical properties reflect this fact. It may, therefore, be extremely difficult to assess and interpret the mechanical behavior of a whole bone *in vivo* because, in addition to the anisotropic and nonhomogeneous character of the various bony structures, many bones contain a viscoelastic marrow and their shape is irregular. However, great progress has been made in carefully preparing dead bone samples for testing in various modes of mechanical loading, and some generally accepted conclusions can be made. For example, although bone is a viscoelastic material, a standard tensile test (even when ultrasonic wave propagation experiments are performed) of a *wet cortical bone sample* at a quasi-static strain rate, at room temperature, will result in an almost linear (elastic) stress–strain curve, approximating cortical bone as an *anisotropic, linear elastic solid*.[7,8] Brittle fracture occurs at low strains between 0.5% and 3%.[8] Values of the Young's modulus, E, indicating the stiffness of the cortical bone, range from 7 to 30 GPa depending on the position, orientation, and age. These values are intermediate between the two main components of the composite cortical bone material, that of the linear elastic ceramic reinforcement (hydroxyapatite) having $E \cong 160$ GPa and that of the compliant, ductile polymer matrix (collagen) with a tangential modulus $E \cong 1.2$ GPa. Of interest is also the fact that the ultimate compressive strength of compact bone is higher than the tensile strength (typical values 180 MPa and 135 MPa, respectively).[5] For comparison, austenitic stainless steel has Young's modulus around 200 GPa (one order of magnitude higher than that of bone) and an ultimate tensile strength from 540 to 1000 MPa (4–7 times higher than that of bone).

Table 11.1 gives examples of the composition and some properties of different tissues.

Another interesting point is that the variations within the body of some of the mechanical parameters is indeed impressive. For example, the maximum elongation (elasticity) at failure of human organs and tissues (adult) varies from less than 1% (spongy bone, vertebra) to

TABLE 11.1   Examples of Tissue Compositions and Selected Properties

Tissue	Composition[a]		Properties[a,b]
Dentin	$Ca_5OH(PO_4)$	50%	$\rho = 1.9\,g/cm^3$
	Collagen	32%	$E = 3 \times 10^2\,MPa$
	Liquids	10%	$\sigma_c = 19\,GPa$
Enamel	$Ca_5OH(PO_4)_3$	99%	$\rho = 2.2\,g/cm^3$
			$E = 84\,GPa$
			$\sigma_c = 4 \times 10^2\,MPa$
Bone	$Ca_5OH(PO_4)_3$	40–45%	$\rho = 1.6$–$1.7\,g/cm^3$
	Collagen	40%	$E = 10\,GPa$
	Fluids	20%	$\sigma_c = 100\,MPa$
Aorta	Collagen	20%	$E = 0.1$–$0.2\,MPa$ (at 20% strain)
	Elastin	60%	$= 1$–$2\,MPa$ (at 80% strain)
	Smooth muscle, liquids	20%	
Tenden	Collagen	95%	$E = 8 \times 10^2\,MPa$
Ligaments	Elastin	85%	$E = 0.3\,MPa$
	Collagen	10%	
Connective	Skin Collagen		$E = 0.02$–$12\,MPa$
Tissue			Ultimate tensile strength $= 1\,MPa$

[a] Values vary according to species, site and definition.
[b] E = Young's modulus; σ_c = compressive strength; ρ = density.

10–20% (carilage, heart valves, nerves) to 50–100% (skin, muscle, arteries) to 200% (urinary bladder), and to over 300% (tracheal membrane, erythrocyte membrane).

11.5 Biomaterials

When the need arises for the alleviation of disease, injury, or handicap for the replacement or modification of the anatomy or of a physiological process or for the control of conception, by processes applying using instruments, apparatus, appliances, materials, or other articles to the human body, we refer to those devices as *medical devices*.

A major part of such medical devices is the *biomaterials* used. Biomaterials include any synthetic or natural polymers, metals, alloys, ceramics, composites, or other nonviable substances including tissue rendered nonviable.[9] For legal and commercial reasons as well as scientific ones, much effort has been made to give more precise definitions of what constitutes a biomaterial. For example, there have been two "Consensus Conferences" of the European Society of Biomaterials in Chester, UK (1987, 1991) for precisely this purpose.

It is futile at the present to try to generalize theoretically approaches for developing an ideal "biomaterial" because there is an enormous variety of applications, there are quite different types of biomaterials, and the complexity of tissue reactions is overwhelming. The general guidelines provide for a biomaterial to possess a set of physicochemical, mechanical, and biological properties that will make it suitable for safe, effective, and reliable use within (or in contact with) a physiological environment that, by its nature, attacks and attempts to expel any foreign intruder whether a bacterium or a foreign surface.

The question of the selection of materials for specific devices is of course very important, and for each application the characteristics and properties of the material—such as chemical, physical, toxicological, electrical, morphological and mechanical—must be taken into account. As there is an enormous variety of medical devices, it is necessary to categorize them somehow, so as to screen out irrelevancies and focus on the most relevant parameters to be considered for the selection of materials for use in manufacturing a particular medical device and to apply biological tests that are pertinent for the specific case.

In this sense, we can generally distinguish the following:

- *Surface-contacting devices*, such as electrodes, bandages, etc. that contact the skin, or contact lenses, urinary catheters, dental prostheses, etc., that contact mucosal membranes, or other devices used to heal compromised surfaces (ulcers, burn).

- *External communicating devices*, as when circulating blood comes out of the body and returns back after passing through such devices as dialysers, extracorporeal oxygenators, or hemoadsorbents, along with their accessory tubings and so on, or when other tissues/bone/dentin come into contact with laparoscopes, arthroscopes, dental filling materials.

- *Implant devices*, which include bone prostheses, pacemakers, neuromuscular sensors, breast implants, heart valves, and ventricular assist devices.

Another criterion for categorization is the duration of use of the device, since obviously the various phenomena associated with the interaction of materials and living tissues depend on the length of time of exposure. One such categorization[9] is into *limited exposure*, contact time, up to 24 h; *prolonged exposure*, from 1 to 30 days; and *permanent* contact for longer periods.

Examples of applications of biomaterials

1. Total joint replacements (hip, knee, fingers, shoulder, elbow); arthroplasty
2. Orthoses (intermedullary rods, intertrochanteric nails, bone plates, screws, etc.)

3. Bone cements

4. Dental restoration, prosthetic bridges, and dentures

5. Alveolar bone replacements, cements, and adhesives

6. Heart pacemakers

7. Cardiac heart valves

8. Artificial heart components, heart assist devices, arterial and vascular prostheses

9. Chronic catheters and shunts

10. Artificial lens, cornea, vitreous humor

11. Neural electrodes (hearing, vision, bladder, brain)

12. Replacement of missing soft and hard tissues (in ear, nose, cheek, maxillofacial, etc.)

13. Artificial skin, tendons

14. Artificial trachea, intestinal wall, bladder, etc.

15. Sutures

16. Surface electrodes (ECG, EMG, EEG)

17. Encapsulation of implanted devices

18. Neurosurgery clips

19. Diagnostic catheters, blood bags, blood needles, blood tubing

20. Kidney dialysis, plasmapheresis, heart–lung machines

21. Drug carriers

11.5.1 Interface

As was suggested earlier, for a biomaterial to serve its purpose it must possess suitable physical and mechanical properties. However, the most critical aspect of its fate will be its contact with the living tissues, which means that the long-term stability of the *interface* formed between living and nonliving substances will determine the success or failure of a biomaterial (or a device fabricated from biomaterials).

At the outset it must be stressed that the formed interface is in a dynamic state. Whether the implantation procedure involves, as it usually does, a wound or simply comes into intimate contact with the living tissue, there will be reparative processes (including immunological ones), histochemical and cellular changes associated with the "foreign invader". These complex tissue reactions will be changing with time, involving kinetics that depend on the type of tissue–implant interface.

Similarly, the performance of an implanted biomaterial depends strongly on the dynamic changes of surface composition and structure of the material at the time of implantation and thereafter.

There are four major types of biomaterials in terms of interfacial response of tissues, according to Hench:[10]

Type 1: Nearly inert, smooth surface

Type 2: Nearly inert, microporous surface

Type 3: Controlled reactive surface

Type 4: Resorbable

Most biomaterials in use today are of type 1, but research has been going on strongly also in developing biomaterials with surfaces of the other types.

Of great importance in the development of the interface between a biomaterial and the living tissue are the initial events that take place at the interface, largely concerned with the physicochemical phenomena that take place within seconds or minutes following contact between biomaterial and tissues. As a corollary of the above considerations, one may expect that the specific set of biological, mechanical, or other tests for the development of, say, a heart valve will be different from those for the development of an intravenous catheter, and quite different again for a hip prosthesis.

As the scope of this chapter is to give only a brief overview of this subject, we shall attempt to describe the major events that take place when nonliving materials come into contact with only one living tissue, the blood. First we shall briefly outline other important biological tests that are recommended by all standards organizations.[9]

- *Cytotoxicity.* These tests involve cell culture techniques and determine the fate of specific cells (lysis, inhibition of growth, etc.) caused by contacting the cells with the material or extracts from the material.

- *Allergic reactions.* These are important because even minute amounts of leachables from the material may result in allergic or sensitization reactions.

- *Irritation.* It may be that some materials in contact with (or implanted in) as the skin, the eyes, etc. may irritate the particular tissue by surface contact or through leachables.

- *Intracutaneous reactivity.* These tests assess the localized reaction of tissue to material extracts in situations in which the irritation tests are not applicable.

- *Acute systemic toxicity.* These tests are carried out on *animal* models, for periods of less than 24 h, to detect possible harmful effects due to leachables, degradation products, or simple exposure to the materials. This category includes the *pyrogenicity* tests. Analogously, for exposure times greater than 1 day we have the *sub-chronic toxicity* tests.

- *Genotoxicity.* These also use specialized cell cultures in contact with the material and/or its extracts to determine possible gene mutations, changes in chromosome structure, etc.

- *Implantation.* Similarly to the sub-chronic toxicity tests, these assess local pathological effects on living tissue by implanting a sample of a material at the actual site.

- *Carcinogenicity.* Involving appropriate animal models, these tests last for the total life-span of the animal (*chronic tests*) and determine the tumorigenic potential of the device, materials, and/or their extracts.

- *Reproductive and developmental toxicity.* These tests evaluate the potential effects of devices, materials, and/or their extracts on reproductive function, embryonic development (teratogenicity), and so on.

11.6   Blood–Material Interactions

In order to understand the enormous variability found in the literature regarding *hemocompatibility*, i.e., the compatibility with blood, of various materials, one should keep in mind that in reality we are dealing with a three-party system: the device, the material, and the blood.[11] This means that the same blood (of exactly the same composition) will interact with the same material surface in different ways depending on the device that incorporates the material.

The blood is a complex tissue consisting mainly of water, cells (erythrocytes, platelets, various leukocytes), ions, hundreds of plasma proteins, and other circulating molecules. Activation of plasma proteins and/or blood cells leads to systemic inflammatory reactions involving many known or suggested pathways and the generation of new molecules along with the release of substances from a multitude of cells.

The material, with biomolecules (such as albumin or heparin) grafted on its surface or not, should have the biomechanical properties necessary for the specific application and should be able to be processed to forms, shapes, and sizes appropriate to its final use. Of primary importance are of course its surface properties, as it is the surface that contacts the elements of blood and will interact dynamically with them. Parameters such as hydrophobicity or hydrophilicity, surface charge, polarity, heterogeneity in the distribution of reactive chemical groups (domains), mobility of the surface molecules, smoothness, etc., may be important both initially and as they change with time, temperature, and evolving physicochemical environment.

The device, or the final product incorporating the biomaterial, will be used in the patient. It could be used for a relatively short time (minutes to hours), as with a catheter, a hemodialyser, a blood oxygenator, or blood tubes used in extracorporeal devices, or it could be incorporated into the cardiovascular system permanently in the form of an artificial vessel, a heart valve, a left ventricular assist device, a total artificial heart, and in the future perhaps other artificial organs (lung, liver, etc.).

Thus, the flow conditions (shear rates, turbulence, secondary flows, etc.), duration of contact, size of the contact surface area, and actual placement site in the cardiovascular system are very important parameters to be considered in addition to the surface finish due to fabrication and to sterilization effects.

It is generally accepted that the first step of the interaction process between blood and artificial surfaces is the adsorption of plasma proteins. Subsequent events involve platelet and leukocyte adhesion and aggregation, and activation of the coagulation system, as well as the fibrinolysis and complement systems, resulting eventually in the formation of a *thrombus* on the artificial (e.g., polymer) surface. More specifically, the nature of the artificial surface, under specific hemodynamic conditions, may promote:

- specific plasma protein *adsorption* with resultant activation, denaturation and/or desorption;

- platelet *adhesion* and *activation* with release of active substances that are important in platelet *aggregation* leading to thrombus formation;

- hemolysis;

- leukocyte adhesion, *spreading* and activation, with further involvement of the inflammatory and complement systems;

- interaction with other blood components such as lipoproteins, trace plasma proteins, *inhibitors* of activated plasma proteins, and adhesive proteins (fibronectin, thrombospondin).

It should be noted that each of the above systems is quite complex in its own interaction with foreign surfaces and, as all the plasmic components, proteins, and cell interact among themselves, the complexity of the events become enormous.

It should once more be emphasized that here we deal only with the complex interactions during the *initial events* at the *interface* and not the long-term effects at the site of implantation or elsewhere in the body.

11.6.1 Protein adsorption at the solid–solution interface

It is almost axiomatic that when protein solutions are put into contact with solid surfaces, adsorption occurs. However, in blood there are about 150 different plasmic proteins (totaling about 70–80 g/l) with great variations in size, concentration, and purpose. It is therefore logical to ask: What is the composition of the protein layer that is laid down on different surfaces in contact with blood? Is there a preference for specific proteins to be absorbed depending on the type of surface? Does the composition on a particular surface change as a function of time? And what happens to the proteins that get adsorbed: do they simply bind to the surface, do they undergo (after binding) different types of transformation, for example denaturation, changes in their biological activities, etc.?

Among the general conclusions that have come out of protein adsorption studies are the following:[12]

- Most proteins are adsorbed on most surfaces.

- The relative amounts of the different proteins vary from surface to surface.

- Many proteins (when whole blood is used) become degraded. This may be due to proteolytic enzymes derived from damaged cells.

- The Vroman effect is seen on most surfaces. This is the turnover at the same surface site of one protein by another as time progresses, which also may later be replaced by a third.

- Many proteins may be deposited at their own times (and conditions of flow) that have not yet been investigated.

11.6.2 Blood coagulation and fibrinolysis

When blood is exposed to damaged tissue or to an artificial surface, blood *coagulation* takes place, which comprises a series of consecutive enzymatic reactions (a cascade) that leads to the conversion of *fibrinogen* into a *fibrin clot*. Intimately associated with coagulation is the process of *fibrinolysis*, the enzymatic breakdown of fibrin (usually by *plasmin*). The dynamic steady state of coagulation and fibrinolysis maintains the patency of the circulatory system. Fibrin formation is a major contributor to thrombosis. There are two pathways for coagulation, which at a certain level merge into a common pathway.

The intrinsic pathway is triggered by collagen or other subendothelial components (or negatively charged surfaces) that on contacting

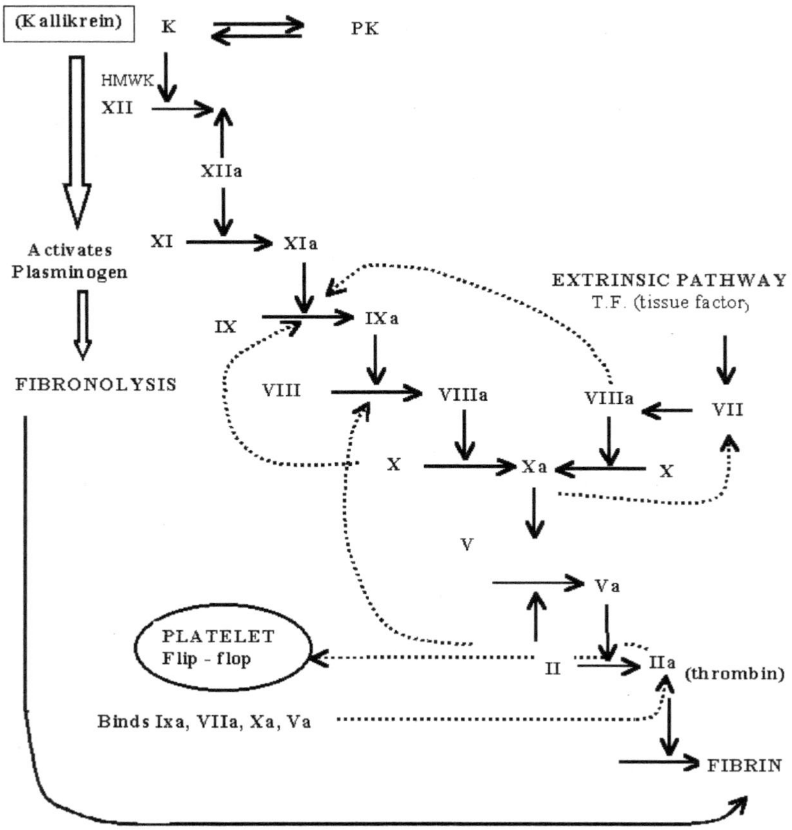

Figure 11.9 Partial set of interactions leading to coagulation (fibrin formation) and fibrinolysis, starting either by the generation of tissue factor (due to injury, for example) or by blood contacting a (negatively charged) artificial surface. Inhibitors and complement system (and other systems) that also interact are not shown. Both pathways need Ca^{2+} and phospholipids to enhance acceleration. The subscript α indicates the activated state of the molecule.

factor XII activate it into factor XIIa. In this process two other plasmatic components are involved as well, high molecular weight kininogen (HMWK) and kallikrein. Factor XIIa activates factor XI into XIa, which in turn, in the presence of Ca^{2+} activates factor IX into IXa. Then IXa in the presence of platelet membrane phospholipids, Ca^{2+}, and factor VIII converts factor X into Xa. This activated factor Xa, together with factor V and in the presence of Ca^{2+} binds to platelet membrane phoshpolipids. The complex formed (sometimes referred as prothrombinase) catalyzes factor II (prothrombin) to *thrombin* (factor IIa). There are positive feedback mechanisms involving thrombin and the previous reactions where factors V and VIII are involved.

The extrinsic pathway is triggered by tissue thromboplastin (made available either by subendothelium or by stimulated leukocytes, or by "activated" endothelial cells). Thromboplastin in the presence of Ca^{2+} complexes with factor VII to activate factor X into Xa. This is a point of convergence with the intrinsic pathway.

As is the case in most situations, the process of activation of the circulating clotting factors has a counteraction. Indeed, there are a number of naturally occurring protease inhibitors, including antithrombin-III (AT- III). Deficiency of ATIII in a patient is associated with an increased risk of venous thrombosis.

Fibrinolysis also follows intrinsic or extrinsic pathways and depends on the plasma protein *plasminogen.*

In summary if an artificial or natural surface, negatively charged, comes in contact with blood, coagulation is initiated. This contact system of coagulation consists of four proteins: factors XII, XI, prekallikrein, and HMWK. Activation of the contact system initiates the intrinsic and extrinsic coagulation pathways, intrinsic fibrinolysis, and the complement, kinin, and renin–angiotension systems.

The plasma proteins involved in contact activation have been identified by studying the reaction in rare individuals with specific protein deficiencies. Figure 11.9 shows all the above cascading reactions in summary.

11.7 Materials Used in Implants

A wide variety of materials are in use as biomaterials, including metals and alloys, glasses and ceramics, polymers, composite materials, and biologically derived biomaterials.

The following list shows important materials along with their composition and main applications.

1. *Pure metals* such as C, Au, Ag, Pt, Ta, W, and Ti, usually more than

99.9% pure. Orthopedics, dental applications, neurosurgery, electrodes, heart valves.

2. *Alloys* based on Au, Pt, Ag, Ni, Co, Ti, and stainless steels of various compositions. Dental, orthopedic, and neurosurgical applications, electrodes.

3. *Glasses* (mainly SiO_2 but including various proportions of Al_2O_3, P_2O_5, CaO, Na_2O, K_2O, ZnO, B_2O_3, ZrO_2, Li_2O). Orthopedic coatings, dental applications, electronic implant packaging.

4. *Ceramics* (alumina, zirconia, calcium hydroxyapatite, TiO_2, ZnO). Orthopedic and dental applications (dental cements included).

5. *Carbons* including graphite, diamond, and glassy, vitreous, and pyrolytic carbons. High purity. Carbon fibers produced by pyrolysis of spun polymeric fibers (or from the vapor phase) are used in composite structures.

6. *Synthetic polymers* (most of which include various amounts of additives such as plasticizers, antioxidants, etc. in addition to the repeating units of the basic unit). Table 11.2 shows different polymers (their monomer) and their main applications.

7. *Natural polymers.* Table 11.3 shows selected natural polymers and their main applications.

11.8 Composite Biomaterials

Composites in general consist of at least two different types of materials (fibers or fillers and matrix) or of one material with different structures (self-reinforced material) at a scale larger than the atomic. There are various hierarchical levels of organization (molecular, microscopic, or macroscopic) of the structures that make up the composite material. To generalize, one might suggest that foamy materials are also composites in which one phase is empty space.[13]

Most composites are fabricated, in principle, to provide better mechanical properties such as strength, stiffness, toughness, and fatigue resistance. This principle is based on the interaction of the fibers and/or fillers with the matrix. The geometry of the reinforcement, therefore, strongly affects the improvement of the mechanical properties. The direct contribution of the matrix, which is the continuous phase and is softer than the discontinuous phase (fibers or fillers), to the mechanical properties of the composite material is rather small. The function of matrix is to transfer the loads from matrix to fibers (or fillers) over the matrix–fiber (filler) interface, to protect the fibers (fillers), and to give the desired shape to the device. It is, therefore, very important to have excellent adhesion between the

TABLE 11.2   **Polymers in Biomaterials and Applications**

Poly(vinyl chloride) (PVC)	$-CH_2-CH-$ $\qquad \vert$ $\qquad Cl$	Blood bags, tubing catheters
Polyethylene (PE)	$-CH_2-CH_2-$	Orthopedics
Poly(ethylene terephthalate) (Dacron, Mylar)	$-O-CH_2\text{-}CH_2\text{-}O\text{-}\overset{O}{\overset{\Vert}{C}}-\bigcirc-\overset{O}{\overset{\Vert}{C}}-$	Vascular prostheses heart valve stents
Polyacrylonitrile (PAN)	$-CH_2-CH-$ $\qquad \vert$ $\qquad CN$	Dialysis filters
Polytetrafluoroethylene (PTFE, Teflon)	$-CF_2-CF_2-$	Orthopedics, neurosurgery
Polycyanoacrylates (PCA)	CN $\vert$ $-CH_2-C-$ $\vert$ C $O^{\nearrow} \;\; OR$	Tissue adhesives
Poly(methyl methacrylate) (PMMA)	CH_3 $\vert$ $-CH_2-C-$ $\vert$ C $O^{\nearrow} \;\; OCH_3$	Orthopedics, dental, ophthalmology neurosurgery
Silicone	CH_3 $\vert$ $-O-Si-$ $\vert$ R where R: CH_3 or $\quad -CH=CH_2$	Heart valves, tubing plastic surgery, oxygenators
Polyurethanes	$A_x B_y$ A: polyether $-(CH_2)_4-O-$ or polyester $-(CH_2)_4-O-\overset{O}{\overset{\Vert}{C}}-(CH_2)_4-\overset{O}{\overset{\Vert}{C}}-$ B: "hard" segment $-\overset{O}{\overset{\Vert}{C}}-N\underset{\diagup}{\diagdown}\quad N-\overset{O}{\overset{\Vert}{C}}-$ or other types	Catheters, tubing, artificial heart ventricules, pumps, heart valves

(*continued*)

Table 11.2   Polymers in Biomaterials and Applications (*continued*)

Poly(glycolic acid) (PGA)	$-CH_2-\overset{\overset{\textstyle O}{\|\|}}{C}-O-$	Resorbable sutures
Copolymers of silicone with polycarbonates or polyurethanes		Oxygenators, intra-aortic baloon pumps, LVAD
Polyamides (Nylon 6)	$-\overset{\overset{\textstyle O}{\|\|}}{C}-(CH_2)_6-\overset{\overset{\textstyle H}{\|}}{N}-$	Sutures
Polyformaldehyde (Delrin)	$-CH_2-O-$	Heart valves occluders
Polycarbonate		Heart valves occluders, oxygenators, cranioplasty
Polypropylene	$-\overset{}{\underset{\underset{\textstyle CH_3}{\|}}{CH}}-CH_2-$	Syringes, catheters, sutures, occluders, oxygenators, dialysis

TABLE 11.3   Natural Polymers

Polyisoprene (*cis*-1,4) (latex)	$-CH_2-\overset{}{\underset{\underset{\textstyle CH_3}{\|}}{C}}=\overset{}{\underset{\underset{\textstyle CH_3}{\|}}{C}}-CH_2-$	Artificial heart ventricules, shunts
Polyisoprene (*trans*-1,4) (gutta-percha)		Dental, casts
Collagen Arteries, veins, pericardium, valves fascia lata	Reconstituted Glutaraldehyde treated	Dialysis Grafts

fiber (filler) and the matrix for the composite to perform effectively. The adhesion between the fiber and the matrix should be good during the whole period of application (implantation of an artificial hip, for example) and not only in the beginning. The surface of the fibers should be thoroughly cleaned, or a special treatment of the surface should be applied so as to ensure that agents on the fiber surface such as extrusion lubricants (silicone oils) or textile finishes are removed and cannot interfere with the integrity contact.

TABLE 11.4   Materials for Composites

Matrix	Reinforcement	
	Fibers	Fillers
Resins	**Polymeric**	**Organic**
Epoxy	Polyesters	Polyacrylate
Polyacrylates	Polyolefins	Polymethacrylate
Polymethacrylates	Poly(ether ketones)	Composite
Polyesters		organic/inorganic
Silicones	**Resorbable Polymers**	
	Polyglycolide and its	**Inorganic**
	copolymers	Silicon dioxide
Thermoplastics	Polylactide and its	Silanes
	copolymers	Glass
Polyolefins	Polylactide-coglycolide	Yttrium trifluoride
Polycarbonates	copolymer	Yttrium fluoride
Polysulfones	Polydioxanone	Other metal oxides
Poly(ether ketones)	Aromatic polyamides	Hydroxyapatite
Polyesters		Tricalcium phosphate
	Others	
Resorbable polymers	Carbon	
Poly(glycolide) and its	Glass	
copolymers	Resorbable glasses	
Polylactide and its		
copolymers	**Ceramics**	
Polylactide-coglycolide	Hydroxyapatite	
copolymer	Tricalcium phosphate	
Polydioxanone		
Poly(hydroxy butyrate)		
Natural origin		
Others		
Carbons		
Ceramics		
Hydroxyapatite		
Glass ceramics		
Calcium carbonate		
ceramics		
Calcium phosphate		
ceramics		

Table 11.4 lists the various materials that are used either as matrix or as reinforcing fibers or filler for biomedical applications.[9] Some composites are intended to be nondegradable and stable for their lifetime, while others are designed to be resorbable at a preset rate. Obviously, the requirements in these cases are different and, especially in the second case, apart from the general biocompatibility requirements outlined previously, care should be taken regarding the biocompatibility of the degradation products.

In the following, the main reinforcing materials for biomedical composites will be presented and then the matrix materials. Resorbable composites will be dealt separately. Finally, examples from the dental, orthopedic, and cardiovascular areas will be given.

11.8.1 Reinforcing systems

The main reinforcing materials that have been used in biomedical composites are carbon fibers, polymer fibers, ceramics, and glasses. These reinforcements have been either inert or resorbable.

Carbon fiber for biomedical use is produced from polyacrylonitrile (PAN) precursor fiber in a three-step process: (1) stabilization, (2) carbonization, and (3) graphitization.[14] In the stabilization stage, the PAN fibers are first stretched to align the fibrillar networks within each fiber parallel to the fiber axis; they are then oxidized in air at about 200–220°C while held in tension. The second stage in the production of high-strength carbon fibers is carbonization. In this process, the stabilized PAN-based fibers are pyrolyzed (heated) in a controlled environment until they become transformed into carbon fibers by the elimination of O, H, and N from the precursor fiber. The carbonization heat treatment is usually carried out in an inert atmosphere in the 1000–1500°C temperature range. During the graphitization process, turbostratic graphitelike fibrils or ribbons are formed within each fiber, greatly increasing the tensile strength of the material.

Carbon fibers have been used to reinforce porous polytetrafluoroethylene for soft-tissue augmentation. Because of the complex shapes of facial tissues that need to be augmented (temples, forehead, cheek, middle ear, nasal dorsum, etc.) the implant is manufactured as a sponge with about 70% voids so as to be easily "carved" into custom shape and size.[15] Another application has been that of a carbon fiber stent coated with polylactide for the repair of the anterior cruciate ligament (of the knee) and other joint structures, in reality acting as a scaffold for deposition of new collagen.[16] However, possible problems with the fate of carbon particles have raised concerns about this device.

Other applications where carbon fibers have been used as reinforcement include bearing surfaces in total joint prostheses, repairs of tendons and ligaments, fracture fixation devices, and total joint replacement components. In all these cases the matrix is ultra-high-molecular-weight (UHMW) polyethylene. Other matrices have been tried, including poly(ether ether ketone) (PEEK) and the composite has been used to manufacture femoral stems,[17] polysulfones,[18] and epoxy resins.[19]

Carbon fibers have also been used as reinforcements of carbon matrix manufactured on the basis of phenol–formaldehyde resin. The composite material was covered by pyrolytic carbon. The materials are used as internal fixation devices and in hip endoprothesis.[20]

11.8.2 Polymer fibers

Most polymer fibers are not strong enough or stiff enough to be used to reinforce other polymers. Possible exceptions are aromatic polyamide fibers (commercially known as Kevlar) and UHMW polyethylene fibers.[14] Kelvar composites are used commercially where high strength and stiffness, damage resistance, and resistance to fatigue and stress rupture are important. Potential biomedical applications are in hip prosthesis stems, fracture fixation devices, and ligament and tendon prostheses. UHMW polyethylene fibers have recently become available. To date, these fibers have not seen extensive use in biomedical composites. Bulk UHMW polyethylene demonstrates excellent biocompatibility, but, there are preliminary data suggesting a less favorable response to UHMW polyethylene fibers.[21] Questions are always raised when bulk and fiber properties are compared. While in theory the basic materials should be the same, differences associated with surface characteristics and the details of manufacturing and processing can be significant.

An area in which polymeric membranes are used and in some cases a reinforcement is necessary is the wound healing. When the skin is damaged due to burns or from other causes, epithelialization (towards skin regeneration) is accelerated by keeping the wounds wet with a dressing that retains moisture (and allows for gas exchange). One such dressing is the *Biobrane*; this is made of nylon fibers embedded in silastic membrane and the composite is coated with collagen-derived peptides.[22]

In the cardiovascular domain, the clinically accepted vascular grafts of either knitted or woven polyester (Dacron) or of expanded polytetrafluoroethylene (Teflon) fail for small-diameter grafts (<6 mm). An attempt to use polyurethanes and to construct the tube by forming a fiber spun on a rotating mandrill,[23] gave disappointing clinical results. In an attempt to rectify this problem, the known anisotropy in the mechanical properties of blood vessels was taken into account and two different types of fibers were used, polyurethane ($E = 0.3$ MPa] and Dacron ($E = 500$ MPa). These were woven in the appropriate directions (longitudinally only Dacron fibers were used, but both types of fibers were used circumferentially).[24]

11.8.3 Collagen fibers

As already indicated, collagen is the most abundant protein in the mammalian species. This class of proteins, collectively termed collagen (from the Greek synthetic word κολλα-γονο, meaning giving birth to glue, because thermal denaturation of collagen transforms it into glue) is a rich source of material that can be made into various forms, sizes, and shapes. Indeed, native collagen molecules are organized in tissues in specific orders. Many forms of rather pure collagen can be reconstituted from the collagen molecules obtained either from enzymatic digestion of native collagenous tissues or by extracting the tissues with salt solutions. The types of fibrils formed in this process, whether the so-called *segment-long-spacing* in which all collagen molecules are aligned in tail-to-head orientation (thus giving the characteristic striation picture in transmission electron microscopy) or the totally random *fibrous-long-spacing*, depends on the environment of the reconstitution medium (salt concentration, etc.). There are two ways to isolate and purify the collagen material, a molecular technology and a fibrillar technology.[25] After further processing, the purified collagen materials thus obtained can be fabricated into different forms: membranes, porous structures, gels, tubular matrix, and filaments.

Collagen fibers are resorbable and have been used to construct composite materials by embedding them in a collagen matrix or in various polylactides[26] to be used as resorbable anterior cruciate ligament prostheses. Another very important area of application is that of *biological adhesives*. Fibrin sealants (recall that fibrin is the polymerization product of fibrinogen during blood clotting) are the most successful tissue adhesives. However, a major drawback is their limited strength. To overcome this problem in part, fibrillar collagen was added into the fibrinogen solution to yield a composite adhesive with increased strength.[27]

11.8.4 Ceramics

Many ceramic materials have been used as fillers to reinforce biomedical composites. They have been used as fillers because they are rather weak and brittle when loaded in tension or shear, which excludes them from use as (long) fibers. These reinforcements have included various calcium phosphates, aluminum- and zinc-based phosphates, glass and glass ceramics, and bone minerals. Minerals in bone are numerous: in the past, bone has been defatted, ground, and calcined or heated to yield a relatively pure mix of the naturally occurring bone minerals. It was recognized early that this mixture of natural bone mineral was poorly defined and extremely variable. Consequently, its use as an implant material was limited.

The calcium phosphate ceramic system has been the most intensely studied ceramic system. Of particular interest are the calcium phosphates having calcium-to-phosphorus ratios of 1.5–1.67. Tricalcium phosphate and hydroxyapatite form the boundaries of this compositional range. At present, these two materials are used clinically for dental and orthopedic applications. Tricalcium phosphate (the common name for which is whitlockite) has a nominal composition of $Ca_3(PO_4)_2$. It exists in two crystallographic forms, α- and β-whitlockite; in general, it has been used in the β-form.

11.8.5 Hydroxyapatite

Hydroxyapatite, with nominal composition $Ca_{10}(PO_4)_6(OH)_2$, which is the major mineral composition of bone, has been used extensively in biocomposites. The major applications are dental and orthopedic. Hydroxyapatipe particles have been embedded in collagen matrix, in UHMW polyethylene, in glass matrix, mixed with zirconia or alumina, embedded in polylactides, and associated with elastin-derived fibers. Selective applications will be illustrated later.

Silica particles (SiO_2) also have been used to give strength to rubber, used in catheters, gloves, and the like (in very fine particles), and as spherical silica fillers with resin to produce a composite that goes by the tradename Palfique Clear used in dental applications.[28]

11.8.6 Matrix systems

There are stable nonresorbable matrices and resorbable one of synthetic or of natural origin. The stable ones are mainly synthetic polymers. By far the largest literature exists for the use of polysulfone, (UHMW) polyethylene, polytetrafluoroethylene, poly(ether ether ketone), and poly(methyl methacrylate). These matrices, reinforced with carbon fibers and ceramics, have been used as prosthetic hip stems, fracture fixation devices, artificial joint bearing surfaces, artificial tooth roots, and bone cements.

11.8.7   Resorbable matrices

Resorbable matrices are produced from polyester materials such as polylactides (PLA) and poly(glycolic acid) (PGA) or from copolymers of these two, among others.

More detail about these polyesteric resorbable polymers will be presented later. Polyhydroxybutyrate has also been used, reinforced with hydroxyapatite,[29] as a composite bone-analogue material; a copolymer of hydroxybutyrate and hydroxyvalerate formed the resorbable matrix in which calcium metaphosphate fibers were embedded[30] to form a composite to be used as a totally resorbable fracture fixation device. In addition to these, polydioxanone, polycarbonates, polycaprolactones, polyorthoesters and polyesteramides have been used.

Resorbable polymers of natural origin have also been used to form the matrices of composite biomaterials. Collagen, as indicated previously, can be reconstituted not only in fibrillar form but also in membranous matrix. A simple method of production of collagen films is to dry a collagen solution or a fibrillar collagen dispersion cast on a nonadhesive surface. One can control the thickness of the collagen film (thicknesses up to 0.5 mm can easily be formed). However, to stabilize this film it must be crosslinked (usually with glutaraldehyde). The crosslinking of collagen films that are intended to be used as biomaterials serves two purposes. One is to decelerate the well-known degradation of collagen by the body's enzymes (collagenases). The other is to reduce the immunogenicity of the collagen: as the main source of it is from animals, it would be rejected, that is, would elicit the production of antibodies in the human organism.

A very important aspect of the preparation is the resulting porosity. It is important to have pores in a collagenous implant because they allow cells to pass through from neighboring tissues into the bulk of the implant. In practice, pore sizes between 1 and 800 μm have been formed depending on the use of the implant (e.g., 20–125 μm pore size is necessary for skin regeneration, and less than 10 μm for sciatic nerve regeneration).[31]

Highly porous graft copolymers of collagen and various glycosaminoglycans (also known as mucopolysaccharides or ground substance, they form the matrix in which collagen and elastin are embedded in forming the natural soft tissues) have also been prepared[31] with applications in the skin, the nervous tissue and the meniscus of the knee.

11.8.8   Resorbable composites

There are many situations in which the matrix of a biomedical composite or even the whole composite (i.e., including the fibers) is desired to be resorbed in a time-controlled way by the body. One application is release of drugs such as antibiotics or growth factors by exposing them to the biological environment. The other two applications that will be outlined in some detail are fracture fixation devices and tissue engineering.

11.8.9   Fracture fixation

Today, the fracture fixation devices such as osteosynthesis plates, screws, wires, pins, etc. that are used to secure and stabilize bone fractures for the purpose of synthesis of new bone that will join the fractured pieces together and reestablish the function of the particular bony structure are made mainly from metals or alloys. In the early stages of fracture healing, the rigid devices serve a useful purpose of aligning the fractured bone and stabilizing it by compression. However as the healing progresses, but especially after the healing is complete, the metallic plates in place offer negative service to bone, even leading to atrophy. The phenomenon is called *stress shielding* and is based on the fact that while the elastic modulus of cortical bone is of the order of 10–30 MPa, that of the commonly used metals and alloys is between 100 and 200 MPa. In loading, therefore, because of the disproportionate load sharing due to the difference in stiffness, disadvantageous mechanical interactions may take place, such as relative micromotion between the implant and bone, high stress concentrations, etc. Furthermore, in the long run, corrosion reactions may lead to adverse complications. When ions are released as a result of corrosion, the clinical problems may be enormous. They may cause local reactions, allogenic responses, tumor formation, or improper bone mineralization, and may affect the disease state of certain patients (Alzheimer's, etc.). For this reason a second surgical operation is performed to remove the implant. Apart from the economic cost, a second episode of surgery, is a very serious matter, especially for aged people (who suffer more broken bones and are osteoporotic).

The use of resorbable osteosynthesis devices, instead of the metallic ones, is therefore advantageous for at least two reasons. First, the devices will degrade with time so that their mechanical properties will change with time, reducing the stress shielding and allowing for stronger bone healing. Second, there will be no need for another surgical operation. This will be true provided that the controlled decrease in the mechanical properties follows the prescribed path and that the products of the biodegradation are biocompatible, i.e., either they are absorbed by the body or they are excreted without harm to the organism. Several resorbable polymers have been tried for this purpose. Invariably the unreinforced ones have compromised mechanical properties, so that the use of composites is a necessity. The major polymer used is polylactic acid, because its degradation in the body can be controlled (it depends mainly on the molecular weight of the polymer) and the degradation products are biocompatible, nontoxic, and easily excreted. PLA is hydrolyzed into lactic acid, which is a human metabolite, that ultimately ends up as CO_2 and H_2O.

PLA polymer reinforced with randomly oriented chopped carbon fiber was used to produce partially degradable bone plates. When implanted, the PLA matrix degraded and the plates lost rigidity, gradually transferring load to the healing bone. However, the mechanical properties of such chopped fiber plates were relatively low and consequently the plates were only adequate for low-load situations. An improved design was necessary if composite plates of these materials were to be successful in high-load situations. Hence, the possibility was investigated of using a long-fiber, angle-ply-laminated composite of carbon fiber and PLA.[32] Composite theory was used to determine an optimum fiber lay-up for a composite bone plate. Composite analysis suggested the mechanical superiority of a $0°/\pm45°$ laminae lay-up. Although the $0°/\pm45°$ carbon/PLA composite possessed adequate initial mechanical properties, water absorption and subsequent delamination degraded the properties rapidly in an aqueous environment.

In an attempt to develop a totally resorbable composite biomaterial, poly(L-lactic acid) matrix was reinforced with continuous calcium sodium metaphosphate glass fibers, which were pretreated using a silane coupling agent to protect the fiber–matrix interface from early dissolution.[33]

Another approach is to use fibers from the same material. The use of continuous fibers necessitates processing by pultrusion or compression molding, which are more expensive and more difficult than the injection molding used for chopped fibers. The matrix can be fed to the fibers in the form of powder, or as fibres by solution or by melt impregnation. As poly(glycolic acids) have a very short life-time, the polyesters of the lactic acid were used both as fibers and matrix.[34] The reinforcement fibers were made from pure poly(L-lactide) (P-L-LA) with inherent viscosity in the range 1.5–3.5 dl/g. As a matrix, an amorphous copolymer was produced that was a 70 : 30 mixture of P-L-LA and poly(DL-lactide), noted as P-L-DL-LA 70 : 30, with inherent viscosity of 1.5–2.5 dl/g. The fibers were produced by melt spinning and their elastic modulus was $E = 7$–$10\,GPa$. For the composite, hybrid yarn technology was utilized whereby reinforcement fibers and matrix fibers were intermingled. Although the initial mechanical characteristics were quite promising, the degradation kinetics were unsatisfactory mainly because of the existence of large quantities of monomer (1–3%), leading to an autocatalytic degradation reaction.

11.8.10   Mechanisms of degradation of biomedical composites

The degradation mechanisms for the component parts of the composites, i.e., polymers, ceramics, or metals, apply, of course, for the whole composite. However, of special interest are the specific mechanisms at the interfaces due to interaction of different materials. The processing method for fibre preparation may affect the crystallinity, the molecular weight, etc., and thus affect the degradation. Specific chemical and biological mechanisms include the following: adsorption of liquids either at the surface of the device or at the matrix–fiber (filler) interface; adsorption and swelling of either one; chemical reaction between the components and the biological environment; or reactions between one phase and the degradation products of the other. Local stresses can be generated in reinforced materials as a result of different coefficients of thermal expansion or of swelling, and the local stresses lead to enhanced degradation. Mechanical processes include surface abrasion because of micromotional friction due to the use of the implant (hip joints, dental implants, etc.) or friction at the implant–tissue interface, and interface abrasion by internal motion of the reinforcing material with respect to the matrix. Abrasion may expose the reinforcing material and as a consequence the total chemical–biological–mechanical performance of the implant may change.

Both real-time and accelerated tests are performed to test the degradation of composites. Time, pH, and temperature are important parameters to be considered. The composition of the test medium is also very important as the simulation of in vivo conditions is crucial. Parameters that are usually measured include (the list is not exhaustive): molecular weight changes (measured by viscometry or gel permeation chromatography); mass changes; thermal properties (by differential scanning calorimetry); chemical structure alterations (by infrared spectroscopy, NMR, ESCA, SIMS, etc.); changes in shape due to erosion; evidence of cracks and voids; quality of adherence between reinforcement and matrix (samples must be broken with the material frozen in liquid nitrogen and observed with light microscopy or SEM); changes in specific mechanical properties (ISO DP527, ISO 178, ISO 4049, ISO CD604, ASTM D 3846); solubility and water adsorption; and color stability (for dental implants, ISO 4049).

11.8.11   Tissue engineering scaffolds

Tissue and organ generation is an important research field in recent years. The concept is that if autologous cells are provided with the necessary means to sustain their life, as is done with cell cultures in glass or plastic flasks or dishes at the right temperature and with nutrient solutions, the cells proliferate and survive as if they were in the body. Furthermore, if they are provided with a scaffold which they can adhere, move and produce extracellular material in such a way as to copy the three-dimensional structure of their tissue or organ, then tissues and organs may be generated. A lot of research has been directed at finding the suitable biomaterial in terms of biocompatibility, processability, porosity, etc., that will permit cells to "work" as if they are in their natural environment. In most cases, but not in all, the scaffold material should be able to be resorbed at a controlled rate and substituted by biological material generated by the cells. It is not surprising that most research has been on skin regeneration as, among other reasons, skin is a layered flat membrane and can be considered two-dimensional for practical purposes.

As outlined previously,[31] crosslinked collagen by itself or together with glucosaminoglycans has been used as a substrate for skin regeneration. Further strength can be provided to this biomaterial by further crosslinking with carbodiimides and diamines.[35] However, this treatment inhibited fibroblast proliferation and therefore is not recommended for artificial skin applications. Synthetic resorbable biopolymers based on polylactides, polyglycolides, or their copolymers have also been used.

Tissue engineering of liver or kidney requires more stringent conditions for development. For example, the properties on the two sides of a membrane or a tube may be different in many ways, since on the one side hepatocytes or kidney cells should adhere and proliferate while the other side is to be exposed to blood. Thus, porosity, texture, hemocompatibility, micropattern, etc. might have to be different.

Both degradable and nondegradable synthetic polymer matrices have been evaluated for the development of a hybrid liver, i.e. a hepatocyte–polymer construction. Poly(vinyl alcohol) (PVA), which is nondegradable, in spongy form can be made into a noncollapsible scaffold on which hepatocytes survived in vivo.[36] However, as PVA does not degrade it could be a cause of infection and chronic inflamation. For this reason, again, polylactides, polyglycolides, and their copolymers are the materials of choice.

Peripheral nerve regeneration is a major problem that, if solved, will relieve many neurological diseases. Efforts have been concentrated on constructing nerve guidance channels that, when placed so as to bridge the gap of a severed nerve, will guide cells and other biological materials into regenerating nervous tissue. Different biomaterials used for this purpose, with minimal attention to their mechanical properties, include silicon elastomer, PVC, PE, acrylonitrile/vinyl chloride copolymer, PLLA, PGA, collagens, ethylene/vinyl acetate copolymer, and others.[37] Matters in the central nervous system are more difficult due to both its size and its complexity. As electrical activity is the major mode of operation of the nervous tissue, current efforts are directed to combining the physical guidance channels, matrices, growth factors, and electrically charged moieties to produce three-dimensional neural circuits.

Cartilage has a limited capacity for self-repair as it is avascular, i.e., does not have a blood supply. It may be damaged by trauma, by congenital abnormalities, or by arthritis. Engineered cartilage is constructed by growing isolated cells, the chondrocytes, on biomaterial scaffolds in vitro. Another approach has been to implant the cells in the body at the correct site directly or by implanting a biomaterial scaffold in the site. Both nondegradable and degradable polymer and composite matrices have been used for this purpose.[38] These include naturally occurring biodegradable materials like collagen sponges and gels, collagen and glycosaminoglycan composites and hyaluronic acid, naturally occurring nonbiodegradable materials like agarose, synthetic biodegradable materials like polyesters (e.g., polylactides) or their copolymers (e.g., polyglycolic–colactic acid with the trade name Vicryl), Dacron, composites of polyurethanes with Dacron and synthetic nonbiodegradable materials like PVA (with the trade name Ivalon), carbon fibers, Teflon, and Teflon/polyurethane composites.[38] Many problems of biomechanical nature, of biodegradation kinetics and of cellular function remain to be solved for clinical applications of tissue- engineered cartilage.

11.8.12 Nanotechnology in biomaterials

In the preceding sections the underlying assumption of cell–material interactions was that the cells need the right physicochemical environment for their movement and function. It has been shown on several occasions that cells prefer a given stereometry over another for their growth and function.[39] For this reason it may be necessary to use nanotechnological techniques to manufacture these biomaterials to produce an optimized biomaterial surface on which particular cells will grown and function.

Techniques that may be used for this purpose are photolithography, which can be applied to silanes, peptides, etc., and using electron-beams can introduce surface features at the nanometer scale, or microcontact printing which offers resolutions of $1\,\mu m$ so far.

11.8.13 Nonresorbable matrix composites

Such composites usually employ fillers in particulate or chopped-fiber form to improve the mechanical properties over those of the nonresorbable matrix. Bone cements, bearing surfaces, and dental prostheses are examples of applications. For fracture fixation, carbon fiber reinforced epoxy osteosynthesis plates have been used clinically as they are less stiff than the corresponding metallic ones. Bone analogues have also been researched extensively involving polymeric matrices and hydroxyapatite or bioglass reinforcing particles.

Three major classes of nonresorbable composite implants will be discussed in the following sections.

11.9 Dental Implants

Many people have been fitted with one type or another of artificial dental material. Dentures (false teeth), cavity fillings, root forms, blades, endodontic stabilizers, bone plates, and endosteal screws are some of the many dental devices in use. Polymers, alloys, and ceramics, along with composites of these, are all used in dentistry. Polymeric materials are used as impression materials, denture-base and restorative resins, cements, and materials that promote repair of periodontal defects. Metal alloys are used for preparation of crowns and bridges, for inlays, and for repair of cavities. Ceramics are used as impression materials, in cements, and in artificial teeth.

One important consideration in choosing dental materials is the knowledge that chemical and biological conditions within the mouth make it a particularly hostile environment so that hydrolytic degradation processes are accelerated.

Regarding the dental composites, in particular those that are used to repair defects in teeth involving either the filling of cavities or the replacement of grinding surfaces, different considerations apply for the front teeth (anterior composites) and for those designed for the back teeth (posterior composites). The anterior composites should have aesthetic qualities, such as variable shades and degrees of translucency to match the teeth of the patient; they should retain their gloss and be resistant to wear with everyday toothbrushing. For the posterior composites the prime requirement is to withstand the chewing forces over a long period in the hostile oral environment without wear and, of course, fracture. The magnitude of chewing stresses varies considerably, depending on the diet of the person and on the position of the tooth; the farther back in the mouth, the greater the load imposed on the tooth during the chewing process. Some data on physical characteristics during the masticatory cycle are as follows:[40]

Rate of chewing: 60–80 cycles/min

Normal force on single tooth: 3–18 N

Maximum biting force on single tooth: 265 N

Contact time of maximum stress: 0.07 s

Mean sliding distance: 1.0 mm

Frequency of application of maximum stress: 3000 times/day

The data show that dental composites used in fillings must withstand a large number of cyclic stresses and indicate that fatigue mechanisms, and in particular slow crack growth, may be involved in the wear processes.

Cavities are filled using a composite made from an organic resin matrix in which the inorganic particulate reinforcement is dispersed. The resin normally contains a polymer such as BIG-GMA, which is a derivative of methyl methacrylate or urethane dimethacrylate. This polymer is mixed with peroxide to initiate crosslinking after mixing with another liquid or paste composed of dimethacrylate monomer and amine, which is also required to complete the crosslinking. In addition, UV and visible light curing systems are also available for resin crosslinking. For the posterior composites an extra requirement is that they should be X-ray opaque for clear radiographic visualization.

The inorganic filler particles may be 70–80% of the total volume, with sizes from 20 nm to 60 μm. Crystalline quartz and/or lithium glass ceramics, calcium silicate, glass beads, glass fibers, and calcium fluoride may be used. Elements such as barium, strontium, and lanthanum are incorporated into the glass structure to make it X-ray opaque. These elements, because of their high atomic number, increase the refractive index of the glass and, because of their relatively high atomic radius, disrupt the glass network and increase solubility. When the filler is composed of very fine particles, 20–60 nm, there is improvement in surface smoothness and wear. By varying the quantity and size of the fillers, different mechanical properties are obtained. Three main groups can be categorized as follows:[40]

- *Conventional.* These contain glass fillers in the range 1–60 μm in volumes up to 60%. These composites have been superseded in recent years by the other two groups because they failed to fulfill the majority of the performance requirements, in particular aesthetics, surface gloss, and wear resistance. They were the first dental composites and were developed in the early 1960s.

- *Microfilled.* These contain fumed silica particles only and, because of their high surface area, can only be incorporated in volumes up to 40% in the resin. To produce a handlable paste, some manufacturers produce prepolymerized blocks of composite containing agglomerated silica particles. These are fractured and ground down to 10 μm silica-impregnated polymer particles that are then redispersed into the silica filled resin paste. Because these composites contain very fine particles with dimensions below the wavelength of light, they are relatively translucent and similar in appearance to tooth structure and can easily be polished to a high gloss. The mechanical properties of these composites are too low for use in back teeth but they are acceptable as restorative materials in the front teeth.

- *Hydrid.* These contain up to three different size distributions of particles. The fine mode is made up from fumed silica with particle sizes ranging from 20 nm to 0.5–3 μm, and the large mode is made up of glass particles in the size range 5–15 μm. The total volume of the fillers in the composite paste is typically between 60% and 80%. The main benefit of these composites is that a lower viscosity is obtained along with higher moduli.

Table 11.5 gives some comparative data on these different composites.

Many different materials have been evaluated with regard to composites for purposes other than fillings. A tripartite composite made from hydroxyapatite, glass, and titanium has been developed for use in cementless artificial joints and in dental implants.[41] The surface of the substrate, titanium, was roughened and a ceramic layer comprising a continuous glass phase with dispersed hydroxyapatite particles was coated onto it. Electrochemical stability was demonstrated and positive results were reported with tests in vitro (in a simulated body fluid) and in vivo (rods inserted in the femur of dogs). Another composite has been examined for the purpose of increasing the strength in dental applications (bending strength 600–850 MPa).[42] For this, tetragonal zirconia polycrystals were dispersed in bioactive glass-ceramics which were the matrix of the composite. The glass-ceramic was apatite and wollastonite with CaO, P_2O_5, SiO_2, MgO, and CaF_2. Monolithic zirconia was used in another way. In another application monolithic zircoma and a hydroxyapatite/zirconia compo-

TABLE 11.5 Mechanical Properties of Conventional, Microfilled Anterior and Hybrid Posterior Composites

Type	Water Adsorption (%)	Compressive Strength (MPa)	Modulus (GPa)	Surface Hardness (Vickers)
Unfilled resin	2	69	2.4	
Conventional composite	0.6	235	13.7	
Microfilled composite	1.4	276	5–10	30–50
Hybrid composite			14–25	60–120

site were joined by cold isostatic pressing and, subsequently, the resulting 2-phase material was glass encapsulated by hot isostatic pressing at 1225°C and 200 MPa. The resulting composite structure had high strength (860 MPa) and was considered a good combination of a strong, inert ceramic with a bioactive ceramic composite.[43] A similar attempt to produce an alumina and hydroxyapatite/alumina composite failed, however. A different type of composite made of polyethylene reinforced with hydroxyapatite, under the trademark name Hapex,[44] has been developed for dental orthodontic brackets because of isotropic reinforcement, easy molding, and good bonding to dental enamel, possibily due to the presence of hydroxyapatite. A monotonic increase in both Young's modulus and shear modulus was measured with increasing the filler content.

11.10 Artificial Bone

Several different composite systems have been (and still are being) investigated for the development of mechanically compatible artificial bone material. The polymers have much lower Young's modulus than bone, while the metals/alloys have a higher moduli and the ceramics are brittle. Hence, a combination of a biocompatible polymer with a reinforcing material is a logical approach. However, a well demonstrated general effect for polymers is that a progressive increase in the volume fraction of the second phase will correspondingly increase the Young's modulus but decrease the fracture toughness, so that, starting with a ductile polymer, an eventual transition into brittle behavior is observed. As PMMA is brittle, while polyethylene is ductile with an elongation to fracture of 100%, polyethylene provides the more suitable starting matrix material and has comparable mechanical behavior to collagen, the natural bone matrix. Various materials could be considered as additions to polyethylene to stiffen the material at the expense of fracture toughness, but the selection of hydroxyapatite, which is the reinforcing constituent of natural bone and hence biocompatible, proved particularly appropriate in the development of artificial bone. Hydroxyapatite, in the form of calcined bone, is in fact already a significant tonnage engineering material as a constituent of bone china and alternatively can readily be prepared in synthetic form in the laboratory.

The group of Bonfield has extensively worked on this system and the following details are taken mainly from their published work.[8,45,46] Starting with polyethylene granules (of average molecular weight of ~400 000) and hydroxyapatite particles (~0.5–20 μm), composites of hydroxyapatite-reinforced polyethylene are prepared to various volume fractions by a compounding and molding techniques designed to minimize polymer degradation. As no coupling agents are used, all the components of the composite are biocompatible. The composite artifacts produced are pore free with a homogeneous distribution of hydroxyapatite particles, with various volume fractions from 0.1 to 0.5. It was demonstrated that the Young's modulus, as predicted, increases progressively with increase in volume fraction from a starting value of ~1 GPa to ~9 GPa at the 0.5 level. This increase in modulus is accompanied by a decrease in percentage elongation to fracture. However, the composite remains unusually ductile, and hence fracture tough, up to a hydroxyapatite volume fraction of 0.4, at which the elongation to fracture is still ~30% (and hence fracture mechanics analysis is inappropriate). From 0.4 to 0.5 volume fraction, the elongation to fracture decreases markedly, with brittle fracture observed at the 0.45 and 0.5 volume fractions.

Hence in terms of cortical bone replacement, a polyethylene–hydroxyapatite composite with 0.5 volume fraction hydroxyapatite has a Young's modulus within the lower band of the values for bone and a comparable fracture toughness, while a polyethylene–hydroxyapatite composite with 0.4 volume fraction hydroxyapatite is less stiff $(E \sim 5\,\text{GPa})$, but has a superior fracture toughness. Composites with both volume fractions, or intermediate levels, would have applications as artificial bone, depending on the nature of the bone being replaced and the applied physiological loading.

For application as a prosthetic material it is also essential to demonstrate that the hydroxyapatite-reinforced polyethylene composites are biocompatible with bone. Following sterilization by γ-irradiation, a satisfactory zero cytotoxic response of fibroblast cells placed on the composite surface in a cell culture was demonstrated. In addition, the presence of bone-making cells (osteoblasts) in the cell culture was found to promote formation of osteoid (the precursor of bone) at the composite surface, which indicates a favorable response of the natural bone to hydroxyapatite-reinforced polyethylene. This result was confirmed in vivo by implantation of sterilized machined composite pins, 2.5 mm in diameter, in the lateral femoral condyle of adult rabbits for periods up to 6 months. It was established by histological sectioning and microprobe analysis, for both 0.4 and 0.5 hydroxyapatite volume fraction composites, that an initial fibrous encapsulation was succeeded by localized bone apposition at the implant surface to create a secure bond between the natural bone and the implant. As the limb was not immobilized, this favorable bone response occurred during physiological loading and demonstrates the absence of significant relative movement at the bone–implant interface resulting from the mechanical compatibility of the natural and artificial bone. The stable interface and the presence of hydroxyapatite provide the necessary factors for a favorable condition of bioactivity in which bone growth around the implant is encouraged.

Extensive mechanical testing, employing dynamic mechanical analysis, verified that an increase in hydroxyapatite volume fraction led to increases of both strength and modulus of the Hapex composite with a simultaneous reduction in strain to failure.[45] On comparison of two different sizes (roughly 4 μm and 7 μm), it was observed that the composite with the smaller particles possessed higher torsional modulus, Young's modulus, and tensile strength, but lower strain to failure. Finite element analysis with mechanical material properties as accurate as possible provides a predictive tool to help explain phenomenologically unexpected results.[46] The large differences between the material properties of the filler and the matrix were shown to be the cause of high concentration of direct stresses within the composite that resulted in lack of ductility at the required high volume fraction of fillers.

Another composite reported recently[47] as used in the manufacturing of orthopedic implants consists of polyurethane matrix (60%) and granules (63–200 μm) of a calcium phosphosilicate glass ceramic (40%). Mechanical and histological observations in a loaded situation in the tibia of sheep gave evidence of very good biocompatibility and much better osseous integration compared to pure titanium implants.

Other fillers have also been used within the polyethylene matrix. An example is bioglass, which is a class of glasses with active surface that have been shown to bond to tissues. A classical bioglass is composed of 45% SiO_2, 6% P_2O_5, 24.5% CaO, and 24.5% Na_2O and is found in the market with the trademark Bioglass. By blending, compounding, and compression molding, Bioglass 45S5 particles were incorporated into high-density polyethylene to produce composite plates.[48] After immersing them in a simulated body fluid and examining them with various surface analytical techniques and biological testing, it was concluded that the composite was bioactive and biocompatible.

Other composites for orthopedic implants include titania/hydroxyapatite,[49] vapor plasma sprayed titanium, or preroughened carbon fiber reinforced poly(ether ether ketone) (PEEK).[50] A second coating of hydroxyapatite has also been incorporated on such coated composites.[51]

11.11 Bone Cements

Whenever a bone prosthesis is to be implanted in the hip, tibia, knee or other bony structure in such a way that the implant and the remaining bone will make up the new functional unit, there is a choice of either direct press-fitting of the prosthesis into the surgically formed bony cavity (and therefore the need for porous biocompatible surfaces for osseointegration), or a third material to form the intermediate phase to join the two parts (the implant and the bone), i.e., a bone cement. Bone cement is made primarily of poly(methyl methacrylate) (PMMA) powder and methyl methacrylate monomer liquid. Hydroquinone is added to prevent premature polymerization of the monomer, and N,N-dimethyl-p-toluidine is added to promote or accelerate curing of the polymerizing compound.

Bone cement fixation creates two interfaces, cement–bone and cement–implant, with different possible loosening mechanisms in the two interfaces. Sometimes precoating of the implant improves the situation, but the relatively high strains or stresses on the cement at the tip of the prosthesis and elsewhere necessitate the reinforcement of the PMMA to avoid loosening of the prosthesis with concomitant catastrophic results.

Several reinforcements of bone cements have been investigated. Some examples are given below.

Wire-reinforced bone cement.[52]   PMMA incorporating metal wires has been used clinically in the stabilization of the cervical spine. Investigations of the effect of metal wire on the tensile, bending, and shear properties of PMMA have shown that wire reinforcement significantly increases the failure stress of PMMA in all these modes.

Vitalium and stainless steel wires have been utilized and an increase in the tensile stress of between 25% and 85% has been reported, compared with an unreinforced control. As the use of metal wire is not practicable in total joint replacement, due to the narrow annular space between the prosthesis and the bone, carbon or other fiber reinforcement may be used in such cases.

Fiber reinforced bone cement.[52]   Several authors have shown that graphite or carbon fiber reinforcement increases the tensile, compressive, and shear strengths of bone cement and increases its fatigue life. The modulus of elasticity of fiber reinforced PMMA also increases with increasing fiber content when the fiber volume percentage is kept small. Carbon fiber reinforcement of bone cement can also reduce the deformation due to creep. Several authors have also experimented with aramid and Kevlar fiber reinforcement of bone cement. They demonstrated that addition of these fibers can also improve the mechanical properties of PMMA. The improved mechanical properties of graphite fiber reinforced PMMA would allow the clinical use of this material in more diverse applications than is presently possible with normal bone cement.

11.12   Summary

The subject of composites in biomaterials and in medical devices is enormous. In this chapter a brief outline was presented for composite materials and systems in mammalian applications, starting with the cellular membrane as exemplified by the erythrocyte, then with examples of soft tissues such as the vascular vessels, the heart valves, and skin, and continuing with bone and other hard tissues. The major classes of biomaterials were presented with emphasis on the interaction of the material with the living tissue, and in particular with blood, and focusing on composite biomaterials. We followed the scheme of first describing the reinforcing materials such as calbon/polymer/collagen fibers and ceramic fillers, then the matrix systems, with emphasis on the resorbable ones, and gave examples of the fracture fixation plates and touched upon the degradation mechanisms. Finally, more details were given for nonresorbable systems: dental implants, artificial bone, and bone cement.

Acknowledgments

The assistance of my graduate students Anastasia Apostolaki and Nikoletta Katsala in producing the drawings and in proofreading the text is greatly appreciated.

References

1. R. Lipowsky and E. Sackmann (eds.) (1995) *Structure and Dynamics of Membranes*, North-Holland, 5.
2. G. Athanasiou, N., Zoubos, and Y. Missirlis (1991) Erythrocyte membrane deformability in patients with thalassemia syndromes, *Nouv. Rev. Fr. Hematol.*, **33**, 15–20.
3. F. H. Silver (1994) *Biomaterials, Medical Devices and Tissue Engineering*, Chapman & Hall, London, 160.
4. Y. F. Missirlis (1973) In-vitro studies of human aortic valve mechanics, Ph.D. thesis, Rice University.
5. H. Yamada (1973) *Strength of Biological Materials*, Krieger.
6. Y. F. Missirlis (1977) Use of enzymolysis techniques in studying the mechanical properties of connective tissue components, *J. Bioeng.*, **1**, 215–222.
7. J. L. Katz (1995) Mechanics of hard tissue, in *The Biomedical Engineering Handbook* (J. D. Bronzino, ed.), CRC Press, Boca Raton, 273–290.
8. W. Bonfield (1994) Artificial bone, in *Concise Encyclopedia of Composite Materials* (A. Kelly, ed.), Pergamon Press, 15–17.
9. *Biological Evaluation of Medical Devices*, ISO 10993, 1992.
10. L. L. Hench and E. C. Ethridge (1982) *Biomaterials. An Interfacial Approach*, Academic Press, New York.
11. Y. F. Missirlis (1992) How to deal with the complexity of the blood-polymer interactions, *Clin. Mater.*, **11**, 9–12.
12. Y. F. Missirlis and W. Lemm (eds.) (1991) *Modern Aspects of Protein Adsorption on Biomaterials*, Kluwer, Dordrecht.
13. R. Lakes (1995) Composite biomaterials, in *The Biomedical Engineering Handbook* (J. D. Bronzino, ed.), CRC Press, Boca Raton, 598–610.

14. H. Alexander (1996) Composites, in *Biomaterials Science* (B. D. Ratner et al., eds.), Academic Press, San Diego, 94–105.
15. J. S. Adams (1991) Facial augmentation with solid alloplastic implants: a rational approach to material selection, in *Applications of Biomaterials in Facial Plastic Surgery* (A. I. Glasgold and F. H. Silver, eds.), CRC Press, Boca Raton, chapter 17.
16. R. M. Rusch, E. F. Nelson, and D. Noel (1988) Integraft anterior cruciate ligament reconstruction, in *Prosthetic Ligament Reconstruction of the Knee* (M. J. Friedman and R. D. Ferkel, eds.), W.B. Saunders, Philadelphia, 59.
17. M. Spector, et al. (1992) Histological response to composite stems in dogs, *Transactions of the 4th World Biomaterials Congress*, Berlin, 263.
18. F. P. Magee, et al. (1988) A canine composite femoral stem, *Clin. Orthop. Relat. Res.*, **235**, 237.
19. J. S. Bradley, G. W. Hastings, and C. Johnson-Nurce (1980) Carbon fiber reinforced epoxy as a high strength, low modulus material for internal fixation plates, *Biomaterials*, **1**, 38–40.
20. M. Lewandowska-Szumiel, et al. (1997) Fixation of carbon fibre-reiforced carbon composite implanted into bone, *J. Mater. Sci.: Mater. Med.*, **8**, 485–488.
21. S.-J. Shieh, M. C. Zimmerman, and J. R. Parson (1990) Preliminary characterization of bioresorbable and nonresorbable synthetic fibers for the repair of soft tissue injuries, *J. Biomed. Mater. Res.*, **24**, No. 7, 789–808.
22. B. G. MacMillan (1984) Present status of bioadherent materials, barrier dressings, and biosynthetics as skin substitutes, in *Burn Wound Coverings*, vol. 1 (D. L. Wise, ed.), CRC Press, Boca Raton, 115.
23. D. Annis, et al. (1978) An elastomeric vascular prosthesis, *Trans. Am. Soc. Artif. Intern. Org.*, **14**, 209.
24. V. Kasyanov, et. al. (1992) Biomechanical properties of novel compliant textile grafts, *Transactions of the 4th World Biomaterials Congress*, Berlin, 42.
25. S.-T. Li (1995) Tissue-derived biomaterials (collagen), in *The Biomedical Engineering Handbook* (J. D. Bronzino, ed.), CRC Press, Boca Raton, 627–647.
26. M. G. Dunn, et al. (1992) Preliminary development of a collagen fiber/PLA matrix composite prosthesis for ligament reconstruction, in *Transactions of the 4th World Biomaterials Congress*, Berlin, 103.
27. D. Sierra, et al. (1992) Physico-chemical characteristics of a fibrin- based composite tissue adhesive, in *Transactions of the 4th World Biomaterials Congress*, Berlin, 336.
28. Y. Tani (1992) Clinical application of a new colorless and transparent composite resin, in *Transactions of the 4th World Biomaterials Congress*, Berlin, 371.
29. Z. B. Luklinska and W. Bonfield (1997) Morphology and ultrastracture of the interface between hydroxyapatite-polyhydroxybutyrate composite implant and bone, *J. Mater. Sci.: Mater. Med.*, **8**, 379–383.
30. A. M. Babayan, R. M. Pilliar, and M. D. Grnypas (1992) Resorbable short-fibre reinforced composites for fracture fixation, in *Transactions of the 4th World Biomaterials Congress*, Berlin, 245.
31. I. V. Yannas (1996) Natural materials, in *Biomaterials Science* (B. D. Ratner et al., eds.), Academic Press, San Diego, 84–94.
32. M. C. Zimmerman, J. R. Parsons, and H. Alexander (1987) The design and analysis of laminated partially degraded composite bone plate for fracture fixation, *J. Biomed. Mater. Res. Appl. Biomater.*, **21A**, No. 3, 345.
33. K. P. Andriano, A. U. Daniels, and J. Heller (1992) Mechanical properties of composites reinforced with surface modified absorbable calcium-sodium-metaphosphate microfibers, *Transactions of the 4th World Biomaterials Congress*, Berlin, 418.
34. M. Dauner, M., et al. (1998) Resorbable continuous-fibre reinforced polymers for osteosynthesis, *J. Mater. Sci.: Mater. Med.*, **9**, 173–179.
35. C. S. Osborne, et al. (1998) Investigation into the tensile properties of collagen/chondroitin-6-sulphate gels: the effect of crosslinking agents and diamines, *Med. Biol. Eng. Comput.*, **6**, 129–134.
36. S. Uyama, T. Takeda, and J. P. Vacanti (1993) Delivery of whole liver equivalent hepatic mass using polymer devices and heterotrophic stimulation, *Transplantation*, **55**, No. 4, 932.

37. R. Bellamkoda and P. Aebischer (1995) Tissue engineering in the nervous system, in *The Biomedical Engineering Handbook* (J. D. Bronzino, ed.), CRC Press, Boca Raton, 1754–1773.
38. L. E. Freed and G. Vunjak-Novakovic (1995) Tissue engineering of cartilage, in *The Biomedical Engineering Handbook* (J. D. Bronzino, ed.), CRC Press, Boca Raton, 1788–1806.
39. A. S. G. Curtis and C. D. W. Wilkinson (1998) Reactions of cells to topography, *J. Biomater. Sci. Polym. Edn.*, **9**, No. 12, 1313–1329.
40. T. A. Roberts (1994) Dental composites, in *Concise Encyclopedia of Composite Materials* (A. Kelly, ed.), Pergamon Press, 71–74.
41. S. Ban, et al. (1992) Calcium phosphate formation on the surface of HA-G-Ti composite in-vivo and in-vitro, *Transactions of the 4th World Biomaterials Congress*, Berlin, 6.
42. M. Yoshiba, et al. (1992) Mechanical properties and bioactivity of zirconia-toughened glass-ceramics, in *Transactions of the 4th World Biomaterials Congress*, Berlin, 236.
43. J. Li, L. Hermansson, and R. Soeremark (1992) Joining of ceramics of different biofunctions by hot isostatic pressing, in *Transactions of the 4th World Biomaterials Congress*, Berlin, 345.
44. M. Wang, et al. (1998) Young's and shear moduli of ceramic particle filled polyethylene, *J. Mater Sci.: Mater. Med.*, **9**, 621–624.
45. M. Wang, R. Joseph, and W. Bonfield (1997) Influence of hydroxyapatite particle size and morphology on Hapex, in *Bioceramics*, vol. 10 (L. Sedel and C. Rey, eds.), Pergamon Press, 15–18.
46. F. J. Guild and W. Bonfield (1998) Predictive modeling of the mechanical properties and failure processes in hydroxyapatite-polyethylene (Hapex) composite, *J. Mater. Sci.: Mater. Med.*, **9**, 497–502.
47. A. Ignatius, et al. (1997) A new composite made of poyurethane and glass ceramic in a loaded implant model: a biomechanical and histological analysis, *J. Mater. Sci.: Mater. Med.*, **8**, 753–756.
48. J. Huang, et al. (1997) Evaluation of in-vitro bioactivity and biocompatibility of Bioglass-reinforced polyethylene composite, *J. Mater. Sci.: Mater. Med.*, **8**, 809–813.
49. H. Liao (1997) Responses of bone to titania-hydroxypatite composite and nacreous implants: a preliminary comparison by in situ hydridization, *J. Mater. Sci.: Mater. Med.*, **8**, 823–827.
50. S. W. Ha (1997) Topographical characterization and microstructural interface analysis of vascuum-plasma-sprayed titanium and hydroxyapatite coatings on carbon fibre-reinforced poly (etheretherketone), *J. Mater. Sci.: Mater. Med.*, **8**, 891–896.
51. S. W. Ha, et al. (1997) Vacuum plasma sprayed titanium and hydroxyapatite coatings on carbon fiber reinforced polyetheretherketone (PEEK), in *Bioceramics*, vol. 10, Pergamon Press, 203–206.
52. S. Saha and S. Pal (1994) Solid fiber composites as biomedical materials, in *Concise Encyclopedia of Composite Materials* (A. Kelly, ed.), Pergamon Press, 266–271.

Index

ABOUT THE EDITORS

PROFESSOR E. E. GDOUTOS is Director of the Laboratory of Applied Mechanics and Director of the Section of Design and Construction of Structures, Department of Civil Engineering, the Democritus University of Thrace. He is visiting Professor at the University of California at Santa Barbara and Davis; the Michigan Technological University; the National Technical University of Athens; and the University of Toledo. He is a member of the editorial board of the journals *Theoretical and Applied Fracture Mechanics, Applied Composite Materials*, and *Advanced Composite Letters*. Gdoutos is Fellow of ASME, and received the Distinguished Visiting Professorship Award from Toledo University in 1992. A member of the Structures and Materials Panel of AGARD since 1992, he serves on the Research Board of Advisors; is a Fellow of the American Biographical Institute; and is Vice-Chairman of the Research Committee of the Democritus University of Thrace. He is listed in *Who's Who in the World*, 8th-14th editions.

DR. K. PILAKOUTAS is Reader in Civil Engineering at Sheffield University and Manager of the Centre for Cement and Concrete. He is a Registered Engineer in Cyprus, and is a member of the Society for Earthquake and Civil Engineering Dynamics; Institution of Civil Engineers; American Concrete Institute; and the Committee Euro International du Baton.

DR. C. A. RODOPOULOS is recognized worldwide as an expert in the failure analysis of composite materials. He is Research Associate at the Structural Integrity Research Institute, University of Sheffield, (SIRIUS), and fellow member of the Composite Research Group. He received two research awards from the EU, and holds memberships in ASME, SAE, ASTM, AIAA and ESIS. The author of dozens of scientific and technical papers published in national and international journals, he is on the advisory board of National Technical Publications of Greece and is listed in *Who's Who in the World*.

www.ingramcontent.com/pod-product-compliance
Lightning Source LLC
Chambersburg PA
CBHW060419220326
41598CB00021BA/2229